Studies in Systems, Decision a

CW01249609

Volume 385

Series Editor

Janusz Kacprzyk, Systems Research Institute, Polish Academy of Sciences, Warsaw, Poland

The series "Studies in Systems, Decision and Control" (SSDC) covers both new developments and advances, as well as the state of the art, in the various areas of broadly perceived systems, decision making and control–quickly, up to date and with a high quality. The intent is to cover the theory, applications, and perspectives on the state of the art and future developments relevant to systems, decision making, control, complex processes and related areas, as embedded in the fields of engineering, computer science, physics, economics, social and life sciences, as well as the paradigms and methodologies behind them. The series contains monographs, textbooks, lecture notes and edited volumes in systems, decision making and control spanning the areas of Cyber-Physical Systems, Autonomous Systems, Sensor Networks, Control Systems, Energy Systems, Automotive Systems, Biological Systems, Vehicular Networking and Connected Vehicles, Aerospace Systems, Automation, Manufacturing, Smart Grids, Nonlinear Systems, Power Systems, Robotics, Social Systems, Economic Systems and other. Of particular value to both the contributors and the readership are the short publication timeframe and the world-wide distribution and exposure which enable both a wide and rapid dissemination of research output.

Indexed by SCOPUS, DBLP, WTI Frankfurt eG, zbMATH, SCImago.

All books published in the series are submitted for consideration in Web of Science.

More information about this series at http://www.springer.com/series/13304

Xiaojie Su · Yao Wen · Yue Yang · Peng Shi

Intelligent Control, Filtering and Model Reduction Analysis for Fuzzy-Model-Based Systems

Springer

Xiaojie Su
College of Automation
Chongqing University
Chongqing, China

Yao Wen
College of Automation
Chongqing University
Chongqing, China

Yue Yang
College of Automation
Chongqing University
Chongqing, China

Peng Shi
School of Electrical and Electronic Engineering
The University of Adelaide
Adelaide, SA, Australia

ISSN 2198-4182 ISSN 2198-4190 (electronic)
Studies in Systems, Decision and Control
ISBN 978-3-030-81216-4 ISBN 978-3-030-81214-0 (eBook)
https://doi.org/10.1007/978-3-030-81214-0

This Springer imprint is published by the registered company Springer Nature Switzerland AG
The registered company address is: Gewerbestrasse 11, 6330 Cham, Switzerland

To My Family

X. Su

To My Family

Y. Wen

To My Family

Y. Yang

To My Family

P. Shi

Preface

Problem formulations of physical systems and processes can often lead to complex nonlinear systems, which may cause analysis and synthesis difficulties. Study of nonlinear systems is often problematic due to their complexities. One effective way of representing a complex nonlinear dynamic system is the so-called Takagi-Sugeno (T-S) fuzzy model, which is governed by a family of fuzzy IF-THEN rules that represent local linear input-output relations of the system. It incorporates a family of local linear models that smoothly blend together through fuzzy membership functions. This, in essence, is a multi-model approach in which simple sub-models (typically linear models) are fuzzily combined to describe the global behavior of a nonlinear system. Fuzzy logic method has been studied and developed for decades. It is known to be an effective control approach to some ill-defined and complex control processes. Thanks to fuzzy logic, expert knowledge about the control processes can be employed to heuristically design fuzzy controllers with some linguistic IF-THEN rules. Practically, human knowledge can be represented as linguistic statements and incorporated into the fuzzy logic controller. As a result, the design method can operate with intelligence.

Analysis and synthesis including state-feedback control, output-feedback control, tracking control, optical control, filtering, fault detection, and model reduction for a class of T-S fuzzy systems are all thoroughly studied. Fresh novel techniques including the Linear Matrix Inequality (LMI) techniques, the slack matrix method, and so on, are applied to such systems. This monograph is divided into four sections. First, we focus on stabilization synthesis and controller design for T-S fuzzy systems. The following problems are investigated in this book: (1) the problem of stability analysis and stabilization for T-S fuzzy systems with the time-varying delay; (2) the problem of Hankel-norm output feedback controller design for a class of T-S fuzzy stochastic systems; (3) the problem of $\mathcal{L}_2-\mathcal{L}_\infty$ dynamic output feedback controller design for nonlinear switched systems with nonlinear perturbations. Secondly, the reliable filtering and fault detection problems are solved for fuzzy systems. The below problems are studied: (1) the problem of the dissipativity-based filtering problem for fuzzy switched systems with stochastic perturbation; (2) the fault detection filtering problem for nonlinear switched stochastic system; (3) the problem of reliable filter design with strictly dissipativity for discrete-time T-S fuzzy time-delay systems.

Then the theories and techniques developed in the previous part are extended to the model reduction and model approximation of T-S fuzzy systems. The below problems are studied: (1) the reduced-order model approximation problem for discrete-time hybrid switched nonlinear systems; (2) the model approximation problem for dynamic systems with time-varying delays under the fuzzy framework; (3) the model approximation problem for T-S fuzzy switched systems with stochastic disturbance; (4) the $\mathcal{H}_\infty$ reduced-order filter design problem for discrete-time fuzzy delayed systems with stochastic perturbation. Finally, two real applications are proposed to demonstrate the feasibility and effectiveness of the fuzzy control design presented in the previous parts. The first application is the dissipative event-triggered fuzzy control of truck-trailer system. In view of the fuzzy model, the stability of the resulting system is analyzed in terms of Lyapunov stability theory. Additionally, the explicit expression of the desired controller is given in view of Linear Matrix Inequalities (LMIs), which ensures the resulting closed-loop system is asymptotically stable and strictly (X, Y, Z)–θ–*dissipative*. The second one is the event-triggered fuzzy control of inverted pendulum systems. By employing the parallel distributed compensation law, sufficient conditions for the resulting fuzzy system and the event-triggered fuzzy controller are presented for the nonlinear inverted pendulum system.

The main contents are suitable for a one-semester graduate course. This publication is a research reference whose intended audience includes researchers, postgraduate students.

Chongqing, China Xiaojie Su
Chongqing, China Yao Wen
Chongqing, China Yue Yang
Adelaide, Australia Peng Shi
July 2021

Acknowledgements

There are numerous individuals without whose help this book will not have been completed. Special thanks go to Prof. Ligang Wu from Harbin Institute of Technology, Prof. Yong-Duan Song from Chongqing University, Prof. Michael V. Basin from the Autonomous University of Nuevo Leon, Prof. Hamid Reza Karimi from University of Agder, Dr. Hak Keung Lam from King's College London, Prof. Rongni Yang from Shandong University, Prof. Jianxing Liu from Harbin Institute of Technology, for their valuable suggestions, constructive comments, and support.

Our acknowledgments also go to our fellow colleagues who have offered invaluable support and encouragement throughout this research effort. Thanks go to our students, Hongying Zhou, Fengqin Xia, Xinxin Liu, Bingna Qiao, Yaoyao Tan, Feng Hu, and Qianqian Chen for their commentary. The authors are especially grateful to their families for their encouragement and never-ending support when it was most required. Finally, we would like to thank the editors at Springer for their professional and efficient handling of this project.

The writing of this book was supported in part by the National Key R&D Program of China under Grant (2019YFB1312002), the Key-Area Research and Development Program of Guangdong Province under Grant (2020B0909020001), the National Natural Science Foundation of China (617 72095), and Chongqing Science Fund for Outstanding Young Scholars (cstc2019jcyjjqX0015).

Contents

Notations and Acronyms

$\triangleq$	is defined as
$\in$	belongs to
$\forall$	for all
$\sum$	sum
$\mathbf{R}$	field of real numbers
$\mathbf{R}^n$	space of n -dimensional real vectors
$\mathbf{R}^{n\times m}$	space of $n \times m$ real matrices
$\mathbf{Z}$	field of integral numbers
$\mathbf{Z}^+$	field of positive integral numbers
$\mathbf{E}\{\cdot\}$	mathematical expectation operator
$\mathbf{He}(A)$	$A + A^T$
lim	limit
max	maximum
min	minimum
sup	supremum
inf	infimum
$\text{rank}(\cdot)$	rank of a matrix
$\text{trace}(\cdot)$	trace of a matrix
$\lambda_{\min}(\cdot)$	minimum eigenvalue of a real symmetric matrix
$\lambda_{\max}(\cdot)$	maximum eigenvalue of a real symmetric matrix
I	identity matrix
I_n	$n \times n$ identity matrix
0	zero matrix
$0_{n\times m}$	zero matrix of dimension $n \times m$
X^T	transpose of matrix X
X^*	conjugate transpose of matrix X
X^{-1}	inverse of matrix X
$X > (<)0$	X is real symmetric positive (negative) definite
$X \geq (\leq)0$	X is real symmetric positive (negative) semi-definite
$\mathcal{L}_2\{[0, \infty), [0, \infty)\}$	space of square summable sequences on $\{[0, \infty), [0, \infty)\}$ (continuous case)

$\ell_2\{[0,\infty),[0,\infty)\}$	space of square summable sequences on $\{[0,\infty),[0,\infty)\}$ (discrete case)
$\lvert\cdot\rvert$	Euclidean vector norm
$\Vert\cdot\Vert$	Euclidean matrix norm (spectral norm)
$\Vert\cdot\Vert_2$	$\mathcal{L}_2$ -norm: $\sqrt{\int_0^\infty \lvert\cdot\rvert^2 dt}$ (continuous case) ℓ_2 -norm: $\sqrt{\sum_0^\infty \lvert\cdot\rvert^2}$ (discrete case)
$\Vert\cdot\Vert_E$	$\mathbf{E}\{\Vert\cdot\Vert_2\}$
$\Vert\mathbf{T}\Vert_\infty$	$\mathcal{H}_\infty$ norm of transfer function $\mathbf{T}$: $\sup_{\omega\in[0,\infty)}\Vert\mathbf{T}(j\omega)\Vert$ (continuous case) $\sup_{\omega\in[0,2\pi)}\Vert\mathbf{T}(e^{j\omega})\Vert$ (discrete case)
diag	block diagonal matrix with blocks $\{X_1,\ldots,X_m\}$
*	symmetric terms in a symmetric matrix
ADT	average dwell time
CCL	cone complementary linearization
DOF	dynamic output feedback
DOFC	dynamic output feedback control
FLC	fuzzy-logic-control
LKF	Lyapunov–Krasovskii function
LMI(s)	linear matrix inequality (inequalities)
MIMO	multiple-input multiple-output
NCSs	networked control systems
OFC	output feedback control
PDC	parallel distributed compensation
SISO	single-input single-output
SOFC	state-output feedback control
SOS	sum-of-squares
T-S	Takagi–Sugeno
TSK	Takagi–Sugeno–Kang

List of Figures

List of Tables

Chapter 1
Introduction

1.1 Background

Many real-world systems are nonlinear and complex and thus challenging to synthesize and analyse using the conventional theory and approach. However, with the advent of fuzzy modelling and introduction of the "fuzzy sets" theory in [272], the Takagi–Sugeno (T-S) fuzzy model has emerged as an effective tool to analyse complex nonlinear systems [118, 153, 194, 202, 357]. By using IF-THEN rules, nonlinear equations can be approximated as a set of local linear input–output relations, and the complete fuzzy model can be obtained by mixing local linear models with piecewise fuzzy membership functions. Consequently, the results based on traditional linear system techniques can be extended to nonlinear systems. The resulting fuzzy system is described by a family of IF-THEN rules to express the local input–output relations of the nonlinear systems and realized by smoothly blending these local linear models together through the membership functions. Owing to the promising approximation ability of T-S fuzzy systems, the research concerning fuzzy systems has attracted increasing interest, and many researchers have focused on T-S fuzzy systems. For example, a stability analysis was conducted and stabilization issues were investigated in [115, 143, 321]; filtering problems were investigated in [10, 254, 337]; fault detection issues were reported in [83, 284, 341]; and model reduction problems were examined in [255, 256, 294].

The basic structure of a fuzzy system consists of four conceptual components: knowledge base, fuzzification interface, interface engine, and defuzzification interface. Figure 1.1 shows the block diagram of a fuzzy system. The fuzzification interface maps the crisp inputs into fuzzy inputs through membership functions. The knowledge base is a collection of rules in the IF-THEN format, which describes the expert knowledge using linguistic rules. The interface engine performs reasoning based on the fuzzy inputs and rules to generate the fuzzy outputs. Subsequently, the defuzzification interface converts the fuzzy outputs into crisp outputs. The fuzzy system shown in Fig. 1.1 has been employed as a fuzzy controller, as proposed by

X. Su et al., *Intelligent Control, Filtering and Model Reduction Analysis for Fuzzy-Model-Based Systems*, Studies in Systems, Decision and Control 385,
https://doi.org/10.1007/978-3-030-81214-0_1

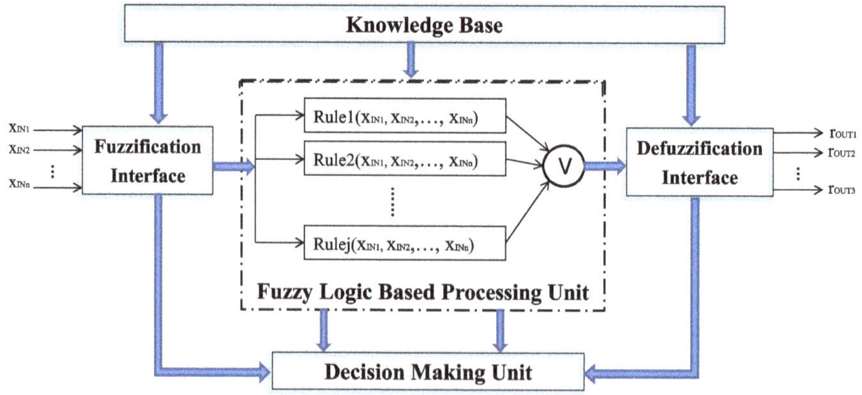

Fig. 1.1 Basic structure of a fuzzy system

Prof. Ebrahim Mamdani. By incorporating the knowledge of the control experts in the knowledge base, fuzzy controllers demonstrating human spirits have been successfully applied in various engineering applications such as industrial processes [119, 150, 170, 181], sludge waste water treatment [159, 229, 264], and construction engineering [13, 62, 71, 182].

Fuzzy control has been demonstrated to be a successful control approach for complex nonlinear systems. Moreover, the use of fuzzy control has been suggested as an alternative to conventional control techniques [117, 129, 216, 320]. In general, fuzzy sets can effectively capture the system nonlinearities, and the system dynamics of the nonlinear systems can be represented as an average weighted sum of certain local linear subsystems, with the weights characterized by the membership functions. In this context, stability analyses represent a key aspect in the field of fuzzy control systems [8, 148, 222].

1.2 Fuzzy-Model-Based Systems

Fuzzy-model-based control [19, 70, 312] is a powerful approach to address mathematically ill-defined nonlinear systems. To investigate the system stability, the T-S fuzzy model (also known as the Takagi–Sugeno–Kang (TSK) model) [59, 93, 235, 355] was proposed to provide a general and systematic framework to represent a nonlinear plant as a weighted sum of several linear subsystems. Each linear subsystem effectively models the dynamics of the nonlinear plant in the local operating domain. As the linear and nonlinear parts of the nonlinear plant are extracted, the T-S fuzzy model exhibits a semi-linear characteristic that can facilitate the realization of the stability analysis and controller synthesis. Based on the T-S fuzzy model, a fuzzy

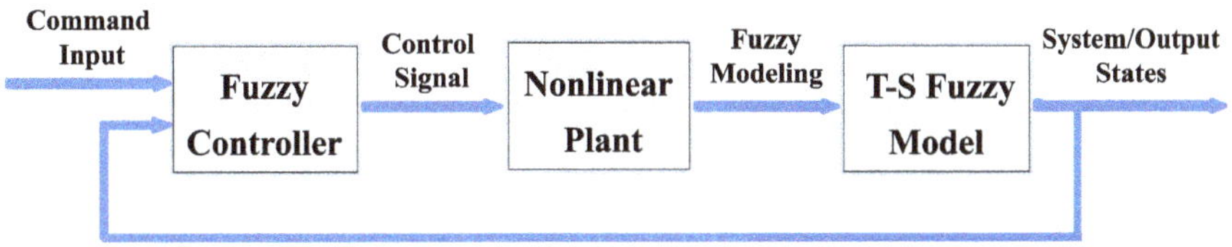

Fig. 1.2 A block diagram of the fuzzy-model-based control system

controller can be designed to close the feedback loop and form a fuzzy-model-based control system, as shown in Fig. 1.2.

In recent decades, fuzzy control has undergone rapid development in both theoretical research and industrial practice. In particular, since 1985, when Japanese scholars Takagi and Sugeno proposed the T-S fuzzy models [272], the design and analysis problem of fuzzy systems has evolved. Specifically, the T-S fuzzy model has emerged as the main method to address nonlinear problems because it can combine the strict mathematics theory with fuzzy logic theory to accurately approach nonlinear systems [2, 220, 286]. The dynamics of a nonlinear system can be expressed as a weighted average of linear subsystems. The linear and nonlinear characteristics of the nonlinear plant are extracted and expressed as the linear subsystems and nonlinear weights, respectively. In this manner, the T-S fuzzy model exhibits a favourable semi-linear property, enabling the use of certain linear analysis and design methods in performing the system analysis.

Due to its simple modelling process, T-S fuzzy models have received considerable attention from domestic and foreign scholars, and many relevant studies have been conducted. In [63, 131, 213, 232], the stability analysis of a T-S fuzzy model was performed using the linear matrix inequality (LMI) toolbox. In [66, 136, 162, 273, 352], certain researchers designed observer-based controllers and output feedback controllers for a case involving unmeasurable system states. Furthermore, in [4, 30, 86], fault detection problems were considered to detect failures in real time. Moreover, in [51, 121, 292], the $\mathcal{H}_\infty$ filter problems were investigated; in particular, the filter error system was rendered asymptotically stable, and a certain performance was achieved in the presence of interfering signals.

1.2.1 T-S Fuzzy Dynamic Model

The T-S fuzzy model is a useful mathematical tool to model nonlinear plants. This model provides a fixed framework to represent nonlinear systems through several linguistic rules to facilitate the stability analysis and control synthesis by using the LMI or sum-of-squares (SOS) -based analysis approach. In particular, a complex nonlinear system can be modelled as the following T-S fuzzy system:

♦ **Plant Form**:

Rule i: IF $\theta_1(x(t))$ is $\mathcal{M}_{i1}$ and $\cdots$ and $\theta_p(x(t))$ is $\mathcal{M}_{ip}$, THEN

$$\delta x(t) = A_i x(t) + B_i u(t), \quad i = 1, 2, \ldots, r,$$

where $x(t) \in \mathbf{R}^n$ is the state vector, δ above denotes the derivative operator in continuous time $\big(\text{i.e., } \delta x(t) = \dot{x}(t)\big)$ and the shift forward operator in discrete time $\big(\text{i.e., } \delta x(t) = x(t+1)\big)$; $u(t) \in \mathbf{R}^m$ is the input vector. $\mathcal{M}_{im}$ is the fuzzy set of rule i corresponding to the function $\theta_m(x(t))$, $i = 1, 2, \ldots, r$, $m = 1, 2, \ldots, p$, and r is the number of IF-THEN rules; $A_i \in \mathbf{R}^{n\times n}$ and $B_i \in \mathbf{R}^{n\times m}$ are system parameter matrices.

The premise variables are assumed to be independent of the input variables $u(t)$. This assumption is implemented to avoid the complex defuzzification process of the fuzzy controllers [275]. Given a pair $(x(t), u(t))$, the T-S fuzzy dynamic model can be derived:

$$\delta x(t) = \sum_{i=1}^{r} h_i(x(t)) \left[A_i x(t) + B_i u(t)\right], \tag{1.1}$$

where $h_i(x(t))$ are the normalized grades of membership function with

$$h_i(x(t)) = \frac{\nu_i(x(t))}{\sum_{i=1}^{r} \nu_i(x(t))}, \quad \nu_i(x(t)) = \prod_{m=1}^{p} \mu_{\mathcal{M}_{im}}\big(\theta_m(x(t))\big),$$

where $\mu_{\mathcal{M}_{im}}\big(\theta_m(x(t))\big)$ are the membership functions corresponding to the fuzzy set $\mathcal{M}_{im}$. It is assumed that

$$\nu_i(x(t)) \geq 0, \quad i = 1, 2, \ldots, r,$$
$$\sum_{i=1}^{r} \nu_i(x(t)) > 0, \quad \forall t \geq 0.$$

Therefore,

$$h_i(x(t)) \geq 0, \quad i = 1, 2, \ldots, r; \quad \sum_{i=1}^{r} h_i(x(t)) = 1. \tag{1.2}$$

In general, the T-S fuzzy model can be established using two approaches: (1) By using certain system identification algorithms [235, 272] based on the input–output data. This approach is suitable for nonlinear systems for which mathematical models are not available, but input–output data are available. (2) If the mathematical model of the nonlinear system is available, the T-S fuzzy model can be derived from the mathematical model by using the concept of sector nonlinearity or local approximation [275, 299]. Notably, in the second approach, the grades of the membership may be uncertain if they are in terms of uncertain system parameters. In this case, a nonlinear plant subject to parameter uncertainties can be represented as a T-S fuzzy model with uncertain grades of membership.

1.2.2 Fuzzy-Model-Based Control System

The most widely used approach to design the fuzzy controller pertains to the state-feedback fuzzy controller [122, 226, 338], which has a structure similar to the T-S fuzzy model and is a weighted sum of several linear state-feedback sub-controllers. The control action is described by certain linguistic rules. Let us consider the following state-feedback fuzzy controller:

♦ **Controller Form**:

Rule j: IF $\vartheta_1(x(t))$ is $\mathcal{N}_{j1}$ and ... and $\vartheta_q(x(t))$ is $\mathcal{N}_{jq}$, THEN

$$u(t) = K_j x(t), \quad j = 1, 2, \ldots, s,$$

where $\mathcal{N}_{jn}$ is the fuzzy set of rule j corresponding to the function $\vartheta_n(x(t))$, $j = 1, 2, \ldots, s$, $n = 1, 2, \ldots, q$, and s is the number of IF-THEN rules. $K_j \in \mathbf{R}^{m \times n}$ is the gain matrix of the state feedback controller in each rule, and a compact form of the controller is given by

$$u(t) = \sum_{j=1}^{s} g_j(x(t)) K_j x(t), \tag{1.3}$$

where

$$g_j(x(t)) = \frac{\nu_j(x(t))}{\sum_{j=1}^{s} \nu_j(x(t))}, \quad \nu_j(x(t)) = \prod_{n=1}^{q} \mu_{\mathcal{N}_{jn}}\big(\theta_n(x(t))\big),$$

with

$$g_j(x(t)) \geq 0, \quad j = 1, 2, \ldots, s; \quad \sum_{j=1}^{s} g_j(x(t)) = 1, \tag{1.4}$$

$g_j(x(t))$ are the normalized grades of membership function, and $\mu_{\mathcal{N}_{jn}}\big(\vartheta_n(x(t))\big)$ are the membership functions corresponding to the fuzzy set $\mathcal{N}_{jn}$.

A fuzzy-model-based control system consists of a nonlinear plant represented by the T-S fuzzy model (1.1) and fuzzy controller (1.3) connected in a closed loop.

Throughout this book, as derived from (1.2) and (1.4), the following property is used during the system performance analysis:

$$\sum_{i=1}^{r} h_i(x(t)) = \sum_{j=1}^{s} g_j(x(t)) = \sum_{i=1}^{r}\sum_{j=1}^{s} h_i(x(t))g_j(x(t)) = 1. \tag{1.5}$$

Consider (1.1), (1.3) and (1.5), the closed-loop fuzzy-model-based control system can be described by

$$\begin{aligned}\delta x(t) &= \sum_{i=1}^{r} h_i(x(t))\big[A_i x(t) + B_i \sum_{j=1}^{s} g_j(x(t))K_j x(t)\big]\\ &= \sum_{i=1}^{r}\sum_{j=1}^{s} h_i(x(t))g_j(x(t))A_{ij}x(t),\end{aligned}$$

where $A_{ij} \triangleq A_i + B_i K_j$, $x(t) \in \mathbf{R}^n$ is the state vector, δ denotes the derivative operator in continuous time $\big($i.e., $\delta x(t) = \dot{x}(t)\big)$ and the shift forward operator in discrete time $\big($i.e., $\delta x(t) = x(t+1)\big)$.

1.2.3 *Stability Analysis of Fuzzy Control Systems*

The stability analysis and control synthesis are essential aspects in fuzzy-model-based control problems. The most popular approach to investigate the stability of fuzzy-model-based control systems is based on the Lyapunov method [151, 225, 324]. The stability analysis process can be realized using the following steps:

1. Construct a fuzzy model representing the nonlinear plant.
2. Select the type of fuzzy controller for the control process.
3. Formulate a fuzzy-model-based control system by connecting the fuzzy model and fuzzy controller in a closed loop, as displayed in Fig. 1.2.
4. Define a Lyapunov function candidate, which is a scalar positive function.
5. Set the stability conditions based on the Lyapunov stability method.

Moreover, in the stability analysis of fuzzy-model-based control systems, the conservativeness is related to several factors:

1. **Types of Lyapunov Functions**: A Lyapunov function is a mathematical tool to investigate the stability problem for fuzzy-model-based control systems. By employing different types or forms of Lyapunov function candidates to approximate the domain of the feasible solution, different stability conditions pertaining to different levels of conservativeness are obtained.
2. **Types of Stability Analyses**: The type of stability analysis determines the information of the membership functions to be considered, which influences

the conservativeness of the stability conditions. In particular, in membership-function-independent and membership-function-dependent stability analyses, the information of the membership functions is not considered and considered to set the stability conditions, respectively. In the latter case, the stability analysis results depend on the considered nonlinear model and often correspond to relaxed stability conditions.
3. **Methods of Stability Analyses**: The methods used to realize the stability analysis, such as those for managing and incorporating the membership functions influence the degree of conservativeness for the stability conditions.

For T-S fuzzy-model-based control systems, as shown in Fig. 1.3, the approaches to realize the stability analysis based on the Lyapunov stability theory can be classified into two types, specifically, membership-function-independent and membership-function-dependent approaches depending on whether the information of the membership functions is considered in the stability analysis. As the membership-function-independent stability analysis does not consider the membership functions, the stability analysis results are often more conservative than those based on the membership-function-dependent approach, in which the membership functions are considered in the stability analysis.

1.2.3.1 Membership-Function-Independent Stability Analysis

In the membership-function-independent stability analysis, the information of the membership functions is not considered, and only the local control subsystems of the fuzzy-model-based control systems are managed. Thus, the stability conditions do not involve any membership functions. Once there exists a feasible solution to the stability conditions, the fuzzy-model-based control system is guaranteed to be stable for any shape of the membership functions. However, when membership functions are not considered in the stability analysis, certain information of the nonlinearity is ignored. Therefore, the membership-function-independent stability results are potentially conservative.

1.2.3.2 Membership-Function-Dependent Stability Analysis

The membership-function-dependent stability conditions take into account the membership functions of the fuzzy model and fuzzy controller in the stability analysis. The obtained stability conditions include the information of the membership functions. In general, more relaxed stability conditions are obtained compared with those in the membership-function-independent stability analysis as more information of the fuzzy-model-based control system is considered. However, as the information of the membership functions is implemented in the form of slack matrices in the stability analysis and the number of stability conditions is generally high, the computational demand to determine a feasible solution for the stability conditions is high. Moreover,

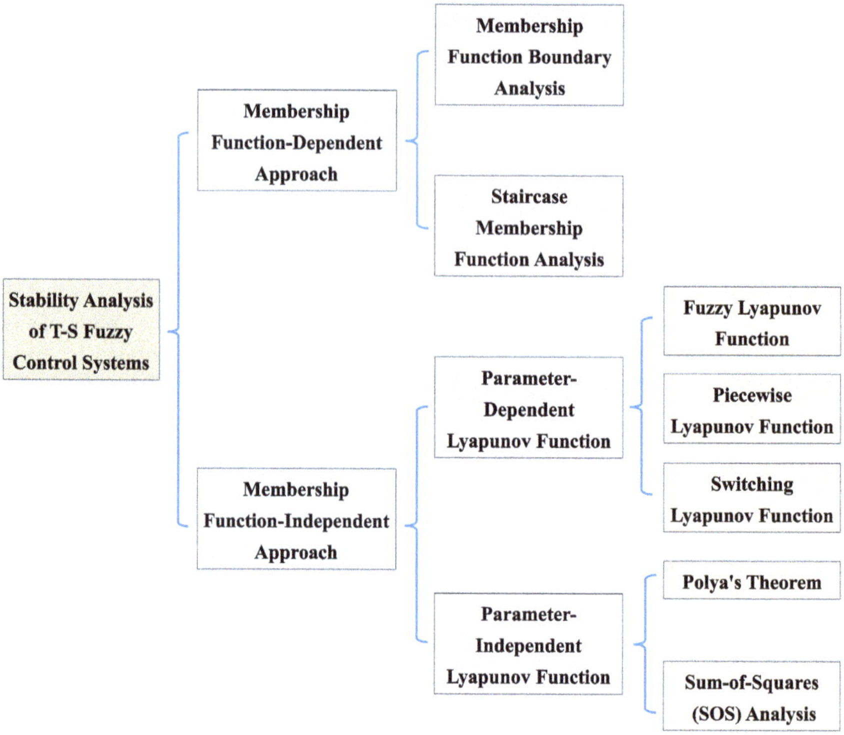

Fig. 1.3 Stability analysis approaches for T-S fuzzy-model-based control systems

the obtained stability conditions are specific to the fuzzy-model-based control system to be controlled and not generalized for all shapes of the membership functions.

1.3 Intelligent Control of Nonlinear Systems

1.3.1 Intelligent Control

Since the 20th century, the requirements for control systems have evolved with the development of science and technology [107, 112, 114, 309]. From linear to nonlinear systems and from single-input single-output (SISO) control systems to multiple-input multiple-output (MIMO) control systems, multiple control approaches have been integrated to complement the strengths of different techniques [46, 68, 178, 271, 313]. Intelligent control, as a novel technology, can help realize the preset control tasks of a system autonomously without human intervention [60, 75, 85, 201, 236]. Through this control implementation, the system control mode evolves from ordinary automatic control to a more advanced intelligent control mode. Intelligent control strategies transform the control model from certain to uncertain and provide a more convenient path for the information exchange between the input and output devices of the control system and the external environment. Moreover, when using intelligent control schemes, the control task of the system changes from a single task to a more complex control task. Thus, a more ideal solution exists for the control problem of nonlinear systems, which cannot be easily solved using ordinary automatic control systems. Intelligent control strategies enable an automatic control system to achieve self-adaptation, self-organization, self-learning, and self-coordination [197, 279, 301, 317]. Intelligent control represents the development trend of the control theory, which can effectively solve complex control problems, and several related techniques can be applied to industry, agriculture, service industry, military aviation, and other fields beyond control pertaining to finance, management, civil engineering, and design, among other domains.

Most control systems are designed under the assumption of perfect data transmission in both the sensor-to-controller and controller-to-actuator channels. This assumption is valid for most point-to-point control structures, but not for the widely used networked control systems (NCSs), in which the control loop is closed through a certain form of communication networks. Compared with traditional point-to-point control systems, the main advantages of NCSs are the low cost, flexibility and easy reconfigurability, inherent reliability and robustness to failure, and adaptation capability [20, 101, 108, 155, 265]. Consequently, NCSs have been applied in a broad range of areas such as power grids, water distribution networks, transportation networks, haptics collaboration over the Internet, mobile sensor networks, and unmanned aerial vehicles [78, 160, 234, 344, 348]. However, the introduction of communication channels in the control loop induces several network-induced critical issues or constraints such as variable transmission delays, data-packet dropouts, packet disorder, and quantization errors, which can significantly degrade the system performance and even destabilize the system in certain conditions. In recent years, the issues induced by the NCSs have posed considerable challenges to conventional control and communication theory and have attracted considerable attention from researchers of multiple disciplines including control, communication, and mathematics [65, 92, 174, 354]. The typical research topics pertaining to NCSs include the stability of NCSs under various network constraints, state estimation over lossy networks, controller/filter design of NCSs with guaranteed stability, and performance optimization [134, 198, 263, 339].

Moreover, recently, the benefit of using wireless communication technology in large-scale industrial processes has become evident [64, 145, 282, 283, 314], especially in the form of cyber-physical systems. The utilization of wireless networks in industrial process control enables the realization of new system architectures and designs. Nevertheless, many industrial control processes involve severe nonlinear characteristics, which renders the analysis and design highly challenging. In recent decades, fuzzy-logic-control (FLC) has received considerable attention from both academic and industrial communities. Notably, the FLC has been demonstrated to be a simple and powerful strategy for the analysis and synthesis of many complex nonlinear systems and nonanalytic systems [224, 227, 233, 336]. Significant research efforts have been devoted for both theoretical advances and implementation techniques for fuzzy controllers, and many industrial applications of the FLC have been reported in the existing literature [203, 228, 274]. Among various model-based fuzzy control methods, the approach based on the T-S model is well suited to design model-based nonlinear system controller [80, 126, 149, 244]. The research interest in the systematic analysis and design of networked nonlinear systems via T-S fuzzy dynamic models has increased, and multiple significant results have been reported [88, 165, 215, 346, 356].

1.3.2 Fuzzy Control

The mathematical modelling of physical systems and processes often generates complex nonlinear systems, the synthesis and analysis of which is highly difficult. The research on nonlinear systems is often problematic due to their complexities. One effective way of representing a complex nonlinear dynamic system is by using the T-S fuzzy model [5, 16, 27, 126, 142], which is governed by a family of fuzzy IF-THEN rules that represent the local linear input–output relations of the system. This model incorporates a family of local linear models smoothly blended through fuzzy membership functions. This approach, in essence, is a multi-model approach, in which simple sub-models (typically linear models) are fuzzily combined to describe the global behaviour of a nonlinear system [7, 21, 23, 260]. Within these fuzzy models, the local dynamics in different state space regions are represented by linear models [24, 29, 37, 41, 48]. An overall fuzzy model of the system is created by fuzzily 'blending' these linear models. Based on the fuzzy model, the control design can be realized based on the parallel distributed compensation (PDC) scheme. In particular, a linear state-feedback controller is designed for each local linear model. The obtained overall controller is usually nonlinear and represents a fuzzy 'blending' of each individual linear controller [33, 44, 56, 61, 69, 72].

Nevertheless, because only a part of the state information may be known in real-world engineering, the state-feedback control is not entirely effective to ensure the desired performance level [11, 12, 189, 257]. Consequently, intensive research has been conducted in the area of output feedback control (OFC) design. Although the state-output feedback control (SOFC) technique (see, for instance, [100, 110, 188]) can be used to address the dynamic output feedback control (DOFC) problem, the process involves several analytical difficulties. Nonetheless, several solutions have been obtained for DOFC problems. For instance, the DOFC problems for T-S fuzzy systems were addressed in [40, 128, 147, 191, 345], the corresponding results for Markovian hybrid jump systems were reported in [9, 156, 177, 240**?**], and the feasibility conditions for stochastic switched systems were defined in [54, 90, 91, 175, 361]. In addition, the mixed $\mathcal{H}_2/\mathcal{H}_\infty$ fuzzy OFC design techniques were introduced in [49, 81, 221, 305]. However, to the best of our knowledge, only a few studies have focused on the fuzzy output feedback controller design with the $\mathcal{L}_2$–$\mathcal{L}_\infty$ performance for nonlinear switched systems. Furthermore, the previously obtained results in this domain must be further investigated in terms of multiple aspects, for instance, the mechanism of selecting the fuzzy piecewise Lyapunov functions for nonlinear complex systems to reduce conservativeness and the design strategy for switched DOFC to ensure the system stability and achieve $\mathcal{L}_2$–$\mathcal{L}_\infty$ performance level. These problems are the motivation for the current research.

Practical systems, especially those pertaining to chemical processes and communication, commonly involve time delays, which reduce the system performance and may lead to instability. The prevalent use of stochastic systems has led to the widespread application of stochastic modelling in science and engineering domains [247, 303]. Many key results have been reported for the T-S fuzzy model [42, 239, 262, 360], switched systems [34, 158, 252, 291], and Markovian jumping systems [14, 36, 241, 248, 332]. The general control synthesis methodologies cannot satisfy the requirements for T-S fuzzy systems that incorporate intelligent control method, filtering, and model reduction analysis. Considering these aspects, this monograph presents the innovative research developments and methodologies pertaining to the synthesis and analysis of T-S fuzzy systems in a unified matrix inequality setting. Researchers exploring the problems of intelligent control, filtering, and model reduction for fuzzy-model-based systems can find valuable reference material in this text. The aspects of stability analysis and stabilization, dynamic output feedback (DOF) control, full- and reduced-order filter design, fault detection, and model reduction problems for a class of T-S fuzzy systems are thoroughly investigated. Moreover, novel techniques are applied to systems, including the delay-partitioning method, slack matrix method, reciprocally convex approach, and event-triggered strategy, [15, 22, 31, 288, 349].

1.4 Reduced-Order Method Synthesis

1.4.1 Model Reduction

With the increasing demands of higher security, reliability, and performance in several engineering domains, fault detection techniques have attracted considerable attention. Model-based fault diagnosis represents an effective approach to solve the fault detection problems in technical processes [45, 87, 102, 231, 269, 277]. The key strategy for fault detection involves two parts, specifically, the construction of a residual signal and computation of a residual evaluation function to be compared against a predefined threshold. A fault alarm is generated when the residual evaluation function exceeds the threshold. To detect faults in a timely manner and avoid false alarms, the residual signal should be sensitive to faults and robust against modelling errors or disturbances for a fault detection problem [26, 35, 38, 79, 280]. Recently, several model-based approaches for fault detection problems have been reported for dynamic systems, along with several significant results. For example, the authors of

[192] used a dynamic event-triggered strategy to address the fault detection problem for nonlinear stochastic systems. Certain researchers proposed a novel reliable decentralized control method for interconnected discrete delay systems, in [176]. Moreover, a novel model-based fault detection and prediction scheme for dynamic T-S fuzzy systems was established in [267]. In [95], the sensor fault reconfigurable control problem was investigated for switched systems. In [304], the fault detection problem for two-dimensional Markovian jump systems was addressed. Furthermore, the fault detection issue for uncertain sampled-data systems using deterministic learning was solved in [50]. Despite these achievements, only a few results pertaining to nonlinear switched models with stochastic disturbances have been reported, although considerable effort was expended in the past decade to examine filtering problems due to their theoretical and practical significance in control engineering and signal processing [47, 105, 199, 259, 292].

The nonlinear dynamic systems in various engineering fields commonly correspond to high-order mathematical models [113, 144, 214], and thus, the performance evaluation and system stability analysis are highly difficult and complex. Hence, methods to simplify the original system with lower-order filters based on certain criteria have been focused on [6, 96, 111, 157]. The goal of reduced-order filtering is to incorporate a filter of a lower order than that of the original system based on specific standards. In recent years, many techniques have been introduced to process mathematical models by using reduced-order filters, such as $\mathcal{H}_\infty$ filtering [3, 67, 138], $\mathcal{H}_2$ filtering [84, 185, 246], and $\mathcal{L}_2 - \mathcal{L}_\infty$ filter designs [32, 173, 211]. In [238], the reduced-order model approximation problem for discrete-time hybrid switched nonlinear systems was addressed via T-S fuzzy modelling. In [211], the problem of $\mathcal{L}_2 - \mathcal{L}_\infty$ filtering for discrete-time T-S fuzzy systems with stochastic perturbation was considered. Model order reduction for an electronic circuit design was described in [270]. Moreover, in [221], the problem of stabilization and mixed $\mathcal{H}_2/\mathcal{H}_\infty$ reduced-order dynamic OFC of discrete systems was solved. The authors in [327] addressed the robust fuzzy $\mathcal{L}_2 - \mathcal{L}_\infty$ filtering problem for uncertain discrete-time Markov jump systems with nonhomogeneous jump processes. Furthermore, the dissipative filtering issues for a class of nonlinear delayed systems approximated by T-S fuzzy models were considered in [17]. In addition, the authors of [255] investigated the model approximation for switched systems with stochastic perturbation. In general, the reduced-order filter design can be flexibly and simply implemented in practical applications, which is a motivation for the current study.

1.4.2 Reduced Filtering and Control

Among different filtering techniques, the $\mathcal{H}_\infty$ filter has garnered notable interest as it does not involve any statistical assumptions regarding the exogenous noises and can consider the uncertainty in the system model [116, 330, 335, 362]. The $\mathcal{H}_\infty$ filter design approach considers the worst case from the process noise to the minimization of the estimation error and is thus of considerable value in many practical applications. Significant results pertaining to $\mathcal{H}_\infty$ filtering have been reported. Specifically, the $\mathcal{H}_\infty$ filtering problem for linear systems with uncertain disturbances ([43, 94, 97]), hybrid systems ([57, 168, 308]), singular systems ([172, 187, 289]), and T-S fuzzy systems ([106, 193, 254]) was solved. However, the measurement outputs of a dynamic system contain incomplete observations in practice, as contingent failures of all sensors may occur in a system [40, 130, 152], which may deteriorate the filter performances and pose certain hazard risks, see for example, [163, 164, 290, 318]. In this regard, the problem of reliable filter design is a key issue and has attracted considerable attention, for instance, in [249, 315, 331], reliable filtering problems for discrete-time-delay systems were investigated. Furthermore, for T-S fuzzy systems ([127, 151, 296, 297]), problems of reliable filtering in the presence of sensor faults have attracted research interest [171, 276, 306]. Moreover, the event-driven reliable dissipative filtering issue for a class of T-S fuzzy systems was considered in [140]. The problem of dissipativity-based reliable fuzzy filter design for uncertain nonlinear systems was investigated in [364] using the interval type-2 fuzzy method. The authors in [281] focused on the reliable $\mathcal{L}_2 - \mathcal{L}_\infty$ filter design for continuous-time Markov jump systems, based on the T-S fuzzy model. In addition, the reliable exponential $\mathcal{H}_\infty$ filtering problem for singular Markovian jump systems with time-varying delays subject to sensor failures was resolved in [167]. The authors of [123] investigated the robust passive reliable filtering issue for T-S fuzzy systems with time-varying delays, random uncertainties, and missing measurements.

In practice, many modern industrial systems involve complex features, such as system parameter jumps and the chattering phenomenon [28, 243, 266, 304]. To accurately describe such phenomena, a concept of switched systems has been introduced to overcome the drawbacks of discrete or continuous dynamics derived from the traditional control theory. Switched systems are composed of a finite number of continuous or discrete dynamic subsystems, as well as a switching strategy, indicating the active subsystem at each time instant [98, 124, 300]. Moreover, hybrid stochastic switched systems, which represent a significant component of stochastic jump systems, are composed of a finite number of independent control subsystems, including discrete-time or continuous-time dynamics, and a switching signal governing the activation of these concerned subsystems [200, 206, 208, 212]. A large class of practical systems and processes, including advanced transportation management systems, automated highway systems, communication systems, and network control systems, can be characterized as hybrid stochastic switched systems [154, 319, 322]. Moreover, several intelligent control strategies have been introducing by hybrid switching controllers, which can effectively overcome the limitations of the traditionally adopted single controller and considerably enhance the resulting closed-loop control system performance. In this manner, the corresponding closed-loop control systems can be transformed to typical hybrid stochastic switched systems [210, 223, 261]. Recently, several researchers focused on hybrid stochastic switched systems and achieved notable achievements. For example, the switched controller design problems of hybrid switched systems were addressed in [166, 307, 328], model reduction/approximation approaches for stochastic switched time-delay systems were established in [39, 305, 316], mixed $\mathcal{H}_\infty$ and passive filtering issues were

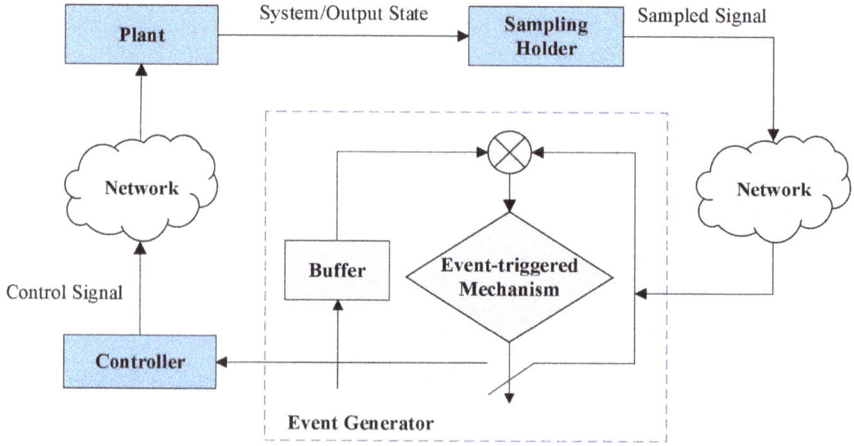

Fig. 1.4 A structure diagram of the event-triggered control system

investigated in [52, 262, 363], and stability analysis and stabilization problems for switched linear systems were discussed in [141, 196, 302].

1.5 Event-Triggered Strategy

At present, two primary communication schemes, specifically, the time-triggered scheme and event-triggered scheme dominate the given research field. In general, the time-triggered communication scheme, proposed for stability analysis and system modelling, may result in transmitting certain "unnecessary" sampling data and wasting the limited space and communication bandwidth in the network [179, 310, 325]. Consequently, the event-triggered scheme was established to gradually replace the time-triggered scheme. The event-triggered technique can mitigate the unnecessary wastage of limited resources while ensuring a satisfactory system performance and has thus garnered increasing research interest [285, 326, 334]. The overall concept of the event-triggered communication scheme is that the control process is completed only if the pre-specified triggering criteria is satisfied, owing to which the state signals are not constantly triggered and transmitted in each sampling period. In this manner, the requirement for the communication bandwidth and computation resources during transmission can be considerably reduced. The structure of the event-triggered control system is shown in Fig. 1.4.

Moreover, NCSs have attracted considerable research attention owing to their advantages of reduced weight, low cost, high reliability, and flexible installation and maintenance [253, 283, 287]. Nevertheless, in NCSs, the sharing of the limited network bandwidth of the communication channels often leads to network time delays, stochastic processes, data-packet dropouts, and the quantization effect, which may lead to the instability and divergence of the NCSs [53, 77, 218]. Time delays generally lead to oscillation, divergence, and even instability in many practical dynamic systems such as electronic circuits, telecommunication, chemical processes, and neural networks [195, 253, 311]. The recently proposed techniques of stability analysis can be classified as delay-dependent [176, 251, 358] and delay-independent approaches [1, 139, 242]. The former approach, which considers the size of the time-delay information, is widely preferred over delay-independent technique. Accordingly, the establishment of methods to reduce the conservativeness by adopting a delay-dependent technique is of key interest in the research on time-varying delay systems.

In addition, it is essential to alleviate the negative impacts of these relevant issues to achieve the ideal system performance, and the primary task to reduce the data transmission. In this regard, it is necessary to employ the event-triggered communication scheme. Notable achievements has been realized the research on event-triggered approach research. In particular, solutions for event-triggered $\mathcal{H}_\infty$ filtering issues were introduced in [333, 342]; the fault detection problems were considered in [133, 295]; $\mathcal{H}_\infty$ stabilization problems were presented in [104, 268]; optimal control methods for discrete event systems were proposed in [180, 258]; and maximum likelihood state estimation approaches were described in [230, 347]. In general, T-S fuzzy systems and dissipativity have been widely recognized and utilized [137, 351].

As another active research domain, many developed methods have been expounded to enhance the effectiveness of the delay-dependent stability condition. Among these advanced methods, the delay-partitioning method, as specified in [184, 217, 245], has been recognized as an effective approach to reduce the conservativeness of the stability conditions for T-S fuzzy delayed systems. This approach initially divides the lower bound of the time-varying delay into several uniform components. Subsequently, the corresponding Lyapunov–Krasovskii function (LKF) is constructed for each delay subinterval. The obtained stability conditions are less conservative than those obtained previously [55, 323]. However, these conditions lead to an increased computational complexity during optimization. Conversely, the reciprocally convex approach discussed in [205] established a lower bound lemma, the key concept of which is to address the information of each delay subinterval and achieve results identical to those of other inequality lemmas with considerably less conservativeness and reduced requirements for state decision variables. This approach has been extensively applied to different dynamic time-varying delay systems such as nonlinear dynamic systems with time-varying delay constraints [249, 343, 353] and continuous-time/discrete-time linear delayed systems [74, 190, 209]. However, there remains considerable scope to enhance the existing results for T-S fuzzy dynamic systems involve uncertain time-delay constraints [109, 350]. A complex and challenging task pertains to further reducing the conservativeness and optimizing the results affected by time-varying delays, particularly when the interval of the time-varying delay increases and becomes uncertain. Thus, based on the considered fuzzy control method, the control approach with an efficient event-triggered scheme [135, 161, 253] is proposed to reduce the burden of communication and save bandwidth resources.

1.6 Publication Contribution

This book is aimed at highlighting the state-of-the-art research for realizing stability/performance analysis and addressing optimal synthesis problems for fuzzy-model-based systems. The contents of this book can be divided into four parts. The book describes the stability analysis and controller design of fuzzy systems. Certain sufficient conditions for the stability and controller design for the considered fuzzy systems with different performances are established. The adopted methodologies include the Lyapunov stability approach and LMI technique. The key objective of using these advanced approaches is to reduce the conservativeness of the obtained results, thereby facilitating the subsequent design. Furthermore, several optimal synthesis problems, including the stabilization, OFC with different system performances, $\mathcal{L}_2$–$\mathcal{L}_\infty$ OFC, dissipative fuzzy control, switched OFC, switched dissipative filtering, fault detection in the presence of sensor nonlinearities, and reduced-order model approximation, are investigated based on the analysis results. Moreover, relevant simulation examples and applications are described to validate the effectiveness and applicability of the design methods.

The key features of this book can be summarized as follows. (1) A unified framework is established for the analysis and synthesis of T-S fuzzy systems involving parameter uncertainties such as external perturbations and faults. (2) A series of problems are solved using novel approaches to analyse and synthesize continuous- and discrete-time fuzzy systems, including stability/performances analysis and stabilization, OFC, tracking control, filtering, fault detection, and model reduction. (3) A set of newly developed techniques (e.g., the Lyapunov stability theory, LMI technique, and convex optimization) are exploited to address the emerging mathematical/computational challenges.

This publication is a timely reflection of the developing new domain of system analysis and synthesis theories for T-S fuzzy systems. The book can be considered to be a collection of a series of latest research results and therefore serves as a useful textbook for senior and/or graduate students who are interested in gaining knowledge regarding the (1) state-of-the-art of fuzzy systems and fuzzy control area; (2) recent advances in nonlinear systems; (3) recent advances in stability/performance analysis, stabilization, OFC, tracking control, fault-tolerant control, filtering, fault detection, and reduced-order model approximation problems. The readers can also familiarize themselves with several new concepts, models, and methodologies with theoretical significance in system analysis and control synthesis. Moreover, the book can be used as a practical research reference for engineers working on stabilization, intelligent control, and filtering problems for T-S fuzzy systems. The objective is to fill the gaps in the literature by providing a unified and structured framework for stability/performances analysis and synthesis of T-S fuzzy systems.

Generally, this advanced publication is aimed at 3rd/4th-year undergraduates, postgraduates and academic researchers. Prerequisite knowledge includes that of fuzzy sets, linear algebra, matrix analysis, and linear control system theory.

The target readership includes (1) control engineers working on nonlinear control, fuzzy control and optimal control; (2) system engineers working on intelligent control systems; (3) mathematicians and physicians working on uncertain systems; (4) postgraduate students majoring in control engineering, system sciences and applied mathematics. This publication is also a useful reference for (1) mathematicians and physicians working on intelligent systems and nonlinear systems; (2) computer scientists working on algorithms and computational complexities; (3) 3rd/4th-year students who are interested in advanced control theories and its applications.

1.7 Publication Outline

The content of this monograph is divided into four parts. Part one focuses on the stability analysis and controller design problem for T-S fuzzy systems. The second part corresponds to the filter design and fault detection problems for fuzzy switched systems. Part three focuses on the model reduction problem for fuzzy systems. Part four presents several applications of fuzzy control approaches. The organization structure of this monograph is shown in Fig. 1.5, and the main contents of this monograph are illustrated in Fig. 1.6.

This chapter describes the research backgrounds, motivations, and research problems, including stabilization synthesis, output feedback control design, filter design, fault detection design, and reduced-order model approximation for fuzzy systems. Subsequently, the outline of the monograph is presented.

Part One focuses on the stability analysis and controller design problem for T-S fuzzy systems. Part One begins with Chap. 2 and consists of three chapters.

Chapter 2 investigates the issues regarding the stability analysis and stabilization for T-S fuzzy systems with time-varying delays. By appropriately choosing the LKF, together with the reciprocally convex method, a new sufficient stability condition employing the delay-partitioning approach is proposed for the delayed T-S fuzzy systems, which can significantly reduce the conservativeness compared to that pertaining to the existing results. Based on the obtained stability condition and PDC law, the state-feedback fuzzy controller is designed for the overall fuzzy systems. Furthermore, the parameters of the designed fuzzy controller are derived in terms of LMIs, which can be easily obtained by applying optimization techniques.

Chapter 3 focuses on the issue of the Hankel-norm output feedback controller design for T-S fuzzy stochastic systems. The full-order controller design scheme with the Hankel-norm performance is established by using the fuzzy-basis-dependent Lyapunov function method and transforming the Hankel-norm controller parameters. Sufficient conditions are presented to design the controllers

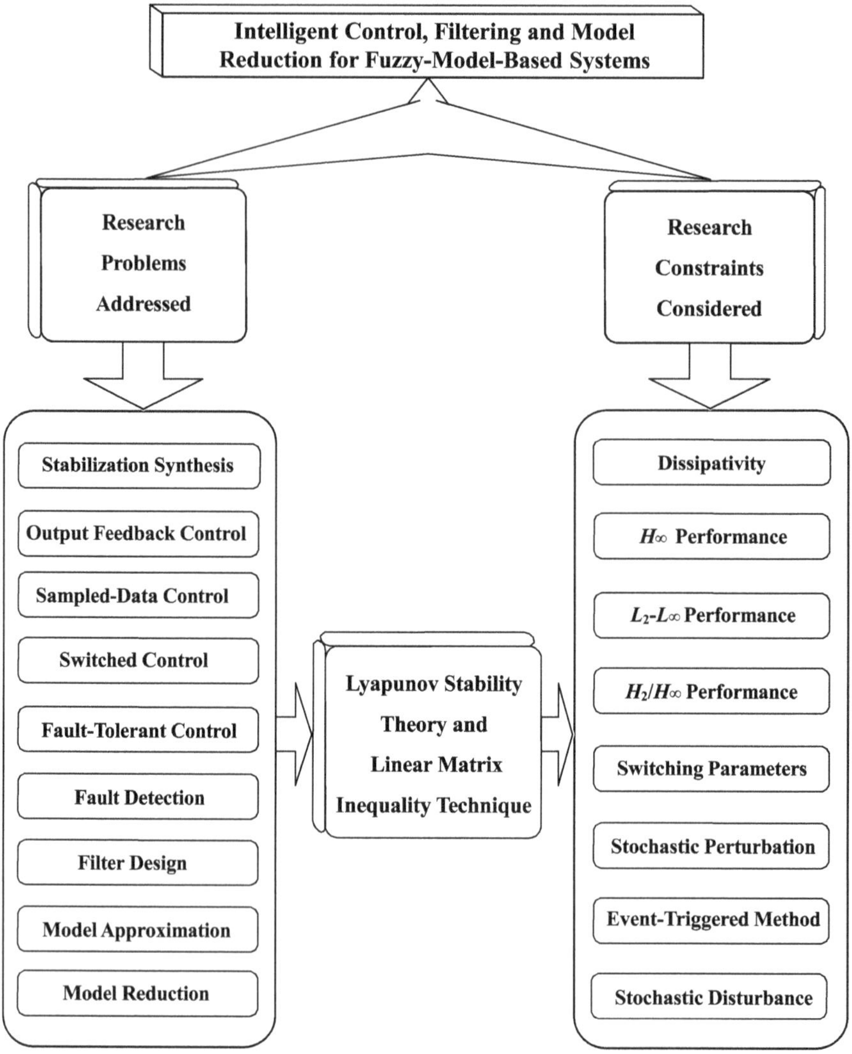

Fig. 1.5 Organizational structure of this publication

such that the resulting closed-loop system is stochastically stable and satisfies the specified performance. By solving a convex optimization issue, we can obtain the desired controller, which can be promptly resolved using standard numerical algorithms.

Chapter 4 address the $\mathcal{L}_2$–$\mathcal{L}_\infty$ output feedback controller design problem for switched systems with nonlinear perturbations in the T-S fuzzy framework. First, the average dwell time (ADT) method is considered to stabilize the non-

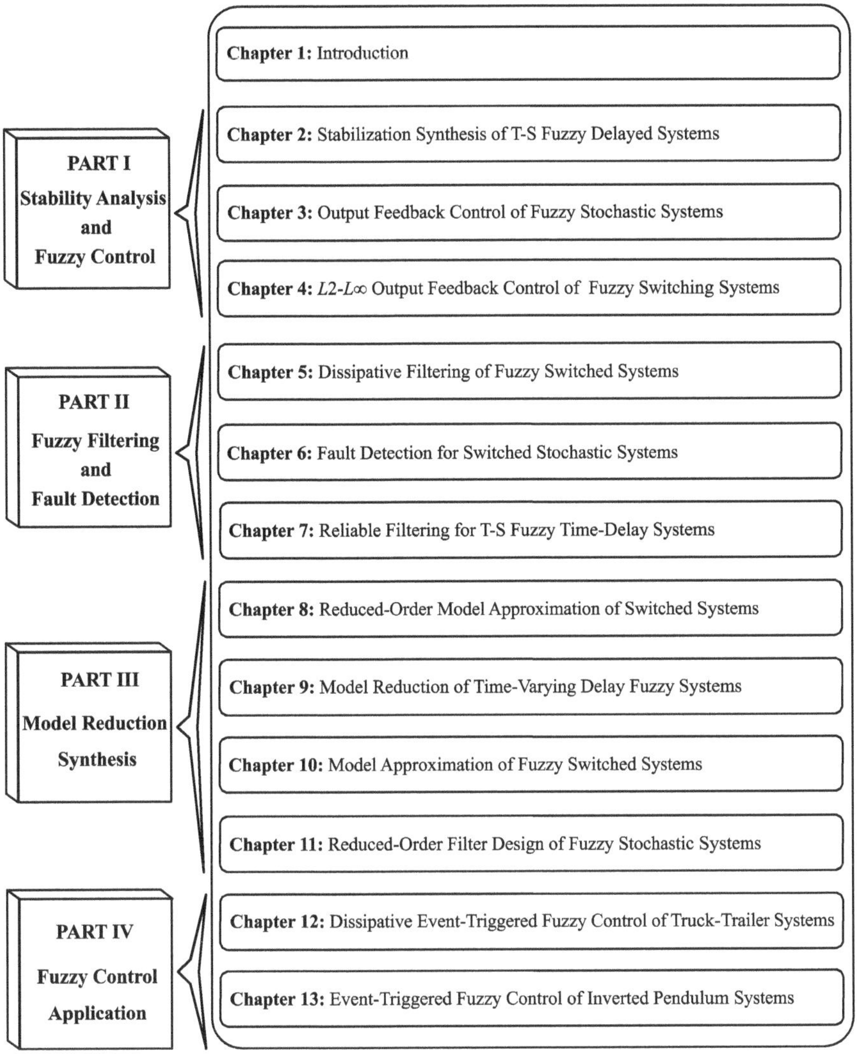

Fig. 1.6 Main contents of this publication

linear switched system exponentially through an arbitrary switching law. Subsequently, based on the piecewise Lyapunov functions, a fuzzy-rule-dependent output feedback controller is proposed to ensure that the closed-loop system is exponentially stable with a weighted $\mathcal{L}_2$–$\mathcal{L}_\infty$ performance (γ, α). The solvable conditions of the desired dynamic controller are derived by employing the linearization approach. The controller matrices can be obtained in terms of several strict LMIs, which can be resolved numerically through efficient standard software.

Part Two focuses on the filter design and fault detection problem for fuzzy switched systems. Part Two begins with Chap. 5 and consists of three chapters.

Chapter 5 investigates the dissipativity-based filtering problem for T-S fuzzy switched systems with stochastic perturbation. First, sufficient conditions are established to ensure the mean-square exponential stability of T-S fuzzy switched systems. Next, we design a filter for the system of interest, subject to Brownian motion. The piecewise Lyapunov function approach and ADT method are combined, and valid fuzzy filters are presented such that the filter error dynamics exhibit mean-square exponential stability and the prescribed dissipative property. Moreover, the solvability conditions of the designed filter are specified through linearization.

Chapter 6 considers the problem of fault detection filtering for the nonlinear switched stochastic system in the form of the T-S fuzzy model. The objective is to establish a robust fault detection approach for a nonlinear switched system with Brownian motion. Based on the observer-based fuzzy filter as a residual generator, the fault detection problem can be reformulated as a fuzzy filtering problem. By using the piecewise Lyapunov functions and ADT approach, a fuzzy-rule-dependent fault detection filter is established to ensure that the overall dynamic system is mean-square exponential stable and exhibits weighted $\mathcal{H}_\infty$ performance. Moreover, the solvable conditions of the fuzzy filter are established through linearization.

Chapter 7 addresses the reliable filtering problem with the requirement of strict dissipative performance for discrete-time T-S fuzzy delayed systems. A reliable filter design is prepared to ensure strict dissipativity for dynamic filtering systems. First, sufficient conditions for the reliable dissipativity analysis are established using a reciprocally convex method for T-S fuzzy systems subject to sensor failures. In addition, a reliable filter is established based on the convex optimization technique, which can be readily resolved through the standard numerical toolbox.

Part Three focuses on the model reduction problem for fuzzy switched systems and fuzzy stochastic systems. Part Three begins with Chap. 8 and consists of four chapters.

Chapter 8 considers the reduced-order model approximation issue for discrete hybrid switched nonlinear systems through T-S fuzzy modelling. We attempt to establish a reduced-order model for a high-dimension hybrid switched system, which can approximate the original high-order model subject to the prescribed system performance index. First, the mean-square exponential stability conditions are formulated to ensure the specific $\mathcal{H}_\infty$ performance for the resulting dynamic error system via the efficient Lyapunov stability method and ADT approach. The solutions of the relevant model reduction problems are derived using the projection technique, through which the algorithms of the reduced-order model parameters are established using a cone complementary linearization (CCL) method.

Chapter 9 presents a novel solution for the model approximation problem for dynamic time-varying delay systems based on fuzzy modelling. We attempt to establish a reduced-order model, which can approximate the original high-order model under a specific system performance index and ensure the asymptotic stability for the overall system. A less conservative stability condition for the overall error system with a specific performance level is derived using the reciprocally convex strategy. The reduced-order model can be ultimately established via the projection strategy, which transfers the model approximation work to a sequential minimization issue under LMI constraints based on the CCL algorithms.

Chapter 10 examines the model approximation issue of a T-S fuzzy switched system with stochastic disturbances. Considering a high-order system, we constructing a reduced-order model, which can approximate the original system with a Hankel-norm performance and convert it to a lower-order switched system. Sufficient conditions are derived using the ADT method and piecewise Lyapunov function to ensure the resulting error system is mean-square exponentially stable and satisfies the Hankel-norm performance sense. Based on a linearization process, the preceding model approximation can be transformed to a convex optimization issue.

Chapter 11 reports on the design of the $\mathcal{H}_\infty$ reduced-order filter for T-S fuzzy delayed systems with stochastic perturbation. Using a novel Lyapunov function and the reciprocally convex approach, the fuzzy-dependent conditions are established to ensure that the resulting filtering system is mean-square asymptotically stable and satisfies the prescribed $\mathcal{H}_\infty$ performance. Next, feasible solutions of the designed reduced-order filter are specified, which can be converted to a convex optimization problem via the convex linearization strategy. Finally, simulation examples, including that of an inverted pendulum, are explained to demonstrate the validity and superiority of the presented $\mathcal{H}_\infty$ reduced-order filtering scheme.

Part Four is aimed at examining two applications of fuzzy control approaches. Part Four begins with Chap. 12 and consists of two chapters.

Chapter 12 addresses the dissipative fuzzy control issue for fuzzy dynamic systems using an event-triggered approach, which is adopted to reduce the transmissions and ensure the stability of the closed-loop system. Furthermore, the dissipative performance is considered in the design of the fuzzy controller to ensure that the resulting closed-loop system is asymptotically stable and strictly (X, Y, Z)–θ–$dissipative$. Based on the fuzzy model, the stability is analysed by using the Lyapunov stability theory. In addition, the designed controller is compactly described in the form of LMIs. An illustrative example is presented to validate the proposed design approach.

Chapter 13 focuses on the event-based fuzzy control design and its application to an inverted pendulum system. First, the time-varying delays in the inverted pendulum system and event-based method are considered in the system stability analysis. The interval of the time-delay is segmented to l non-uniform sub-intervals by applying an efficient delay-partition approach. The information in every subinterval is handled using the reciprocally convex technique. The stability conditions of the inverted pendulum model are established to be less conservative than existing research achievements. The reduction in the conservativeness is more notable when the number of delay partitions l is smaller. In addition, through the PDC rule, feasible conditions for the designed event-triggering fuzzy controller are derived for the considered inverted pendulum system.

Chapter 14 presents the concluding remarks and highlights the future research directions for the presented work.

Part I
Stability Analysis and Fuzzy Control

Chapter 2
Stabilization Synthesis of T-S Fuzzy Delayed Systems

2.1 Introduction

This chapter investigates the issues regarding the stability analysis and stabilization for T-S fuzzy systems with time-varying delays. By appropriately choosing the LKF, together with the reciprocally convex method, a new sufficient stability condition employing the delay-partitioning approach is proposed for the delayed T-S fuzzy systems, which can significantly reduce the conservativeness compared to that pertaining to the existing results. Based on the obtained stability condition and PDC law, the state feedback fuzzy controller is designed for the overall fuzzy systems. Furthermore, the parameters of the designed fuzzy controller are derived in terms of LMIs, which can be easily obtained by applying optimization techniques.

2.2 System Description and Preliminaries

Introduce a class of continuous-time nonlinear systems which can be expressed by the following T-S fuzzy time-varying delay model:

♦ **Plant Form**:
Rule i: IF $\theta_1(t)$ is μ_{i1} and $\theta_2(t)$ is μ_{i2} and $\cdots$ and $\theta_p(t)$ is μ_{ip}, THEN

$$\begin{cases} \dot{x}(t) = A_i x(t) + A_{di} x(t - d(t)) + B_i u(t), \\ x(t) = \varphi(t), \quad t \in \left[-\bar{d}, 0\right], \quad i = 1, 2, \ldots, r, \end{cases}$$

where $x(t) \in \mathbb{R}^n$ denotes the state vector; $u(t) \in \mathbb{R}^q$ denotes the control input vector; $d(t)$ denotes the time-varying delay which satisfies $0 \leqslant \underline{d} \leqslant d(t) \leqslant \bar{d} < \infty, \dot{d}(t) \leqslant \tau$, where $\underline{d}$, $\bar{d}$ and τ are real constant scalars. Partition the delay interval into m fractions, $[\underline{d}, \bar{d}] = \bigcup_{i=1}^{m} [d_{i-1}, d_i]$, with $d_0 = \underline{d}$, $d_m = \bar{d}$. Let η_i be the length of the

X. Su et al., *Intelligent Control, Filtering and Model Reduction Analysis for Fuzzy-Model-Based Systems*, Studies in Systems, Decision and Control 385,
https://doi.org/10.1007/978-3-030-81214-0_2

subinterval, $\eta_i = d_i - d_{i-1}$, $i = 1, \ldots, m$, with $d_{-1} = 0$. A_i, A_{di}, B_i are system matrices with suitable dimensions; $\theta_1(t), \theta_2(t), \ldots, \theta_p(t)$ are the premise variables; $\mu_{i1}, \ldots, \mu_{ip}$ denote the fuzzy sets; r is the number of IF-THEN rules; $\varphi(t)$ represents the initial condition.

Given a pair of $(x(t), u(t))$, the complete T-S fuzzy time-varying delay system is described as

$$\dot{x}(t) = \sum_{i=1}^{r} h_i(\theta(t)) \left[A_i x(t) + A_{di} x(t - d(t)) + B_i u(t)\right], \tag{2.1}$$

with

$$h_i(\theta(t)) = \frac{\nu_i(\theta(t))}{\sum_{i=1}^{r} \nu_i(\theta(t))}, \quad \nu_i(\theta(t)) = \prod_{j=1}^{p} \mu_{ij}(\theta_j(t)),$$

where $\mu_{ij}(\theta_j(t))$ represents the grade of membership of $\theta_j(t)$ in μ_{ij}. Thus, for all t, we can see $\nu_i(\theta(t)) \geqslant 0$, $i = 1, 2, \ldots, r$, $\sum_{i=1}^{r} h_i(\theta(t)) = 1$. For notational simplicity, define

$$A(t) \triangleq \sum_{i=1}^{r} h_i(\theta(t)) A_i, \quad A_d(t) \triangleq \sum_{i=1}^{r} h_i(\theta(t)) A_{di}, \quad B(t) \triangleq \sum_{i=1}^{r} h_i(\theta(t)) B_i.$$

Suppose the premise variable of the fuzzy model $\theta(t)$ is available for feedback, which means $h_i(\theta(t))$ is available for feedback control. Assume the controller's premise variables are the same as those in the plant. The PDC strategy is applied and the state feedback fuzzy controller follows the rules as below:

♦ **Fuzzy Controller**:
Rule i: IF $\theta_1(t)$ is μ_{i1} and $\theta_2(t)$ is μ_{i2} and $\cdots$ and $\theta_p(t)$ is μ_{ip}, THEN

$$u(t) = K_i x(t), \quad i = 1, 2, \ldots, r,$$

where K_i denotes the gain matrix of the state feedback controller.

The controller can be further constructed as follows:

$$u(t) = \sum_{i=1}^{r} h_i(\theta(t)) K_i x(t). \tag{2.2}$$

The compact form of the fuzzy controller is expressed as

$$u(t) = K(t)x(t),$$

where $K(t) = \sum_{i=1}^{r} h_i(\theta(t)) K_i$.

Therefore, the closed-loop system can be obtained as

$$\dot{x}(t) = \sum_{i=1}^{r}\sum_{j=1}^{r} h_i(\theta(t))h_j(\theta(t))\left[A_i x(t) + A_{di}x(t-d(t)) + B_i K_j x(t)\right], \quad (2.3)$$

or expressed in the compact form as

$$\dot{x}(t) = [A(t) + B(t)K(t)]\,x(t) + A_d(t)x(t-d(t)). \quad (2.4)$$

To propose the main results, the following lemmas are introduced.

Lemma 2.1 *([206]) Let $f_1, f_2, \ldots, f_N$: $\mathbb{R}^m \to R$ have positive values in an open subset D of $\mathbb{R}^m$. And the reciprocally convex combination of f_i over D meets*

$$\min_{\{\beta_i | \beta_i > 0, \sum_i \beta_i = 1\}} \sum_i \frac{1}{\beta_i} f_i(t) = \sum_i f_i(t) + \max_{g_{i,j}(t)} \sum_{i \neq j} g_{i,j}(t),$$

which subject to

$$\left\{ g_{i,j} : \mathbb{R}^m \to \mathbb{R},\, g_{j,i}(t) = g_{i,j}(t), \begin{bmatrix} f_i(t) & g_{i,j}(t) \\ g_{j,i}(t) & f_j(t) \end{bmatrix} \geqslant 0 \right\}.$$

Lemma 2.2 *([238]) Given any constant matrix $M > 0$, scalars $b > a > 0$, vector function $w : [a, b] \to \mathbb{R}^n$, then*

$$-(b-a)\int_{t-b}^{t-a} w^T(s)Mw(s)ds \leqslant -\left(\int_{t-b}^{t-a} w(s)ds\right)^T M \left(\int_{t-b}^{t-a} w(s)ds\right),$$

$$-\frac{b^2-a^2}{2}\int_{-b}^{-a}\int_{t+\theta}^{t} w^T(s)Mw(s)dsd\theta$$

$$\leqslant -\left(\int_{-b}^{-a}\int_{t+\theta}^{t} w(s)dsd\theta\right)^T M \left(\int_{-b}^{-a}\int_{t+\theta}^{t} w(s)dsd\theta\right).$$

Remark 2.3 During the subsequent derivation of the LKF, the key is how to handle the amplification derived from the negative integral terms of quadratic quantities. Thus, Lemma 2.1 is an effective approach to deal with these terms for the concerned delayed systems. This processing method has been used for time-varying delay systems in [74] and [206], which will be utilized in our later derivation.

2.3 Main Results

2.3.1 Stability Analysis

We partition the delay interval into two parts in this section to describe our scheme more clearly. By using the reciprocally convex method to handle the Lyapunov function, we present a new stability criterion for fuzzy systems with time-varying delays.

Theorem 2.4 *Assume that the controller gains in (2.2) are known in advance. Given scalars d_0, d_2 and τ, the delayed fuzzy system (2.3) is asymptotically stable, if there exist matrices $S_N > 0$, $R_N > 0$, $Q_N > 0$, $N = 0, 1, 2$, $P > 0$, $Y > 0$, $M(t)$, Z_1 and Z_2 which meet the following conditions:*

$$\Phi(t) + \Phi_1(t) < 0, \tag{2.5}$$

$$\Phi(t) + \Phi_2(t) < 0, \tag{2.6}$$

$$\begin{bmatrix} S_1 & Z_1 \\ \star & S_1 \end{bmatrix} \geqslant 0, \tag{2.7}$$

$$\begin{bmatrix} S_2 & Z_2 \\ \star & S_2 \end{bmatrix} \geqslant 0, \tag{2.8}$$

where

$$\begin{aligned}
\Phi(t) \triangleq\ & d_0^2 W_9^T S_0 W_9 - (W_1 - W_2)^T S_0 (W_1 - W_2) + \eta_1^2 W_9^T S_1 W_9 \\
& + \eta_2^2 W_9^T S_2 W_9 + \frac{d_0^4}{4} W_9^T R_0 W_9 - (\eta_0 W_1 - W_6)^T R_0 (\eta_0 W_1 - W_6) \\
& + W_1^T Q_0 W_1 + \frac{(d_1^2 - d_0^2)^2}{4} W_9^T R_1 W_9 + \frac{(d_2^2 - d_1^2)^2}{4} W_9^T R_2 W_9 \\
& - (\eta_1 W_1 - W_7)^T R_1 (\eta_1 W_1 - W_7) - W_2^T Q_0 W_2 + W_1^T Y W_1 \\
& + (\tau - 1) W_5^T Y W_5 + sym(W_1^T P W_9) - (\eta_2 W_1 - W_8)^T R_2 (\eta_2 W_1 - W_8) \\
& + sym(M(t) W_A(t)) - W_3^T Q_1 W_3 + W_3^T Q_2 W_3 - W_4^T Q_2 W_4 + W_2^T Q_1 W_2, \\
\Phi_1(t) \triangleq & - (W_5 - W_2)^T S_1 (W_5 - W_2) - (W_5 - W_3)^T Z_1^T (W_5 - W_2) \\
& - (W_5 - W_2)^T Z_1 (W_5 - W_3) - (W_5 - W_3)^T S_1 (W_5 - W_3) \\
& - (W_3 - W_4)^T S_2 (W_3 - W_4), \\
\Phi_2(t) \triangleq & - (W_5 - W_3)^T S_2 (W_5 - W_3) - (W_5 - W_4)^T Z_2^T (W_5 - W_3) \\
& - (W_5 - W_3)^T Z_2 (W_5 - W_4) - (W_5 - W_4)^T S_2 (W_5 - W_4) \\
& - (W_2 - W_3)^T S_1 (W_2 - W_3), \\
W_1 \triangleq\ & [I_n \ \ 0_{n,8n}], \quad W_2 \triangleq \begin{bmatrix} 0_{n,n} & I_n & 0_{n,7n} \end{bmatrix}, \\
W_3 \triangleq\ & \begin{bmatrix} 0_{n,2n} & I_n & 0_{n,6n} \end{bmatrix}, \quad W_4 \triangleq \begin{bmatrix} 0_{n,3n} & I_n & 0_{n,5n} \end{bmatrix}, \\
W_5 \triangleq\ & \begin{bmatrix} 0_{n,4n} & I_n & 0_{n,4n} \end{bmatrix}, \quad W_6 \triangleq \begin{bmatrix} 0_{n,5n} & I_n & 0_{n,3n} \end{bmatrix}, \\
W_7 \triangleq\ & \begin{bmatrix} 0_{n,6n} & I_n & 0_{n,2n} \end{bmatrix}, \quad W_8 \triangleq \begin{bmatrix} 0_{n,7n} & I_n & 0_{n,n} \end{bmatrix}, \quad W_9 \triangleq \begin{bmatrix} 0_{n,8n} & I_n \end{bmatrix}, \\
W_A(t) \triangleq\ & [A(t) + B(t)K(t) \ \ 0_{n,3n} \ \ A_d(t) \ \ 0_{n,3n} \ \ -I_n].
\end{aligned}$$

Proof Firstly, we construct the Lyapunov function as follows:

$$V(t) \triangleq \sum_{i=1}^{5} V_i(t), \tag{2.9}$$

where

$$\begin{aligned}
V_1(t) &\triangleq x^T(t)Px(t),\\
V_2(t) &\triangleq \sum_{i=0}^{2}\int_{t-d_i}^{t-d_{i-1}} x^T(s)Q_ix(s)ds,\\
V_3(t) &\triangleq \int_{t-d(t)}^{t} x^T(s)Yx(s)ds + d_0\int_{-d_0}^{0}\int_{t+\theta}^{t}\dot{x}^T(s)S_0\dot{x}(s)dsd\theta,\\
V_4(t) &\triangleq \sum_{i=1}^{2}\eta_i\int_{-d_i}^{-d_{i-1}}\int_{t+\theta}^{t}\dot{x}^T(s)S_i\dot{x}(s)dsd\theta,\\
V_5(t) &\triangleq \sum_{i=0}^{2}\frac{d_i^2-d_{i-1}^2}{2}\int_{-d_i}^{-d_{i-1}}\int_{\theta}^{0}\int_{t+\lambda}^{t}\dot{x}^T(s)R_i\dot{x}(s)dsd\lambda d\theta.
\end{aligned}$$

The time derivative of the Lyapunov function is obtained as

$$\begin{aligned}
\dot{V}_1(t) &\triangleq x^T(t)P\dot{x}(t) + \dot{x}^T(t)Px(t),\\
\dot{V}_2(t) &\triangleq \sum_{i=0}^{2}\left\{x^T(t-d_{i-1})Q_ix(t-d_{i-1}) - x^T(t-d_i)Q_ix(t-d_i)\right\},\\
\dot{V}_3(t) &\triangleq d_0^2\dot{x}^T(t)S_0\dot{x}(t) - d_0\int_{t-d_0}^{t}\dot{x}^T(s)S_0\dot{x}(s)ds + x^T(t)Yx(t)\\
&\quad - (1-\dot{d}(t))x^T(t-d(t))Yx(t-d(t)),\\
\dot{V}_4(t) &\triangleq \sum_{i=1}^{2}\left\{\eta_i^2\dot{x}^T(t)S_i\dot{x}(t) - \eta_i\int_{t-d_i}^{t-d_{i-1}}\dot{x}^T(s)S_i\dot{x}(s)ds\right\},\\
\dot{V}_5(t) &\triangleq -\sum_{i=0}^{2}\frac{d_i^2-d_{i-1}^2}{2}\int_{-d_i}^{-d_{i-1}}\int_{t+\theta}^{t}\dot{x}^T(s)R_i\dot{x}(s)dsd\theta\\
&\quad - \sum_{i=0}^{2}\frac{(d_i^2-d_{i-1}^2)^2}{4}\dot{x}^T(t)R_i\dot{x}(t).
\end{aligned} \tag{2.10}$$

Based on Lemma 2.2, it follows that

$$\dot{V}_3(t) \leqslant d_0^2\dot{x}^T(t)S_0\dot{x}(t) - (x(t) - x(t-d_0))^T S_0(x(t) - x(t-d_0)) + x^T(t)Yx(t) - (1-\tau)x^T(t-d(t))Yx(t-d(t)), \quad (2.11)$$

$$\dot{V}_5(t) \leqslant \sum_{i=0}^{2} \frac{(d_i^2 - d_{i-1}^2)^2}{4}\dot{x}^T(t)R_i\dot{x}(t) - \sum_{i=0}^{2}\left(\eta_i x(t) - \int_{t-d_i}^{t-d_{i-1}} x(s)ds\right)^T R_i \left(\eta_i x(t) - \int_{t-d_i}^{t-d_{i-1}} x(s)ds\right). \quad (2.12)$$

For $\dot{V}_4(t)$, if $d_0 < d(t) < d_1$, according to Lemmas 2.1 and 2.2, it can obtain that

$$\begin{aligned}\dot{V}_4(t) =& \eta_1^2\dot{x}^T(t)S_1\dot{x}(t) - \eta_1\int_{t-d(t)}^{t-d_0}\dot{x}^T(s)S_1\dot{x}(s)ds + \eta_2^2\dot{x}^T(t)S_2\dot{x}(t)\\ &- \eta_1\int_{t-d_1}^{t-d(t)}\dot{x}^T(s)S_1\dot{x}(s)ds - \eta_2\int_{t-d_2}^{t-d_1}\dot{x}^T(s)S_2\dot{x}(s)ds\\ \leqslant& \eta_1^2\dot{x}^T(t)S_1\dot{x}(t) + \eta_2^2\dot{x}^T(t)S_2\dot{x}(t)\\ &- (x(t-d_1) - x(t-d_2))^T S_2(x(t-d_1) - x(t-d_2))\\ &- \frac{\eta_1}{d(t)-d_0}(x(t-d(t)) - x(t-d_0))^T S_1(x(t-d(t)) - x(t-d_0))\\ &- \frac{\eta_1}{d_1-d(t)}(x(t-d(t)) - x(t-d_1))^T S_1(x(t-d(t)) - x(t-d_1))\\ \leqslant& \eta_1^2\dot{x}^T(t)S_1\dot{x}(t) + \eta_2^2\dot{x}^T(t)S_2\dot{x}(t)\\ &- (x(t-d_1) - x(t-d_2))^T S_2(x(t-d_1) - x(t-d_2))\\ &- \begin{bmatrix} x(t-d(t)) - x(t-d_0)\\ x(t-d(t)) - x(t-d_1)\end{bmatrix}^T \begin{bmatrix} S_1 & Z_1\\ \star & S_1\end{bmatrix}\begin{bmatrix} x(t-d(t)) - x(t-d_0)\\ x(t-d(t)) - x(t-d_1)\end{bmatrix}. \end{aligned} \quad (2.13)$$

When $d(t) = d_0$ or $d(t) = d_1$, (2.13) still holds because of $x(t-d(t)) - x(t-d_0) = 0$ or $x(t-d(t)) - x(t-d_1) = 0$. In view of the T-S fuzzy systems in (2.3), it has

$$\Delta = 2\zeta^T(t)M(t)[(A(t) + B(t)K(t))x(t) + A_d(t)x(t-d(t)) - \dot{x}(t)] = 0,$$

where $M(t) = \sum_{i=1}^{r} h_i(\theta(t))M_i$.

Hence, considering (2.10), (2.11), (2.12) and (2.13), when $\zeta(t) \neq 0$, it is easy to obtain

$$\dot{V}(t) = \dot{V}(t) + \Delta \leq \zeta^T(t)(\Phi(t) + \Phi_1(t))\zeta(t) < 0,$$

where

$$\zeta(t) \triangleq \Big[x^T(t)\ \ x^T(t-d_0)\ \ x^T(t-d_1)\ \ x^T(t-d_2)\ \ x^T(t-d(t))$$
$$\int_{t-d_0}^{t} x^T(s)ds\ \ \int_{t-d_1}^{t-d_0} x^T(s)ds\ \ \int_{t-d_2}^{t-d_1} x^T(s)ds\ \ \dot{x}^T(t)\Big]^T.$$

For $\dot{V}_4(t)$, if $d_1 < d(t) < d_2$, utilizing the same method as above, it is not difficult to get that

$$\begin{aligned}
\dot{V}_4(t) = &\ \eta_1^2 \dot{x}^T(t) S_1 \dot{x}(t) - \eta_1 \int_{t-d_1}^{t-d_0} \dot{x}^T(s) S_1 \dot{x}(s) ds \\
&+ \eta_2^2 \dot{x}^T(t) S_2 \dot{x}(t) - \eta_2 \int_{t-d(t)}^{t-d_1} \dot{x}^T(s) S_2 \dot{x}(s) ds \\
&- \eta_2 \int_{t-d_2}^{t-d(t)} \dot{x}^T(s) S_2 \dot{x}(s) ds \\
\leqslant &\ \eta_1^2 \dot{x}^T(t) S_1 \dot{x}(t) + \eta_2^2 \dot{x}^T(t) S_2 \dot{x}(t) \\
&- (x(t-d_0) - x(t-d_1))^T S_1 (x(t-d_0) - x(t-d_1)) \\
&- \begin{bmatrix} x(t-d(t)) - x(t-d_1) \\ x(t-d(t)) - x(t-d_2) \end{bmatrix}^T \begin{bmatrix} S_2 & Z_2 \\ \star & S_2 \end{bmatrix} \begin{bmatrix} x(t-d(t)) - x(t-d_1) \\ x(t-d(t)) - x(t-d_2) \end{bmatrix}.
\end{aligned} \tag{2.14}$$

When $d(t) = d_1$ or $d(t) = d_2$, (2.14) still holds due to $x(t-d(t)) - x(t-d_1) = 0$ or $x(t-d(t)) - x(t-d_2) = 0$. Likewise, if $d_1 < d(t) < d_2$, considering (2.10), (2.12) and (2.14), when $\zeta(t) \neq 0$, we can obtain

$$\dot{V}(t) = \dot{V}(t) + \Delta \leqslant \zeta^T(t)(\Phi(t) + \Phi_2(t))\zeta(t) < 0.$$

Therefore, it is shown that the fuzzy system (2.3) is asymptotically stable.

It is noticed that Theorem 2.4 cannot be directly solved by LMIs for stability analysis. The next goal is to transform the inequalities (2.5) and (2.6) to finite LMIs, which can be readily solved by the standard numerical software. Then we propose the following theorem.

Theorem 2.5 *Assuming the controller gains K_i in (2.2) are known in advance. Given scalars d_0, d_2 and τ, the delayed fuzzy system (2.3) is asymptotically stable, if there exist matrices $S_N > 0$, $R_N > 0$, $Q_N > 0$, $N = 0, 1, 2$, $P > 0$, $Y > 0$, M_i, Z_1 and Z_2 which satisfy (2.7), (2.8) and the following inequalities for $i, j, k = 1, 2, \ldots, r$:*

$$\Phi_{iik} + \Phi_{1iik} < 0, \tag{2.15}$$
$$\Phi_{ijk} + \Phi_{1ijk} + \Phi_{jik} + \Phi_{1jik} < 0, \quad 1 \leqslant i < j \leqslant r, \tag{2.16}$$
$$\Phi_{iik} + \Phi_{2iik} < 0, \tag{2.17}$$
$$\Phi_{ijk} + \Phi_{2ijk} + \Phi_{jik} + \Phi_{2jik} < 0, \quad 1 \leqslant i < j \leqslant r, \tag{2.18}$$

where

$$\begin{aligned}
\Phi_{ijk} \triangleq & d_0^2 W_9^T S_0 W_9 - (W_1 - W_2)^T S_0 (W_1 - W_2) \\
& + \eta_1^2 W_9^T S_1 W_9 + \eta_2^2 W_9^T S_2 W_9 + \frac{d_0^4}{4} W_9^T R_0 W_9 \\
& - (\eta_0 W_1 - W_6)^T R_0 (\eta_0 W_1 - W_6) + \frac{(d_1^2 - d_0^2)^2}{4} W_9^T R_1 W_9 \\
& - (\eta_1 W_1 - W_7)^T R_1 (\eta_1 W_1 - W_7) + \frac{(d_2^2 - d_1^2)^2}{4} W_9^T R_2 W_9 \\
& + sym(W_1^T P W_9) - (\eta_2 W_1 - W_8)^T R_2 (\eta_2 W_1 - W_8) \\
& + W_1^T Y W_1 + (\tau - 1) W_5^T Y W_5 + W_1^T Q_0 W_1 - W_2^T Q_0 W_2 \\
& + W_2^T Q_1 W_2 - W_3^T Q_1 W_3 + W_3^T Q_2 W_3 - W_4^T Q_2 W_4 \\
& + sym(M_i W_{Ajk}), \\
\Phi_{1ijk} \triangleq & - (W_5 - W_2)^T S_1 (W_5 - W_2) - (W_5 - W_3)^T Z_1^T (W_5 - W_2) \\
& - (W_5 - W_2)^T Z_1 (W_5 - W_3) - (W_5 - W_3)^T S_1 (W_5 - W_3) \\
& - (W_3 - W_4)^T S_2 (W_3 - W_4), \\
\Phi_{2ijk} \triangleq & - (W_5 - W_3)^T S_2 (W_5 - W_3) - (W_5 - W_4)^T Z_2^T (W_5 - W_3) \\
& - (W_5 - W_3)^T Z_2 (W_5 - W_4) - (W_5 - W_4)^T S_2 (W_5 - W_4) \\
& - (W_2 - W_3)^T S_1 (W_2 - W_3), \\
W_{Ajk} \triangleq & [A_j + B_j K_k \ \ 0_{n,3n} \ \ A_{dj} \ \ 0_{n,3n} \ \ -I_n].
\end{aligned}$$

Proof On account of the fuzzy basis functions, the inequalities (2.5) and (2.6) can be expressed as

$$\Phi(t) + \Phi_1(t) \triangleq \sum_{i=1}^{r} h_i(\theta(t)) \sum_{j=1}^{r} h_j(\theta(t)) \sum_{k=1}^{r} h_k(\theta(t))(\Phi_{ijk} + \Phi_{1ijk}) < 0, \quad (2.19)$$

$$\Phi(t) + \Phi_2(t) \triangleq \sum_{i=1}^{r} h_i(\theta(t)) \sum_{j=1}^{r} h_j(\theta(t)) \sum_{k=1}^{r} h_k(\theta(t))(\Phi_{ijk} + \Phi_{2ijk}) < 0. \quad (2.20)$$

It is obvious that if the conditions (2.15), (2.16), (2.17) and (2.18) are met, the above-mentioned inequalities (2.19) and (2.20) hold. In other words, the T-S fuzzy time-delay system (2.3) is asymptotically stable.

For the sake of reducing the conservatism, the results are extended to the case whose delay interval is partitioned into arbitrarily m parts and the Lyapunov function is chosen as follows:

$$V(t) \triangleq \sum_{i=1}^{5} V_i(t), \quad (2.21)$$

where

$$
\begin{aligned}
V_1(t) &\triangleq x^T(t)Px(t),\\
V_2(t) &\triangleq \sum_{i=0}^{m}\int_{t-d_i}^{t-d_{i-1}} x^T(s)Q_i x(s)ds,\\
V_3(t) &\triangleq \int_{t-d(t)}^{t} x^T(s)Yx(s)ds + d_0\int_{-d_0}^{0}\int_{t+\theta}^{t}\dot{x}^T(s)S_0\dot{x}(s)dsd\theta,\\
V_4(t) &\triangleq \sum_{i=1}^{m}\eta_i\int_{-d_i}^{-d_{i-1}}\int_{t+\theta}^{t}\dot{x}^T(s)S_i\dot{x}(s)dsd\theta,\\
V_5(t) &\triangleq \sum_{i=0}^{m}\frac{d_i^2-d_{i-1}^2}{2}\int_{-d_i}^{-d_{i-1}}\int_{\theta}^{0}\int_{t+\lambda}^{t}\dot{x}^T(s)R_i\dot{x}(s)dsd\lambda d\theta.
\end{aligned}
$$

On the basis of the aforementioned Lyapunov function, it can easily obtain the following results, which can be proved with similar arguments as those in the proof of Theorems 2.4 and 2.5.

Theorem 2.6 *Assuming the controller gains K_i in (2.2) are known. Given scalars d_0, d_m, $m \geq 2$ and τ, the delayed fuzzy system (2.3) is asymptotically stable, if there exist matrices $S_N > 0$, $R_N > 0$, $Q_N > 0$, $N = 0, 1, \ldots, m$, Z_l, $l = 1, \ldots, m$, $P > 0$, $Y > 0$ and M_i which satisfy the following conditions for $i, j, k = 1, 2, \ldots, r$:*

$$
\Phi_{iik} + \Phi_{liik} < 0, \tag{2.22}
$$

$$
\Phi_{ijk} + \Phi_{lijk} + \Phi_{jik} + \Phi_{ljik} < 0, \quad 1 \leqslant i < j \leqslant r, \tag{2.23}
$$

$$
\begin{bmatrix} S_l & Z_l \\ \star & S_l \end{bmatrix} \geqslant 0, \quad l = 1, \ldots, m, \tag{2.24}
$$

where

$$
\begin{aligned}
\Phi_{ijk} \triangleq{}& \sum_{N=0}^{m}\frac{(d_N^2-d_{N-1}^2)^2}{4}W_{2m+5}^T R_N W_{2m+5}\\
&-\sum_{N=0}^{m}(\eta_N W_1 - W_{m+4+N})^T R_N(\eta_N W_1 - W_{m+4+N})\\
&+\sum_{N=0}^{m}(W_{N+1}^T Q_N W_{N+1} - W_{N+2}^T Q_N W_{N+2}) + W_1^T Y W_1\\
&+\sum_{N=0}^{m}\eta_N^2 W_{2m+5}^T S_N W_{2m+5} + sym(W_1^T P W_{2m+5})\\
&-(W_1 - W_2)^T S_0(W_1 - W_2) + (\tau - 1)W_{m+3}^T Y W_{m+3}\\
&+ sym(M_i W_{Ajk}),\\
\Phi_{lijk} \triangleq{}& -\begin{bmatrix} W_{m+3} - W_{l+1} \\ W_{m+3} - W_{l+2}\end{bmatrix}^T \begin{bmatrix} S_l & Z_l \\ \star & S_l \end{bmatrix}\begin{bmatrix} W_{m+3} - W_{l+1} \\ W_{m+3} - W_{l+2}\end{bmatrix}
\end{aligned}
$$

$$- \sum_{h=1,\ h\neq l}^{m} (W_{h+1} - W_{h+2})^T S_h (W_{h+1} - W_{h+2}),$$

$$W_{Ajk} \triangleq \left[A_j + B_j K_k \;\; 0_{n,(m+1)n} \;\; A_{dj} \;\; 0_{n,(m+1)n} \;\; - I_n \right],$$
$$W_t \triangleq \left[0_{n,(t-1)n} \;\; I_n \;\; 0_{n,(2m+5-t)n} \right], t = 1, 2, \ldots, 2m + 5.$$

Remark 2.7 The presented stability conditions do not require the limited upper bound for the delay derivative, which are more common than the results with strict restraints $\tau < 1$. It is noticeable that our approach is are better suited for real systems.

Remark 2.8 In Theorem 2.6, we utilized the reciprocally convex approach combined with the delay partitioning method and a novel Lyapunov function (2.21) to analyze the stability of the T-S fuzzy delayed system. The key point is on reducing the conservativeness so that the controller design issues have workable solutions. To demonstrate the effectiveness of our proposed delay division method, the sufficient conditions are proposed firstly when the delay interval is partitioned into 2 parts. Then, to further reduce the conservatism, this case is extended to a general one, and we partition the delay interval into arbitrary $m(m \geq 2)$ parts.

2.3.2 State Feedback Fuzzy Control

In this section, we focus on the state feedback controller design for the concerned T-S fuzzy systems (2.3).

Theorem 2.9 *Given scalars $\lambda_{i1}, \lambda_{i2}, \ldots, \lambda_{i(2m+5)}, d_0, d_m, m \geq 2$ and τ, a state feedback controller (2.2) exists such that the closed-loop fuzzy system in (2.3) is asymptotically stable, if there exist matrices $\hat{S}_N > 0$, $\hat{R}_N > 0$, $\hat{Q}_N > 0$, $N = 0, 1, \ldots, m$, X, $\hat{Z}_l$, $l = 1, 2, \ldots, m$, $\hat{P} > 0$, $\hat{Y} > 0$ and G_i which satisfy the following conditions for $i, j = 1, 2, \ldots, r$:*

$$\Pi_{ii} + \Pi_{lii} < 0, \tag{2.25}$$
$$\Pi_{ij} + \Pi_{lij} + \Pi_{ji} + \Pi_{lji} < 0, \quad 1 \leqslant i < j \leqslant r, \tag{2.26}$$
$$\begin{bmatrix} \hat{S}_l & \hat{Z}_l \\ \star & \hat{S}_l \end{bmatrix} \geqslant 0, \quad l = 1, \ldots, m, \tag{2.27}$$

where

$$
\begin{aligned}
\Pi_{ij} \triangleq & \sum_{N=0}^{m} \frac{(d_N^2 - d_{N-1}^2)^2}{4} W_{2m+5}^T \hat{R}_N W_{2m+5} \\
& - \sum_{N=0}^{m} (\eta_N W_1 - W_{m+4+N})^T \hat{R}_N (\eta_N W_1 - W_{m+4+N}) \\
& + \sum_{N=0}^{m} \eta_N^2 W_{2m+5}^T \hat{S}_N W_{2m+5} + \sum_{N=0}^{m} (W_{N+1}^T \hat{Q}_N W_{N+1} \\
& - W_{N+2}^T \hat{Q}_N W_{N+2}) + W_1^T \hat{Y} W_1 + sym(W_1^T \hat{P} W_{2m+5}) \\
& - (W_1 - W_2)^T \hat{S}_0 (W_1 - W_2) + (\tau - 1) W_{m+3}^T \hat{Y} W_{m+3} \\
& + sym(\hat{M}_i \hat{W}_{Aij}), \\
\Pi_{lij} \triangleq & - \begin{bmatrix} W_{m+3} - W_{l+1} \\ W_{m+3} - W_{l+2} \end{bmatrix}^T \begin{bmatrix} \hat{S}_l & \hat{Z}_l \\ \star & \hat{S}_l \end{bmatrix} \begin{bmatrix} W_{m+3} - W_{l+1} \\ W_{m+3} - W_{l+2} \end{bmatrix} \\
& - \sum_{h=1,\ h \neq l}^{m} (W_{h+1} - W_{h+2})^T \hat{S}_h (W_{h+1} - W_{h+2}), \\
\hat{W}_{Aij} \triangleq & \begin{bmatrix} A_i X + B_i G_j & 0_{n,(m+1)n} & A_{di} X & 0_{n,(m+1)n} & -X \end{bmatrix}, \\
\hat{M}_i \triangleq & \begin{bmatrix} \lambda_{i1} I_n & \lambda_{i2} I_n & \dots & \lambda_{i(2m+5)} I_n \end{bmatrix}^T .
\end{aligned}
$$

If the above conditions are solvable, the feedback gains of the fuzzy controllers can be represented as

$$
K_i = G_i X^{-1}, \quad i = 1, 2, \dots, r. \tag{2.28}
$$

Proof For the sake of brevity, define $Z = X^{-T}$ and introduce the following matrix:

$$
E \triangleq diag\{\underbrace{Z, Z, \dots, Z}_{2m+5}\}.
$$

Pre- and post-multiplying (2.25) and (2.26) with E and E^T respectively, which yields

$$
\begin{aligned}
E(\Pi_{ii} + \Pi_{lii})E^T &< 0, \\
E(\Pi_{ij} + \Pi_{lij})E^T + E(\Pi_{ji} + \Pi_{lji})E^T &< 0, \quad 1 \leqslant i < j \leqslant r,
\end{aligned}
$$

where

$$
\begin{aligned}
E\Pi_{ij}E^T \triangleq & \sum_{N=0}^{m} \frac{(d_N^2 - d_{N-1}^2)^2}{4} W_{2m+5}^T Z\hat{R}_N Z^T W_{2m+5} \\
& + \sum_{N=0}^{m} \eta_N^2 W_{2m+5}^T Z\hat{S}_N Z^T W_{2m+5} + W_1^T Z\hat{Y} Z^T W_1 \\
& - \sum_{N=0}^{m} (\eta_N W_1 - W_{m+4+N})^T Z\hat{R}_N Z^T (\eta_N W_1 - W_{m+4+N}) \\
& + \sum_{N=0}^{m} (W_{N+1}^T Z\hat{Q}_N Z^T W_{N+1} - W_{N+2}^T Z\hat{Q}_N Z^T W_{N+2}) \\
& + sym(W_1^T Z\hat{P} Z^T W_{2m+5}) - (W_1 - W_2)^T Z\hat{S}_0 Z^T (W_1 - W_2) \\
& + (\tau - 1) W_{m+3}^T Z\hat{Y} Z^T W_{m+3} + sym(E\hat{M}_i \hat{W}_{Aij} E^T), \\
E\Pi_{lij}E^T \triangleq & - \begin{bmatrix} W_{m+3} - W_{l+1} \\ W_{m+3} - W_{l+2} \end{bmatrix}^T \begin{bmatrix} Z\hat{S}_l Z^T & Z\hat{Z}_l Z^T \\ \star & Z\hat{S}_l Z^T \end{bmatrix} \begin{bmatrix} W_{m+3} - W_{l+1} \\ W_{m+3} - W_{l+2} \end{bmatrix} \\
\triangleq & - \sum_{h=1,\ h\neq l}^{m} (W_{h+1} - W_{h+2})^T Z\hat{S}_h Z^T (W_{h+1} - W_{h+2}).
\end{aligned}
$$

Define

$$
\begin{aligned}
& S_N \triangleq Z\hat{S}_N Z^T, \quad R_N \triangleq Z\hat{R}_N Z^T, \quad Q_N \triangleq Z\hat{Q}_N Z^T, \quad Y \triangleq Z\hat{Y} Z^T, \\
& P \triangleq Z\hat{P} Z^T, \quad Z_l \triangleq Z\hat{Z}_l Z^T, \quad M_i \triangleq E\hat{M}_i.
\end{aligned}
$$

Then we have

$$
\begin{aligned}
E\hat{M}_i \hat{W}_{Aij} E^T = & \left[\lambda_{i1} Z^T \ \ \lambda_{i2} Z^T \ \ \ldots \ \ \lambda_{i(2m+5)} Z^T\right]^T \\
& \times \left[A_i X + B_i G_j \ 0_{n,(m+1)n} \ A_{di} X \ 0_{n,(m+1)n} \ -X\right] E^T \\
= & M_i \left[A_j + B_j K_k \ 0_{n,(m+1)n} \ A_{dj} \ \ 0_{n,(m+1)n} \ -I_n\right], \\
& i, j, k = 1, 2, \ldots, r.
\end{aligned}
$$

Thus, it can see that the results obtained by pre- and post-multiplying (2.25) and (2.26) are equivalent to (2.22) and (2.23). In a similar way, pre- and post-multiplying (2.27) by $diag\{Z, Z\}$ and $diag\{Z^T, Z^T\}$, we can get (2.24). Therefore, from the results in Theorem 2.6, the T-S fuzzy delayed system with our designed controller is asymptotically stable. The proof is then completed.

2.4 Illustrative Example

Example 2.10 Consider the inverted pendulum system in [132] and its schematic diagram is shown in Fig. 2.1. To design a fuzzy controller, we set up a T-S fuzzy model for the corresponding nonlinear system. So first of all, construct a T-S fuzzy model of the pendulum system with approximation methods, and then stabilize the inverted pendulum system with the proposed controller.

Some parameters in the pendulum system are given as

M	mass of the cart, 1.378 kg;
m	mass of the pendulum, 0.051 kg;
l	length of the pendulum, 0.325 m;
g_r	coefficient of the delayed resonator, 0.7 kg/s;
g	acceleration due to gravity, 9.8 m/s^2;
c_r	coefficient of the damper, 5.98 kg/s;
$\theta(t)$	angle the pendulum makes with the top vertical;
$y(t)$	displacement of the cart;
$d(t)$	time-varying delay;
$u(t)$	force applied to the cart.

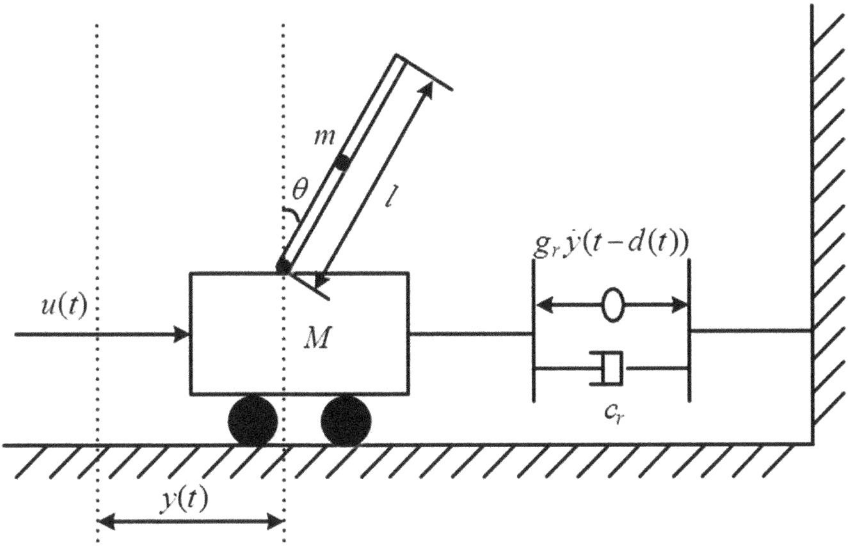

Fig. 2.1 Inverted pendulum on a cart with a delayed resonator

For alignbrevity's sake, the notation "(t)" in some places will be omitted. Suppose the pendulum can be modeled as a thin rod and consider Newton's law, the following equations of the system motion can be obtained:

$$
\begin{aligned}
M\frac{d^2 y}{dt^2} + m\frac{d^2}{dt^2}(y + l\sin\theta) &= u - F_r,\\
m\frac{d^2}{dt^2}(y + l\sin\theta)l\cos\theta &= mgl\sin\theta,
\end{aligned}
$$

where $F_r(t) = g_r\dot{y}(t - d(t)) + c_r\dot{y}(t)$ represents the force of the damper and delayed resonator. Based on the above equations and the state variables $x_1 = y$, $x_2 = \theta$, $x_3 = \dot{y}$, $x_4 = \dot{\theta}$, the achieved state-space equations of the concerned system are as below:

$$
\begin{aligned}
\dot{x}_1 &= x_3,\\
\dot{x}_2 &= x_4,\\
\dot{x}_3 &= \frac{-mg\sin x_2}{M\cos x_2} - \frac{c_r x_3 + g_r x_3(t - d(t)) - u}{M},\\
\dot{x}_4 &= \frac{(M+m)g\sin x_2}{Ml\cos^2 x_2} + \frac{x_4^2\sin x_2}{\cos x_2} + \frac{c_r x_3 + g_r x_3(t - d(t)) - u}{Ml\cos x_2}.
\end{aligned}
$$

Then introduce the following T-S fuzzy model which represents the inverted pendulum system.

♦ **Plant Form**:
Rule 1: IF x_2 is about 0 rad, THEN $\dot{x}(t) = A_1x(t) + A_{d1}x(t - d(t)) + B_1u(t)$.
Rule 2: IF x_2 is near γ $(0 < |\gamma| < 1.57rad)$, THEN $\dot{x}(t) = A_2x(t) + A_{d2}x(t - d(t)) + B_2u(t)$,

$$
A_1 = \begin{bmatrix} 0 & 0 & 1 & 0\\ 0 & 0 & 0 & 1\\ 0 & -\frac{mg}{M} & -\frac{c_r}{M} & 0\\ 0 & \frac{(M+m)g}{Ml} & \frac{c_r}{Ml} & 0 \end{bmatrix},\quad
A_{d1} = \begin{bmatrix} 0 & 0 & 0 & 0\\ 0 & 0 & 0 & 0\\ 0 & 0 & -\frac{g_r}{M} & 0\\ 0 & 0 & \frac{g_r}{Ml} & 0 \end{bmatrix},\quad
B_1 = \begin{bmatrix} 0\\ 0\\ \frac{1}{M}\\ -\frac{1}{Ml} \end{bmatrix},
$$

$$
A_2 = \begin{bmatrix} 0 & 0 & 1 & 0\\ 0 & 0 & 0 & 1\\ 0 & -\frac{mg\beta}{M\alpha} & -\frac{c_r}{M} & 0\\ 0 & \frac{(M+m)g\beta}{Ml\alpha^2} & \frac{c_r}{Ml\alpha} & 0 \end{bmatrix},\quad
A_{d2} = \begin{bmatrix} 0 & 0 & 0 & 0\\ 0 & 0 & 0 & 0\\ 0 & 0 & -\frac{g_r}{M} & 0\\ 0 & 0 & \frac{g_r}{Ml\alpha} & 0 \end{bmatrix},\quad
B_2 = \begin{bmatrix} 0\\ 0\\ \frac{1}{M}\\ -\frac{1}{Ml\alpha} \end{bmatrix},
$$

where $x(t) = [x_1^T(t)\ \ x_2^T(t)\ \ x_3^T(t)\ \ x_4^T(t)]^T$, set $|\gamma| = 0.52$ rad, $\alpha = \cos\gamma$, $\beta = (\sin\gamma)/\gamma$. The fuzzy basis functions $h_1(x_2(t))$ and $h_2(x_2(t))$ are chosen as triangular ones:

$$
h_1(x_2(t)) = 1 - \frac{|x_2(t)|}{|\gamma|},\quad h_2(x_2(t)) = \frac{|x_2(t)|}{|\gamma|}.
$$

Table 2.1 Comparison of upper bounds and controller feedback gains for different cases

Method	d_0	$\bar{d}$	K_1	K_2
Theorem 2.9, $m = 2$	0.1	1.4527	[0.0013 30.2859 7.5837 5.1625]	[0.0011 29.5536 7.3600 4.4054]
	0.3	1.5321	[0.0026 31.6369 7.6595 5.3417]	[0.0022 30.6965 7.4303 4.5661]
	0.5	1.3224	[0.0028 29.7307 7.4063 5.0375]	[0.0026 29.0849 7.2264 4.3056]
Theorem 2.9, $m = 3$	0.1	1.4575	[0.0134 30.7909 7.6601 5.2958]	[0.0112 30.0020 7.4255 4.4940]
	0.3	1.5797	[0.0059 31.5425 7.6834 5.3664]	[0.0052 30.6082 7.4267 4.5654]
	0.5	1.4074	[0.0054 30.7601 7.6066 5.2111]	[0.0048 29.9372 7.3605 4.4352]
Theorem 2.9, $m = 4$	0.1	1.5456	[0.1696 31.2491 7.7618 5.4203]	[0.1422 30.2531 7.4861 4.5755]
	0.3	1.5985	[0.0103 31.8430 7.6205 5.4085]	[0.0088 30.8357 7.3804 4.6171]
	0.5	1.4615	[0.0107 30.8056 7.4986 5.2148]	[0.0095 29.9472 7.2809 4.4471]

Firstly, we make a comparison on the upper bounds obtained from the inverted pendulum system with different lower bound d_0. Table 2.1 lists the upper bounds and the fuzzy controller gains, it's easily to conclude that the allowable upper bound of the controller is enhanced along with the number of partitioned segments m raising. That is to say, the conservatism of our designed scheme is decreased when the number of fractioning is increased. In the mean time, the unknown variables increase. Thus, there is a tradeoff between conservatism reduction and computation complexity. Set the partitioning fractions $m = 2$, $d_0 = 0.1$, $\tau = 0.3$, and the initial condition is set as $x(t) = [0\ \ 0.4\ \ 0\ \ 0]^T$. The dynamics of the original inverted pendulum system is presented in Fig. 2.2, from which we can see the system is unstable. By employing Theorem 2.9, we stabilize the system with the allowable upper bound of time-varying delay $\bar{d} = 1.3894$. Letting $\bar{d} = 1.2$, the following solutions can be obtained which satisfy the conditions in Theorem 2.9.

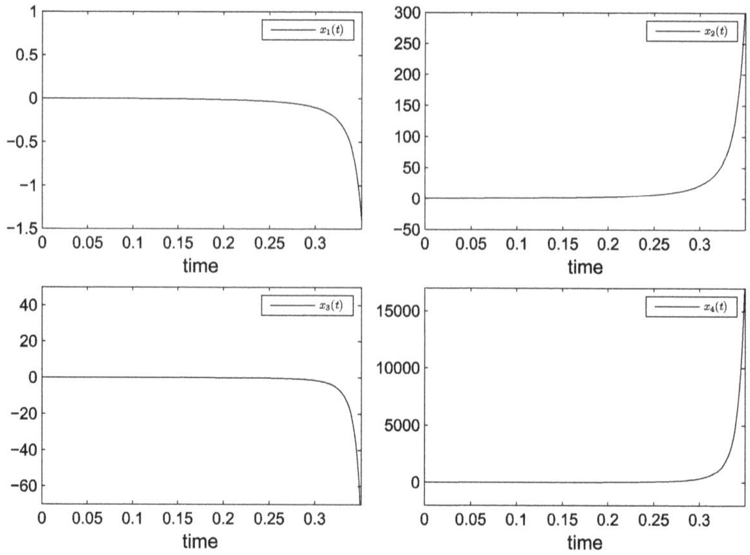

Fig. 2.2 States of the original system without control

$$G_1 = \begin{bmatrix} -0.0265 & 0.0132 & -0.0291 & 0.0370 \end{bmatrix},$$
$$G_2 = \begin{bmatrix} -0.0272 & 0.0189 & -0.0301 & 0.0093 \end{bmatrix},$$
$$X = \begin{bmatrix} 0.0664 & -0.0011 & 0.0008 & -0.0011 \\ 0.0000 & 0.0017 & -0.0035 & -0.0040 \\ -0.0053 & 0.0018 & 0.0084 & -0.0108 \\ 0.0014 & -0.0109 & 0.0020 & 0.0498 \end{bmatrix},$$
$$K_1 = \begin{bmatrix} 0.0983 & 30.0079 & 7.7666 & 4.8405 \end{bmatrix},$$
$$K_2 = \begin{bmatrix} 0.0851 & 29.4028 & 7.5489 & 4.1888 \end{bmatrix}.$$

Consequently, the proposed fuzzy controller in (2.2) is given by

$$\begin{aligned} u(t) = & h_1(x_2(t)) \begin{bmatrix} 0.0983 & 30.0079 & 7.7666 & 4.8405 \end{bmatrix} x(t) \\ & + h_2(x_2(t)) \begin{bmatrix} 0.0851 & 29.4028 & 7.5489 & 4.1888 \end{bmatrix} x(t). \end{aligned}$$

Figure 2.3 shows that the obtained fuzzy state feedback system makes the closed-loop states converge to zero combined with the above controller.

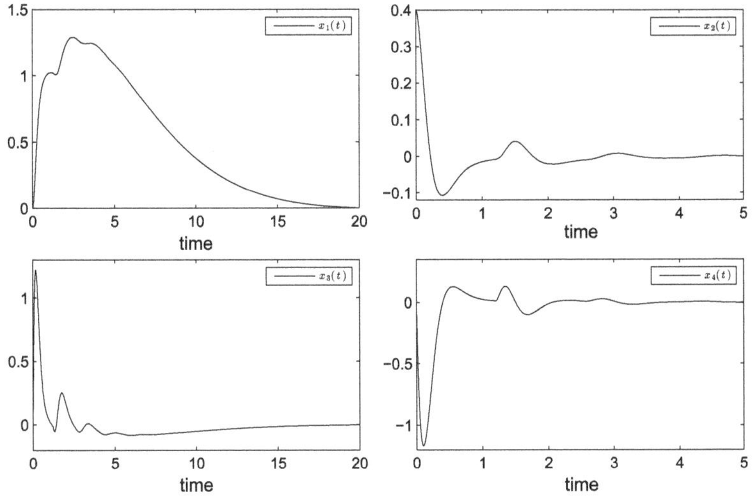

Fig. 2.3 States of the controlled fuzzy system

2.5 Conclusion

The stability and stabilization issues for continuous-time T-S fuzzy systems with time-varying delays were considered. Using the delay-partitioning approach and reciprocally convex method, new sufficient conditions, which could reduce the conservativeness against that of the existing results, were established. Next, an effective primary domain controller was developed to stabilize the fuzzy closed-loop systems. Finally, an illustrative example was given to verify the effectiveness of the approaches presented in this chapter.

Chapter 3
Output Feedback Control of Fuzzy Stochastic Systems

3.1 Introduction

This chapter focuses on the issue of the Hankel-norm output-feedback controller design for T-S fuzzy stochastic systems. The full-order controller design scheme with the Hankel-norm performance is established by using the fuzzy-basis-dependent Lyapunov function method and transforming the Hankel-norm controller parameters. Sufficient conditions are presented to design the controllers such that the resulting closed-loop system is stochastically stable and satisfies the specified performance. By solving a convex optimization issue, we can obtain the desired controller, which can be promptly resolved using standard numerical algorithms.

3.2 System Description and Preliminaries

Some stochastic nonlinear systems can be described with a set of linear systems in local regions. The T-S fuzzy stochastic models with r fuzzy rules are given by

♦ **Plant Form**:

Rule i: IF $\theta_1(k)$ is M_{i1} and $\theta_2(k)$ is M_{i2} and ... and $\theta_p(k)$ is M_{ip}, THEN

$$\begin{cases} x(k+1) = A_i x(k) + B_i u(k) + D_i \omega(k) + L_i x(k) \varpi(k), \\ y(k) = C_i x(k) + F_i \omega(k) + N_i x(k) \varpi(k), \\ z(k) = E_i x(k) + G_i u(k) + H_i \omega(k), \end{cases} \tag{3.1}$$

where $x(k) \in \mathbb{R}^n$ denotes the state vector; $y(k) \in \mathbb{R}^p$ denotes the measured output; $u(k) \in \mathbb{R}^s$ denotes the control input; $z(k) \in \mathbb{R}^q$ denotes the controlled output; and $\omega(k) \in \mathbb{R}^l$ denotes the disturbance input vector which is assumed to belong to $\ell_2[0, \infty)$. M_{ij} represents the fuzzy set; r is the number of IF-THEN rules;

X. Su et al., *Intelligent Control, Filtering and Model Reduction Analysis for Fuzzy-Model-Based Systems*, Studies in Systems, Decision and Control 385,
https://doi.org/10.1007/978-3-030-81214-0_3

$\theta(k) = [\theta_1(k), \theta_2(k), \ldots, \theta_p(k)]$, denoted by θ for simplicity, represents the premise variables vector. $\varpi(k)$ is a scalar Brownian motion defined on the probability space $(\Omega, \mathcal{F}, \mathcal{P})$ relative to an increasing family $(\mathcal{F}_k)_{k\in\mathbb{N}}$ of σ-algerbras $\mathcal{F}_k \subset \mathcal{F}$ generated by $(\varpi(k))_{k\in\mathbb{N}}$. The stochastic process $\varpi(k)$ is independent, and it's supposed that $\mathscr{E}\{\varpi(k)\} = 0$, $\mathscr{E}\{\varpi(k)^2\} = \delta^2$. A_i, B_i, C_i, D_i, E_i, F_i, G_i, H_i, L_i and N_i are known constant matrices subject to suitable dimensions. The fuzzy membership functions are expressed as $h_i(\theta) \triangleq \frac{\prod_{j=1}^{p} M_{ij}(\theta_j)}{\sum_{i=1}^{r}\prod_{j=1}^{p} M_{ij}(\theta_j)}$, $i = 1, \ldots, r$, with $M_{ij}(\theta_j)$ representing the grade of membership of θ_j in M_{ij}. Thus, for all k we have

$$h_i(\theta) \geq 0, i = 1, 2, \ldots, r, \quad \sum_{i=1}^{r} h_i(\theta) = 1. \tag{3.2}$$

Let Θ be a set of basis functions which satisfy (3.2). A more compact presentation of the nonlinear systems in the discrete-time T-S fuzzy model is written as

$$\begin{cases} x(k+1) = A(h)x(k) + B(h)u(k) + D(h)\omega(k) \\ \qquad\qquad +L(h)x(k)\varpi(k), \\ y(k) = C(h)x(k) + F(h)\omega(k) + N(h)x(k)\varpi(k), \\ z(k) = E(h)x(k) + G(h)u(k) + H(h)\omega(k), \end{cases} \tag{3.3}$$

where

$$\begin{array}{ll} A(h) \triangleq \sum_{i=1}^{r} h_i(\theta)A_i, & B(h) \triangleq \sum_{i=1}^{r} h_i(\theta)B_i, \\ C(h) \triangleq \sum_{i=1}^{r} h_i(\theta)C_i, & E(h) \triangleq \sum_{i=1}^{r} h_i(\theta)E_i, \\ D(h) \triangleq \sum_{i=1}^{r} h_i(\theta)D_i, & F(h) \triangleq \sum_{i=1}^{r} h_i(\theta)F_i, \\ G(h) \triangleq \sum_{i=1}^{r} h_i(\theta)G_i, & H(h) \triangleq \sum_{i=1}^{r} h_i(\theta)H_i, \\ L(h) \triangleq \sum_{i=1}^{r} h_i(\theta)L_i, & N(h) \triangleq \sum_{i=1}^{r} h_i(\theta)N_i, \end{array}$$

with $h \triangleq (h_1, h_2, \ldots, h_r) \in \Theta$.

The fuzzy-basis-dependent output feedback controller is designed as

$$(K): \quad \begin{cases} x_c(k+1) = A_c(h)x_c(k) + B_c(h)y(k), \\ u(k) = C_c(h)x_c(k) + D_c(h)y(k), \end{cases} \tag{3.4}$$

where $x_c(k) \in \mathbb{R}^n$ denotes the state vector of the controller; the fuzzy-basis-function-dependent matrices $A_c(h)$, $B_c(h)$, $C_c(h)$ and $D_c(h)$ are controller parameters with appropriate dimensions.

The resulting closed-loop system which combines the proposed controller (3.4) with original system (3.3) is represented as

$$\begin{cases} \xi(k+1) = \bar{A}(h)\xi(k) + \bar{D}(h)\omega(k) + \bar{L}(h)\xi(k)\varpi(k), \\ z(k) = \bar{E}(h)\xi(k) + \bar{H}(h)\omega(k) + \bar{M}(h)\xi(k)\varpi(k), \end{cases} \tag{3.5}$$

where $\xi(k) \triangleq \begin{bmatrix} x^T(k) & x_c^T(k) \end{bmatrix}^T$, and

$$\begin{aligned}
\bar{A}(h) &\triangleq \begin{bmatrix} A(h) + B(h)D_c(h)C(h) & B(h)C_c(h) \\ B_c(h)C(h) & A_c(h) \end{bmatrix}, \\
\bar{D}(h) &\triangleq \begin{bmatrix} D(h) + B(h)D_c(h)F(h) \\ B_c(h)F(h) \end{bmatrix}, \\
\bar{E}(h) &\triangleq \begin{bmatrix} E(h) + G(h)D_c(h)C(h) & G(h)C_c(h) \end{bmatrix}, \\
\bar{H}(h) &\triangleq H(h) + G(h)D_c(h)F(h), \\
\bar{L}(h) &\triangleq \begin{bmatrix} L(h) + B(h)D_c(h)N(h) & 0 \\ B_c(h)N(h) & 0 \end{bmatrix}, \\
\bar{M}(h) &\triangleq \begin{bmatrix} G(h)D_c(h)N(h) & 0 \end{bmatrix}.
\end{aligned}$$

Definition 3.1 The closed-loop system in (3.5) is mean-square asymptotically stable if $\omega(k) = 0$, $\lim_{k\to\infty} \mathscr{E}\{|\xi(k)|\} = 0$.

For a mean-square asymptotically stable closed-loop system in (3.5), we have $z = \{z(k)\} \in \ell_2[0, \infty)$ when $\omega = \{\omega(k)\} \in \ell_2[0, \infty)$.

Definition 3.2 Given a scalar $\gamma > 0$, the closed-loop system in (3.5) is mean-square asymptotically stable with a Hankel-norm error performance level γ if $\mathscr{E}\left\{\sum_{i=T}^{\infty} z^T(i)z(i)\right\} < \gamma^2 \sum_{i=0}^{T-1} \omega^T(i)\omega(i)$, for all $\omega \in \ell_2[0, \infty)$ with $\omega(k) = 0$, $\forall k \geq T$ and satisfy the zero initial condition.

Remark 3.3 In this chapter, the Hankel-norm measure for the desired stable systems is proposed. This criterion possesses a series of important characteristics. Firstly, it is well known that the Hankel-norm is to lie between the $\mathcal{L}_2$–$\mathcal{L}_2$ (for continuous-time systems) or ℓ_2–ℓ_2 (for discrete-time systems) norm and the $\mathcal{L}_2$–$\mathcal{L}_\infty$ (for continuous-time systems) or ℓ_2–ℓ_∞ (for discrete-time systems) norm in the frequency domain and thus appears as a tradeoff between the two common error criteria. Secondly, this norm is aimed to the singular values of Hankel matrices, and the singular values of such matrices are rather insensitive to perturbations, which lead to the obtained model fairly robust to handle the uncertainties. Thirdly, the principal advantage of the Hankel-norm index is its ability to quantify frequency dependent interactions and it can be utilized for input-output pairing. In consideration of these features, the Hankel-norm measure has been widely applied to system control and analysis during some practical applications.

3.3 Main Results

3.3.1 State-Feedback Control

In this section, by employing fuzzy-basis-dependent Lyapunov functions, some new sufficient conditions are proposed for the resulting closed-loop system in (3.5), which is to be mean-square asymptotically stable with Hankel-norm error performance level γ. First of all, the fuzzy-basis-dependent Lyapunov functions are constructed:

$$\begin{cases} \mathscr{V}(\xi(k), k) \triangleq \xi^T(k)[\mathscr{P}^{-1}(h)]\xi(k), \\ \mathcal{V}(\xi(k), k) \triangleq \xi^T(k)[\mathcal{P}^{-1}(h)]\xi(k), \end{cases} \tag{3.6}$$

where

$$\mathscr{P}(h) \triangleq \sum_{i=1}^{r} h_i(\theta)\mathscr{P}_i, \quad \mathcal{P}(h) \triangleq \sum_{i=1}^{r} h_i(\theta)\mathcal{P}_i. \tag{3.7}$$

On the strength of fuzzy-basis-dependent Lyapunov functions in (3.6), we propose the following results.

Theorem 3.4 *The closed-loop system in (3.5) is mean-square asymptotically stable with Hankel-norm error performance level γ, if there exist fuzzy-basis-dependent matrices $\mathscr{P}(h)$ and $\mathcal{P}(h)$ which satisfy*

$$0 < \mathscr{P}(h) < \mathcal{P}(h), \tag{3.8}$$

for any $h \in \Theta$ and a constant matrix Φ such that for any $h, h^+ \in \Theta$,

$$\begin{bmatrix} -\mathscr{P}(h^+) & 0 & \delta\bar{L}(h)\Phi & 0 \\ \star & -\mathscr{P}(h^+) & \bar{A}(h)\Phi & \bar{D}(h) \\ \star & \star & \mathscr{P}(h) - \Phi - \Phi^T & 0 \\ \star & \star & \star & -\gamma^2 I \end{bmatrix} < 0, \tag{3.9}$$

$$\begin{bmatrix} -\mathcal{P}(h^+) & 0 & 0 & 0 & \delta\bar{L}(h)\Phi \\ \star & -\mathcal{P}(h^+) & 0 & 0 & \bar{A}(h)\Phi \\ \star & \star & -I & 0 & \delta\bar{M}(h)\Phi \\ \star & \star & \star & -I & \bar{E}(h)\Phi \\ \star & \star & \star & \star & \mathcal{P}(h) - \Phi - \Phi^T \end{bmatrix} < 0, \tag{3.10}$$

where $h^+ \triangleq \{h_1[\theta(k+1)], h_2[\theta(k+1)], \ldots, h_r[\theta(k+1)]\}$.

Proof Assume there exist a constant matrix Φ and the matrices $\mathscr{P}(h)$, $\mathcal{P}(h)$ satisfying (3.8) such that inequalities (3.9) and (3.10) hold. Based on (3.9) and (3.10), there exist scalars $\beta_2 > \beta_1 > 0$ and

$$\beta_1 I \leq \mathscr{P}(h) < \Phi + \Phi^T, \quad \mathcal{P}(h) < \Phi + \Phi^T \leq \beta_2 I. \tag{3.11}$$

On account of (3.8), the matrix $\mathscr{P}(h)$ is invertible and for any $h \in \Theta$,

$$0 < \frac{1}{\beta_2} I \leq \mathcal{P}^{-1}(h) < \mathscr{P}^{-1}(h) \leq \frac{1}{\beta_1} I. \tag{3.12}$$

Then we can obtain

$$\mathscr{V}(k) > \mathcal{V}(k). \tag{3.13}$$

Based on (3.12), the following inequalities

$$\begin{cases} [\mathscr{P}(h) - \Phi]^T \mathscr{P}^{-1}(h) [\mathscr{P}(h) - \Phi] \geq 0, \\ [\mathcal{P}(h) - \Phi]^T \mathcal{P}^{-1}(h) [\mathcal{P}(h) - \Phi] \geq 0, \end{cases}$$

signify that

$$\begin{cases} \Phi^T \mathscr{P}^{-1}(h)\Phi \geq \Phi + \Phi^T - \mathscr{P}(h), \\ \Phi^T \mathcal{P}^{-1}(h)\Phi \geq \Phi + \Phi^T - \mathcal{P}(h). \end{cases} \tag{3.14}$$

Combining with (3.9) and (3.10), we obtain

$$\begin{bmatrix} -\mathscr{P}(h^+) & 0 & \delta\bar{L}(h) & 0 \\ \star & -\mathscr{P}(h^+) & \bar{A}(h) & \bar{D}(h) \\ \star & \star & -\mathscr{P}^{-1}(h) & 0 \\ \star & \star & \star & -\gamma^2 I \end{bmatrix} < 0, \tag{3.15}$$

$$\begin{bmatrix} -\mathcal{P}(h^+) & 0 & 0 & 0 & \delta\bar{L}(h) \\ \star & -\mathcal{P}(h^+) & 0 & 0 & \bar{A}(h) \\ \star & \star & -I & 0 & \delta\bar{M}(h) \\ \star & \star & \star & -I & \bar{E}(h) \\ \star & \star & \star & \star & -\mathcal{P}^{-1}(h) \end{bmatrix} < 0. \tag{3.16}$$

Along the trajectory of closed-loop system in (3.5) and in view of the Lyapunov functions in (3.6), it follows that

$$\begin{aligned} &\mathscr{E}\{\Delta\mathscr{V}(\xi(k+1), k+1)\} \\ &\triangleq \mathscr{E}\{\mathscr{V}(\xi(k+1), k+1) - \mathscr{V}(\xi(k), k)\} \\ &= \mathscr{E}\Big\{\big[\bar{A}(h)\xi(k) + \bar{D}(h)\omega(k) + \bar{L}(h)\xi(k)\varpi(k)\big]^T \mathscr{P}^{-1}(h^+) \\ &\times \big[\bar{A}(h)\xi(k) + \bar{D}(h)\omega(k) + \bar{L}(h)\xi(k)\varpi(k)\big] \\ &- \xi^T(k)\mathscr{P}^{-1}(h)\xi(k)\Big\} \\ &= \begin{bmatrix} \xi(k) \\ \omega(k) \end{bmatrix}^T \begin{bmatrix} \Pi_1(k) & \Pi_2(k) \\ \star & \Pi_3(k) \end{bmatrix} \begin{bmatrix} \xi(k) \\ \omega(k) \end{bmatrix}, \end{aligned} \tag{3.17}$$

where

$$\begin{aligned}\Pi_1(k) &\triangleq \bar{A}^T(h)\mathscr{P}^{-1}(h^+)\bar{A}(h) - \mathscr{P}^{-1}(h) + \delta\bar{L}^T(h)\mathscr{P}^{-1}(h^+)\bar{L}(h),\\ \Pi_2(k) &\triangleq \bar{A}^T(h)\mathscr{P}^{-1}(h^+)\bar{D}(h),\\ \Pi_3(k) &\triangleq \bar{D}^T(h)\mathscr{P}^{-1}(h^+)\bar{D}(h).\end{aligned}$$

On the basis of the inequalities (3.15) and $\omega(k) = 0$, we have $\Delta\mathscr{V}(k) < 0$, and the resulting closed-loop system in (3.5) is mean-square asymptotically stable.

Based on Definition 3.2, introduce the performance index as

$$\mathscr{E}\left\{\sum_{i=T}^{\infty} z^T(i)z(i)\right\} < \gamma^2 \sum_{i=0}^{T-1} \omega^T(i)\omega(i), \tag{3.18}$$

for all $\omega \in \ell_2[0, \infty)$ with $\omega(k) = 0, \forall k \geq T$. Then establish two inequalities:

$$\begin{cases}\mathscr{E}\left\{\sum_{i=T}^{\infty} z^T(i)z(i)\right\} < \mathcal{V}(\xi(T), T),\\ \quad\quad \mathscr{V}(\xi(T), T) < \gamma^2\sum_{i=0}^{T-1}\omega^T(i)\omega(i).\end{cases} \tag{3.19}$$

From (3.17), we can obtain

$$\mathscr{E}\{\Delta\mathscr{V}(\xi(k+1), k+1)\} - \gamma^2\omega^T(k)\omega(k) = \begin{bmatrix}\xi(k)\\ \omega(k)\end{bmatrix}^T \begin{bmatrix}\bar{\Pi}_1(k) & \bar{\Pi}_2(k)\\ \star & \bar{\Pi}_3(k)\end{bmatrix}\begin{bmatrix}\xi(k)\\ \omega(k)\end{bmatrix},$$

where

$$\begin{aligned}\bar{\Pi}_1(k) &\triangleq \bar{A}^T(h)\mathscr{P}^{-1}(h^+)\bar{A}(h) - \mathscr{P}^{-1}(h) + \delta\bar{L}^T(h)\mathscr{P}^{-1}(h^+)\bar{L}(h),\\ \bar{\Pi}_2(k) &\triangleq \bar{A}^T(h)\mathscr{P}^{-1}(h^+)\bar{D}(h),\\ \bar{\Pi}_3(k) &\triangleq \bar{D}^T(h)\mathscr{P}^{-1}(h^+)\bar{D}(h) - \gamma^2 I.\end{aligned}$$

We can obtain the first inequality in (3.19) by summing up two sides of the inequality from 0 to $T - 1$.

Then propose $\mathcal{V}(k)$ in (3.6). Similarly, it's easy to get $\Delta\mathcal{V}(k) < 0$ with zero initial condition. Consider $\omega(k) = 0, \forall k \geq T$ and $\Delta\mathcal{V}(k) < 0$, for any $k \geq T$, (3.16) ensures

$$\mathscr{E}\{\Delta\mathcal{V}(\xi(k+1), k+1)\} + z^T(k)z(k) = \xi^T(k)\hat{\Pi}(k)\xi(k) < 0, \tag{3.20}$$

where

$$\begin{aligned}\hat{\Pi}(k) \triangleq{}& \bar{A}^T(h)\mathscr{P}^{-1}(h^+)\bar{A}(h) + \delta\bar{L}^T(h)\mathscr{P}^{-1}(h^+)\bar{L}(h)\\ &- \mathscr{P}^{-1}(h) + \bar{E}^T(h)\bar{E}(h) + \delta\bar{M}^T(h)\bar{M}(h).\end{aligned}$$

Summing up two sides of (3.20) from T to ∞ that results in

$$\mathcal{V}(\xi(\infty),\infty) - \mathcal{V}(\xi(T),T) + \sum_{i=T}^{\infty} z^T(i)z(i) < 0. \tag{3.21}$$

Since $\mathcal{V}(\infty) \geq 0$, we have the second inequality in (3.19). By considering (3.13) and (3.19), it yields (3.18), and the proof is completed.

Remark 3.5 The obtained results of T-S fuzzy stochastic systems further reduce the conservativeness because of the universality of the Lyapunov functions utilized which include the fuzzy-basis-independent one as a particular case [7, 191, 250]. Next, we are going to demonstrate the full-order output feedback Hankel-norm controller design issues, and these issues can be converted into linear matrices inequalities optimization issues, which can be resolved numerically readily.

3.3.2 Hankel-Norm Output Feedback Control

Theorem 3.6 *Based on the overall closed-loop system in (3.5), and given a constant scalar* $\gamma > 0$*, if there exist fuzzy-basis-dependent matrices* $\mathcal{P}_1(h)$, $\mathcal{P}_2(h)$, $\mathcal{P}_3(h)$, $\mathscr{P}_1(h)$, $\mathscr{P}_2(h)$, $\mathscr{P}_3(h)$, $\mathcal{R}(h)$, $\mathcal{S}(h)$, $\mathscr{A}(h)$, $\mathscr{B}(h)$, $\mathscr{C}(h)$, $\mathscr{D}(h)$, *and constant matrices* $\mathcal{X}$, $\mathcal{Y}$ *and* $\mathcal{S}$ *such that for any* $h, h^+ \in \Theta$,

$$\begin{bmatrix} -\tilde{\mathscr{P}}(h^+) & 0 & \delta\tilde{\Pi}_L(h) & 0 \\ \star & -\tilde{\mathscr{P}}(h^+) & \tilde{\Pi}_A(h) & \tilde{\Pi}_D(h) \\ \star & \star & \tilde{\mathscr{P}}(h) - \tilde{\Pi}_\Phi - \tilde{\Pi}_\Phi^T & 0 \\ \star & \star & \star & -\gamma^2 I \end{bmatrix} < 0, \tag{3.22}$$

$$\begin{bmatrix} -\tilde{\mathcal{P}}(h^+) & 0 & 0 & 0 & \delta\tilde{\Pi}_L(h) \\ \star & -\tilde{\mathcal{P}}(h^+) & 0 & 0 & \tilde{\Pi}_A(h) \\ \star & \star & -I & 0 & \delta\tilde{\Pi}_M(h) \\ \star & \star & \star & -I & \tilde{\Pi}_E(h) \\ \star & \star & \star & \star & \tilde{\mathcal{P}}(h) - \tilde{\Pi}_\Phi - \tilde{\Pi}_\Phi^T \end{bmatrix} < 0, \tag{3.23}$$

$$\begin{bmatrix} \mathscr{P}_1(h) & \mathscr{P}_2(h) \\ \star & \mathscr{P}_3(h) \end{bmatrix} - \begin{bmatrix} \mathcal{P}_1(h) & \mathcal{P}_2(h) \\ \star & \mathcal{P}_3(h) \end{bmatrix} < 0, \tag{3.24}$$

$$\begin{bmatrix} -\mathscr{P}_1(h) & -\mathscr{P}_2(h) \\ \star & -\mathscr{P}_3(h) \end{bmatrix} < 0, \tag{3.25}$$

where

$$\begin{aligned}
\tilde{\mathscr{P}}(h^+) &\triangleq \begin{bmatrix} \mathscr{P}_1(h^+) & \mathscr{P}_2(h^+) \\ \star & \mathscr{P}_3(h^+) \end{bmatrix}, \ \tilde{\mathcal{P}}(h) \triangleq \begin{bmatrix} \mathcal{P}_1(h) & \mathcal{P}_2(h) \\ \star & \mathcal{P}_3(h) \end{bmatrix}, \\
\tilde{\mathcal{P}}(h^+) &\triangleq \begin{bmatrix} \mathcal{P}_1(h^+) & \mathcal{P}_2(h^+) \\ \star & \mathcal{P}_3(h^+) \end{bmatrix}, \ \tilde{\mathscr{P}}(h) \triangleq \begin{bmatrix} \mathscr{P}_1(h) & \mathscr{P}_2(h) \\ \star & \mathscr{P}_3(h) \end{bmatrix}, \\
\tilde{\Pi}_L(h) &\triangleq \begin{bmatrix} L(h)\mathcal{X} + B(h)\mathcal{R}(h) & 0 \\ \mathcal{S}(h) & 0 \end{bmatrix}, \ \tilde{\Pi}_\Phi \triangleq \begin{bmatrix} \mathcal{X} & I \\ \mathcal{S} & \mathcal{Y} \end{bmatrix}, \\
\tilde{\Pi}_A(h) &\triangleq \begin{bmatrix} A(h)\mathcal{X}+B(h)\mathscr{C}(h) & A(h)+B(h)\mathscr{D}(h)C(h) \\ \mathscr{A}(h) & \mathcal{Y}A(h)+\mathscr{B}(h)C(h) \end{bmatrix}, \\
\tilde{\Pi}_D(h) &\triangleq \begin{bmatrix} D(h)+B(h)\mathscr{D}(h)F(h) \\ \mathcal{Y}D(h)+\mathscr{B}(h)F(h) \end{bmatrix}, \\
\tilde{\Pi}_M(h) &\triangleq \begin{bmatrix} G(h)\mathcal{R}(h) & 0 \end{bmatrix}, \\
\tilde{\Pi}_E(h) &\triangleq \begin{bmatrix} E(h)\mathcal{X}+G(h)\mathscr{C}(h) & E(h)+G(h)\mathscr{D}(h)C(h) \end{bmatrix},
\end{aligned}$$

then there exists a dynamic output feedback controller in (3.4) such that the closed-loop system in (3.5) is mean-square asymptotically stable subject to Hankel-norm error performance level γ. Furthermore, we can obtain two nonsingular matrices $\mathcal{U}$ and $\mathcal{V}$ which satisfy

$$\mathcal{V}\mathcal{U} = \mathcal{S} - \mathcal{Y}\mathcal{X}, \tag{3.26}$$

and the corresponding controller matrix can be written as

$$\begin{aligned}
\mathscr{K}(h) \triangleq \begin{bmatrix} A_c(h) & B_c(h) \\ C_c(h) & D_c(h) \end{bmatrix} &= \begin{bmatrix} \mathcal{V}^{-1} & -\mathcal{V}^{-1}\mathcal{Y}B(h) \\ 0 & I \end{bmatrix} \begin{bmatrix} \mathscr{A}(h)-\mathcal{Y}A(h)\mathcal{X} & \mathscr{B}(h) \\ \mathscr{C}(h) & \mathscr{D}(h) \end{bmatrix} \\
&\times \begin{bmatrix} \mathcal{U}^{-1} & 0 \\ -C(h)\mathcal{X}\mathcal{U}^{-1} & I \end{bmatrix}.
\end{aligned} \tag{3.27}$$

Proof Consider (3.23)–(3.25), it follows that

$$\begin{aligned}
\begin{bmatrix} \mathcal{P}_1(h) & \mathcal{P}_2(h) \\ \star & \mathcal{P}_3(h) \end{bmatrix} &< \begin{bmatrix} \mathcal{X} + \mathcal{X}^T & \mathcal{S}^T + I \\ \star & \mathcal{Y} + \mathcal{Y}^T \end{bmatrix}, \\
0 < \begin{bmatrix} \mathscr{P}_1(h) & \mathscr{P}_2(h) \\ \star & \mathscr{P}_3(h) \end{bmatrix} &< \begin{bmatrix} \mathcal{P}_1(h) & \mathcal{P}_2(h) \\ \star & \mathcal{P}_3(h) \end{bmatrix},
\end{aligned}$$

which means matrices $\mathcal{X}$ and $\mathcal{Y}$ are nonsingular. Pre-multiplying $\begin{bmatrix} \mathcal{X}^{-T} & -I \end{bmatrix}$ and post-multiplying $\begin{bmatrix} \mathcal{X}^{-1} \\ -I \end{bmatrix}$ to the above inequalities, it can get

$$\begin{aligned}
&\begin{bmatrix} \mathcal{X}^{-T} & -I \end{bmatrix} \begin{bmatrix} \mathcal{X} + \mathcal{X}^T & \mathcal{S}^T + I \\ \star & \mathcal{Y} + \mathcal{Y}^T \end{bmatrix} \begin{bmatrix} \mathcal{X}^{-1} \\ -I \end{bmatrix} \\
&= (\mathcal{Y}\mathcal{X} - \mathcal{S})\mathcal{X}^{-1} + \mathcal{X}^{-T}(\mathcal{Y}\mathcal{X} - \mathcal{S})^T > 0,
\end{aligned}$$

which signifies $\mathcal{Y}\mathcal{X} - \mathcal{S}$ is also nonsingular, thus there are nonsingular matrices $\mathcal{U}$ and $\mathcal{V}$ such that (3.26) holds.

Define

$$\begin{cases} \Xi \triangleq \begin{bmatrix} I & \mathcal{Y}^T \\ 0 & \mathcal{V}^T \end{bmatrix}, \quad \Phi \triangleq \begin{bmatrix} \mathcal{X} & (I - \mathcal{X}\mathcal{Y}^T)\mathcal{V}^{-T} \\ \mathcal{U} & -\mathcal{U}\mathcal{Y}^T\mathcal{V}^{-T} \end{bmatrix}, \\ \mathcal{P}(h) \triangleq \Xi^{-T}\tilde{\mathcal{P}}(h)\Xi^{-1}, \quad \mathscr{P}(h) \triangleq \Xi^{-T}\tilde{\mathscr{P}}(h)\Xi^{-1}, \\ \mathcal{S}(h) \triangleq \mathcal{Y}L(h)\mathcal{U} + \mathcal{Y}B(h)D_c(h)N(h)\mathcal{U} + \mathcal{V}B_c(h)N(h)\mathcal{U}, \\ \mathcal{R}(h) \triangleq D_c(h)N(h)\mathcal{X}, \end{cases} \tag{3.28}$$

and

$$\begin{aligned} &\begin{bmatrix} \mathscr{A}(h) & \mathscr{B}(h) \\ \mathscr{C}(h) & \mathscr{D}(h) \end{bmatrix} \\ &\triangleq \begin{bmatrix} \mathcal{V} & \mathcal{Y}B(h) \\ 0 & I \end{bmatrix} \begin{bmatrix} A_c(h) & B_c(h) \\ C_c(h) & D_c(h) \end{bmatrix} \begin{bmatrix} \mathcal{U} & 0 \\ C(h)\mathcal{X} & I \end{bmatrix} + \begin{bmatrix} \mathcal{Y}A(h)\mathcal{X} & 0 \\ 0 & 0 \end{bmatrix} \\ &= \begin{bmatrix} \Omega(h) & \mathcal{V}B_c(h) + \mathcal{Y}B(h)D_c(h) \\ C_c(h)\mathcal{U} + D_c(h)C(h)\mathcal{X} & D_c(h) \end{bmatrix}, \end{aligned} \tag{3.29}$$

where

$$\begin{aligned} \Omega(h) \triangleq\ & \mathcal{V}A_c(h)\mathcal{U} + \mathcal{Y}B(h)B_c(h)\mathcal{U} + \mathcal{V}B_c(h)C(h)\mathcal{X} \\ & + \mathcal{Y}B(h)D_c(h)C(h)\mathcal{X} + \mathcal{Y}A(h)\mathcal{X}. \end{aligned}$$

Performing congruence transformations to (3.8)–(3.10) by Ξ, diag $\{\Xi, \Xi, \Xi, I,\}$ and diag$\{\Xi, \Xi, I, I, \Xi\}$, respectively. From (3.28)–(3.29), one have (3.22)–(3.24) and (3.8) yields (3.25). The proof is completed.

Due to the conditions (3.22)–(3.25) in Theorem 3.6 can't be solved immediately for the proposed controller design. The next goal is to convert the inequalities to finite LMIs, which can be resolved with standard MATLAB toolbox. Then the following results are given.

Theorem 3.7 *If there exist matrices $\mathcal{X}$, $\mathcal{Y}$, $\mathcal{S}$, $\mathcal{P}_{1i} > 0$, $\mathcal{P}_{3i} > 0$, $\mathscr{P}_{1i} > 0$, $\mathscr{P}_{3i} > 0$, $\mathcal{P}_{2i}$, $\mathscr{P}_{2i}$, $\mathcal{R}_i$, $\mathcal{S}_i$, $\mathscr{A}_i$, $\mathscr{B}_i$, $\mathscr{C}_i$ and $\mathscr{D}_i$ for all $i, j, g, l \in \{1, \ldots, r\}$ such that*

$$\begin{bmatrix} -\tilde{\mathscr{P}}_j & 0 & \delta\tilde{\Pi}_{Lig} & 0 \\ \star & -\tilde{\mathscr{P}}_j & \tilde{\Pi}_{Aig} & \tilde{\Pi}_{Dig} \\ \star & \star & \tilde{\mathscr{P}}_i - \tilde{\Pi}_{\Phi} - \tilde{\Pi}_{\Phi}^T & 0 \\ \star & \star & \star & -\gamma^2 I \end{bmatrix} < 0, \tag{3.30}$$

$$\begin{bmatrix} -\tilde{\mathcal{P}}_j & 0 & 0 & 0 & \delta\tilde{\Pi}_{Lig} \\ \star & -\tilde{\mathcal{P}}_j & 0 & 0 & \tilde{\Pi}_{Aig} \\ \star & \star & -I & 0 & \delta\tilde{\Pi}_{Mig} \\ \star & \star & \star & -I & \tilde{\Pi}_{Eig} \\ \star & \star & \star & \star & \tilde{\mathcal{P}}_i - \tilde{\Pi}_{\Phi} - \tilde{\Pi}_{\Phi}^T \end{bmatrix} < 0, \tag{3.31}$$

$$\begin{bmatrix} \mathscr{P}_{1i} & \mathscr{P}_{2i} \\ \star & \mathscr{P}_{3i} \end{bmatrix} - \begin{bmatrix} \mathcal{P}_{1j} & \mathcal{P}_{2j} \\ \star & \mathcal{P}_{3j} \end{bmatrix} < 0, \tag{3.32}$$

$$\begin{bmatrix} -\mathscr{P}_{1i} & -\mathscr{P}_{2i} \\ \star & -\mathscr{P}_{3i} \end{bmatrix} < 0, \tag{3.33}$$

where

$$\begin{aligned}
\tilde{\mathscr{P}}_i &\triangleq \begin{bmatrix} \mathscr{P}_{1i} & \mathscr{P}_{2i} \\ \star & \mathscr{P}_{3i} \end{bmatrix}, \quad \tilde{\mathcal{P}}_i \triangleq \begin{bmatrix} \mathcal{P}_{1i} & \mathcal{P}_{2i} \\ \star & \mathcal{P}_{3i} \end{bmatrix}, \\
\tilde{\Pi}_{Aig} &\triangleq \begin{bmatrix} A_i\mathcal{X} + B_i\mathscr{C}_g & A_i + B_i\mathscr{D}_g C_i \\ \mathscr{A}_g & \mathcal{Y}A_i + \mathscr{B}_g C_i \end{bmatrix}, \\
\tilde{\Pi}_{Lig} &\triangleq \begin{bmatrix} L_i\mathcal{X} + B_i\mathcal{R}_g & 0 \\ \mathcal{S}_g & 0 \end{bmatrix}, \quad \tilde{\Pi}_{Dig} \triangleq \begin{bmatrix} D_i + B_i\mathscr{D}_g F_i \\ \mathcal{Y}D_i + \mathscr{B}_g F_i \end{bmatrix}, \\
\tilde{\Pi}_{Eig} &\triangleq \begin{bmatrix} E_i\mathcal{X} + G_i\mathscr{C}_g & E_i + G_i\mathscr{D}_g C_i \end{bmatrix}, \quad \tilde{\Pi}_{Mig} \triangleq \begin{bmatrix} G_i\mathcal{R}_g & 0 \end{bmatrix},
\end{aligned}$$

where $\tilde{\Pi}_{\Phi}$ is given in Theorem 3.6. There exists a Hankel-norm output feedback controller as (3.4) such that the closed-loop system in (3.5) is mean-square asymptotically stable with Hankel-norm performance level γ. In addition, there are two nonsingular matrices $\mathcal{U}$ and $\mathcal{V}$ such that

$$\mathcal{V}\mathcal{U} = \mathcal{S} - \mathcal{Y}\mathcal{X},$$

and the designed fuzzy controller can be obtained as

$$A_c(h) \triangleq \sum_{i=1}^{r}\sum_{g=1}^{r} h_i(\theta)h_g(\theta)\big(\mathcal{V}^{-1}\mathcal{Y}B_i\mathscr{D}_gC_i\mathcal{X}\mathcal{U}^{-1} + \mathcal{V}^{-1}\mathscr{A}_g\mathcal{U}^{-1} - \mathcal{V}^{-1}\mathcal{Y}B_i\mathscr{C}_g\mathcal{U}^{-1} - \mathcal{V}^{-1}\mathcal{Y}A_i\mathcal{X}\mathcal{U}^{-1} - \mathcal{V}^{-1}\mathscr{B}_gC_i\mathcal{X}\mathcal{U}^{-1}\big),$$

$$B_c(h) \triangleq \sum_{i=1}^{r}\sum_{g=1}^{r} h_i(\theta)h_g(\theta)\big(\mathcal{V}^{-1}\mathscr{B}_g - \mathcal{V}^{-1}\mathcal{Y}B_i\mathscr{D}_g\big),$$

$$C_c(h) \triangleq \sum_{i=1}^{r}\sum_{g=1}^{r} h_i(\theta)h_g(\theta)\big(\mathscr{C}_g\mathcal{U}^{-1} - \mathscr{D}_gC_i\mathcal{X}\mathcal{U}^{-1}\big),$$

$$D_c(h) \triangleq \sum_{g=1}^{r} h_g(\theta)\mathscr{D}_g. \tag{3.34}$$

Proof For the given fuzzy-based function $h \in \Theta$ and matrices $\mathcal{P}_{1i}$, $\mathcal{P}_{2i}$, $\mathcal{P}_{3i}$, $\mathscr{P}_{1i}$, $\mathscr{P}_{2i}$, $\mathscr{P}_{3i}$, $\mathcal{R}_i$, $\mathcal{S}_i$, $\mathscr{A}_i$, $\mathscr{B}_i$, $\mathscr{C}_i$ and $\mathscr{D}_i$ which satisfy (3.22)–(3.25), define

$$\begin{cases} \mathcal{P}_1(h) \triangleq \sum_{i=1}^{r} h_i(\theta)\mathcal{P}_{1i}, & \mathcal{P}_2(h) \triangleq \sum_{i=1}^{r} h_i(\theta)\mathcal{P}_{2i}, \\ \mathscr{P}_1(h) \triangleq \sum_{i=1}^{r} h_i(\theta)\mathscr{P}_{1i}, & \mathscr{P}_2(h) \triangleq \sum_{i=1}^{r} h_i(\theta)\mathscr{P}_{2i}, \\ \mathcal{R}(h) \triangleq \sum_{i=1}^{r} h_i(\theta)\mathcal{R}_i, & \mathcal{S}(h) \triangleq \sum_{i=1}^{r} h_i(\theta)\mathcal{S}_i, \\ \mathcal{P}_3(h) \triangleq \sum_{i=1}^{r} h_i(\theta)\mathcal{P}_{3i}, & \mathscr{P}_3(h) \triangleq \sum_{i=1}^{r} h_i(\theta)\mathscr{P}_{3i}, \\ \mathscr{A}(h) \triangleq \sum_{i=1}^{r} h_i(\theta)\mathscr{A}_i, & \mathscr{B}(h) \triangleq \sum_{i=1}^{r} h_i(\theta)\mathscr{B}_i, \\ \mathscr{C}(h) \triangleq \sum_{i=1}^{r} h_i(\theta)\mathscr{C}_i, & \mathscr{D}(h) \triangleq \sum_{i=1}^{r} h_i(\theta)\mathscr{D}_i. \end{cases} \tag{3.35}$$

It is not difficult to have

$$\sum_{i=1}^{r}\sum_{i=j}^{r}\sum_{i=g}^{r} h_i(\theta)h_j(\theta)h_g(\theta) \times \begin{bmatrix} -\tilde{\mathscr{P}}_j & 0 & \delta\tilde{\Pi}_{Lig} & 0 \\ \star & -\tilde{\mathscr{P}}_j & \tilde{\Pi}_{Aig} & \tilde{\Pi}_{Dig} \\ \star & \star & \tilde{\mathscr{P}}_i - \tilde{\Pi}_{\Phi} - \tilde{\Pi}_{\Phi}^T & 0 \\ \star & \star & \star & -\gamma^2 I \end{bmatrix} < 0, \tag{3.36}$$

$$\sum_{i=1}^{r}\sum_{i=j}^{r}\sum_{i=g}^{r} h_i(\theta)h_j(\theta)h_g(\theta) \times \begin{bmatrix} -\tilde{\mathcal{P}}_j & 0 & 0 & 0 & \delta\tilde{\Pi}_{Lig} \\ \star & -\tilde{\mathcal{P}}_j & 0 & 0 & \tilde{\Pi}_{Aig} \\ \star & \star & -I & 0 & \delta\tilde{\Pi}_{Mig} \\ \star & \star & \star & -I & \tilde{\Pi}_{Eig} \\ \star & \star & \star & \star & \tilde{\mathcal{P}}_i - \tilde{\Pi}_{\Phi} - \tilde{\Pi}_{\Phi}^T \end{bmatrix} < 0, \tag{3.37}$$

$$\sum_{i=1}^{r}\sum_{i=j}^{r} h_i(\theta)h_j(\theta)\begin{bmatrix} \mathscr{P}_{1i} - \mathcal{P}_{1j} & \mathscr{P}_{2i} - \mathcal{P}_{2j} \\ \star & \mathscr{P}_{3i} - \mathcal{P}_{3j} \end{bmatrix} < 0, \tag{3.38}$$

$$\sum_{i=1}^{r} h_i(\theta)\begin{bmatrix} -\mathscr{P}_{1i} & -\mathscr{P}_{2i} \\ \star & -\mathscr{P}_{3i} \end{bmatrix} < 0. \tag{3.39}$$

It is clear that (3.36)–(3.39) are satisfied if the conditions in (3.30)–(3.33) hold. Hence, it completes the proof.

Remark 3.8 Based on the fuzzy basis functions, Theorem 3.7 is presented from parameter-dependent matrix inequalities' conditions in Theorem 3.6, which cannot be directly implemented for the output feedback controller design. This converts the obtained sufficient conditions to some finite LMIs, which can be readily solved using standard numerical software.

Remark 3.9 Theorem 3.7 provides the sufficient solvability conditions for Hankel-norm output feedback controller design issue of T-S fuzzy stochastic system, therefore, a desired controller can be obtained by resolving the optimization issue as follow:

$$\min\ \varphi \quad \text{subject to } ((3.30) - (3.33)) \text{ with } \quad \varphi = \gamma^2.$$

3.4 Illustrative Example

Example 3.10 Consider the modified Henon mapping system with stochastic disturbances:

$$\begin{cases} x_1(k+1) = -[\varepsilon x_1(k)]^2 + [0.01x_1(k) + 0.02x_2(k)]\,\varpi(k) \\ \qquad\qquad\quad +0.3x_2(k) + 1.4, \\ x_2(k+1) = \varepsilon x_1(k) + 0.01x_2(k)\varpi(k), \end{cases} \tag{3.40}$$

where $\varpi(k)$ denotes the stochastic process and $\delta^2 = \mathscr{E}\{\varpi(k)^2\}$. The constant $\varepsilon \in [0, 1]$ is the retarded coefficient. Let $\theta = \varepsilon x_1(k)$. Suppose $\theta(k) \in [-\varrho, \varrho]$, $\varrho > 0$. With similar methods in [275], the nonlinear term θ^2 can be described as $\theta^2 = h_1(\theta)(-\varrho)\theta + h_2(\theta)\varrho\theta$, where $h_1(\theta), h_2(\theta) \in [0, 1]$, and $h_1(\theta) + h_2(\theta) = 1$. Then the fuzzy basis functions $h_1(\theta)$ and $h_2(\theta)$ are expressed by

$$h_1(\theta) = \frac{1}{2}\left(1 - \frac{\theta}{\varrho}\right), \quad h_2(\theta) = \frac{1}{2}\left(1 + \frac{\theta}{\varrho}\right).$$

It is noted that $h_1(\theta) = 1$ and $h_2(\theta) = 0$ when θ is $-\varrho$ and that $h_1(\theta) = 0$ and $h_2(\theta) = 1$ when θ is ϱ. And for the given fuzzy control input $u(k)$, the nonlinear stochastic system in (3.40) is constructed with the following T-S fuzzy model:

♦ **Plant Form**:

Rule 1: IF θ is $-\varrho$, THEN

$$x(k+1) = A_1 x(k) + B_1 u^\star(k) + L_1 x(k)\varpi(k).$$

Rule 2: IF θ is ϱ, THEN

$$x(k+1) = A_2 x(k) + B_2 u^\star(k) + L_2 x(k)\varpi(k),$$

where $u^\star(k) = 1.4 + u(k)$ and

$$A_1 = \begin{bmatrix} \varepsilon\varrho & 0.3 \\ \varepsilon & 0 \end{bmatrix}, \quad L_1 = \begin{bmatrix} 0.01 & 0.02 \\ 0 & 0.01 \end{bmatrix}, \quad B_1 = \begin{bmatrix} 1 \\ 0 \end{bmatrix},$$
$$A_2 = \begin{bmatrix} -\varepsilon\varrho & 0.3 \\ \varepsilon & 0 \end{bmatrix}, \quad L_2 = \begin{bmatrix} 0.01 & 0.02 \\ 0 & 0.01 \end{bmatrix}, \quad B_2 = \begin{bmatrix} 0 \\ 1 \end{bmatrix}.$$

The regulated output and disturbance terms are added, then (3.40) turns as

♦ **Plant Form**:

Rule 1: IF $\theta(k)$ is $-\varrho$, THEN

$$\begin{cases} x(k+1) = A_1 x(k) + B_1 u^\star(k) + D_1 \omega(k) + L_1 x(k)\varpi(k), \\ y(k) = C_1 x(k) + F_1 \omega(k) + N_1 x(k)\varpi(k), \\ z(k) = E_1 x(k). \end{cases}$$

Rule 2: IF $\theta(k)$ is ϱ, THEN

$$\begin{cases} x(k+1) = A_2 x(k) + B_2 u^\star(k) + D_2 \omega(k) + L_2 x(k)\varpi(k), \\ y(k) = C_2 x(k) + F_2 \omega(k) + N_2 x(k)\varpi(k), \\ z(k) = E_2 x(k), \end{cases}$$

where

$$C_1 = \begin{bmatrix} 1 - \varepsilon & 0 \end{bmatrix}, \; N_1 = \begin{bmatrix} 0.01 & 0.02 \end{bmatrix}, \; D_1 = \begin{bmatrix} 1 & 0 \end{bmatrix}^T, \; E_1 = \begin{bmatrix} 1 & 0 \end{bmatrix}, \; F_1 = 1,$$
$$C_2 = \begin{bmatrix} \varepsilon & 0 \end{bmatrix}, \; N_2 = \begin{bmatrix} 0.03 & 0.06 \end{bmatrix}, \; D_2 = \begin{bmatrix} 1 & 0 \end{bmatrix}^T, \; E_2 = \begin{bmatrix} 1 & 0 \end{bmatrix}, \; F_2 = 0.5.$$

Moreover, $x(k) = \begin{bmatrix} x_1(k) \\ x_2(k) \end{bmatrix}$, $\varepsilon = 0.9$ and $\varrho = 0.9$. Assumed the initial condition as $\varphi(k) = \begin{bmatrix} 1 \\ -1 \end{bmatrix}$. Here, the Hankel-norm error performance level is given by $\gamma = 25.2684$, and the corresponding controller parameters are obtained as

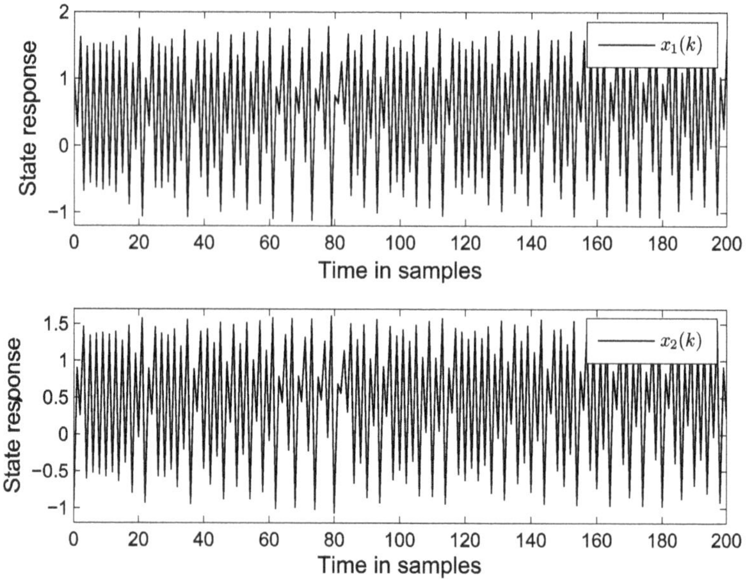

Fig. 3.1 States of the open-loop system

$$\begin{aligned}
A_c(h) &= h_1(\theta)h_2(\theta)\begin{bmatrix} 9.5385 & 62.4944 \\ -1.6038 & -10.5558 \end{bmatrix} + h_1^2(\theta)\begin{bmatrix} 10.5735 & 61.9773 \\ -1.7835 & -10.4730 \end{bmatrix} \\
&\quad + h_1^2(\theta)\begin{bmatrix} -1.0350 & 0.5171 \\ 0.1797 & -0.0828 \end{bmatrix}, \\
B_c(h) &= h_1^2(\theta)\begin{bmatrix} -17.1856 \\ 3.0213 \end{bmatrix} + h_2^2(\theta)\begin{bmatrix} 3.3442 \\ -0.5781 \end{bmatrix} + h_1(\theta)h_2(\theta)\begin{bmatrix} -13.8414 \\ 2.4432 \end{bmatrix}, \\
C_c(h) &= h_1(\theta)h_2(\theta)\begin{bmatrix} -0.0764 & -0.5225 \end{bmatrix} + h_1^2(\theta)\begin{bmatrix} -1.0974 & -5.9260 \end{bmatrix} \\
&\quad + h_2^2(\theta)\begin{bmatrix} 1.0210 & 5.4035 \end{bmatrix}, \\
D_c(h) &= -1.8821.
\end{aligned}$$

The disturbance input $\omega(k)$ is set as $\omega(k) = 2e^{(-0.18k)}\sin(5k)$. The simulation results are shown in Figs. 3.1, 3.2, 3.3. Thereinto, Fig. 3.1 draws the state response $x_1(k)$ and $x_2(k)$ of the open-loop system, and Fig. 3.2 plots the state response $x_1(k)$ (solid line) and $x_2(k)$ (dash-dot line) of the closed-loop system. The control input $u(k)$ is plotted in Fig. 3.3. It can be seen that the proposed controller guarantees the mean-square asymptotic stability with a Hankel-norm performance level of the concerned system.

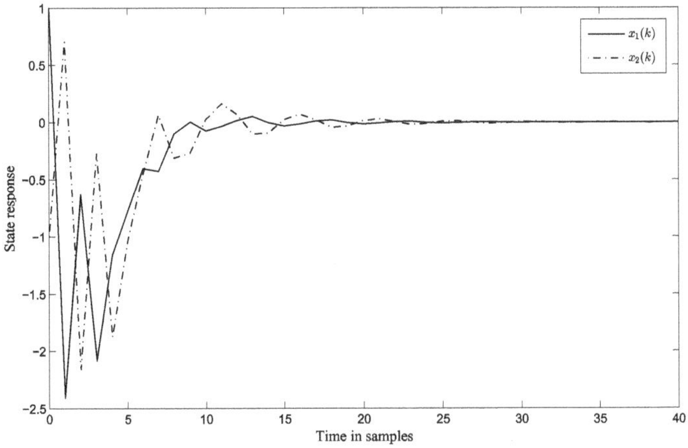

Fig. 3.2 States of the closed-loop system

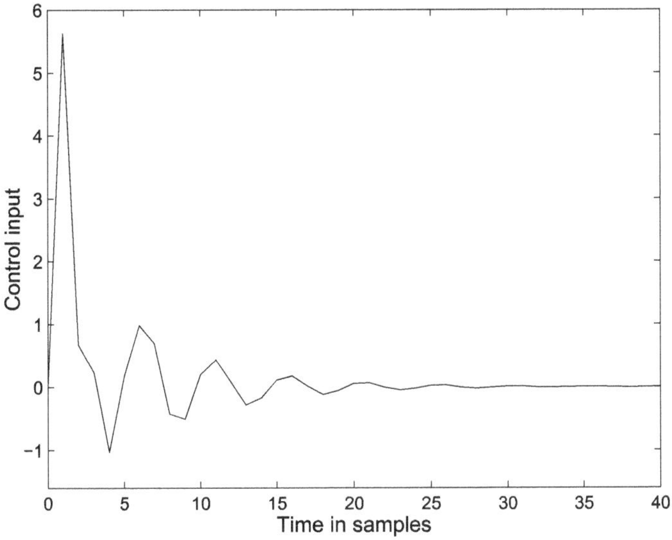

Fig. 3.3 Control input $u(k)$

3.5 Conclusion

In this chapter, the DOF Hankel-norm controller design issue was examined for T-S fuzzy stochastic systems. First, the Hankel-norm controller design scheme was established using certain additional matrix variables, which decoupled the Lyapunov functions and rendered the controller design feasible. Sufficient conditions with less conservativeness were obtained through the use of fuzzy-basis-dependent Lyapunov functions. Furthermore, a full-order OFC problem could be transformed to an optimization problem through the parameter transformation. Finally, a numerical simulation was provided to verify the validity of our controller design method.

Chapter 4
$\mathcal{L}_2$–$\mathcal{L}_\infty$ Output Feedback Control of Fuzzy Switching Systems

4.1 Introduction

This chapter address the $\mathcal{L}_2$–$\mathcal{L}_\infty$ output-feedback controller design problem for switched systems with nonlinear perturbations in the T-S fuzzy framework. First, the ADT method is considered to stabilize the nonlinear switched system exponentially through an arbitrary switching law. Subsequently, based on the piecewise Lyapunov functions, a fuzzy-rule-dependent output-feedback controller is proposed to ensure that the closed-loop system is exponentially stable with a weighted $\mathcal{L}_2$–$\mathcal{L}_\infty$ performance (γ, α). The solvable conditions of the desired dynamic controller are derived by employing the linearization approach. The controller matrices can be obtained in terms of several strict LMIs, which can be resolved numerically through efficient standard software.

4.2 System Description and Preliminaries

By applying the following T-S fuzzy modelling, a dynamic nonlinear system can be described by a class of switched fuzzy linear systems:

♦ **Plant Form**:
Rule $\mathcal{R}_i^{[j]}$: IF $\theta_1^{[j]}(t)$ is $\mu_{i1}^{[j]}$ and $\theta_2^{[j]}(t)$ is $\mu_{i2}^{[j]}$ and $\cdots$ and $\theta_p^{[j]}(t)$ is $\mu_{ip}^{[j]}$, THEN

$$\dot{x}(t) = A_i^{[j]}x(t) + B_{1i}^{[j]}u(t) + B_{2i}^{[j]}\omega(t) + F_i^{[j]}f\big(x(t), t\big), \tag{4.1a}$$

$$y(t) = C_i^{[j]}x(t) + D_{1i}^{[j]}u(t) + D_{2i}^{[j]}\omega(t) + G_i^{[j]}g\big(x(t), t\big), \tag{4.1b}$$

$$z(t) = L_i^{[j]}x(t) + K_i^{[j]}u(t), \quad i = 1, 2, \ldots, r, \tag{4.1c}$$

where $x(t) \in \mathbb{R}^n$ represents the state vector; $u(t) \in \mathbb{R}^m$ represents the control input; $\omega(t) \in \mathbb{R}^l$ represents an exogenous disturbance that $\mathcal{L}_2[0, \infty)$; $y(t) \in \mathbb{R}^p$ denotes the measurement output; $z(t) \in \mathbb{R}^q$ represents the controlled output; r

X. Su et al., *Intelligent Control, Filtering and Model Reduction Analysis for Fuzzy-Model-Based Systems*, Studies in Systems, Decision and Control 385,
https://doi.org/10.1007/978-3-030-81214-0_4

is the number of IF-THEN rules; $\mu_{i1}^{[j]}, \ldots, \mu_{ip}^{[j]}$ denote the fuzzy sets; $\theta^{[j]}(t) = \left[\theta_1^{[j]}(t), \theta_2^{[j]}(t), \ldots, \theta_p^{[j]}(t)\right]$ are premise variables; the positive integer N denotes the number of subsystems.

$$\sigma_j(t) : [0, \infty) \to \{0, 1\}, \quad \sum_{j=1}^{\mathcal{N}} \sigma_j(t) = 1, \; t \in [0, \infty),$$

$j \in \mathcal{N} = \{1, 2, \ldots, N\}$, is a switching signal assigning which subsystem is activated at a switching instant.

$$\left\{ \left(A_i^{[j]}, B_{1i}^{[j]}, B_{2i}^{[j]}, F_i^{[j]}, C_i^{[j]}, D_{1i}^{[j]}, D_{2i}^{[j]}, G_i^{[j]}, L_i^{[j]}, K_i^{[j]} \right) \right\}$$

$j \in \mathcal{N}$ is a set of matrices parameterized by an index set $\mathcal{N} = \{1, 2, \ldots, N\}$.

$$\left\{ A_i^{[j]},\ B_{1i}^{[j]},\ B_{2i}^{[j]},\ F_i^{[j]},\ C_i^{[j]},\ D_{1i}^{[j]},\ D_{2i}^{[j]},\ G_i^{[j]},\ L_i^{[j]},\ K_i^{[j]} \right\}$$

are real constant matrices. $f\left(x(t), t\right) \in \mathbb{R}^f$ and $g\left(x(t), t\right) \in \mathbb{R}^g$ are known nonlinear functions which satisfy the assumption as below.

Remark 4.1 This chapter focuses on the nonlinear switched system which is modeled by T-S fuzzy rules such as (4.1a)–(4.1c). It is observed that Karer [120] has proposed the hybrid fuzzy systems in some cases, i.e., as a hybrid fuzzy model satisfying particular conditions as (4.1a)–(4.1c). Then employing the ADT method and piecewise Lyapunov functions, DOFC is proposed to guarantee the closed-loop system is exponentially stable subject to a weighted $\mathcal{L}_2$–$\mathcal{L}_\infty$ performance level. Hence, we need additional theoretical verifications to see whether the results for switched fuzzy systems can be generalized to hybrid fuzzy systems, which prompts us to carry on the research in this area in the future.

Assumption 4.1 The nonlinear functions $f\left(x(t), t\right)$ and $g\left(x(t), t\right)$ satisfy the zero initial condition ($f\left(0, 0\right) = 0$ and the Lipschitz condition, that is there exist known real matrices M and N which satisfy

$$\left\| f\left(x(t), t\right) - f\left(y(t), t\right) \right\| \leqslant \left\| M(x - y) \right\|,$$
$$\left\| g\left(x(t), t\right) - g\left(y(t), t\right) \right\| \leqslant \left\| N(x - y) \right\|.$$

Suppose the control input $u(t)$ cannot influence the premise variables $\theta^{[j]}(t)$. For a pair of $\left(x(t), u(t)\right)$, the resulting system output is described by

$$\dot{x}(t)=\sum_{j=1}^{N}\sigma_j(t)\sum_{i=1}^{r}h_i^{[j]}\left(\theta^{[j]}(t)\right)\Big[A_i^{[j]}x(t)+B_{1i}^{[j]}u(t)$$
$$+B_{2i}^{[j]}\omega(t)+F_i^{[j]}f(t)\Big], \tag{4.2a}$$

$$y(t)=\sum_{j=1}^{N}\sigma_j(t)\sum_{i=1}^{r}h_i^{[j]}\left(\theta^{[j]}(t)\right)\Big[C_i^{[j]}x(t)+D_{1i}^{[j]}u(t)$$
$$+D_{2i}^{[j]}\omega(t)+G_i^{[j]}g(t)\Big], \tag{4.2b}$$

$$z(t)=\sum_{j=1}^{N}\sigma_j(t)\sum_{i=1}^{r}h_i^{[j]}\left(\theta^{[j]}(t)\right)\Big[L_i^{[j]}x(t)+K_i^{[j]}u(t)\Big], \tag{4.2c}$$

where

$$h_i^{[j]}\left(\theta^{[j]}(t)\right)=\frac{\nu_i^{[j]}\left(\theta^{[j]}(t)\right)}{\sum_{i=1}^{r}\nu_i^{[j]}\left(\theta^{[j]}(t)\right)},\quad \nu_i^{[j]}\left(\theta^{[j]}(t)\right)=\prod_{l=1}^{p}\mu_{il}^{[j]}\left(\theta_l^{[j]}(t)\right),$$

and $\mu_{il}^{[j]}\left(\theta_l^{[j]}(t)\right)$ represents the grade of membership of $\theta_l^{[j]}(t)$ in $\mu_{il}^{[j]}$. Hence, it follows that $\nu_i^{[j]}\left(\theta^{[j]}(t)\right)\geqslant 0$, $h_i^{[j]}\left(\theta^{[j]}(t)\right)\geqslant 0$ for $i=1,2,\ldots,r$, and $\sum_{i=1}^{r}h_i^{[j]}\left(\theta^{[j]}(t)\right)=1$ for all t.

It is assumed that the premise variables $\theta^{[j]}(t)$ are available for the controller design. Thus, the structure of designed controllers can be described as follows by utilizing the PDC technique.

♦ **Dynamic Output Feedback Control Form**:
Rule $\mathcal{R}_i^{[j]}$: IF $\theta_1^{[j]}(t)$ is $\mu_{i1}^{[j]}$ and $\theta_2^{[j]}(t)$ is $\mu_{i2}^{[j]}$ and $\cdots$ and $\theta_p^{[j]}(t)$ is $\mu_{ip}^{[j]}$, THEN

$$\dot{x}_c(t)=A_{ci}^{[j]}x_c(t)+B_{ci}^{[j]}y(t), \tag{4.3a}$$
$$u(t)=C_{ci}^{[j]}x_c(t),\quad i=1,2,\ldots,r, \tag{4.3b}$$

where $x_c(t)\in\mathbb{R}^r$ is the controller state vector with $r\leqslant n$, $A_{ci}^{[j]}$, $B_{ci}^{[j]}$ and $C_{ci}^{[j]}$ are controller parameters to be determined later. A complete form of the dynamic output feedback controller (4.3) is expressed as

$$\dot{x}_c(t)=\sum_{j=1}^{N}\sigma_j(t)\sum_{i=1}^{r}h_i^{[j]}\left(\theta^{[j]}(t)\right)\Big[A_{ci}^{[j]}x_c(t)+B_{ci}^{[j]}y(t)\Big], \tag{4.4a}$$

$$u(t)=\sum_{j=1}^{N}\sigma_j(t)\sum_{i=1}^{r}h_i^{[j]}\left(\theta^{[j]}(t)\right)C_{ci}^{[j]}x_c(t). \tag{4.4b}$$

Thus, combining the system model (4.2) and designed controller (4.4), the resulting closed-loop system can be given by

$$\begin{aligned}\dot{\xi}(t) = &\sum_{j=1}^{N}\sigma_j(t)\sum_{i=1}^{r}h_i^{[j]}\big(\theta^{[j]}(t)\big)\sum_{l=1}^{r}h_l^{[j]}\big(\theta^{[j]}(t)\big)\\ &\times\Big[\tilde{A}_{il}^{[j]}\xi(t) + \tilde{B}_{il}^{[j]}\omega(t) + \tilde{F}_{il}^{[j]}\eta(t)\Big],\end{aligned} \tag{4.5a}$$

$$z(t) = \sum_{j=1}^{N}\sigma_j(t)\sum_{i=1}^{r}h_i^{[j]}\big(\theta^{[j]}(t)\big)\sum_{l=1}^{r}h_l^{[j]}\big(\theta^{[j]}(t)\big)\tilde{C}_{il}^{[j]}\xi(t), \tag{4.5b}$$

where $\xi(t) \triangleq \begin{bmatrix} x(t) \\ x_c(t)\end{bmatrix}$, $\eta(t) \triangleq \begin{bmatrix} f\big(x(t), t\big) \\ g\big(x(t), t\big)\end{bmatrix}$ and

$$\begin{cases} \tilde{A}_{il}^{[j]} \triangleq \begin{bmatrix} A_i^{[j]} & B_{1i}^{[j]}C_{cl}^{[j]} \\ B_{cl}^{[j]}C_i^{[j]} & A_{cl}^{[j]} + B_{cl}^{[j]}D_{1i}^{[j]}C_{cl}^{[j]}\end{bmatrix}, \\ \tilde{F}_{il}^{[j]} \triangleq \begin{bmatrix} F_i^{[j]} & 0 \\ 0 & B_{cl}^{[j]}G_i^{[j]}\end{bmatrix},\ \tilde{B}_{il}^{[j]} \triangleq \begin{bmatrix} B_{2i}^{[j]} \\ B_{cl}^{[j]}D_{2i}^{[j]}\end{bmatrix}, \\ \tilde{C}_{il}^{[j]} \triangleq \begin{bmatrix} L_i^{[j]} & K_i^{[j]}C_{cl}^{[j]}\end{bmatrix}. \end{cases} \tag{4.6}$$

Define

$$\begin{cases} \tilde{A}\big(t, \sigma_j(t)\big) \triangleq \sum_{j=1}^{N}\sigma_j(t)\sum_{i=1}^{r}h_i^{[j]}\big(\theta^{[j]}(t)\big)\sum_{l=1}^{r}h_l^{[j]}\big(\theta^{[j]}(t)\big)\tilde{A}_{il}^{[j]}, \\ \tilde{B}\big(t, \sigma_j(t)\big) \triangleq \sum_{j=1}^{N}\sigma_j(t)\sum_{i=1}^{r}h_i^{[j]}\big(\theta^{[j]}(t)\big)\sum_{l=1}^{r}h_l^{[j]}\big(\theta^{[j]}(t)\big)\tilde{B}_{il}^{[j]}, \\ \tilde{C}\big(t, \sigma_j(t)\big) \triangleq \sum_{j=1}^{N}\sigma_j(t)\sum_{i=1}^{r}h_i^{[j]}\big(\theta^{[j]}(t)\big)\sum_{l=1}^{r}h_l^{[j]}\big(\theta^{[j]}(t)\big)\tilde{C}_{il}^{[j]}, \\ \tilde{F}\big(t, \sigma_j(t)\big) \triangleq \sum_{j=1}^{N}\sigma_j(t)\sum_{i=1}^{r}h_i^{[j]}\big(\theta^{[j]}(t)\big)\sum_{l=1}^{r}h_l^{[j]}\big(\theta^{[j]}(t)\big)\tilde{F}_{il}^{[j]}. \end{cases}$$

Figure 4.1 plots the compact presentation of the closed-loop system (4.5). Before proposing the main results, the following definitions are introduced.

Definition 4.2 [306] When $\omega(t) = 0$, the equilibrium $\xi^\star(t) = 0$ of the closed-loop system (4.5) is exponentially stable under the switching parameter $\sigma_j(t)$, if $\xi(t)$ satisfies

$$\|\xi(t)\|^2 \leqslant \mu\,\|\xi(t_0)\|^2\, e^{-\lambda(t-t_0)}, \quad \forall t \geqslant t_0,$$

for any constants $\mu \geqslant 1$ and $\lambda > 0$.

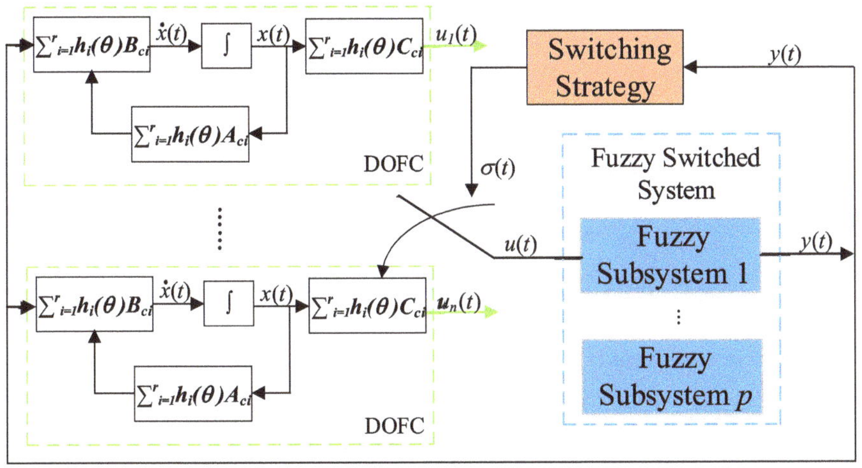

Fig. 4.1 Block diagram of the resulting closed-loop system

Definition 4.3 [306] For the scalars $\gamma > 0$ and $\alpha > 0$, the closed-loop system in (4.5) is exponentially stable with a weighted $\mathcal{L}_2$–$\mathcal{L}_\infty$ performance level (γ, α), if it is exponentially stable for any switching signal $\sigma_j(t)$ when $\omega(t) = 0$, and under zero initial condition $\big(\xi(0) = 0\big)$, for all nonzero $\omega(t) \in \mathcal{L}_2[0, \infty)$, the following inequality holds:

$$\sup_{\forall t} e^{-\alpha t} z^T(t) z(t) < \gamma^2 \int_0^\infty \omega^T(t)\omega(t)dt.$$

4.3 System Performance Analysis

Assume that for given matrices $A_{ci}^{[j]}$, $B_{ci}^{[j]}$, and $C_{ci}^{[j]}$ in (4.4), the sufficient criteria are established to ensure the dynamic closed-loop system (4.5) is exponentially stable with a weighted $\mathcal{L}_2$–$\mathcal{L}_\infty$ performance level (γ, α).

Theorem 4.4 *Given scalars $\gamma > 0$, $\alpha > 0$, if there exist a scalar $\varepsilon > 0$ and matrices $\mathcal{P}^{[j]} > 0$ such that the following inequalities hold for $j \in \mathcal{N}$:*

$$\phi_{ii}^{[j]} < 0, \quad i = 1, 2, \ldots, r, \tag{4.7}$$

$$\frac{1}{r-1}\phi_{ii}^{[j]} + \frac{1}{2}\left(\phi_{il}^{[j]} + \phi_{li}^{[j]}\right) < 0, \quad 1 \leqslant i < l \leqslant r, \tag{4.8}$$

$$\varphi_{ii}^{[j]} < 0, \quad i = 1, 2, \ldots, r, \tag{4.9}$$

$$\frac{1}{r-1}\varphi_{ii}^{[j]} + \frac{1}{2}\left(\varphi_{il}^{[j]} + \varphi_{li}^{[j]}\right) < 0, \quad 1 \leqslant i < l \leqslant r, \tag{4.10}$$

with

$$\phi_{il}^{[j]} \triangleq \begin{bmatrix} \phi_{11il}^{[j]} & \mathcal{P}^{[j]}\tilde{B}_{il}^{[j]} & \mathcal{P}^{[j]}\tilde{F}_{il}^{[j]} \\ \star & -I & 0 \\ \star & \star & -\varepsilon I \end{bmatrix},$$

$$\varphi_{il}^{[j]} \triangleq \begin{bmatrix} -\mathcal{P}^{[j]} & \left(\tilde{C}_{il}^{[j]}\right)^T \\ \star & -\gamma^2 I \end{bmatrix}, \quad \begin{array}{l} \mathcal{M} \triangleq M^T M + N^T N, \\ \mathcal{K} \triangleq \begin{bmatrix} I & 0 \end{bmatrix}, \end{array}$$

$$\phi_{11il}^{[j]} \triangleq \mathcal{P}^{[j]}\tilde{A}_{il}^{[j]} + \left(\tilde{A}_{il}^{[j]}\right)^T \mathcal{P}^{[j]} + \alpha\mathcal{P}^{[j]} + \varepsilon\mathcal{K}^T\mathcal{M}\mathcal{K}.$$

For any switching signal, the dynamic system (4.5) is exponentially stable with a weighted $\mathcal{L}_2$–$\mathcal{L}_\infty$ performance level (γ, α) if $T_a > T_a^\star = \frac{\ln\rho}{\alpha}$, $\rho \geqslant 1$, and the following inequality holds:

$$\mathcal{P}^{[j]} \leqslant \rho\mathcal{P}^{[s]}, \quad \forall j, s \in \mathcal{N}. \tag{4.11}$$

In addition, an estimate of the state decay is described as

$$\|\xi(t)\|^2 \leqslant \mu e^{-\lambda t} \|\xi(0)\|^2, \tag{4.12}$$

where

$$\begin{cases} \lambda = \alpha - \frac{\ln\rho}{T_a} > 0, & \tau = \min_{\forall j \in \mathcal{N}} \lambda_{\min}\left(\mathcal{P}^{[j]}\right), \\ \mu = \frac{\vartheta}{\tau} \geqslant 1, & \vartheta = \max_{\forall j \in \mathcal{N}} \lambda_{\max}\left(\mathcal{P}^{[j]}\right). \end{cases} \tag{4.13}$$

Proof The Lyapunov function is chosen as follow:

$$V\left(\xi(t), \sigma_j(t)\right) = \xi^T(t)\mathcal{P}\left(\sigma_j(t)\right)\xi(t), \tag{4.14}$$

where $\mathcal{P}\left(\sigma_j(t)\right) \triangleq \sum_{j=1}^{N} \sigma_j(t)\mathcal{P}^{[j]}$ $(j \in \mathcal{N})$ are to be determined later. And along the trajectory of (4.5), the derivative of (4.14) is equivalent to

$$\begin{aligned} \dot{V}\left(\xi(t), \sigma_j(t)\right) &= 2\sum_{j=1}^{N} \sigma_j(t) \sum_{i=1}^{r} h_i^{[j]}\left(\theta^{[j]}(t)\right) \sum_{l=1}^{r} h_l^{[j]}\left(\theta^{[j]}(t)\right) \\ &\quad \times \xi^T(t)\mathcal{P}^{[j]}\left[\tilde{A}_{il}^{[j]}\xi(t) + \tilde{F}_{il}^{[j]}\eta(t)\right] \\ &= \sum_{j=1}^{N} \sigma_j(t) \sum_{i=1}^{r} h_i^{[j]}\left(\theta^{[j]}(t)\right) \sum_{l=1}^{r} h_l^{[j]}\left(\theta^{[j]}(t)\right) \end{aligned}$$

$$\times \left\{ \xi^T(t) \left[\mathcal{P}^{[j]} \tilde{A}_{il}^{[j]} + \left(\tilde{A}_{il}^{[j]} \right)^T \mathcal{P}^{[j]} \right] \xi(t) + 2\xi^T(t) \mathcal{P}^{[j]} \tilde{F}_{il}^{[j]} \eta(t) \right\}$$

$$\leqslant \sum_{j=1}^{N} \sigma_j(t) \sum_{i=1}^{r} h_i^{[j]}\left(\theta^{[j]}(t)\right) \sum_{l=1}^{r} h_l^{[j]}\left(\theta^{[j]}(t)\right) \times \left\{ \varepsilon \eta^T(t)\eta(t) + \xi^T(t) \left[\mathcal{P}^{[j]} \tilde{A}_{il}^{[j]} + \left(\tilde{A}_{il}^{[j]} \right)^T \mathcal{P}^{[j]} + \varepsilon^{-1} \mathcal{P}^{[j]} \tilde{F}_{il}^{[j]} \left(\tilde{F}_{il}^{[j]} \right)^T \mathcal{P}^{[j]} \right] \xi(t) \right\}. \tag{4.15}$$

Consider Assumption 4.1, and define M, N that

$$\left\| f\left(x(t), t\right) \right\| \leqslant \left\| Mx(t) \right\|, \quad \left\| g\left(x(t), t\right) \right\| \leqslant \left\| Nx(t) \right\|.$$

It can be seen that

$$\left\| f\left(x(t), t\right) \right\|^2 = f^T\left(x(t), t\right) f\left(x(t), t\right) \leqslant \|Mx(t)\|^2 = x^T(t) M^T M x(t),$$
$$\left\| g\left(x(t), t\right) \right\|^2 = g^T\left(x(t), t\right) g\left(x(t), t\right) \leqslant \|Nx(t)\|^2 = x^T(t) N^T N x(t).$$

Thus, we can get

$$\begin{aligned} \eta^T\left(x(t), t\right)\eta\left(x(t), t\right) &= f^T\left(x(t), t\right) f\left(x(t), t\right) + g^T\left(x(t), t\right) g\left(x(t), t\right) \\ &\leqslant \xi^T(t) \mathcal{K}^T \mathcal{M} \mathcal{K} \xi(t). \end{aligned} \tag{4.16}$$

Based on (4.14)–(4.16), it follows that

$$\begin{aligned} \dot{V}\left(\xi(t), \sigma_j(t)\right) \leqslant & \sum_{j=1}^{N} \sigma_j(t) \sum_{i=1}^{r} h_i^{[j]}\left(\theta^{[j]}(t)\right) \sum_{l=1}^{r} h_l^{[j]}\left(\theta^{[j]}(t)\right) \xi^T(t) \\ & \times \left[\mathcal{P}^{[j]} \tilde{A}_{il}^{[j]} + \varepsilon^{-1} \mathcal{P}^{[j]} \tilde{F}_{il}^{[j]} \left(\tilde{F}_{il}^{[j]} \right)^T \mathcal{P}^{[j]} \right. \\ & \left. + \left(\tilde{A}_{il}^{[j]} \right)^T \mathcal{P}^{[j]} + \varepsilon \mathcal{K}^T \mathcal{M} \mathcal{K} \right] \xi(t). \end{aligned} \tag{4.17}$$

Based on (4.7)–(4.8) and (4.17), employing the Schur complement method, which yields

$$\dot{V}\left(\xi(t), \sigma_j(t)\right) < -\alpha \xi^T(t) \mathcal{P}\left(\sigma_j(t)\right) \xi(t) = -\alpha V\left(\xi(t), \sigma_j(t)\right). \tag{4.18}$$

As for the piecewise switching signal $\sigma_j(t)$ $(t > 0)$, let $0 = t_0 < t_1 < \cdots < t_k < \cdots < t$, $(k = 0, 1, \ldots)$, denote switching points of $\sigma_j(t)$ under the interval $(0, t)$. Thus, the j_kth subsystem is activated when $t \in \big[t_k, t_{k+1}\big)$. Starting with $t^\star \triangleq t_k$ in (4.18), then

$$V\Big(\xi(t), \sigma_j(t)\Big) < e^{-\alpha(t-t_k)} V\Big(\xi(t_k), \sigma_j(t_k)\Big). \tag{4.19}$$

Utilizing (4.11) and (4.14), at switching instant t_k, we have

$$V\Big(\xi(t_k), \sigma_j(t_k)\Big) < \rho V\Big(\xi(t_k^-), \sigma_j(t_k^-)\Big). \tag{4.20}$$

Consequently, from (4.19)–(4.20) and $\phi = N_{\sigma_j}(0, t) \leqslant \frac{t-0}{T_a}$, it follows that

$$\begin{aligned} V\Big(\xi(t), \sigma_j(t)\Big) &\leqslant e^{-\alpha(t-t_k)} \rho V\Big(\xi(t_k^-), \sigma_j(t_k^-)\Big) \\ &\leqslant \cdots \leqslant e^{-\alpha(t-0)} \rho^{\phi} V\Big(\xi(0), \sigma_j(0)\Big) \\ &\leqslant e^{-(\alpha - \frac{\ln \rho}{T_a})t} V\Big(\xi(0), \sigma_j(0)\Big) \\ &= e^{-(\alpha - \frac{\ln \rho}{T_a})t} V\Big(\xi(0), \sigma_j(0)\Big). \end{aligned} \tag{4.21}$$

On the basis of (4.14), it yields

$$V\Big(\xi(t), \sigma_j(t)\Big) \geqslant \tau \|\xi(t)\|^2, \;\; V\Big(\xi(0), \sigma_j(0)\Big) \leqslant \vartheta \|\xi(0)\|^2, \tag{4.22}$$

where τ and ϑ are given in (4.13). Combining (4.21) and (4.22), we have

$$\Big\|\xi(t)\Big\|^2 \leqslant \frac{1}{\tau} V\Big(\xi(t), \sigma_j(t)\Big) \leqslant \frac{\vartheta}{\tau} e^{-(\alpha - \frac{\ln \rho}{T_a})t} \|\xi(0)\|^2. \tag{4.23}$$

When $\omega(t) = 0$, the closed-loop system in (4.5) is exponentially stable on account of Definition 4.2 with $t_0 = 0$.

When $\omega(t) \neq 0$, the $\mathcal{L}_2$–$\mathcal{L}_\infty$ performance of the overall system is analyzed then. Introduce

$$\begin{aligned} \mathcal{J}\big(\xi(t), \sigma_j(t)\big) &\triangleq \dot{V}\big(\xi(t), \sigma_j(t)\big) + \alpha V\big(\xi(t), \sigma_j(t)\big) - \omega(t)\omega(t) \\ &\leqslant \psi^T(t) \phi\big(t, \sigma_j(t)\big) \psi(t), \end{aligned} \tag{4.24}$$

where

$$\phi(t,\sigma_j(t)) \triangleq \begin{bmatrix} \bar{\phi}(t,\sigma_j(t)) & \mathcal{P}(\sigma_j(t))\tilde{B}(t,\sigma_j(t)) \\ \star & -I \end{bmatrix},$$
$$\begin{aligned}\bar{\phi}(t,\sigma_j(t)) \triangleq\ & \mathcal{P}(\sigma_j(t))\tilde{A}(t,\sigma_j(t)) + \tilde{A}^T(t,\sigma_j(t))\mathcal{P}(\sigma_j(t)) \\ & + \varepsilon^{-1}\mathcal{P}(\sigma_j(t))\tilde{F}(t,\sigma_j(t))\tilde{F}^T(t,\sigma_j(t))\mathcal{P}(\sigma_j(t)) \\ & + \alpha\mathcal{P}(\sigma_j(t)) + \varepsilon\mathcal{K}^T\mathcal{M}\mathcal{K},\ \psi(t) \triangleq \begin{bmatrix}\xi(t) \\ \omega(t)\end{bmatrix}.\end{aligned} \tag{4.25}$$

Hence, in view of $\psi(t) \neq 0$ and (4.7), (4.8), we have $\mathcal{J}(\xi(t),\sigma_j(t)) < 0$. Let $\zeta(t) = -\omega^T(t)\omega(t)$, then

$$\dot{V}(\xi(t),\sigma_j(t)) \leqslant -\alpha V(\xi(t),\sigma_j(t)) - \zeta(t). \tag{4.26}$$

Employing a similar way in the proof of exponential stability, (4.26) yields

$$V\Big(\xi(t),\sigma_j(t)\Big) < e^{-\alpha(t-t_k)}V\Big(\xi(t_k),\sigma_j(t_k)\Big) - \int\limits_{t_k}^{t} e^{-\alpha(t-s)}\zeta(s)ds. \tag{4.27}$$

Consider $\phi = N_{\sigma_j}(0,t) \leqslant \frac{t-0}{T_a}$ and (4.20), (4.27), it follows that

$$\begin{aligned} & V\Big(\xi(t),\sigma_j(t)\Big) \\ & \leqslant \rho e^{-\alpha(t-t_k)}V\Big(\xi(t_k^-),\sigma_j(t_k^-)\Big) - \int\limits_{t_k}^{t} e^{-\alpha(t-s)}\zeta(s)ds \\ & \leqslant \rho^{\phi}e^{-\alpha(t-0)}V\Big(\xi(0),\sigma_j(0)\Big) - \rho^{\phi}\int\limits_{0}^{t_1} e^{-\alpha(t-s)}\zeta(s)ds \\ & - \rho^{\phi-1}\int_{t_1}^{t_2} e^{-\alpha(t-s)}\zeta(s)ds - \cdots - \rho^0\int\limits_{t_k}^{t} e^{-\alpha(t-s)}\zeta(s)ds \\ & = e^{-\alpha t - N_{\sigma_j}(0,t)\ln\rho}V\Big(\xi(0),\sigma_j(0)\Big) - \int\limits_{0}^{t} e^{-\alpha(t-s)+N_{\sigma_j}(s,t)\ln\rho}\zeta(s)ds. \end{aligned} \tag{4.28}$$

Provided that $\xi(0) = 0$, (4.28) leads to

$$V\Big(\xi(t),\sigma_j(t)\Big) \leqslant \int\limits_{0}^{t} e^{-\alpha(t-s)+N_{\sigma_j}(s,t)\ln\rho}\omega^T(s)\omega(s)ds. \tag{4.29}$$

Multiplying two sides of (4.29) by $e^{-N_{\sigma_j}(0,t)\ln\rho}$, it yields

$$\begin{aligned} e^{-N_{\sigma_j}(0,t)\ln\rho} V\big(\xi(t), \sigma_j(t)\big) &\leqslant \int_0^t e^{-\alpha(t-s)-N_{\sigma_j}(0,s)\ln\rho} \omega^T(s)\omega(s)ds \\ &\leqslant \int_0^t \omega^T(s)\omega(s)ds. \end{aligned} \tag{4.30}$$

Note that $N_{\sigma_j}(0,t) \leqslant \frac{t}{T_a}$ and $T_a > T_a^\star = \frac{\ln\rho}{\alpha}$, then $N_{\sigma_j}(0,t)\ln\rho \leqslant \alpha t$. Therefore, (4.30) means

$$e^{-\alpha t} V\big(\xi(t), \sigma_j(t)\big) \leqslant \int_0^t \omega^T(s)\omega(s)ds. \tag{4.31}$$

Based on (4.14) and (4.31), it can get

$$e^{-\alpha t}\xi^T(t)\mathcal{P}\big(\sigma_j(t)\big)\xi(t) \leqslant \int_0^t \omega^T(s)\omega(s)ds \leqslant \int_0^\infty \omega^T(t)\omega(t)dt. \tag{4.32}$$

Since $t = T^\star \geqslant 0$ is an arbitrary time moment, then

$$e^{-\alpha T^\star}\xi^T(T^\star)\mathcal{P}\big(\sigma_j(T^\star)\big)\xi(T^\star) \leqslant \int_0^\infty \omega^T(t)\omega(t)dt. \tag{4.33}$$

On account of (4.9)–(4.10),

$$\gamma^{-2}\tilde{C}^T\big(t, \sigma_j(t)\big)\tilde{C}\big(t, \sigma_j(t)\big) < \mathcal{P}\big(\sigma_j(t)\big). \tag{4.34}$$

Combining (4.33) and (4.34) yields

$$\begin{aligned} &\gamma^{-2}e^{-\alpha T^\star}\xi^T(T^\star)\tilde{C}^T\big(T^\star, \sigma_j(T^\star)\big)\tilde{C}\big(T^\star, \sigma_j(T^\star)\big)\xi(T^\star) \\ &\leqslant e^{-\alpha T^\star}\xi^T(T^\star)\mathcal{P}\big(\sigma_j(T^\star)\big)\xi(T^\star) \leqslant \int_0^\infty \omega^T(t)\omega(t)dt. \end{aligned}$$

For any $T^\star \geqslant 0$,

$$e^{-\alpha T^\star} z^T(T^\star)z(T^\star) \leq \gamma^2 \int_0^\infty \omega^T(t)\omega(t)dt.$$

Taking the supremum over $T^\star \geqslant 0$, which signifies

$$\sup_{\forall t} e^{-\alpha t} z^T(t)z(t) < \gamma^2 \int_0^{\infty} \omega^T(t)\omega(t)dt.$$

Thus, the closed-loop system satisfies a given weighted $\mathcal{L}_2$–$\mathcal{L}_\infty$ performance level.

Remark 4.5 In Theorem 4.4, a fuzzy-basis-dependent Lyapunov function $V(t) \triangleq \xi^T(t)\mathcal{P}(\sigma_j(t))\xi(t)$ is established over the switching signal $\sigma_j(t)$. It is proven to be less conservative compared with usual Lyapunov functions (when $\mathcal{P}(\sigma_j(t)) = \mathcal{P}$).

Remark 4.6 If $\rho = 1$ in $T_a > \frac{\ln \rho}{\alpha}$, then $T_a > T_a^\star = 0$, which means the switching signal $\sigma_j(t)$ is arbitrary. It implies a general Lyapunov function is demanded for all subsystems. If $\rho > 1$ and $\alpha \to 0$ in $T_a > \frac{\ln \rho}{\alpha}$, the closed-loop system can be operated at one of the subsystems continuously as $T_a \to \infty$. In addition, based on Assumption 4.1, the nonlinearities $f(x(t), t)$ and $g(x(t), t)$ satisfy Lipschitz conditions. Therefore, the controller design method is useful in some actual applications.

4.4 Dynamic Output Feedback Control

4.4.1 Reduced-Order Controller Design

In this subsection, a criterion is given to handle the reduced-order controller issue for nonlinear switched systems in (4.5).

Theorem 4.7 *For the given scalars $\gamma > 0$, $\alpha > 0$, if there exist a scalar $\varepsilon > 0$ and matrices $\mathscr{P}^{[j]} > 0$, $\mathscr{Q}^{[j]} > 0$, $\mathscr{A}_{cil}^{[j]}$, $\mathscr{B}_{cl}^{[j]}$, $\mathscr{C}_{cl}^{[j]}$, which satisfy the following conditions for $j \in \mathcal{N}$,*

$$\tilde{\phi}_{ii}^{[j]} < 0, \quad i = 1, 2, \ldots, r, \tag{4.35}$$

$$\frac{1}{r-1}\tilde{\phi}_{ii}^{[j]} + \frac{1}{2}\left(\tilde{\phi}_{il}^{[j]} + \tilde{\phi}_{li}^{[j]}\right) < 0, \quad 1 \leqslant i < l \leqslant r, \tag{4.36}$$

$$\tilde{\varphi}_{ii}^{[j]} < 0, \quad i = 1, 2, \ldots, r, \tag{4.37}$$

$$\frac{1}{r-1}\tilde{\varphi}_{ii}^{[j]} + \frac{1}{2}\left(\tilde{\varphi}_{il}^{[j]} + \tilde{\varphi}_{li}^{[j]}\right) < 0, \quad 1 \leqslant i < l \leqslant r, \tag{4.38}$$

with

$$\tilde{\phi}_{il}^{[j]} \triangleq \begin{bmatrix} \tilde{\phi}_{11il}^{[j]} & \tilde{\phi}_{12il}^{[j]} & \tilde{\phi}_{13il}^{[j]} \\ \star & -I & 0 \\ \star & \star & -\varepsilon I \end{bmatrix}, \quad \tilde{\varphi}_{il}^{[j]} \triangleq \begin{bmatrix} -\mathscr{P}^{[j]} & -I & \left(L_i^{[j]}\right)^T \\ \star & -\mathscr{Q}^{[j]} & \tilde{\varphi}_{23il}^{[j]} \\ \star & \star & -\gamma^2 I \end{bmatrix},$$

$$\tilde{\phi}_{11il}^{[j]} \triangleq \begin{bmatrix} \tilde{\phi}_{111il}^{[j]} & \tilde{\phi}_{112il}^{[j]} \\ \star & \tilde{\phi}_{113il}^{[j]} \end{bmatrix}, \quad \tilde{\phi}_{12il}^{[j]} \triangleq \begin{bmatrix} \mathscr{P}^{[j]} B_{2i}^{[j]} + \mathcal{H}\mathscr{B}_{cl}^{[j]} D_{2i}^{[j]} \\ B_{2i}^{[j]} \end{bmatrix},$$

$$\tilde{\phi}_{13il}^{[j]} \triangleq \begin{bmatrix} \mathscr{P}^{[j]} F_i^{[j]} & \mathcal{H}\mathscr{B}_{cl}^{[j]} G_i^{[j]} \\ F_i^{[j]} & 0 \end{bmatrix},$$

$$\tilde{\phi}_{111il}^{[j]} \triangleq \mathscr{P}^{[j]} A_i^{[j]} + \left(\mathscr{P}^{[j]} A_i^{[j]} + \mathcal{H}\mathscr{B}_{cl}^{[j]} C_i^{[j]}\right)^T + \mathcal{H}\mathscr{B}_{cl}^{[j]} C_i^{[j]} + \alpha \mathscr{P}^{[j]} + \varepsilon \mathcal{M},$$

$$\tilde{\varphi}_{23il}^{[j]} \triangleq \left(L_i^{[j]} \mathscr{Q}^{[j]} + K_i^{[j]} \mathscr{C}_{cl}^{[j]} \mathcal{H}^T\right)^T, \quad \tilde{\phi}_{112il}^{[j]} \triangleq \mathscr{A}_{cil}^{[j]} + \left(A_i^{[j]}\right)^T + \alpha I,$$

$$\tilde{\phi}_{113il}^{[j]} \triangleq A_i^{[j]} \mathscr{Q}^{[j]} + \left(A_i^{[j]} \mathscr{Q}^{[j]} + B_{1i}^{[j]} \mathscr{C}_{cl}^{[j]} \mathcal{H}^T\right)^T + B_{1i}^{[j]} \mathscr{C}_{cl}^{[j]} \mathcal{H}^T + \alpha \mathscr{Q}^{[j]}.$$

Then the output feedback controller in (4.4) guarantees that the closed-loop system in (4.5) is exponentially stable subject to a specific weighted $\mathcal{L}_2$–$\mathcal{L}_\infty$ performance level. And the controller parameters are given by

$$\begin{cases} \mathscr{A}_{cil}^{[j]} \triangleq \mathscr{P}^{[j]} B_{1i}^{[j]} C_{cl}^{[j]} \left(\mathcal{H}\mathcal{Q}_2^{[j]}\right)^T + \mathcal{H}\mathcal{P}_2^{[j]} A_{cl}^{[j]} \left(\mathcal{H}\mathcal{Q}_2^{[j]}\right)^T \\ \quad + \mathscr{P}^{[j]} A_i^{[j]} \mathscr{Q}^{[j]} + \mathcal{H}\mathcal{P}_2^{[j]} B_{cl}^{[j]} C_i^{[j]} \mathscr{Q}^{[j]} \\ \quad + \mathcal{H}\mathcal{P}_2^{[j]} B_{cl}^{[j]} D_{1i}^{[j]} C_{cl}^{[j]} \left(\mathcal{H}\mathcal{Q}_2^{[j]}\right)^T, \\ \mathscr{B}_{cl}^{[j]} \triangleq \mathcal{P}_2^{[j]} B_{cl}^{[j]}, \quad \mathscr{C}_{cl}^{[j]} \triangleq C_{cl}^{[j]} \left(\mathcal{Q}_2^{[j]}\right)^T. \end{cases} \tag{4.39}$$

Proof Firstly, $\mathcal{P}^{[j]}$ is constructed as

$$\mathcal{P}^{[j]} \triangleq \begin{bmatrix} \mathcal{P}_1^{[j]} & \mathcal{H}\mathcal{P}_2^{[j]} \\ \star & \mathcal{P}_3^{[j]} \end{bmatrix},$$

so,

$$\mathcal{Q}^{[j]} \triangleq \left(\mathcal{P}^{[j]}\right)^{-1} \triangleq \begin{bmatrix} \mathcal{Q}_1^{[j]} & \mathcal{H}\mathcal{Q}_2^{[j]} \\ \star & \mathcal{Q}_3^{[j]} \end{bmatrix},$$

where $\mathcal{H} = \begin{bmatrix} I_{r\times r} & 0_{r\times(n-r)} \end{bmatrix}^T$, $\mathcal{P}_1 \in \mathbb{R}^{n\times n}$, $\mathcal{P}_2 \in \mathbb{R}^{r\times r}$, and $\mathcal{P}_3 \in \mathbb{R}^{r\times r}$. Without loss of generality, suppose that $\mathcal{P}_2^{[j]}$ and $\mathcal{Q}_2^{[j]}$ are nonsingular (if not, then $\mathcal{P}_2^{[j]}$ and $\mathcal{Q}_2^{[j]}$ can be added with matrices $\triangle\mathcal{P}_2^{[j]}$ and $\triangle\mathcal{Q}_2^{[j]}$, having sufficiently small norms, such that $\mathcal{P}_2^{[j]} + \triangle\mathcal{P}_2^{[j]}$ and $\mathcal{Q}_2^{[j]} + \triangle\mathcal{Q}_2^{[j]}$ are still nonsingular and satisfy (4.7)–(4.10).

Next, define the following matrices:

$$\mathcal{J}_{\mathcal{P}}^{[j]} \triangleq \begin{bmatrix} \mathcal{P}_1^{[j]} & I \\ \left(\mathcal{H}\mathcal{P}_2^{[j]}\right)^T & 0 \end{bmatrix}, \ \mathcal{J}_{\mathcal{Q}}^{[j]} \triangleq \begin{bmatrix} I & \mathcal{Q}_1^{[j]} \\ 0 & \left(\mathcal{H}\mathcal{Q}_2^{[j]}\right)^T \end{bmatrix}. \tag{4.40}$$

Note that

$$\mathcal{P}^{[j]}\mathcal{J}_{\mathcal{Q}}^{[j]} = \mathcal{J}_{\mathcal{P}}^{[j]}, \ \mathcal{Q}^{[j]}\mathcal{J}_{\mathcal{P}}^{[j]} = \mathcal{J}_{\mathcal{Q}}^{[j]}, \ \mathcal{P}_1^{[j]}\mathcal{Q}_1^{[j]} + \mathcal{H}\mathcal{P}_2^{[j]}\left(\mathcal{H}\mathcal{Q}_2^{[j]}\right)^T = I.$$

Performing the congruence transformation to $\phi_{il}^{[j]} < 0$ with diag$\{\mathcal{J}_{\mathcal{Q}}^{[j]}, I, I\}$, it follows that

$$\begin{bmatrix} \left(\mathcal{J}_{\mathcal{Q}}^{[j]}\right)^T \phi_{11il}^{[j]} \mathcal{J}_{\mathcal{Q}}^{[j]} & \left(\mathcal{J}_{\mathcal{Q}}^{[j]}\right)^T \mathcal{P}^{[j]}\tilde{B}_{il}^{[j]} & \left(\mathcal{J}_{\mathcal{Q}}^{[j]}\right)^T \mathcal{P}^{[j]}\tilde{F}_{il}^{[j]} \\ \star & -I & 0 \\ \star & \star & -\varepsilon I \end{bmatrix} < 0. \tag{4.41}$$

Performing the congruence transformation to $\varphi_{il}^{[j]} < 0$ with diag$\{\mathcal{J}_{\mathcal{Q}}^{[j]}, I\}$, we have

$$\begin{bmatrix} -\left(\mathcal{J}_{\mathcal{Q}}^{[j]}\right)^T \mathcal{P}^{[j]}\mathcal{J}_{\mathcal{Q}}^{[j]} & \left(\mathcal{J}_{\mathcal{Q}}^{[j]}\right)^T \left(\tilde{C}_{il}^{[j]}\right)^T \\ \star & -\gamma^2 I \end{bmatrix} < 0. \tag{4.42}$$

Let $\mathscr{P}^{[j]} = \mathcal{P}_1^{[j]}$, $\mathscr{Q}^{[j]} = \mathcal{Q}_1^{[j]}$, and from (4.39) that

$$\begin{aligned}
\left(\mathcal{J}_{\mathcal{Q}}^{[j]}\right)^T \mathcal{P}^{[j]}\mathcal{J}_{\mathcal{Q}}^{[j]} &\triangleq \begin{bmatrix} \mathscr{P}^{[j]} & I \\ I & \mathscr{Q}^{[j]} \end{bmatrix}, \\
\left(\mathcal{J}_{\mathcal{Q}}^{[j]}\right)^T \mathcal{P}^{[j]}\tilde{A}_{il}^{[j]}\mathcal{J}_{\mathcal{Q}}^{[j]} &\triangleq \begin{bmatrix} \mathscr{P}^{[j]}A_i^{[j]} + \mathcal{H}\mathscr{B}_{cl}^{[j]}C_i^{[j]} & \mathscr{A}_{cil}^{[j]} \\ A_i^{[j]} & A_i^{[j]}\mathscr{Q}^{[j]} + B_{1i}^{[j]}\mathscr{C}_{cl}^{[j]}\mathcal{H}^T \end{bmatrix}, \\
\left(\mathcal{J}_{\mathcal{Q}}^{[j]}\right)^T \mathcal{P}^{[j]}\tilde{B}_{il}^{[j]} &\triangleq \begin{bmatrix} \mathscr{P}^{[j]}B_{2i}^{[j]} + \mathcal{H}\mathscr{B}_{cl}^{[j]}D_{2i}^{[j]} \\ B_{2i}^{[j]} \end{bmatrix}, \\
\left(\mathcal{J}_{\mathcal{Q}}^{[j]}\right)^T \mathcal{P}^{[j]}\tilde{F}_{il}^{[j]} &\triangleq \begin{bmatrix} \mathscr{P}^{[j]}F_i^{[j]} & \mathcal{H}\mathscr{B}_{cl}^{[j]}G_i^{[j]} \\ F_i^{[j]} & 0 \end{bmatrix}, \\
\left(\mathcal{J}_{\mathcal{Q}}^{[j]}\right)^T \left(\tilde{C}_{il}^{[j]}\right)^T &\triangleq \begin{bmatrix} \left(L_i^{[j]}\right)^T \\ \left(L_i^{[j]}\mathscr{Q}^{[j]} + K_i^{[j]}\mathscr{C}_{cl}^{[j]}\mathcal{H}^T\right)^T \end{bmatrix}.
\end{aligned} \tag{4.43}$$

On the basis of (4.40)–(4.43), it can be observed that (4.35)–(4.38) hold. Consequently, in consideration of Theorem 4.4, the overall system is exponentially stable subject to a weighted $\mathcal{L}_2$–$\mathcal{L}_\infty$ performance level (γ, α). Besides, the gains of the reduced-order controller can be obtained by resolving the conditions (4.39).

Remark 4.8 The dynamic output feedback controller design approach can be achieved with independent or dependent fuzzy-rules. The premise variable $\theta(k)$ is fully available in fuzzy-rule-dependent method, but the fuzzy-rule-independent method is employed at instances when $\theta(k)$ is inaccessible. That is to say, we can let $[A_{ci}^{[j]}, B_{ci}^{[j]}, C_{ci}^{[j]}] \triangleq [A_{ci}, B_{ci}, C_{ci}]$ or select $[A_{ci}, B_{ci}, C_{ci}] = [A, B, C]$ in (4.4), which result in different non-parameterized controllers with different computational complexity and conservativeness. In this chapter, the fuzzy-rule-dependent method is utilized to design the controller, which is less conservative.

Remark 4.9 The obtained results employing the projection lemma are usually expressed with LMIs plus an additional rank constraint. Nevertheless, due to the rank constraint is non-convex, the obtained conditions are not easy to solve by the numerical software. In this chapter, a linearization technique is put forward to resolve the output feedback controller design issue, which can be implemented with simulation toolbox. In addition, owing to the matrices $\mathcal{P}_2^{[j]}$ and $\mathcal{Q}_2^{[j]}$ can be available in advance, the controller matrices in (4.39) can be obtained by letting $\mathcal{H}\mathcal{P}_2^{[j]}\left(\mathcal{H}\mathcal{Q}_2^{[j]}\right)^T = I - \mathcal{P}_1^{[j]}\mathcal{Q}_1^{[j]}$.

4.4.2 Full-Order Controller Design

The results of full-order $\mathcal{L}_2$–$\mathcal{L}_\infty$ dynamic output feedback controller are derived employing the similar methods in Theorem 4.7.

Theorem 4.10 *Given scalars $\gamma > 0$ and $\alpha > 0$, if there exist a scalar $\varepsilon > 0$ and matrices $\mathscr{P}^{[j]} > 0$, $\mathscr{Q}^{[j]} > 0$, $\mathscr{A}_{cil}^{[j]}$, $\mathscr{B}_{cl}^{[j]}$, $\mathscr{C}_{cl}^{[j]}$, which satisfy the following conditions for $j \in \mathcal{N}$:*

$$\hat{\phi}_{ii}^{[j]} < 0, \quad i = 1, 2, \ldots, r, \tag{4.44}$$

$$\frac{1}{r-1}\hat{\phi}_{ii}^{[j]} + \frac{1}{2}\left(\hat{\phi}_{il}^{[j]} + \hat{\phi}_{li}^{[j]}\right) < 0, \quad 1 \leqslant i < l \leqslant r, \tag{4.45}$$

$$\hat{\varphi}_{ii}^{[j]} < 0, \quad i = 1, 2, \ldots, r, \tag{4.46}$$

$$\frac{1}{r-1}\hat{\varphi}_{ii}^{[j]} + \frac{1}{2}\left(\hat{\varphi}_{il}^{[j]} + \hat{\varphi}_{li}^{[j]}\right) < 0, \quad 1 \leqslant i < l \leqslant r, \tag{4.47}$$

with

$$\hat{\phi}_{il}^{[j]} \triangleq \begin{bmatrix} \hat{\phi}_{11il}^{[j]} & \hat{\phi}_{12il}^{[j]} & \hat{\phi}_{13il}^{[j]} \\ \star & -I & 0 \\ \star & \star & -\varepsilon I \end{bmatrix}, \quad \hat{\varphi}_{il}^{[j]} \triangleq \begin{bmatrix} -\mathscr{P}^{[j]} & -I & \left(L_i^{[j]}\right)^T \\ \star & -\mathscr{Q}^{[j]} & \hat{\varphi}_{23il}^{[j]} \\ \star & \star & -\gamma^2 I \end{bmatrix},$$

$$\hat{\phi}_{11il}^{[j]} \triangleq \begin{bmatrix} \hat{\phi}_{111il}^{[j]} & \tilde{\phi}_{112il}^{[j]} \\ \star & \hat{\phi}_{113il}^{[j]} \end{bmatrix}, \quad \hat{\phi}_{12il}^{[j]} \triangleq \begin{bmatrix} \mathscr{P}^{[j]} B_{2i}^{[j]} + \mathscr{B}_{cl}^{[j]} D_{2i}^{[j]} \\ B_{2i}^{[j]} \end{bmatrix},$$

$$\hat{\phi}_{13il}^{[j]} \triangleq \begin{bmatrix} \mathscr{P}^{[j]} F_i^{[j]} & \mathscr{B}_{cl}^{[j]} G_i^{[j]} \\ F_i^{[j]} & 0 \end{bmatrix}, \quad \hat{\varphi}_{23il}^{[j]} \triangleq \left(L_i^{[j]} \mathscr{Q}^{[j]} + K_i^{[j]} \mathscr{C}_{cl}^{[j]}\right)^T,$$

$$\hat{\phi}_{111il}^{[j]} \triangleq \mathscr{P}^{[j]} A_i^{[j]} + \mathscr{B}_{cl}^{[j]} C_i^{[j]} + \left(\mathscr{P}^{[j]} A_i^{[j]} + \mathscr{B}_{cl}^{[j]} C_i^{[j]}\right)^T + \alpha \mathscr{P}^{[j]} + \varepsilon \mathcal{M},$$

$$\hat{\phi}_{113il}^{[j]} \triangleq A_i^{[j]} \mathscr{Q}^{[j]} + B_{1i}^{[j]} \mathscr{C}_{cl}^{[j]} + \left(A_i^{[j]} \mathscr{Q}^{[j]} + B_{1i}^{[j]} \mathscr{C}_{cl}^{[j]}\right)^T + \alpha \mathscr{Q}^{[j]},$$

where $\tilde{\phi}_{112il}^{[j]}$ are given in Theorem 4.7, and the full-order controller matrices are expressed as

$$\begin{aligned} \mathscr{A}_{cil}^{[j]} &\triangleq \mathscr{P}^{[j]} B_{1i}^{[j]} C_{cl}^{[j]} \left(\mathcal{Q}_2^{[j]}\right)^T + \mathcal{P}_2^{[j]} A_{cl}^{[j]} \left(\mathcal{Q}_2^{[j]}\right)^T + \mathscr{P}^{[j]} A_i^{[j]} \mathscr{Q}^{[j]} \\ &\quad + \mathcal{P}_2^{[j]} B_{cl}^{[j]} C_i^{[j]} \mathscr{Q}^{[j]} + \mathcal{P}_2^{[j]} B_{cl}^{[j]} D_{1i}^{[j]} C_{cl}^{[j]} \left(\mathcal{Q}_2^{[j]}\right)^T, \\ \mathscr{B}_{cl}^{[j]} &\triangleq \mathcal{P}_2^{[j]} B_{cl}^{[j]}, \quad \mathscr{C}_{cl}^{[j]} \triangleq C_{cl}^{[j]} \left(\mathcal{Q}_2^{[j]}\right)^T. \end{aligned}$$

Proof In the first place, set the matrix $\mathcal{P}^{[j]} \triangleq \begin{bmatrix} \mathcal{P}_1^{[j]} & \mathcal{P}_2^{[j]} \\ \star & \mathcal{P}_3^{[j]} \end{bmatrix}$, then

$$\mathcal{Q}^{[j]} = \left(\mathcal{P}^{[j]}\right)^{-1} \triangleq \begin{bmatrix} \mathcal{Q}_1^{[j]} & \mathcal{Q}_2^{[j]} \\ \star & \mathcal{Q}_3^{[j]} \end{bmatrix}.$$

Define the matrices

$$\mathcal{J}_{\mathcal{P}}^{[j]} \triangleq \begin{bmatrix} \mathcal{P}_1^{[j]} & I \\ \left(\mathcal{P}_2^{[j]}\right)^T & 0 \end{bmatrix}, \quad \mathcal{J}_{\mathcal{Q}}^{[j]} \triangleq \begin{bmatrix} I & \mathcal{Q}_1^{[j]} \\ 0 & \left(\mathcal{Q}_2^{[j]}\right)^T \end{bmatrix}, \tag{4.48}$$

and notice that

$$\mathcal{P}^{[j]} \mathcal{J}_{\mathcal{Q}}^{[j]} = \mathcal{J}_{\mathcal{P}}^{[j]}, \quad \mathcal{Q}^{[j]} \mathcal{J}_{\mathcal{P}}^{[j]} = \mathcal{J}_{\mathcal{Q}}^{[j]}, \quad \mathcal{P}_1^{[j]} \mathcal{Q}_1^{[j]} + \mathcal{P}_2^{[j]} \left(\mathcal{Q}_2^{[j]}\right)^T = I.$$

Performing congruence transformations to (4.7) and (4.9) with diag$\{\mathcal{J}_{\mathcal{Q}}^{[j]}, I, I\}$ and diag$\{\mathcal{J}_{\mathcal{Q}}^{[j]}, I\}$, respectively, it follows that (4.44)–(4.47). Therefore, on the basis of

Theorem 4.4, the resulting closed-loop system is exponentially stable subject to a weighted $\mathcal{L}_2$–$\mathcal{L}_\infty$ performance level (γ, α).

Remark 4.11 The feasible conditions for the DOFC design issue are given in Theorem 4.7 (or Theorem 4.10), which can be validly calculated by using the standard optimization toolbox in terms of strict LMIs. The computational complexity mainly depends on the number of fuzzy rules r and the proposed algorithm asks for $2n^2 + mn + n$ free variables. If r is 3 or less, the corresponding complexity is smaller and the result is simpler. Accordingly, if r is larger, the more iterations happen during the simulation, which increases the algorithm complexity. Therefore, it is important to select appropriate fuzzy-based rules and reduce the number of fuzzy rules properly without influencing system performance. The designed reduced-order/full-order DOFC in (4.4) can be obtained by dealing with the convex optimization issue as below:

$$\min \delta \text{ subject to (4.35)–(4.38) or (4.44)–(4.47) with } \delta = \gamma^2.$$

4.5 Illustrative Example

Two examples are provided to illustrate the effectiveness of the proposed design technique in this section. The first one gives numerical results of reduced-order/full-order DOFC. The second one is presented to demonstrate its applicability in cognitive radio (CR) systems.

Example 4.12 Consider the switched system (4.1) with two subsystems and the parameters are given as follows:

Subsystem 1.

$$A_1^{[1]} = \begin{bmatrix} -1.8 & -0.3 & -0.5 \\ 0.2 & -1.8 & 0.3 \\ 0.3 & 0.6 & -1.5 \end{bmatrix}, \ B_{11}^{[1]} = \begin{bmatrix} 1.2 \\ 0.7 \\ 0.7 \end{bmatrix}, \ B_{21}^{[1]} = \begin{bmatrix} 0.3 \\ 0.3 \\ 0.4 \end{bmatrix},$$

$$G_1^{[1]} = 0.4, \ K_1^{[1]} = 1.2, \ D_{11}^{[1]} = 0.4, \ D_{21}^{[1]} = -0.3,$$

$$F_1^{[1]} = \begin{bmatrix} 0.2 & 0.1 & 0.1 \\ 0.1 & 0.1 & 0.0 \\ 0.0 & 0.2 & 0.2 \end{bmatrix}, \quad \begin{array}{l} C_1^{[1]} = \begin{bmatrix} 1.2 & 0.6 & 1.5 \end{bmatrix}, \\ L_1^{[1]} = \begin{bmatrix} 1.2 & -0.8 & -1.2 \end{bmatrix}, \end{array}$$

$$A_2^{[1]} = \begin{bmatrix} -2.1 & 0.2 & 0.4 \\ 0.3 & 0.6 & 0.2 \\ -0.2 & 0.2 & -2.2 \end{bmatrix}, \ B_{12}^{[1]} = \begin{bmatrix} 0.8 \\ 0.7 \\ 0.5 \end{bmatrix}, \ B_{22}^{[1]} = \begin{bmatrix} -0.3 \\ 0.6 \\ -0.5 \end{bmatrix},$$

$$G_2^{[1]} = 0.3, \ K_2^{[1]} = 0.3, \ D_{12}^{[1]} = 0.5, \ D_{22}^{[1]} = 0.4,$$

$$F_2^{[1]} = \begin{bmatrix} 0.2 & 0.1 & 0.1 \\ 0.1 & 0.2 & 0.1 \\ 0.1 & 0.0 & 0.1 \end{bmatrix}, \quad \begin{array}{l} C_2^{[1]} = \begin{bmatrix} 1.3 & 0.6 & 1.1 \end{bmatrix}, \\ L_2^{[1]} = \begin{bmatrix} 0.7 & 0.8 & 1.2 \end{bmatrix}. \end{array}$$

Subsystem 2.

$$A_1^{[2]} = \begin{bmatrix} -1.8 & -0.3 & -0.5 \\ 0.2 & -1.8 & 0.3 \\ 0.3 & 0.6 & -1.5 \end{bmatrix}, \ B_{11}^{[2]} = \begin{bmatrix} 1.0 \\ 0.7 \\ 0.7 \end{bmatrix}, \ B_{21}^{[2]} = \begin{bmatrix} 0.3 \\ 0.3 \\ -0.4 \end{bmatrix},$$

$$G_1^{[2]} = 0.4, \ K_1^{[2]} = 1.2, \ D_{11}^{[2]} = 0.3, \ D_{21}^{[2]} = -0.3,$$

$$F_1^{[2]} = \begin{bmatrix} 0.2 & 0.0 & 0.1 \\ 0.1 & 0.3 & 0.2 \\ 0.1 & 0.1 & 0.2 \end{bmatrix}, \quad \begin{array}{l} C_1^{[2]} = \begin{bmatrix} 1.1 & 0.6 & 1.5 \end{bmatrix}, \\ L_1^{[2]} = \begin{bmatrix} 1.2 & -0.8 & -1.2 \end{bmatrix}, \end{array}$$

$$A_2^{[2]} = \begin{bmatrix} -2.1 & 0.2 & 0.4 \\ 0.3 & 0.6 & 0.2 \\ -0.2 & 0.2 & -2.2 \end{bmatrix}, \ B_{12}^{[2]} = \begin{bmatrix} 0.8 \\ 0.7 \\ 0.5 \end{bmatrix}, \ B_{22}^{[2]} = \begin{bmatrix} -0.3 \\ 0.6 \\ -0.5 \end{bmatrix},$$

$$G_2^{[2]} = 0.3, \ K_2^{[2]} = 0.3, \ D_{12}^{[2]} = 0.3, \ D_{22}^{[2]} = 0.4,$$

$$F_2^{[2]} = \begin{bmatrix} 0.2 & 0.1 & 0.2 \\ 0.1 & 0.2 & 0.1 \\ 0.1 & 0.2 & 0.1 \end{bmatrix}, \quad \begin{array}{l} C_2^{[2]} = \begin{bmatrix} 1.3 & 0.6 & 1.1 \end{bmatrix}, \\ L_2^{[2]} = \begin{bmatrix} 0.7 & 0.8 & 1.2 \end{bmatrix}. \end{array}$$

The nonlinearities $f\big(x(t), t\big)$ and $g\big(x(t), t\big)$ in (4.1) are set as

$$f\big(x(t), t\big) = \begin{bmatrix} 0.2x_1(t) + 0.1x_2(t) \\ 0.2x_1(t) + 0.3x_2(t) + 0.2x_3(t) \\ 0.1x_1(t) + 0.1x_3(t) \end{bmatrix} \sin(t),$$

$$g\big(x(t), t\big) = \begin{bmatrix} 0.1x_1(t) + 0.2x_2(t) + 0.2x_3(t) \end{bmatrix} \sin(t),$$

which satisfy Assumption 4.1 with

$$M \triangleq \begin{bmatrix} 0.2 & 0.1 & 0.0 \\ 0.2 & 0.3 & 0.2 \\ 0.1 & 0.0 & 0.1 \end{bmatrix}, \quad N \triangleq \begin{bmatrix} 0.1 & 0.2 & 0.2 \end{bmatrix}.$$

Case 1. Firstly, consider the full-order DOFC design problem with $r = 3$. In view of the sufficient conditions (4.44)–(4.47) in Theorem 4.10, the minimum feasible γ can be obtained as $\gamma_{\min} = 0.8904$. The corresponding parameters of the full-order DOFC are calculated as

$$A_{c1}^{[1]} = \begin{bmatrix} -57.7172 & -2.2637 & -1.3706 \\ 7.2168 & -4.4209 & 1.7200 \\ 7.5700 & -3.0131 & -4.2822 \end{bmatrix},$$

$$B_{c1}^{[1]} = \begin{bmatrix} -5.4582 \\ -6.4293 \\ -2.0298 \end{bmatrix}, \quad C_{c1}^{[1]} = \begin{bmatrix} -1.3376 \\ -0.1185 \\ -0.1315 \end{bmatrix}^T,$$

$$A_{c2}^{[1]} = \begin{bmatrix} 0.3481 & 8.4100 & 10.5366 \\ 3.0083 & -7.8703 & 0.6900 \\ 9.2645 & -0.8920 & -2.9582 \end{bmatrix},$$

$$B_{c2}^{[1]} = \begin{bmatrix} -5.8390 \\ -1.2570 \\ -7.6154 \end{bmatrix}, \quad C_{c2}^{[1]} = \begin{bmatrix} 1.4527 \\ 0.8489 \\ 0.5020 \end{bmatrix}^T,$$

$$A_{c1}^{[2]} = \begin{bmatrix} -83.6564 & -5.3496 & 13.5642 \\ -9.6521 & -16.9328 & 5.8525 \\ -18.8543 & -6.5568 & -7.4303 \end{bmatrix},$$

$$B_{c1}^{[2]} = \begin{bmatrix} -21.0503 \\ 0.4475 \\ 6.6054 \end{bmatrix}, \quad C_{c1}^{[2]} = \begin{bmatrix} -1.9514 \\ -0.6286 \\ 0.3450 \end{bmatrix}^T,$$

$$A_{c2}^{[2]} = \begin{bmatrix} -69.0979 & 35.8456 & -18.1179 \\ -4.6730 & -4.4102 & -9.2470 \\ -24.6190 & -3.7769 & -11.4027 \end{bmatrix},$$

$$B_{c2}^{[2]} = \begin{bmatrix} -20.3258 \\ 0.3074 \\ 8.7067 \end{bmatrix}, \quad C_{c2}^{[2]} = \begin{bmatrix} -1.1406 \\ 1.5259 \\ -1.7735 \end{bmatrix}^T.$$

Case 2. Introduce the reduced-order DOFC design issue with $r = 2$. Considering (4.35)–(4.38) in Theorem 4.7, we have the minimum feasible γ as $\gamma_{\min} = 0.9030$. Accordingly, the gains of the reduced-order DOFC can be obtained as

$$A_{c1}^{[1]} = \begin{bmatrix} -92.8590 & -5.6818 \\ 7.5781 & -9.3000 \end{bmatrix}, B_{c1}^{[1]} = \begin{bmatrix} 2.1114 \\ -7.1962 \end{bmatrix}, C_{c1}^{[1]} = \begin{bmatrix} -1.5374 \\ -0.4716 \end{bmatrix}^T,$$

$$A_{c2}^{[1]} = \begin{bmatrix} -58.6042 & 34.0300 \\ 14.2169 & 0.9743 \end{bmatrix}, B_{c2}^{[1]} = \begin{bmatrix} -7.2397 \\ -8.5207 \end{bmatrix}, C_{c2}^{[1]} = \begin{bmatrix} -0.0400 \\ 1.8118 \end{bmatrix}^T,$$

$$A_{c1}^{[2]} = \begin{bmatrix} -66.5137 & -10.5625 \\ 10.0811 & -7.7937 \end{bmatrix}, B_{c1}^{[2]} = \begin{bmatrix} -9.8210 \\ -3.3116 \end{bmatrix}, C_{c1}^{[2]} = \begin{bmatrix} -2.1134 \\ -0.5735 \end{bmatrix}^T,$$

$$A_{c2}^{[2]} = \begin{bmatrix} -66.4444 & 29.6925 \\ 20.1712 & -3.8907 \end{bmatrix}, B_{c2}^{[2]} = \begin{bmatrix} -4.3730 \\ -7.4381 \end{bmatrix}, C_{c2}^{[2]} = \begin{bmatrix} -1.3128 \\ 2.4636 \end{bmatrix}^T.$$

It can be seen from the results that the performance level γ increases when the dimension of reduced-order controllers decreases, as shown in Table 4.1.

Table 4.1 Values of γ vs. reduced-order dimension

Reduced-order dimension	γ	ε
Full-order $r = 3$	0.8904	10.9745
Reduced-order $r = 2$	0.9030	10.9119

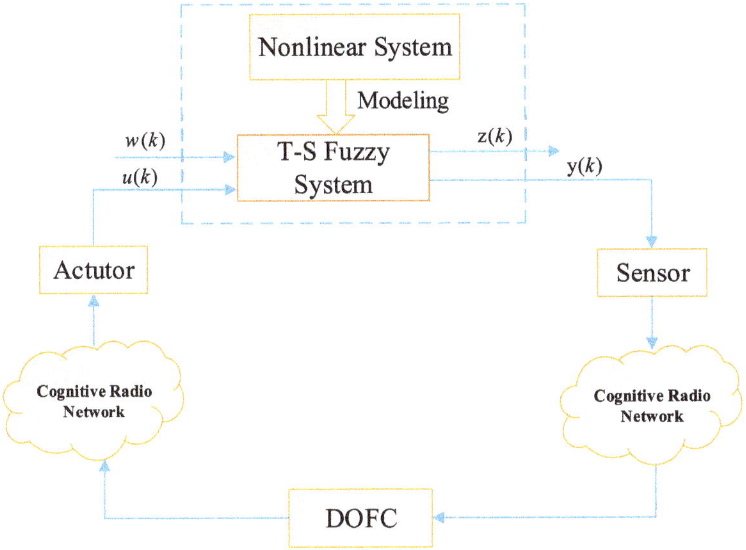

Fig. 4.2 Structure of closed-loop system over CR networks

The initial condition is selected as $x(t) = 0$, $x_c(t) = 0$, the simulation time $T^\star = 6s$, and the switching signal randomly changes from '1' and '2', where '1' and '2' stand for the first and second subsystems, as shown in Fig. 4.2. The fuzzy basis functions are chosen by

$$h_1\big(x_1(t)\big) = \frac{1}{2}\Big[1 - \sin\big(x_1^2(k)\big)\Big],\ h_2\big(x_1(t)\big) = \frac{1}{2}\Big[1 + \sin\big(x_1^2(k)\big)\Big].$$

The disturbance input $\omega(t)$ belongs to $\mathcal{L}_2[0, \infty)$, which makes it energy-bounded (Fig. 4.3). And we choose

$$\omega(t) = \frac{5\sin(0.9t)}{(0.75t)^2 + 3.5}.$$

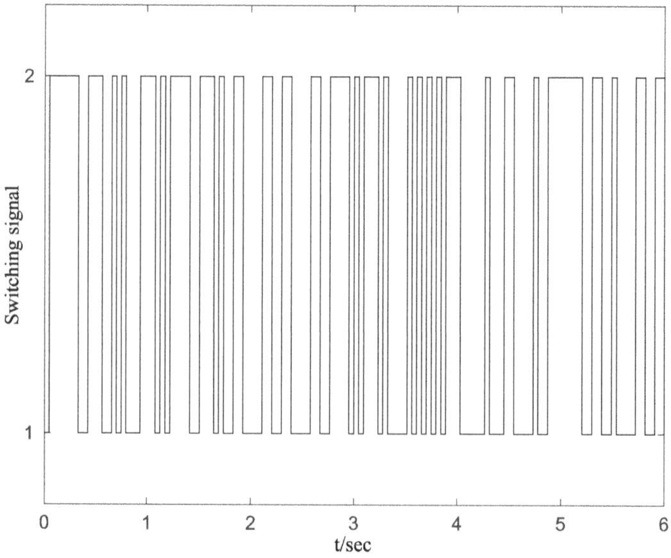

Fig. 4.3 Switching signal

In Case 1, the simulation results of full-order $\mathcal{L}_2$-$\mathcal{L}_\infty$ DOFC are exhibited in Figs. 4.4, 4.5, 4.6, 4.7. Figure 4.4 plots the states of the closed-loop system, and the DOFC states are drawn in Fig. 4.5. The control input $u(k)$ and the controlled output $z(k)$ are depicted in Figs. 4.6 and 4.7, respectively. In Case 2, the simulation results of reduced-order $\mathcal{L}_2$–$\mathcal{L}_\infty$ DOFC are shown in Figs. 4.8, 4.9, 4.10, 4.11.

Example 4.13 In this example, the CR system from [157, 184] is investigated to verify the effectiveness of our proposed scheme, which is plotted schematically in Fig. 4.2. First, suppose that every channel in CR system owes two states (busy and idle), and the channel sojourn times in each state are independent and identically distributed random variables, which follow certain probability distribution functions, that have probable connections to the states to be switched. Then, the sensor first chooses one channel to sense using the sensing strategy; if the channel is idle, the signal is transmitted through it. If not, it stops transmission to avoid collision. Hence, the CR model is approximated by a group of switched fuzzy systems, which denotes the switch between idle and busy states.

Introduce the CR system with following parameters:

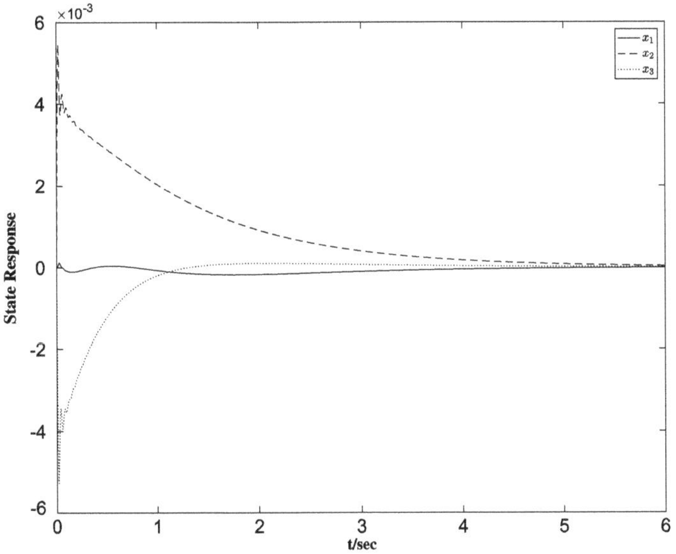

Fig. 4.4 The states of the closed-loop system in Case 1

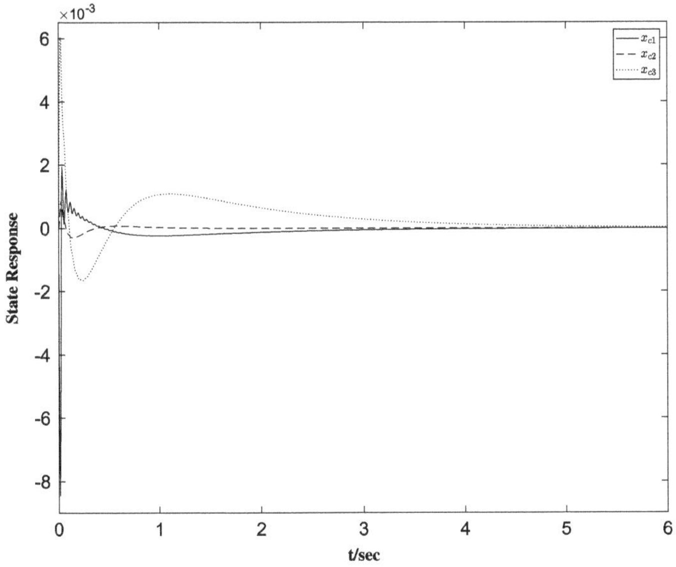

Fig. 4.5 The states of the DOF controller in Case 1

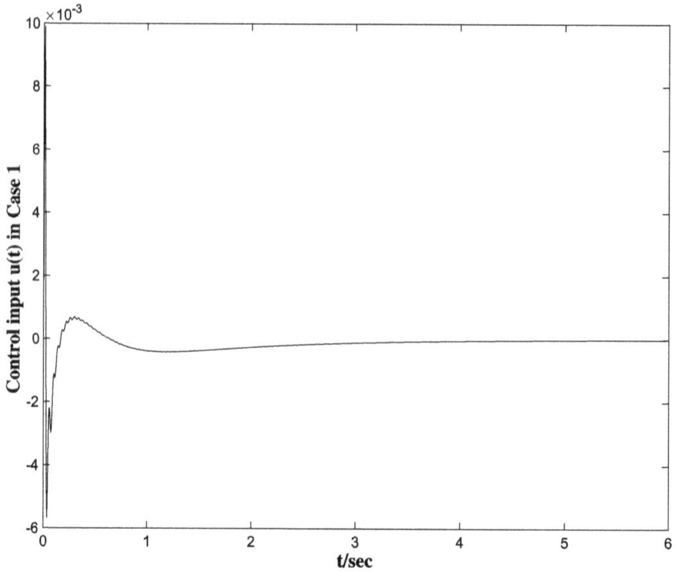

Fig. 4.6 Control input $u(t)$ in Case 1

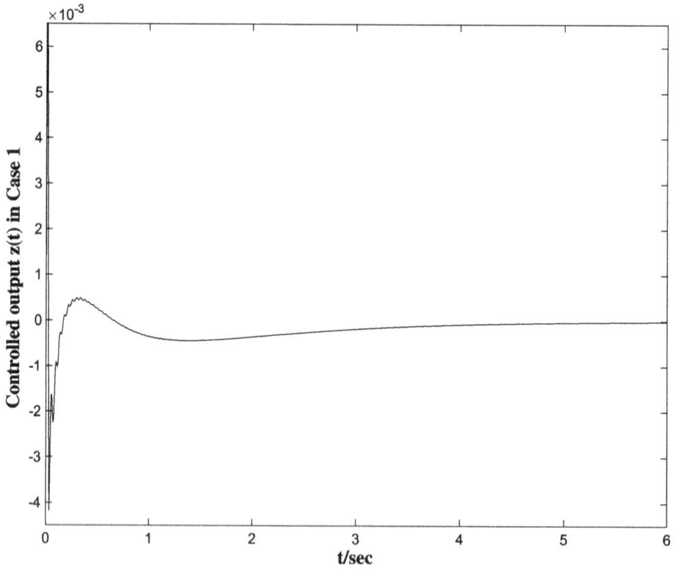

Fig. 4.7 Controlled output $z(t)$ in Case 1

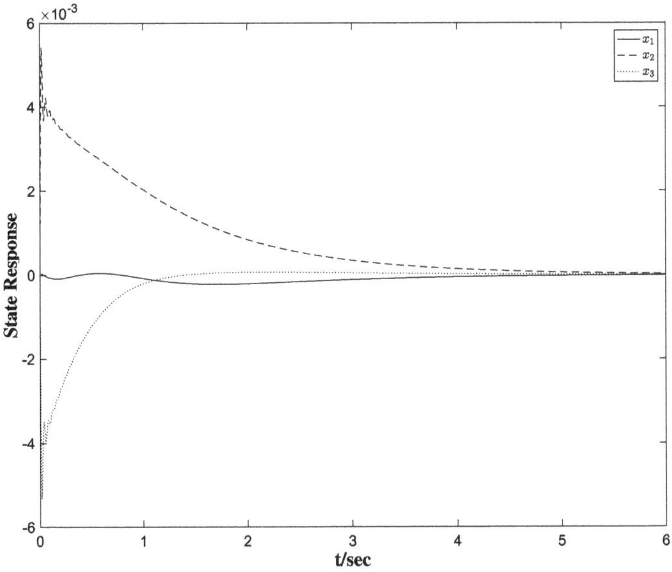

Fig. 4.8 The states of the closed-loop system in Case 2

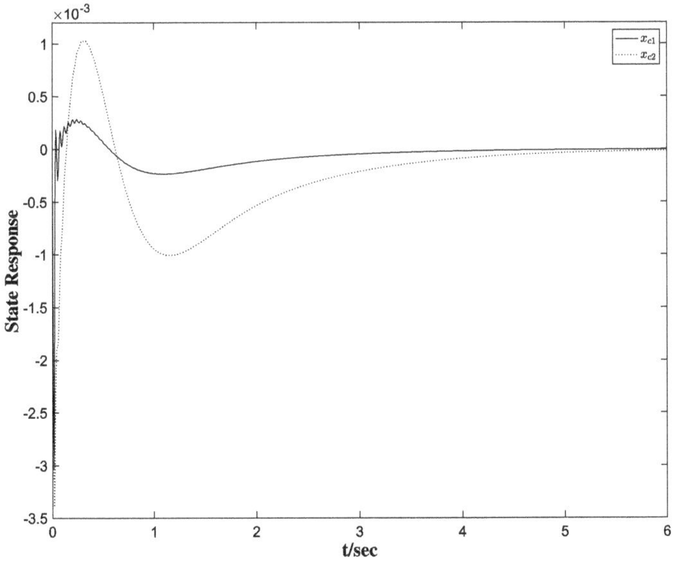

Fig. 4.9 The states of the DOF controller in Case 2

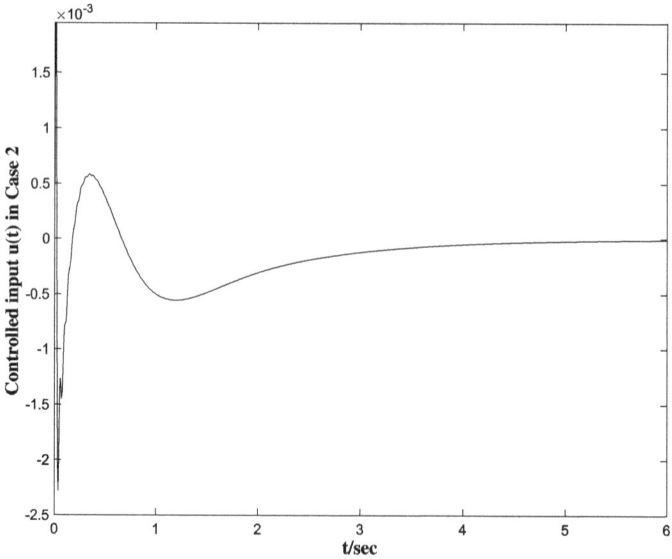

Fig. 4.10 Control input $u(t)$ in Case 2

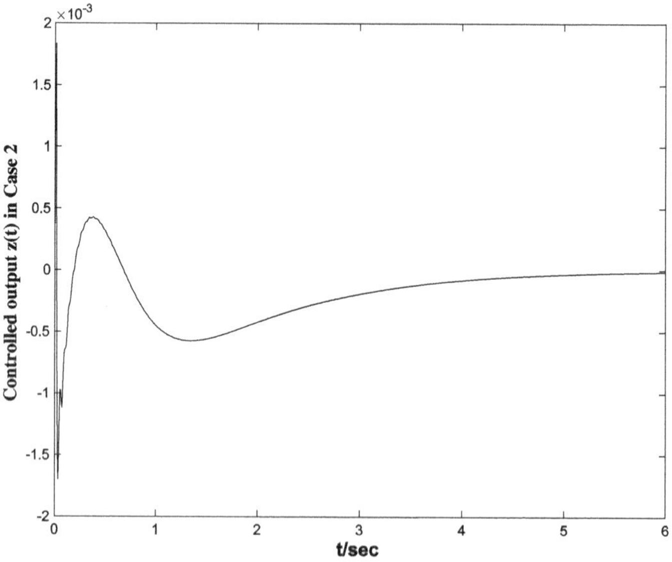

Fig. 4.11 Controlled output $z(t)$ in Case 2

Subsystem 1.

$$A_1^{[1]} = \begin{bmatrix} -1.8 & -0.3 \\ 0.3 & -1.8 \end{bmatrix}, B_{11}^{[1]} = \begin{bmatrix} 1.2 \\ 0.5 \end{bmatrix}, B_{21}^{[1]} = \begin{bmatrix} 0.5 \\ 0.3 \end{bmatrix}, \begin{matrix} G_1^{[1]} = 0.4, & D_{11}^{[1]} = 0.4, \\ K_1^{[1]} = 1.2, & D_{21}^{[1]} = 0.3, \end{matrix}$$

$$F_1^{[1]} = \begin{bmatrix} -1.5 & 1.0 \\ -1.3 & -2.0 \end{bmatrix}, \begin{matrix} C_1^{[1]} = \begin{bmatrix} 1.2 & 0.6 \end{bmatrix}, \\ L_1^{[1]} = \begin{bmatrix} 1.2 & -0.8 \end{bmatrix}, \end{matrix} A_2^{[1]} = \begin{bmatrix} -2.1 & 0.2 \\ 0.5 & -2.1 \end{bmatrix},$$

$$B_{12}^{[1]} = \begin{bmatrix} 0.8 \\ 0.7 \end{bmatrix}, B_{22}^{[1]} = \begin{bmatrix} -0.3 \\ 0.6 \end{bmatrix}, \begin{matrix} G_2^{[1]} = 0.2, & D_{12}^{[1]} = 0.5, \\ K_2^{[1]} = 0.3, & D_{22}^{[1]} = 0.3, \end{matrix}$$

$$F_2^{[1]} = \begin{bmatrix} -2.2 & 1.2 \\ -2.5 & -1.2 \end{bmatrix}, \begin{matrix} C_2^{[1]} = \begin{bmatrix} 1.3 & 0.6 \end{bmatrix}, \\ L_2^{[1]} = \begin{bmatrix} 0.7 & 1.2 \end{bmatrix}. \end{matrix}$$

Subsystem 2.

$$A_1^{[2]} = \begin{bmatrix} -1.8 & -0.3 \\ 0.6 & -1.5 \end{bmatrix}, B_{11}^{[2]} = \begin{bmatrix} 1.0 \\ 0.7 \end{bmatrix}, B_{21}^{[2]} = \begin{bmatrix} 0.5 \\ 0.3 \end{bmatrix}, \begin{matrix} G_1^{[2]} = 0.4, & D_{11}^{[2]} = 0.3, \\ K_1^{[2]} = 1.2, & D_{21}^{[2]} = -0.3, \end{matrix}$$

$$F_1^{[2]} = \begin{bmatrix} -2.1 & 1.5 \\ 1.2 & -2.1 \end{bmatrix}, \begin{matrix} C_1^{[2]} = \begin{bmatrix} 1.1 & 0.6 \end{bmatrix}, \\ L_1^{[2]} = \begin{bmatrix} 1.2 & -1.2 \end{bmatrix}, \end{matrix} A_2^{[2]} = \begin{bmatrix} -2.1 & 0.1 \\ 0.2 & -2.5 \end{bmatrix},$$

$$B_{12}^{[2]} = \begin{bmatrix} 0.8 \\ 0.7 \end{bmatrix}, B_{22}^{[2]} = \begin{bmatrix} -0.3 \\ 0.6 \end{bmatrix}, \begin{matrix} G_2^{[2]} = 0.3, & D_{12}^{[2]} = 0.3, \\ K_2^{[2]} = 0.3, & D_{22}^{[2]} = -0.3, \end{matrix}$$

$$F_2^{[2]} = \begin{bmatrix} -2.1 & 2.5 \\ 2.2 & -2.1 \end{bmatrix}, \begin{matrix} C_2^{[2]} = \begin{bmatrix} 1.3 & 0.6 \end{bmatrix}, \\ L_2^{[2]} = \begin{bmatrix} 0.7 & 1.2 \end{bmatrix}. \end{matrix}$$

Matrices M and N under the Assumption 1 are set to

$$M \triangleq \begin{bmatrix} 0.2 & 0.1 \\ 0.3 & 0.2 \end{bmatrix}, \quad N \triangleq \begin{bmatrix} 0.1 & 0.2 \end{bmatrix}.$$

Then we can calculate the minimum feasible $\gamma_{min} = 0.6655$. The initial condition is $x(0) = [1.1 \ -0.5]^T$, and other system settings are the same as Example 4.12. The simulation results of full-order DFOC are plotted in Figs. 4.12, 4.13, 4.14, 4.15. Figure 4.12 draws the states $x_1(t)$–$x_2(t)$ of the closed-loop system, and the controller states $x_{c1}(t)$–$x_{c2}(t)$ are depicted in Fig. 4.13. The control input $u(k)$ and the controlled output $z(k)$ are displayed in Figs. 4.14 and 4.15, respectively. It can be seen that the system is exponentially stable during every run of the simulation.

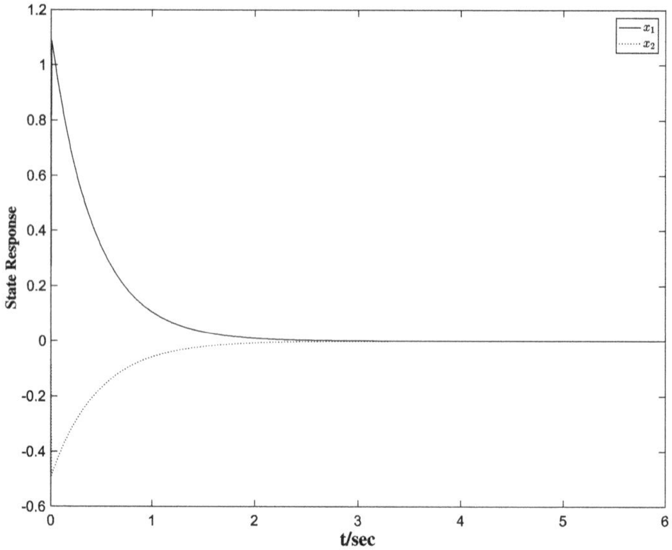

Fig. 4.12 The states of the CR system

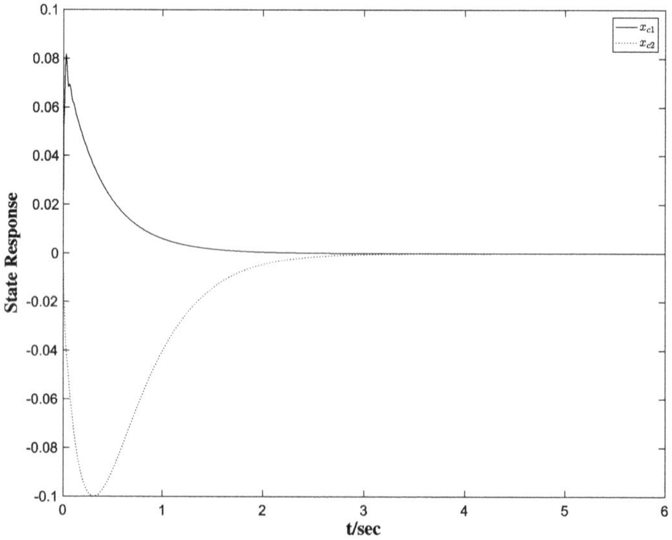

Fig. 4.13 The states of the DOF controller

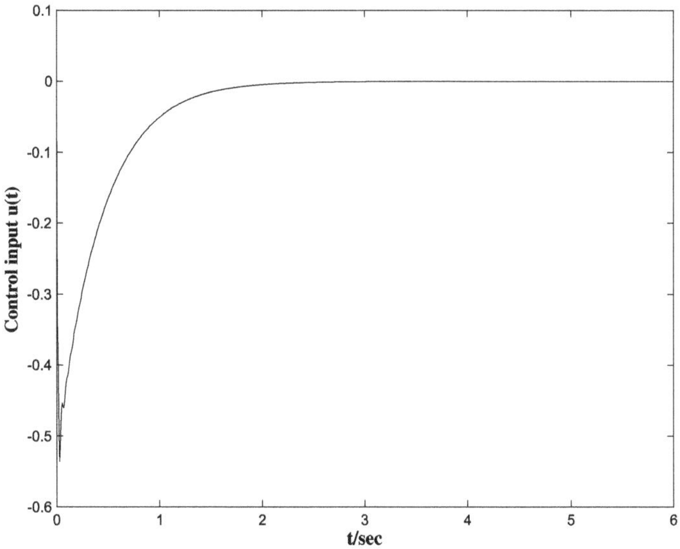

Fig. 4.14 Control input $u(t)$

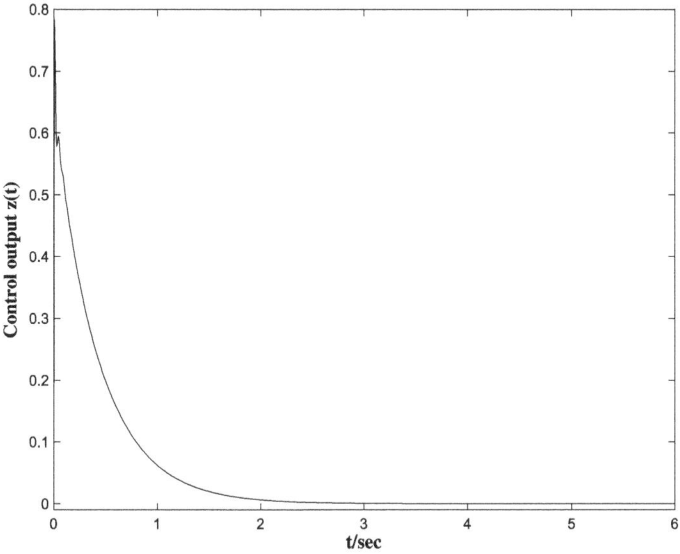

Fig. 4.15 Controlled output $z(t)$

4.6 Conclusion

In this chapter, the DOFC issue for T-S fuzzy switched systems was investigated. The ADT technique was applied to stabilize the switched systems exponentially with an arbitrary switching rule. Next, we constructed a piecewise Lyapunov function and derived the sufficient conditions to ensure that the corresponding closed-loop system was exponentially stable with a specific $\mathcal{L}_2$–$\mathcal{L}_\infty$ performance level γ. Moreover, the feasible conditions of the fuzzy-rule-dependent DOFC were derived through linearization, which could be promptly resolved using the standard toolbox. In the end, two numerical simulations were proposed to illustrate advantages of the developed scheme.

Part II
Fuzzy Filtering and Fault Detection

Chapter 5
Dissipative Filtering of Fuzzy Switched Systems

5.1 Introduction

The dissipativity-based filtering problem is investigated for T-S fuzzy switched systems with stochastic perturbation in this chapter. First, sufficient conditions are established to ensure the mean-square exponential stability of T-S fuzzy switched systems. Next, we design a filter for the system of interest, subject to Brownian motion. The piecewise Lyapunov function approach and ADT method are combined, and valid fuzzy filters are presented such that the filter error dynamics exhibit mean-square exponential stability and the prescribed dissipative property. Moreover, the solvability conditions of the designed filter are specified through linearization.

5.2 System Description and Preliminaries

5.2.1 System Description

In this chapter, we introduce nonlinear switched systems with stochastic perturbation, which are expressed by the T-S fuzzy switched stochastic models as follows:

Rule $\mathcal{R}_i^{[j]}$: IF $\theta_1^{[j]}(t)$ is $\mu_{i1}^{[j]}$ and $\theta_2^{[j]}(t)$ is $\mu_{i2}^{[j]}$ and $\cdots$ and $\theta_p^{[j]}(t)$ is $\mu_{ip}^{[j]}$, THEN

$$dx(t) = \left[A_i^{[j]}x(t) + B_i^{[j]}\omega(t)\right]dt + E_i^{[j]}x(t)d\varpi, \tag{5.1a}$$

$$dy(t) = \left[C_i^{[j]}x(t) + D_i^{[j]}\omega(t)\right]dt + F_i^{[j]}x(t)d\varpi, \tag{5.1b}$$

$$z(t) = L_i^{[j]}x(t), \quad i = 1, 2, \ldots, r, \tag{5.1c}$$

X. Su et al., *Intelligent Control, Filtering and Model Reduction Analysis for Fuzzy-Model-Based Systems*, Studies in Systems, Decision and Control 385,
https://doi.org/10.1007/978-3-030-81214-0_5

where $x(t) \in \mathbb{R}^n$ represents the state vector; $\omega(t) \in \mathbb{R}^q$ represents the disturbance input, and $\omega(t)$ belongs to $\mathcal{L}_2[0, \infty)$; $y(t) \in \mathbb{R}^p$ represents the output; $z(t) \in \mathbb{R}^l$ denotes the signal to be estimated; $\varpi(t)$ denotes the scalar Brownian motion based on the probability space $(\Omega, \mathcal{F}, \{\mathcal{F}_t\}_{t\geq 0}, \mathcal{P})$ satisfying $\mathbf{E}\{d\varpi(t)\} = 0$ and $\mathbf{E}\left\{d\varpi^2(t)\right\} = dt$; the positive integer N is the number of subsystems; $\sigma_j(t) : [0, \infty) \to \{0, 1\}$, and $\sum_{j=1}^{\mathcal{N}} \sigma_j(t) = 1$, $t \in [0, \infty)$, $j \in \mathcal{N} = \{1, 2, \cdots, N\}$, stands for the switching signal specifying which subsystem is activated at the switching moment; r is the number of IF-THEN rules; $\mu_{i1}^{[j]}, \ldots, \mu_{ip}^{[j]}$ denote the fuzzy sets; $\theta_1^j(t)$, $\theta_2^{[j]}(t)$, $\ldots$, $\theta_p^j(t)$ represent the premise variables; $\left\{\left(A_i^{[j]}, B_i^{[j]}, E_i^{[j]}, C_i^{[j]}, D_i^{[j]}, F_i^{[j]}, L_i^{[j]}\right) : j \in \mathcal{N}\right\}$ are a group of matrices parameterized by $\mathcal{N} = \{1, 2, \ldots, N\}$, and $A_i^{[j]}$, $B_i^{[j]}$, $E_i^{[j]}$, $C_i^{[j]}$, $D_i^{[j]}$, $F_i^{[j]}$, $L_i^{[j]}$ are real constant matrices.

Suppose that the premise variables don't depend on the disturbance input $\omega(t)$. Given a set of $(x(t), \omega(t))$, the eventual output of the concerned system is given by

$$
\begin{aligned}
dx(t) =& \sum_{j=1}^{N} \sigma_j(t) \sum_{i=1}^{r} h_i^{[j]}\left(\theta^{[j]}(t)\right)\left\{\left[A_i^{[j]}x(t) + B_i^{[j]}\omega(t)\right]dt \right. \\
&\left. + E_i^{[j]}x(t)d\varpi\right\}, \qquad (5.2a)\\
dy(t) =& \sum_{j=1}^{N} \sigma_j(t) \sum_{i=1}^{r} h_i^{[j]}\left(\theta^{[j]}(t)\right)\left\{\left[C_i^{[j]}x(t) + D_i^{[j]}\omega(t)\right]dt \right. \\
&\left. + F_i^{[j]}x(t)d\varpi\right\}, \qquad (5.2b)\\
z(t) =& \sum_{j=1}^{N} \sigma_j(t) \sum_{i=1}^{r} h_i^{[j]}\left(\theta^{[j]}(t)\right)L_i^{[j]}x(t), \qquad (5.2c)
\end{aligned}
$$

where

$$
h_i^{[j]}\left(\theta^{[j]}(t)\right) = \frac{\nu_i^{[j]}\left(\theta^{[j]}(t)\right)}{\sum_{i=1}^{r} \nu_i^{[j]}\left(\theta^{[j]}(t)\right)}, \quad \nu_i^{[j]}\left(\theta^{[j]}(t)\right) = \prod_{l=1}^{p} \mu_{il}^{[j]}\left(\theta_l^{[j]}(t)\right),
$$

and $\mu_{(il)}^{[j]}\left(\theta_l^{[j]}(t)\right)$ stands for the grade of membership of $\theta_l^{[j]}(t)$ in $\mu_{il}^{[j]}$. Assume $\nu_i^{[j]}\left(\theta^{[j]}(t)\right) \geqslant 0, i = 1, 2, \ldots, r, \sum_{i=1}^{r} \nu_i^{[j]}\left(\theta^{[j]}(t)\right) > 0$ for all t. Hence, $h_i^{[j]}\left(\theta^{[j]}(t)\right) \geqslant 0$ for $i = 1, 2, \ldots, r$ and $\sum_{i=1}^{r} h_i^{[j]}\left(\theta^{[j]}(t)\right) = 1$ for all t.

It is assumed the premise variable of the fuzzy system $\theta^{[j]}(t)$ is available for feedback, which means $h_i^{[j]}\left(\theta(t)\right)$ is available for feedback. Provided that the premise variable of the fuzzy filter is identical to the system. On the basis of PDC technique, the fuzzy-dependent filter is proposed to share the same IF-THEN sections. Subsequently, we propose a fuzzy filter described by

Rule $\mathcal{R}_i^{[j]}$: IF $\theta_1^{[j]}(t)$ is $\mu_{i1}^{[j]}$ and $\theta_2^{[j]}(t)$ is $\mu_{i2}^{[j]}$ and $\cdots$ and $\theta_p^{[j]}(t)$ is $\mu_{ip}^{[j]}$, THEN

$$dx_c(t) = A_{ci}^{[j]} x_c(t)dt + B_{ci}^{[j]} dy(t), \tag{5.3a}$$

$$z_c(t) = L_{ci}^{[j]} x_f(t), \quad i = 1, 2, \cdots, r, \tag{5.3b}$$

where $x_c(t) \in \mathbb{R}^k$ denotes the state vector of the fuzzy filter with $k \leqslant n$; $z_c(t) \in \mathbb{R}^l$ denotes the estimation of $z(t)$; $A_{ci}^{[j]}$, $B_{ci}^{[j]}$, and $L_{ci}^{[j]}$ are fuzzy filter parameters to be designed. The fuzzy filter (5.3) can be further expressed as

$$dx_c(t) = \sum_{j=1}^{N} \sigma_j(t) \sum_{i=1}^{r} h_i^{[j]}(\theta^{[j]}(t)) \Big[A_{ci}^{[j]} x_f(t)dt + B_{ci}^{[j]} dy(t) \Big], \tag{5.4a}$$

$$z_c(t) = \sum_{j=1}^{N} \sigma_j(t) \sum_{i=1}^{r} h_i^{[j]}(\theta^{[j]}(t)) L_{ci}^{[j]} x_c(t). \tag{5.4b}$$

Denoting $e_c(t) \triangleq z(t) - z_c(t)$, $\xi(t) \triangleq \left[x^T(t) \; x_c^T(t) \right]^T$, and augmenting the system (5.2) to contain the information in (5.4), thus the overall fuzzy filtering error dynamic is obtained as

$$\begin{aligned} d\xi(t) = & \sum_{j=1}^{N} \sigma_j(t) \sum_{i=1}^{r} h_i^{[j]}\left(\theta^{[j]}(t)\right) \sum_{l=1}^{r} h_l^{[j]}\left(\theta^{[j]}(t)\right) \\ & \left\{ \left[\tilde{A}_{il}^{[j]} \xi(t) + \tilde{B}_{il}^{[j]} \omega(t) \right] dt + \tilde{E}_{il}^{[j]} \xi(t) d\varpi(t) \right\}, \end{aligned} \tag{5.5a}$$

$$e_c(t) = \sum_{j=1}^{N} \sigma_j(t) \sum_{i=1}^{r} h_i^{[j]}\left(\theta^{[j]}(t)\right) \sum_{l=1}^{r} h_l^{[j]}\left(\theta^{[j]}(t)\right) \tilde{L}_{il}^{[j]} \xi(t), \tag{5.5b}$$

where

$$\tilde{A}_{il}^{[j]} \triangleq \begin{bmatrix} A_i^{[j]} & 0 \\ B_{cl}^{[j]} C_i^{[j]} & A_{cl}^{[j]} \end{bmatrix}, \quad \tilde{B}_{il}^{[j]} \triangleq \begin{bmatrix} B_i^{[j]} \\ B_{cl}^{[j]} D_i^{[j]} \end{bmatrix},$$

$$\tilde{E}_{il}^{[j]} \triangleq \begin{bmatrix} E_i^{[j]} & 0 \\ B_{cl}^{[j]} F_i^{[j]} & 0 \end{bmatrix}, \quad \tilde{L}_{il}^{[j]} \triangleq \begin{bmatrix} L_i^{[j]} & -L_{cl}^{[j]} \end{bmatrix}.$$

Remark 5.1 Actually, there are two methods in the fault detection filter design. One is fuzzy-parameter-independent method, which is fit for the premise variables of the fuzzy model are unavailable. And the other one is fuzzy-parameter-dependent method, which is applied in the case that is available. We solve the fault detection filtering problem with fuzzy-parameter-dependent method in this chapter. Owing to the use of the fuzzy-parameter-dependent approach, information of premise variables are fully taken into account, thus the obtained results are less conservative.

Definition 5.2 For given $T_2 > T_1 \geqslant 0$, $N_{\sigma_j}(T_1, T_2)$ represents the number of switchings of $\sigma_j(t)$ over (T_1, T_2). If $N_{\sigma_j}(T_1, T_2) \leqslant N_0 + (T_2 - T_1)/T_a$ holds for $T_a > 0$, $N_0 \geqslant 0$. Then T_a is named as the ADT.

Definition 5.3 When $\omega(t) = 0$, the equilibrium $\xi^\star(t) = 0$ of fault detection system in (5.5) is said to be mean-square exponentially stable with $\sigma_j(t)$ if $\xi(t)$ satisfies the following condition:

$$\mathbf{E}\left\{\|\xi(t)\|^2\right\} \leqslant \eta \|\xi(t_0)\|^2 e^{-\lambda(t-t_0)}, \quad \forall t \geqslant t_0,$$

where $\eta \geqslant 1$ and $\lambda > 0$.

5.2.2 Dissipativity Definition

Consider [99] and the T-S fuzzy switched system in (5.5), $\mathcal{S}(\omega(t), e_c(t))$ is a real valued function which denotes the following *supply rate*.

Definition 5.4 ***(Supply Rate)*** The supply rate is a real valued function, $\mathcal{S}(\omega(t), e_c(t)) : \Omega \times Z \to \mathbb{R}$, which is locally Lebesgue integrable independently of the initial condition and the input, that is for any $\omega(t) \in \Omega$, $e_c(t) \in Z$, $t^\star \geqslant 0$, it satisfies that $\int_0^{t^\star} |\mathcal{S}(\omega(t), e_c(t))| dt < +\infty$.

Definition 5.5 ***(Dissipative System)*** The T-S fuzzy switched system (5.5) under supply rate $\mathcal{S}(\omega(t), e_c(t))$ is known as dissipative if there is a nonnegative function $V(x(t)) : X \to \mathbb{R}$, namely the storage function, such that the following inequality satisfies:

$$\mathbf{E}\left\{V(x(t^\star)) - V(x(0))\right\} \leqslant \mathbf{E}\left\{\int_0^{T^\star} \mathcal{S}(\omega(t), e_c(t)) dt\right\}, \tag{5.6}$$

for $x_0 \in X$, input $\omega(t) \in \Omega$ and $t^\star \geq 0$ (or, in other words as differently: for admissible inputs $\omega(t)$ that make the state from $x(0)$ to $x(t^\star)$ during $[0, t^\star]$, where $x(t^\star)$ denotes the state variable when $t = t^\star$).

Definition 5.6 Given matrices $\mathscr{Z} \in \mathbb{R}^{l \times l}$, $\mathscr{X} \in \mathbb{R}^{q \times q}$, $\mathscr{Y} \in \mathbb{R}^{l \times q}$ with $\mathscr{Z}$ and $\mathscr{X}$ being symmetric, the T-S fuzzy switched system (5.5) is $(\mathscr{Z}, \mathscr{Y}, \mathscr{X})$-dissipative if there exists the real function $\psi(\cdot)$ with $\psi(0) = 0$, and $\forall t^\star \geqslant 0$,

$$\mathbf{E}\left\{\int_0^{t^\star} \begin{bmatrix} e_c(t) \\ \omega(t) \end{bmatrix}^T \begin{bmatrix} \mathscr{Z} & \mathscr{Y} \\ \star & \mathscr{X} \end{bmatrix} \begin{bmatrix} e_c(t) \\ \omega(t) \end{bmatrix} dt\right\} + \psi(x_0) \geqslant 0. \tag{5.7}$$

Moreover, if the following condition holds for $\delta > 0$,

$$\mathbf{E}\left\{\int_0^{t^\star}\begin{bmatrix}e_c(t)\\ \omega(t)\end{bmatrix}^T\begin{bmatrix}\mathscr{Z} & \mathscr{Y}\\ \star & \mathscr{X}\end{bmatrix}\begin{bmatrix}e_c(t)\\ \omega(t)\end{bmatrix}dt\right\}+\psi(x_0)$$
$$\geqslant \delta\int_0^{t^\star}\omega^T(t)\omega(t)dt,\ \ \forall t^\star\geqslant 0, \tag{5.8}$$

the system (5.5) is called as strictly $(\mathscr{Z},\mathscr{Y},\mathscr{X})$-$\delta$-dissipative.

Remark 5.7 Assume that $\mathscr{Z}\leqslant 0$, it follow that $-\mathscr{Z}=(\mathscr{Z}_-^{\frac{1}{2}})^2$, for some $\mathscr{Z}_-^{\frac{1}{2}}\geqslant 0$. Hence, the index in Definition 5.6 contains some particular cases, such as $\mathcal{H}_\infty$ performance, the positive real performance, and the sector bounded performance.

5.3 Main Results

5.3.1 Dissipativity Performance Analysis

Theorem 5.8 *Given matrices $0\geqslant\mathscr{Z}\in\mathbb{R}^{l\times l}$, $\mathscr{X}\in\mathbb{R}^{q\times q}$, $\mathscr{Y}\in\mathbb{R}^{l\times q}$ with symmetric $\mathscr{Z}$ and $\mathscr{X}$, and scalars $\delta>0$, $\alpha>0$, if there exist $0<P^{[j]}\in\mathbb{R}^{(n+k)\times(n+k)}$ for $j\in\mathcal{N}$,*

$$\Pi_{ii}^{[j]}<0,\quad i=1,2,\ldots,r, \tag{5.9a}$$
$$\Pi_{il}^{[j]}+\Pi_{li}^{[j]}<0,\quad 1\leqslant i<l\leqslant r, \tag{5.9b}$$

where

$$\Pi_{il}^{[j]}\triangleq\begin{bmatrix}\Pi_{11il}^{[j]} & \Pi_{12il}^{[j]} & \left(\tilde{E}_{il}^{[j]}\right)^T P^{[j]} & \left(\tilde{L}_{il}^{[j]}\right)^T\mathscr{Z}_-^{\frac{1}{2}}\\ \star & \delta I-\mathscr{X} & 0 & 0\\ \star & \star & -P^{[j]} & 0\\ \star & \star & \star & -I\end{bmatrix}, \tag{5.10}$$

with $$\begin{cases}\Pi_{11il}^{[j]}\triangleq P^{[j]}\tilde{A}_{il}^{[j]}+\left(\tilde{A}_{il}^{[j]}\right)^T P^{[j]}+\alpha P^{[j]},\\ \Pi_{12il}^{[j]}\triangleq P^{[j]}\tilde{B}_{il}^{[j]}-\left(\tilde{L}_{il}^{[j]}\right)^T\mathscr{Y}^T.\end{cases}$$

Then the concerned system in (5.5) is mean-square exponentially stable subject to the $(\mathscr{Z},\mathscr{Y},\mathscr{X})$-$\delta$-dissipativity for any switching signal with the ADT satisfying $T_a>T_a^*=\dfrac{\ln\mu}{\alpha}$, where $\mu\geqslant 1$ and $P^{[j]}\leqslant\mu P^{[s]}$, $\forall j,s\in\mathcal{N}$.

Proof On account of the switching signal and fuzzy membership functions, it follows from (5.9a)–(5.9b) that

$$\sum_{j=1}^{N}\sigma_j(t)\sum_{i=1}^{r}h_i^{[j]}\left(\theta^{[j]}(t)\right)\sum_{l=1}^{r}h_l^{[j]}\left(\theta^{[j]}(t)\right)$$
$$\begin{bmatrix} \Pi_{11il}^{[j]} & \Pi_{12il}^{[j]} & \left(\tilde{E}_{il}^{[j]}\right)^T P^{[j]} & \left(\tilde{L}_{il}^{[j]}\right)^T \mathscr{Z}_-^{\frac{1}{2}} \\ \star & \delta I - \mathscr{X} & 0 & 0 \\ \star & \star & -P^{[j]} & 0 \\ \star & \star & \star & -I \end{bmatrix} < 0. \tag{5.11}$$

The Lyapunov functional is selected as

$$V(\xi_t,\sigma_j) \triangleq \xi^T(t)\left(\sum_{j=1}^{N}\sigma_j(t)P^{[j]}\right)\xi(t), \tag{5.12}$$

where $P^{[j]} > 0$, $j \in \mathcal{N}$ are to be decided. Along the system in (5.5) for a fixed $\sigma_j(t)$, by using the Itô formula, we can get the stochastic differential as follows:

$$dV(\xi_t,\sigma_j) = \mathscr{L}V(\xi_t,\sigma_j)dt + 2\sum_{j=1}^{N}\sigma_j(t)\sum_{i=1}^{r}h_i^{[j]}\left(\theta^{[j]}(t)\right)$$
$$\sum_{l=1}^{r}h_l^{[j]}\left(\theta^{[j]}(t)\right)\xi^T(t)P^{[j]}\tilde{E}_{il}^{[j]}\xi(t)d\varpi(t), \tag{5.13a}$$
$$\mathscr{L}V(\xi_t,\sigma_j) = \sum_{j=1}^{N}\sigma_j(t)\sum_{i=1}^{r}h_i^{[j]}\left(\theta^{[j]}(t)\right)\sum_{l=1}^{r}h_l^{[j]}\left(\theta^{[j]}(t)\right)$$
$$\left\{\xi^T(t)\left[P^{[j]}\tilde{A}_{il}^{[j]} + \left(\tilde{A}_{il}^{[j]}\right)^T P^{[j]}\right]\xi(t) + \xi^T(t)\right.$$
$$\left.\left(\tilde{E}_{il}^{[j]}\right)^T P^{[j]}\tilde{E}_{il}^{[j]}\xi(t) + 2\xi^T(t)P^{[j]}\tilde{B}_{il}^{[j]}\omega(t)\right\}. \tag{5.13b}$$

Consider (5.13b) and for any nonzero $\omega(t) \in \mathcal{L}_2[0,\infty)$, we have

$$\begin{aligned}\Pi(\xi_t,\sigma_j) &\triangleq \mathscr{L}V(\xi_t,\sigma_j) + \alpha V(\xi_t,t) - e_f^T(t)\mathscr{Z}e_f(t) \\ &\quad - 2\omega^T(t)\mathscr{Y}e_f(t) - \omega^T(t)\left[\mathscr{X} - \delta I\right]\omega(t) \\ &\leqslant \sum_{j=1}^{N}\sigma_j(t)\sum_{i=1}^{r}h_i^{[j]}\left(\theta^{[j]}(t)\right)\sum_{l=1}^{r}h_l^{[j]}\left(\theta^{[j]}(t)\right) \\ &\quad \begin{bmatrix}\xi(t)\\ \omega(t)\end{bmatrix}^T \begin{bmatrix}\tilde{\Pi}_{11il}^{[j]} & \tilde{\Pi}_{12il}^{[j]} \\ \star & \tilde{\Pi}_{22il}^{[j]}\end{bmatrix}\begin{bmatrix}\xi(t)\\ \omega(t)\end{bmatrix},\end{aligned} \tag{5.14}$$

where

$$\begin{cases} \tilde{\Pi}_{11il}^{[j]} \triangleq P^{[j]}\tilde{A}_{il}^{[j]} + \left(\tilde{A}_{il}^{[j]}\right)^T P^{[j]} + \left(\tilde{E}_{il}^{[j]}\right)^T P^{[j]}\tilde{E}_{il}^{[j]} \\ \qquad\qquad +\alpha P^{[j]} - \left(\tilde{L}_{il}^{[j]}\right)^T \mathscr{Z}\tilde{L}_{il}^{[j]}, \\ \tilde{\Pi}_{12il}^{[j]} \triangleq P^{[j]}\tilde{B}_{il}^{[j]} - \left(\tilde{L}_{il}^{[j]}\right)^T \mathscr{Y}^T, \quad \tilde{\Pi}_{22il}^{[j]} \triangleq \delta I - \mathscr{X}. \end{cases} \tag{5.15}$$

Based on (5.11) and employing Schur complement method, it yields $\Pi(\xi_t, t) < 0$, that is $\mathscr{L}V(\xi_t, \sigma_j) < -\alpha V(\xi_t, \sigma_j) + \Psi(t)$, with

$$\Psi(t) \triangleq e_f^T(t)\mathscr{Z}e_f(t) + 2\omega^T(t)\mathscr{Y}e_f(t) + \omega^T(t)\left[\mathscr{X} - \delta I\right]\omega(t).$$

Therefore,

$$\begin{aligned} dV(\xi_t, \sigma_j) <& -\alpha V(\xi_t, \sigma_j)dt + 2\sum_{j=1}^{N}\sigma_j(t)\sum_{i=1}^{r}h_i^{[j]}\left(\theta^{[j]}(t)\right) \\ & \sum_{l=1}^{r}h_l^{[j]}\left(\theta^{[j]}(t)\right)\xi^T(t)P^{[j]}\tilde{E}_{il}^{[j]}\xi(t)d\varpi(t) + \Psi(t)dt. \end{aligned}$$

It can be observed that

$$\begin{aligned} d[e^{\alpha t}V(\xi_t, \sigma_j)] =& \alpha e^{\alpha t}V(\xi_t, \sigma_j)dt + e^{\alpha t}dV(\xi_t, \sigma_j) \\ <& e^{\alpha t}\Bigg[2\sum_{j=1}^{N}\sigma_j(t)\sum_{i=1}^{r}h_i^{[j]}\left(\theta^{[j]}(t)\right)\sum_{l=1}^{r}h_l^{[j]}\left(\theta^{[j]}(t)\right) \\ & \xi^T(t)P^{[j]}\tilde{E}_{il}^{[j]}\xi(t)d\varpi(t) + \Psi(t)dt\Bigg]. \end{aligned} \tag{5.16}$$

Integrate two sides of (5.16) from $t^* > 0$ to t and make expectations. Then it is not difficult to obtain

$$\mathbf{E}\Big\{V(\xi_t, \sigma_j)\Big\} < e^{-\alpha(t-t_k)}\mathbf{E}\Big\{V(\xi_{t_k}, \sigma_j)\Big\} + \mathbf{E}\left\{\int_{t_k}^{t}e^{-\alpha(t-s)}\Psi(s)ds\right\}.$$

Considering $P^{[j]} \leqslant \mu P^{[s]}$and (5.12), at switching instant t_k, it has

$$\mathbf{E}\left\{V(\xi_{t_k}, \sigma_j)\right\} \leqslant \mu\mathbf{E}\left\{V(\xi_{t_k^-}, \sigma_j)\right\}. \tag{5.17}$$

Consequently, based on (5.13)–(5.17) and $\vartheta = N_{\sigma_j}(0, t) \leqslant (t-0)/T_a$,

$$\begin{aligned} \mathbf{E}\Big\{V(\xi_t, \sigma_j)\Big\} \leqslant& \mathbf{E}\left\{\int_0^t e^{-\alpha(t-s)+N_{\sigma_j}(s,t)\ln\mu}\Psi(s)ds\right\} \\ & + e^{-\alpha t+N_{\sigma_j}(0,t)\ln\mu}V\left(\xi_0, \sigma_j\right). \end{aligned} \tag{5.18}$$

Under the zero initial condition $\xi(0) = 0$, (5.18) signifies

$$\mathbf{E}\Big\{V(\xi_t,\sigma_j)\Big\}\leqslant\mathbf{E}\Big\{\int_0^t e^{-\alpha(t-s)+N_{\sigma_j}(s,t)\ln\mu}\Psi(s)ds\Big\}. \tag{5.19}$$

Next, multiplying two sides of (5.18) by $e^{-N_{\sigma_j}(0,t)\ln\mu}$, we can get

$$\mathbf{E}\Big\{e^{-N_{\sigma_j}(0,t)\ln\mu}V(\xi_t,\sigma_j)\Big\}\leqslant\mathbf{E}\Big\{\int_0^t\Psi(s)ds\Big\}. \tag{5.20}$$

Due to $N_{\sigma_j}(0,t)\leq t/T_a$ and $T_a>T_a^*=\ln\mu/\alpha$, we have $N_{\sigma_j}(0,t)\ln\mu\leq\alpha t$. Therefore,

$$\mathbf{E}\Big\{e^{-\alpha t}V(\xi_t,\sigma_j)\Big\}\leqslant\mathbf{E}\Big\{\int_0^t\Psi(s)ds\Big\}. \tag{5.21}$$

It is easily seen that for arbitrary $t^\star\geqslant 0$,

$$\mathbf{E}\Big\{\int_0^t\Psi(s)ds\Big\}\geqslant\mathbf{E}\Big\{e^{-\alpha t}V(\xi_t,\sigma_j)\Big\}\geqslant 0,$$

Then the proof is complete.

On the basis of Theorem 5.8, the following result on the estimation of the state decay can be obtained.

Theorem 5.9 *For $\alpha>0$, assume there are matrices $0<P^{[j]}\in\mathbb{R}^{(n+k)\times(n+k)}$ such that for $j\in\mathcal{N}$,*

$$P^{[j]}\tilde{A}_{ii}^{[j]}+\left(\tilde{A}_{ii}^{[j]}\right)^TP^{[j]}+\alpha P^{[j]}+\left(\tilde{E}_{ii}^{[j]}\right)^TP^{[j]}\tilde{E}_{ii}^{[j]}<0,$$
$$i=1,2,\ldots,r, \tag{5.22a}$$

$$\begin{aligned}&2\alpha P^{[j]}+P^{[j]}\tilde{A}_{il}^{[j]}+\left(\tilde{A}_{il}^{[j]}\right)^TP^{[j]}+\left(\tilde{E}_{il}^{[j]}\right)^TP^{[j]}\tilde{E}_{il}^{[j]}\\&+P^{[j]}\tilde{A}_{li}^{[j]}+\left(\tilde{A}_{li}^{[j]}\right)^TP^{[j]}+\left(\tilde{E}_{li}^{[j]}\right)^TP^{[j]}\tilde{E}_{li}^{[j]}<0,\\&1\leqslant i<l\leqslant r.\end{aligned} \tag{5.22b}$$

Then the dynamic system in (5.5a) is mean-square exponentially stable under any switching signal with ADT which satisfies $T_a>T_a^*=\dfrac{\ln\mu}{\alpha}$, where $\mu\geqslant 1$ and $P^{[j]}\leqslant\mu P^{[s]}$, $\forall j,s\in\mathcal{N}$. In addition, the state decay estimate is expressed as

$$\mathbf{E}\left\{\|\xi(t)\|^2\right\}\leqslant\eta e^{-\lambda t}\|\xi(0)\|^2, \tag{5.23}$$

where

$$\begin{cases} a = \min\limits_{\forall j \in \mathscr{I}} \lambda_{\min}\left(P^{[j]}\right), & \lambda = \alpha - \frac{\ln \mu}{T_a} > 0, \\ b = \max\limits_{\forall j \in \mathscr{I}} \lambda_{\max}\left(P^{[j]}\right), & \eta = \frac{b}{a} \geqslant 1. \end{cases} \tag{5.24}$$

Proof Consider a piecewise Lyapunov function as (5.12). Based on (5.22), we have

$$\sum\nolimits_{j=1}^{N} \sigma_j(t) \sum\nolimits_{i=1}^{r} h_i^{[j]}\left(\theta^{[j]}(t)\right) \sum\nolimits_{l=1}^{r} h_l^{[j]}\left(\theta^{[j]}(t)\right)$$
$$\left[P^{[j]}\tilde{A}_{ii}^{[j]} + \left(\tilde{A}_{ii}^{[j]}\right)^T P^{[j]} + \alpha P^{[j]} + \left(\tilde{E}_{ii}^{[j]}\right)^T P^{[j]}\tilde{E}_{ii}^{[j]}\right] < 0. \tag{5.25}$$

For $\omega(t) = 0$ and a fixed $\sigma_j(t)$, it yields from (5.5a) that

$$\mathscr{L}V(\xi_t, \sigma_j) < -\alpha\xi^T(t)P^{[j]}\xi(t) = -\alpha V(\xi_t, \sigma_j). \tag{5.26}$$

Hence,

$$\begin{aligned} dV(\xi_t, \sigma_j) <& 2\sum_{j=1}^{N} \sigma_j(t) \sum_{i=1}^{r} h_i^{[j]}\left(\theta^{[j]}(t)\right) \sum_{l=1}^{r} h_l^{[j]}\left(\theta^{[j]}(t)\right) \\ & \xi^T(t)P^{[j]}\tilde{E}_{il}^{[j]}\xi(t)d\varpi(t) - \alpha V(\xi_t, \sigma_j)dt. \end{aligned}$$

Note that

$$\begin{aligned} d[e^{\alpha t}V(\xi_t, \sigma_j)] <& 2\sum_{j=1}^{N} \sigma_j(t) \sum_{i=1}^{r} h_i^{[j]}\left(\theta^{[j]}(t)\right) \sum_{l=1}^{r} h_l^{[j]}\left(\theta^{[j]}(t)\right) \\ & \xi^T(t)P^{[j]}\tilde{E}_{il}^{[j]}\xi(t)d\varpi(t). \end{aligned} \tag{5.27}$$

Integrating two sides of (5.27) from $T > 0$ to t and make mathematical expectation, we can get

$$\mathbf{E}\left\{V(\xi_t, \sigma_j)\right\} < e^{-\alpha(t-T)}\mathbf{E}\left\{V(\xi_T, \sigma_j)\right\}. \tag{5.28}$$

Using the similar methods in the proof of Theorem 5.8, set $0 = t_0 < t_1 < \cdots < t_k < \cdots, k = 1, \ldots$, denote the switching points of $\sigma_j(t)$ during $(0, t)$. As in the previous part, the j_kth subsystem is activated when $t \in \left[t_k, t_{k+1}\right)$. When $T = t_k$ in (5.28), then

$$\mathbf{E}\left\{V(\xi_t, \sigma_j)\right\} < e^{-\alpha(t-t_k)}\mathbf{E}\left\{V(\xi_{t_k}, \sigma_j)\right\}. \tag{5.29}$$

Consequently, due to $P^{[j]} \leqslant \mu P^{[s]}$ and (5.29), and $\vartheta = N_{\sigma_j}(0, t) \leqslant (t-0)/T_a$, it can be observed that

$$\mathbf{E}\left\{V(\xi_t, \sigma_j)\right\} \leqslant e^{-(\alpha - \ln\mu/T_a)t}V(\xi_0, \sigma_j). \tag{5.30}$$

Considering (5.12), we can get

$$\mathbf{E}\left\{V(\xi_t,\sigma_j)\right\} \geqslant a\mathbf{E}\left\{\|\xi(t)\|^2\right\}, \quad V(\xi_0,\sigma_j) \leqslant b\|\xi(0)\|^2, \tag{5.31}$$

where a and b are defined in (5.24). Combining (5.30)–(5.31) then

$$\mathbf{E}\left\{\|\xi(t)\|^2\right\} \leqslant \frac{1}{a}\mathbf{E}\left\{V(\xi_t,\sigma_j)\right\} \leqslant \frac{b}{a}e^{-(\alpha-\ln\mu/T_a)t}\,\|\xi(0)\|^2, \tag{5.32}$$

which signifies (5.23). On account of Definition 5.3 with $t_0 = 0$, when $\omega(t) = 0$, the system (5.5a) is mean-square exponentially stable, and thus completing the proof.

5.3.2 Dissipativity-Based Filter Design

Theorem 5.10 *Given scalars $\alpha > 0$, $\delta > 0$, and matrices $0 \geqslant \mathscr{Z} \in \mathbb{R}^{l\times l}$, $\mathscr{X} \in \mathbb{R}^{q\times q}$, $\mathscr{Y} \in \mathbb{R}^{l\times q}$ with symmetric $\mathscr{Z}$ and $\mathscr{X}$, assume there are matrices $0 < \mathscr{U}^{[j]} \in \mathbb{R}^{n\times n}$, $0 < \mathscr{V}^{[j]} \in \mathbb{R}^{k\times k}$, $\mathcal{A}_{ci}^{[j]} \in \mathbb{R}^{k\times k}$, $\mathcal{B}_{ci}^{[j]} \in \mathbb{R}^{k\times p}$, and $\mathcal{L}_{ci}^{[j]} \in \mathbb{R}^{l\times k}$ such that for $j \in \mathcal{N}$, $i = 1, 2, \ldots, r$, $l = 1, 2, \ldots, r$,*

$$\Phi_{ii}^{[j]} < 0, \quad i = 1, 2, \ldots, r, \tag{5.33a}$$

$$\Phi_{il}^{[j]} + \Phi_{li}^{[j]} < 0, \quad 1 \leqslant i < l \leqslant r, \tag{5.33b}$$

$$\begin{bmatrix} \mathscr{U}^{[j]} & \mathscr{V}^{[j]} \\ \star & \mathscr{V}^{[j]} \end{bmatrix} > 0, \tag{5.33c}$$

where

$$\Phi_{il}^{[j]} \triangleq \begin{bmatrix} \Phi_{11il}^{[j]} & \Phi_{12il}^{[j]} & \Phi_{13il}^{[j]} & \left(\Phi_{14il}^{[j]}\right)^T & \left(\Phi_{15il}^{[j]}\right)^T & \Phi_{16il}^{[j]} \\ \star & \Phi_{22il}^{[j]} & \Phi_{23il}^{[j]} & 0 & 0 & \Phi_{26il}^{[j]} \\ \star & \star & \delta I - \mathscr{X} & 0 & 0 & 0 \\ \star & \star & \star & -\mathscr{U}^{[j]} & -\mathscr{V}^{[j]} & 0 \\ \star & \star & \star & \star & -\mathscr{V}^{[j]} & 0 \\ \star & \star & \star & \star & \star & -I \end{bmatrix},$$

with

$$\begin{aligned} \Phi_{11il}^{[j]} &\triangleq \mathscr{U}^{[j]}A_i^{[j]} + \mathcal{B}_{cl}^{[j]}C_i^{[j]} + \left(A_i^{[j]}\right)^T\left(\mathscr{U}^{[j]}\right)^T + \alpha\mathscr{U}^{[j]} \\ &\quad + \left(C_i^{[j]}\right)^T\left(\mathcal{B}_{cl}^{[j]}\right)^T, \quad \Phi_{16il}^{[j]} \triangleq \left(L_i^{[j]}\right)^T\mathscr{Z}_-^{\frac{1}{2}}, \\ \Phi_{12il}^{[j]} &\triangleq \mathcal{A}_{cl}^{[j]} + \left(A_i^{[j]}\right)^T\mathscr{V}^{[j]} + \left(C_i^{[j]}\right)^T\left(\mathcal{B}_{cl}^{[j]}\right)^T + \alpha\mathscr{V}^{[j]}, \end{aligned}$$

$$\Phi_{22l}^{[j]} \triangleq \mathcal{A}_{cl}^{[j]} + \left(\mathcal{A}_{cl}^{[j]}\right)^T + \alpha \mathscr{V}^{[j]}, \quad \Phi_{26il}^{[j]} \triangleq -\left(\mathcal{L}_{cl}^{[j]}\right)^T \mathscr{Z}_-^{\frac{1}{2}},$$

$$\Phi_{13il}^{[j]} \triangleq \mathscr{U}^{[j]} B_i^{[j]} + \mathcal{B}_{cl}^{[j]} D_i^{[j]} - \left(\mathscr{Y} L_i^{[j]}\right)^T,$$

$$\Phi_{23il}^{[j]} \triangleq \mathscr{V}^{[j]} B_i^{[j]} + \mathcal{B}_{fl}^{[j]} D_i^{[j]} + \left(\mathscr{Y} \mathcal{L}_{cl}^{[j]}\right)^T,$$

$$\Phi_{14il}^{[j]} \triangleq \mathscr{U}^{[j]} E_i^{[j]} + \mathcal{B}_{cl}^{[j]} F_i^{[j]}, \quad \Phi_{15il}^{[j]} \triangleq \mathscr{V}^{[j]} E_i^{[j]} + \mathcal{B}_{cl}^{[j]} F_i^{[j]}.$$

In that way, there exists a fuzzy-parameter-dependent filter as (5.3) such that the resulting dynamic system in (5.5) is mean-square exponentially stable subject to a $(\mathscr{Z}, \mathscr{Y}, \mathscr{X})$*-δ-dissipative performance under any switching signal with ADT which satisfies* $T_a > T_a^* = \frac{\ln \mu}{\alpha}$. *Furthermore, if above-mentioned conditions have feasible solutions* $\left\{\mathscr{U}^{[j]}, \mathscr{V}^{[j]}, \mathcal{A}_{ci}^{[j]}, \mathcal{B}_{ci}^{[j]}, \mathcal{L}_{ci}^{[j]}\right\}$, *then the parameters of our designed fuzzy filter (5.3) can be reformulated as*

$$A_{ci}^{[j]} = \left(\mathscr{V}^{[j]}\right)^{-1} \mathcal{A}_{ci}^{[j]}, \quad B_{ci}^{[j]} = \left(\mathscr{V}^{[j]}\right)^{-1} \mathcal{B}_{ci}^{[j]}, \quad L_{ci}^{[j]} = \mathcal{L}_{ci}^{[j]}. \tag{5.34}$$

Proof Partition the matrix $P^{[j]}$ as

$$P^{[j]} \triangleq \begin{bmatrix} P_1^{[j]} & P_2^{[j]} \\ \star & P_3^{[j]} \end{bmatrix} > 0, \quad P_2^{[j]} \triangleq \begin{bmatrix} P_4^{[j]} \\ 0_{(n-k)\times k} \end{bmatrix}, \tag{5.35}$$

where $P_1^{[j]} \in \mathbb{R}^n \times \mathbb{R}^n$, $P_2^{[j]} \in \mathbb{R}^n \times \mathbb{R}^k$, and $P_3^{[j]} \in \mathbb{R}^k \times \mathbb{R}^k$. Without loss of generality, $P_4^{[j]}$ is assumed to be nonsingular.

The following nonsingular matrices are introduced:

$$\begin{cases} \mathscr{F}^{[j]} \triangleq \begin{bmatrix} I & 0 \\ 0 & \left(P_3^{[j]}\right)^{-1} \left(P_2^{[j]}\right)^T \end{bmatrix}, \\ \mathscr{U}^{[j]} \triangleq P_1^{[j]}, \quad \mathscr{V}^{[j]} \triangleq P_2^{[j]} \left(P_3^{[j]}\right)^{-1} \left(P_2^{[j]}\right)^T, \end{cases} \tag{5.36}$$

and

$$\begin{cases} \mathcal{A}_{ci}^{[j]} \triangleq P_2^{[j]} A_{ci}^{[j]} \left(P_3^{[j]}\right)^{-1} \left(P_2^{[j]}\right)^T, \\ \mathcal{L}_{ci}^{[j]} \triangleq L_{ci}^{[j]} \left(P_3^{[j]}\right)^{-1} \left(P_2^{[j]}\right)^T, \quad \mathcal{B}_{ci}^{[j]} \triangleq P_2^{[j]} B_{ci}^{[j]}. \end{cases} \tag{5.37}$$

Performing a congruence transformation to (5.9) with diag$\{\mathscr{F}^{[j]}, I, \mathscr{F}^{[j]}, I\}$, we have

$$\mathrm{diag}\left\{\mathscr{F}^{[j]}, I, \mathscr{F}^{[j]}, I\right\}^T \Pi_{ii}^{[j]} \mathrm{diag}\left\{\mathscr{F}^{[j]}, I, \mathscr{F}^{[j]}, I\right\} < 0, \tag{5.38a}$$

$$\mathrm{diag}\left\{\mathcal{F}^{[j]}, I, \mathcal{F}^{[j]}, I\right\}^T \left(\Pi_{il}^{[j]} + \Pi_{li}^{[j]}\right) \mathrm{diag}\left\{\mathscr{F}^{[j]}, I, \mathscr{F}^{[j]}, I\right\} < 0. \tag{5.38b}$$

In consideration of (5.36)–(5.37), we can get

$$\begin{cases} \left(\mathscr{F}^{[j]}\right)^T P^{[j]} \tilde{E}_{il}^{[j]} \mathscr{F}^{[j]} = \begin{bmatrix} \mathscr{U}^{[j]} E_i^{[j]} + \mathcal{B}_{cl}^{[j]} F_i^{[j]} & 0 \\ \mathscr{V}^{[j]} E_i^{[j]} + \mathcal{B}_{cl}^{[j]} F_i^{[j]} & 0 \end{bmatrix}, \\ \mathscr{Y} \tilde{L}_{il}^{[j]} \mathscr{F}^{[j]} = \begin{bmatrix} \mathscr{Y} L_i^{[j]} & -\mathscr{Y} \mathcal{L}_{cl}^{[j]} \end{bmatrix}, \\ \left(\mathscr{F}^{[j]}\right)^T P^{[j]} \tilde{A}_{il}^{[j]} \mathscr{F}^{[j]} = \begin{bmatrix} \mathscr{U}^{[j]} A_i^{[j]} + \mathcal{B}_{cl}^{[j]} C_i^{[j]} & \mathcal{A}_{cl}^{[j]} \\ \mathscr{V}^{[j]} A_i^{[j]} + \mathcal{B}_{cl}^{[j]} C_i^{[j]} & \mathcal{A}_{cl}^{[j]} \end{bmatrix}, \\ \left(\mathscr{F}^{[j]}\right)^T P^{[j]} \tilde{B}_{il}^{[j]} = \begin{bmatrix} \mathscr{U}^{[j]} B_i^{[j]} + \mathcal{B}_{cl}^{[j]} D_i^{[j]} \\ \mathscr{V}^{[j]} B_i^{[j]} + \mathcal{B}_{cl}^{[j]} D_i^{[j]} \end{bmatrix}, \\ \left(\mathscr{F}^{[j]}\right)^T P^{[j]} \mathscr{F}^{[j]} = \begin{bmatrix} \mathscr{U}^{[j]} & \mathscr{V}^{[j]} \\ \star & \mathscr{V}^{[j]} \end{bmatrix}, \\ \left(\mathscr{Z}_-^{\frac{1}{2}}\right)^T \tilde{L}_{il}^{[j]} \mathscr{F}^{[j]} = \begin{bmatrix} \left(\mathscr{Z}_-^{\frac{1}{2}}\right)^T L_i^{[j]} & -\left(\mathscr{Z}_-^{\frac{1}{2}}\right)^T \mathcal{L}_{cl}^{[j]} \end{bmatrix}. \end{cases} \tag{5.39}$$

Considering (5.39), it yields from (5.38) that (5.33). Besides, (5.37) is equal to

$$\begin{cases} A_{ci}^{[j]} = \left[\left(P_2^{[j]}\right)^{-T} P_3^{[j]}\right]^{-1} \left(\mathscr{V}^{[j]}\right)^{-1} \mathcal{A}_{ci}^{[j]} \left(P_2^{[j]}\right)^{-T} P_3^{[j]}, \\ B_{ci}^{[j]} = \left[\left(P_2^{[j]}\right)^{-T} P_3^{[j]}\right]^{-1} \left(\mathscr{V}^{[j]}\right)^{-1} \mathcal{B}_{ci}^{[j]}, \quad L_{ci}^{[j]} = \mathcal{L}_{ci}^{[j]}. \end{cases} \tag{5.40}$$

Notice that the filter parameters $A_{ci}^{[j]}$, $B_{ci}^{[j]}$ and $L_{ci}^{[j]}$ in (5.3) can be described by (5.40), which means $\left(P_2^{[j]}\right)^{-T} P_3^{[j]}$ can be considered as a similarity transformation on the state-space realization of the filter, which makes no influence in the filter mapping from y to e_f. Without loss of generality, set $\left(P_2^{[j]}\right)^{-T} P_3^{[j]} = I$, then (5.34). As a result, the dissipativity-based fuzzy filter in (5.3) can be obtained with (5.34). This completes the proof then.

Remark 5.11 It can be observed that the obtained conditions in Theorem 5.10 are all in the form of LMIs, and the dissipative fuzzy filter parameters of nonlinear switched stochastic systems can be calculated via resolving the convex optimization issue.

5.4 Illustrative Example

Consider the nonlinear switched stochastic model with $\mathcal{N} = 2$, which can be approximated by the T-S fuzzy model in (5.1). And the model parameters are set as below:

Subsystem 1.

$$A_1^{[1]} = \begin{bmatrix} -2.4 & -0.1 & 0.2 \\ 0.2 & -1.5 & 0.2 \\ -0.3 & 0.0 & -1.8 \end{bmatrix}, \; B_1^{[1]} = \begin{bmatrix} 0.4 \\ 0.3 \\ 0.6 \end{bmatrix}, \; D_1^{[1]} = 0.3,$$

$$A_2^{[1]} = \begin{bmatrix} -2.1 & 0.2 & -0.1 \\ 0.4 & -1.6 & -0.2 \\ 0.3 & 0.2 & -2.8 \end{bmatrix}, \; B_2^{[1]} = \begin{bmatrix} 0.7 \\ 0.8 \\ 0.4 \end{bmatrix}, \; D_2^{[1]} = 0.1,$$

$$E_1^{[1]} = \begin{bmatrix} 1.2 & 0.1 & 0.1 \\ 0.1 & 1.3 & 0.2 \\ 0.1 & 0.1 & 1.1 \end{bmatrix}, \quad E_2^{[1]} = \begin{bmatrix} 1.1 & 0.2 & 0.1 \\ 0.0 & 1.2 & 0.2 \\ 0.3 & 0.1 & 1.3 \end{bmatrix},$$

$$F_1^{[1]} = \begin{bmatrix} 0.2 & 0.1 & 0.3 \end{bmatrix}, \quad F_2^{[1]} = \begin{bmatrix} 0.3 & 0.2 & 0.1 \end{bmatrix},$$

$$C_1^{[1]} = \begin{bmatrix} 1.0 & 0.6 & 1.5 \end{bmatrix}, \quad C_2^{[1]} = \begin{bmatrix} 1.0 & 1.0 & 2.0 \end{bmatrix},$$

$$L_1^{[1]} = \begin{bmatrix} 1.0 & 0.5 & 1.3 \end{bmatrix}, \quad L_2^{[1]} = \begin{bmatrix} 0.8 & 1.5 & 1.0 \end{bmatrix}.$$

Subsystem 2.

$$A_1^{[2]} = \begin{bmatrix} -1.9 & -0.3 & 0.2 \\ 0.4 & -2.1 & 0.2 \\ -0.2 & 0.0 & -1.6 \end{bmatrix}, \; B_1^{[2]} = \begin{bmatrix} 0.3 \\ 0.4 \\ 0.5 \end{bmatrix}, \; D_1^{[2]} = 0.2,$$

$$A_2^{[2]} = \begin{bmatrix} -1.9 & 0.3 & -0.2 \\ 0.3 & -1.7 & -0.2 \\ 0.2 & 0.1 & -2.3 \end{bmatrix}, \; B_2^{[2]} = \begin{bmatrix} 0.6 \\ 0.7 \\ 0.3 \end{bmatrix}, \; D_2^{[2]} = 0.3,$$

$$E_1^{[2]} = \begin{bmatrix} 1.2 & 0.2 & 0.3 \\ 0.2 & 1.2 & 0.2 \\ 0.1 & 0.2 & 1.3 \end{bmatrix}, \quad E_2^{[2]} = \begin{bmatrix} 1.1 & 0.1 & 0.2 \\ 0.0 & 1.2 & 0.2 \\ 0.2 & 0.3 & 1.3 \end{bmatrix},$$

$$F_1^{[2]} = \begin{bmatrix} 0.5 & 0.2 & 0.4 \end{bmatrix}, \quad F_2^{[2]} = \begin{bmatrix} 0.2 & 0.3 & 0.2 \end{bmatrix},$$

$$C_1^{[2]} = \begin{bmatrix} 1.1 & 0.7 & 1.2 \end{bmatrix}, \quad C_2^{[2]} = \begin{bmatrix} 0.9 & 0.8 & 2.2 \end{bmatrix},$$

$$L_1^{[2]} = \begin{bmatrix} 1.1 & 0.4 & 1.2 \end{bmatrix}, \quad L_2^{[2]} = \begin{bmatrix} 0.8 & 1.4 & 1.2 \end{bmatrix}.$$

The dissipative performance parameters in Definition 5.6 are chosen as $\mathscr{Z} = -0.9$, $\mathscr{Y} = 0.3$ and $\mathscr{X} = 0, 7$. By solving the conditions in Theorem 5.10, the fuzzy filter matrices can be computed as

$$A_{c1}^{[1]} = \begin{bmatrix} -4.4135 & -1.1553 & -2.4555 \\ -5.1192 & -4.6978 & -7.6770 \\ -2.0791 & -1.1566 & -5.1215 \end{bmatrix}, \ B_{c1}^{[1]} = \begin{bmatrix} -1.8014 \\ -5.1877 \\ -2.0073 \end{bmatrix},$$
$$C_{c1}^{[1]} = \begin{bmatrix} -0.9768 & -0.5022 & -1.2007 \end{bmatrix},$$
$$A_{c2}^{[1]} = \begin{bmatrix} -5.0918 & -1.3716 & -3.8062 \\ 0.0109 & -6.0679 & -6.7309 \\ -1.6008 & -1.6673 & -7.9648 \end{bmatrix}, \ B_{c2}^{[1]} = \begin{bmatrix} -2.0518 \\ -3.1718 \\ -2.1770 \end{bmatrix},$$
$$C_{c2}^{[1]} = \begin{bmatrix} -0.5810 & -1.4839 & -1.0604 \end{bmatrix},$$
$$A_{c1}^{[2]} = \begin{bmatrix} -3.6636 & -1.0860 & -1.6620 \\ -1.0052 & -3.6918 & -1.4547 \\ -1.7457 & -0.9609 & -3.7415 \end{bmatrix}, \ B_{c1}^{[2]} = \begin{bmatrix} -1.4592 \\ -1.5834 \\ -1.5698 \end{bmatrix},$$
$$C_{c1}^{[2]} = \begin{bmatrix} -1.1016 & -0.3927 & -1.1612 \end{bmatrix},$$
$$A_{c2}^{[2]} = \begin{bmatrix} -4.3988 & -1.2216 & -3.3446 \\ -1.2884 & -3.6354 & -5.0169 \\ -1.2536 & -1.7072 & -8.4814 \end{bmatrix}, \ B_{c2}^{[2]} = \begin{bmatrix} -1.8432 \\ -2.1572 \\ -2.4045 \end{bmatrix},$$
$$C_{c2}^{[2]} = \begin{bmatrix} -0.7637 & -1.3552 & -1.2269 \end{bmatrix}.$$

Figure 5.1 shows the switching signal, which is generated randomly, where '1' and '2' denote the first and the second subsystem, respectively. The fuzzy basis functions

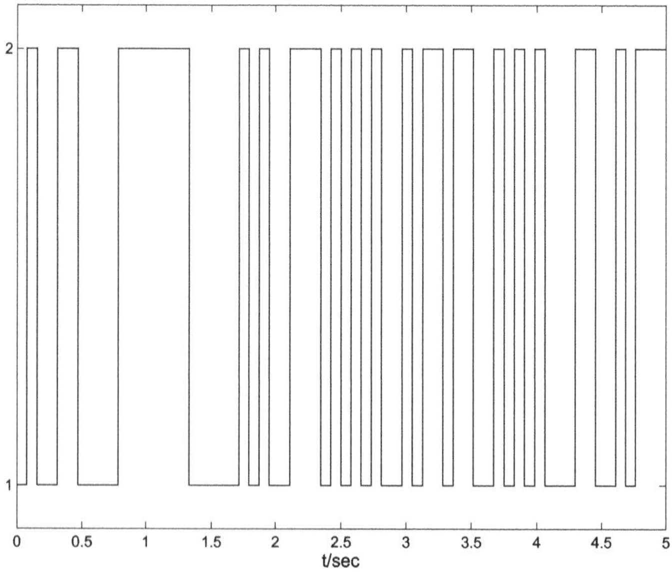

Fig. 5.1 Switching signal $\sigma_j(t)$, $j \in \mathcal{N} = \{1, 2\}$

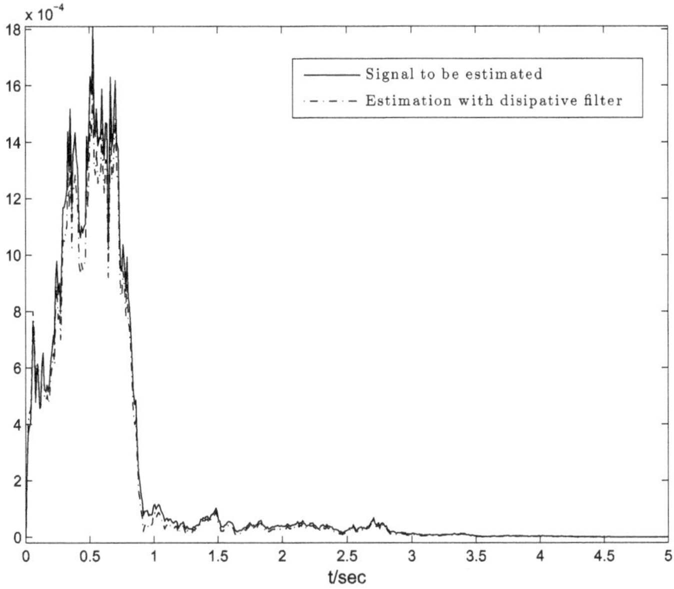

Fig. 5.2 Signal $z(t)$ and its estimation $z_c(t)$

are given by

$$h_1(x_1(t)) \triangleq \exp\left[-\frac{(x_1(t)-\vartheta)^2}{2\sigma^2}\right], \quad h_2(x_1(t)) \triangleq 1 - h_1(x_1(t)).$$

When $x(t) = 0$ ($x_f(t) = 0$), that is the zero initial condition, and the disturbance input $\omega(t)$ is set to $\omega(t) = \frac{\sin(0.1t)}{4t^2+5}$. The simulation results of the dissipative fuzzy filter design scheme are exhibited in Figs. 5.2, 5.3. The signal $z(t)$ (solid line) and its estimation $z_c(t)$ (dash-dot line) are plotted in Fig. 5.2. Figure 5.3 draws the corresponding estimation error $e_c(t)$.

5.5 Conclusion

The dissipativity-based filtering issue was resolved for T-S fuzzy switched systems with stochastic perturbations. Using the ADT approach and piecewise Lyapunov functions, sufficient conditions were established to ensure the mean-square exponential stability with dissipative performance for the corresponding filtering error dynamics. Furthermore, the solvable conditions for the dissipativity-based fuzzy filter were derived through linearization. In the end, a simulation example was presented to demonstrate the validity of the proposed scheme.

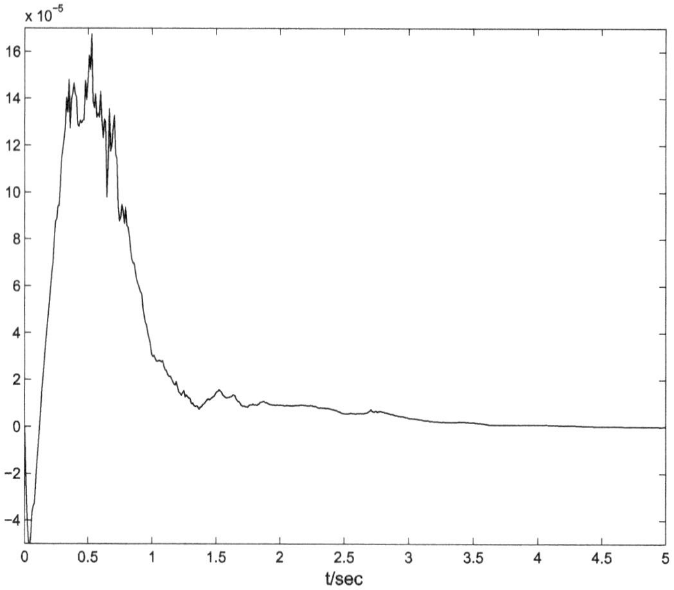

Fig. 5.3 Estimation error $e_c(t)$

Chapter 6
Fault Detection for Switched Stochastic Systems

6.1 Introduction

In this chapter, the problem of fault detection filtering is settled for the nonlinear switched stochastic system in the form of T-S fuzzy model. The objective is to establish a robust fault detection approach for a nonlinear switched system with Brownian motion. Based on the observer-based fuzzy filter as a residual generator, the fault detection problem can be reformulated as a fuzzy filtering problem. By using the piecewise Lyapunov functions and ADT approach, a fuzzy-rule-dependent fault detection filter is established to ensure that the overall dynamic system is mean-square exponential stable and exhibits weighted $\mathcal{H}_\infty$ performance. Moreover, the solvable conditions of the fuzzy filter are established through linearization.

6.2 System Description and Preliminaries

The nonlinear switched stochastic systems are considered as

$$\begin{aligned} dx(t) = \sum\nolimits_{j=1}^{N} \sigma_j(t)\Big[& F_j\left(x(t), u(t), \omega(t), f(t)\right) dt \\ & + G_j\left(x(t), u(t), \omega(t), f(t)\right) d\varpi(t)\Big], \end{aligned} \tag{6.1a}$$

$$\begin{aligned} dy(t) = \sum\nolimits_{j=1}^{N} \sigma_j(t)\Big[& H_j\left(x(t), u(t), \omega(t), f(t)\right) dt \\ & + J_j\left(x(t), u(t), \omega(t), f(t)\right) d\varpi(t)\Big], \end{aligned} \tag{6.1b}$$

where $x(t) \in \mathbb{R}^n$ denotes the state vector; $u(t) \in \mathbb{R}^m$ denotes the known input; $\omega(t) \in \mathbb{R}^q$ is the disturbance input; $f(t) \in \mathbb{R}^l$ represents the fault to be detected; $u(t)$, $\omega(t)$ and $f(t)$ belong to $\mathcal{L}_2[0, \infty)$; $y(t) \in \mathbb{R}^p$ denotes the system output; $\varpi(t)$ represents

X. Su et al., *Intelligent Control, Filtering and Model Reduction Analysis for Fuzzy-Model-Based Systems*, Studies in Systems, Decision and Control 385,
https://doi.org/10.1007/978-3-030-81214-0_6

the Brownian motion defined on the probability space $(\Omega, \mathcal{F}, \{\mathcal{F}_t\}_{t\geq 0}, \mathcal{P})$, which satisfies $\mathbf{E}\{d\varpi(t)\} = 0$ and $\mathbf{E}\left\{d\varpi^2(t)\right\} = dt$; the positive integer N stands for the number of subsystems; $\sigma_j(t) : [0, \infty) \rightarrow \{0, 1\}$, and $\sum_{j=1}^{\mathcal{N}} \sigma_j(t) = 1$, $t \in [0, \infty)$, $j \in \mathcal{N} = \{1, 2, \cdots, N\}$, represents the switching signal specifying which subsystem is activated at the switching instant; $F_j(\cdot), G_j(\cdot), H_j(\cdot)$ and $J_j(\cdot)$ are nonlinear regular functions.

For given t, $\sigma_j(t)$ is denoted as σ_j for simplicity, which may depend on t or $x(t)$, or both, or other hybrid schemes. Suppose σ_j is unknown, while its momentary value is obtainable. Based on the switching signal σ_j, the switching sequence is given by $\{(j_0, t_0), (j_1, t_1), \ldots, (j_k, t_k), \ldots, \mid j_k \in \mathcal{N},\ k = 0, 1, \ldots\}$ with $t_0 = 0$, which implies the j_kth subsystem is activated during $t \in \left[t_k, t_{k+1}\right)$.

The nonlinear switched system is approximated by a T-S fuzzy model, which is utilized to deal with the fault detection filter design issue for the nonlinear interconnected system.

Rule $\mathcal{R}_i^{[j]}$: IF $\theta_1^{[j]}(t)$ is $\mu_{i1}^{[j]}$ and $\theta_2^{[j]}(t)$ is $\mu_{i2}^{[j]}$ and $\cdots$ and $\theta_p^{[j]}(t)$ is $\mu_{ip}^{[j]}$, THEN

$$\begin{aligned} dx(t) &= \left[A_i^{[j]}x(t)+B_{0i}^{[j]}u(t)+B_i\omega(t)+B_{1i}^{[j]}f(t)\right]dt+E_i^{[j]}x(t)d\varpi(t), \\ dy(t) &= \left[C_i^{[j]}x(t)+D_{0i}^{[j]}u(t)+D_i^{[j]}\omega(t)+D_{1i}^{[j]}f(t)\right]dt+F_i^{[j]}x(t)d\varpi(t), \end{aligned}$$

where $i = 1, 2, \ldots, r$, and r denotes the number of IF-THEN fuzzy rules; $\mu_{i1}^{[j]}, \ldots,$ $\mu_{ip}^{[j]}$ represent fuzzy sets; $\theta_1^j(t)$, $\theta_2^{[j]}(t)$, $\ldots$, $\theta_p^j(t)$ are premise variables, denoted by θ_p^j; $\left\{\left(A_i^{[j]}, B_i^{[j]}, B_{0i}^{[j]}, B_{1i}^{[j]}, E_i^{[j]}, C_i^{[j]}, D_i^{[j]}, D_{0i}^{[j]}, D_{1i}^{[j]}, F_i^{[j]}\right) : j \in \mathcal{N}\right\}$ is a class of matrices parameterized by $\mathcal{N} = \{1, 2, \ldots, N\}$, and $A_i^{[j]}$, $B_{0i}^{[j]}$, $B_i^{[j]}$, $B_{1i}^{[j]}$, $E_i^{[j]}$, $C_i^{[j]}$, $D_{0i}^{[j]}$, $D_i^{[j]}$, $D_{1i}^{[j]}$ and $F_i^{[j]}$ represent some constant real matrices.

Provided that the premise variables don't depend on the input $u(t)$. For a group of $(x(t), u(t))$, the resulting output of the fuzzy switched systems is expressed as

$$\begin{aligned} dx(t) = \sum\nolimits_{j=1}^{N} \sigma_j \sum\nolimits_{i=1}^{r} h_i^{[j]}\left(\theta^{[j]}\right)\Big\{ & \Big[A_i^{[j]}x(t) + B_{0i}^{[j]}u(t) \\ & + B_i\omega(t) + B_{1i}^{[j]}f(t)\Big]dt + E_i^{[j]}x(t)d\varpi(t)\Big\}, \end{aligned} \tag{6.2a}$$

$$\begin{aligned} dy(t) = \sum\nolimits_{j=1}^{N} \sigma_j \sum\nolimits_{i=1}^{r} h_i^{[j]}\left(\theta^{[j]}\right)\Big\{ & \Big[C_i^{[j]}x(t) + D_{0i}^{[j]}u(t) \\ & + D_i^{[j]}\omega(t) + D_{1i}^{[j]}f(t)\Big]dt + F_i^{[j]}x(t)d\varpi(t)\Big\}, \end{aligned} \tag{6.2b}$$

where $h_i^{[j]}\left(\theta^{[j]}\right) = \nu_i^{[j]}\left(\theta^{[j]}\right) / \sum_{i=1}^{r} \nu_i^{[j]}\left(\theta^{[j]}\right)$, $\nu_i^{[j]}\left(\theta^{[j]}\right) = \prod_{l=1}^{p} \mu_{il}^{[j]}\left(\theta_l^{[j]}\right)$, with $\mu_{il}^{[j]}\left(\theta_l^{[j]}\right)$ be the grade of membership of $\theta_l^{[j]}$ in $\mu_{il}^{[j]}$. If $\nu_i^{[j]}\left(\theta^{[j]}\right) \geq 0$, $i = 1, 2,$

$\ldots, r$, $\sum_{i=1}^{r} \nu_i^{[j]}\left(\theta^{[j]}\right) > 0$ for all t, then $h_i^{[j]}\left(\theta^{[j]}\right) \geq 0$ for $i = 1, 2, \ldots, r$ and $\sum_{i=1}^{r} h_i^{[j]}\left(\theta^{[j]}\right) = 1$ for all t.

In this note, for the plant (6.2), a fault detection filter is proposed as a residual generator. Assumed that the premise variable of the fault detection filter is the same as the one of the plant. By using the PDC technique, the fuzzy-rule-dependent filter is designed to use the same IF-THEN sections as the following form:

Rule $\mathcal{R}_i^{[j]}$: IF $\theta_1^{[j]}(t)$ is $\mu_{i1}^{[j]}$ and $\theta_2^{[j]}(t)$ is $\mu_{i2}^{[j]}$ and $\cdots$ and $\theta_p^{[j]}(t)$ is $\mu_{ip}^{[j]}$, THEN

$$dx_f(t) = A_{fi}^{[j]} x_f(t)dt + B_{fi}^{[j]} dy(t), \tag{6.3a}$$

$$\chi_f(t) = C_{fi}^{[j]} x_f(t), \quad i = 1, 2, \cdots, r, \tag{6.3b}$$

where $x_f(t) \in \mathbb{R}^n$ stands for the state of the filter; $\chi_f(t) \in \mathbb{R}^l$ denotes the residual signal; $A_f^{[j]}$, $B_f^{[j]}$ and $C_f^{[j]}$ are filter matrices to be determined. The filter (6.3) can be reformulated as

$$dx_f(t) = \sum\nolimits_{j=1}^{N} \sigma_j \sum\nolimits_{i=1}^{r} h_i^{[j]}(\theta^{[j]}) \Big[A_{fi}^{[j]} x_f(t)dt + B_{fi}^{[j]} dy(t) \Big], \tag{6.4a}$$

$$\chi_f(t) = \sum\nolimits_{j=1}^{N} \sigma_j \sum\nolimits_{i=1}^{r} h_i^{[j]}(\theta^{[j]}) C_{fi}^{[j]} x_f(t). \tag{6.4b}$$

To improve the performance in fault detection, a weighting function is combined with the fault $f(s)$, that is to say, $f_w(s) = W(s)f(s)$, where $f(s)$ and $f_w(s)$ represent the Laplace transformations of $f(t)$ and $f_w(t)$, respectively. A state space realization of $f_w(s) = W(s)f(s)$ is

$$\dot{x}_w(t) = A_w x_w(t) + B_w f(t), \tag{6.5a}$$

$$f_w(t) = C_w x_w(t), \tag{6.5b}$$

where $x_w(t) \in \mathbb{R}^k$ stands for the state vector with $x_w(0) = 0$, and A_w, B_w, C_w are constant matrices.

Setting $e_f(t) \triangleq \chi_f(t) - f_w(t)$ and expanding the system (6.2) to contain the information in (6.4) and (6.5), then the resulting fault detection system is reformulated as

$$\begin{aligned} d\varsigma(t) = {} & \sum\nolimits_{j=1}^{N} \sigma_j \sum\nolimits_{i=1}^{r} h_i^{[j]}\left(\theta^{[j]}\right) \sum\nolimits_{l=1}^{r} h_l^{[j]}\left(\theta^{[j]}\right) \\ & \left\{ \left[\tilde{A}_{il}^{[j]} \varsigma(t) + \tilde{B}_{il}^{[j]} \upsilon(t) \right] dt + \tilde{E}_{il}^{[j]} K \varsigma(t) d\varpi(t) \right\}, \end{aligned} \tag{6.6a}$$

$$e_f(t) = \sum\nolimits_{j=1}^{N} \sigma_j \sum\nolimits_{i=1}^{r} h_i^{[j]}\left(\theta^{[j]}\right) \sum\nolimits_{l=1}^{r} h_l^{[j]}\left(\theta^{[j]}\right) \tilde{C}_{fl}^{[j]} \varsigma(t), \tag{6.6b}$$

where

$$\tilde{A}_{il}^{[j]} \triangleq \left[\begin{array}{cc|c} A_i^{[j]} & 0 & 0 \\ B_{fl}^{[j]}C_i^{[j]} & A_{fl}^{[j]} & 0 \\ \hline 0 & 0 & A_w \end{array}\right], \ \zeta(t) \triangleq \begin{bmatrix} x(t) \\ x_f(t) \\ x_w(t) \end{bmatrix},$$

$$\tilde{B}_{il}^{[j]} \triangleq \left[\begin{array}{ccc} B_{0i}^{[j]} & B_i^{[j]} & B_{1i}^{[j]} \\ B_{fl}^{[j]}D_{0i}^{[j]} & B_{fl}^{[j]}D_i^{[j]} & B_{fl}^{[j]}D_{1i}^{[j]} \\ \hline 0 & 0 & B_w \end{array}\right], \ K \triangleq \begin{bmatrix} I \\ 0 \\ 0 \end{bmatrix}^T,$$

$$\tilde{E}_{il}^{[j]} \triangleq \left[\begin{array}{c} E_i^{[j]} \\ B_{fl}^{[j]}F_i^{[j]} \\ \hline 0 \end{array}\right], \ \upsilon(t) \triangleq \begin{bmatrix} u(t) \\ \omega(t) \\ f(t) \end{bmatrix}, \ \tilde{C}_{fl}^{[j]} \triangleq \begin{bmatrix} 0 \\ \left(C_{fl}^{[j]}\right)^T \\ -C_w^T \end{bmatrix}^T.$$

Definition 6.1 The equilibrium $\zeta^{\star}(t) = 0$ of fault detection system in (6.6) with $\upsilon(t) = 0$ is mean-square exponentially stable under $\sigma_j(t)$ if its solution $\zeta(t)$ satisfies $\mathbf{E}\left\{\|\zeta(t)\|^2\right\} \leq \eta \|\zeta(t_0)\|^2 e^{-\lambda(t-t_0)}$, $\forall t \geq t_0$, for constants $\eta \geq 1$ and $\lambda > 0$.

Definition 6.2 [125] Given $T_2 > T_1 \geq 0$, let $N_{\sigma_j}(T_1, T_2)$ represent the number of switchings of $\sigma_j(t)$ over (T_1, T_2). If $N_{\sigma_j}(T_1, T_2) \leq N_0 + (T_2 - T_1)/T_a$ holds for $T_a > 0$, $N_0 \geq 0$. And T_a is named as an ADT.

Definition 6.3 For a scalar $\gamma > 0$, the fault detection system in (6.6) is mean-square exponentially stable subject to a weighted $\mathcal{H}_\infty$ performance level (γ, α) if it is mean-square exponentially stable when $\upsilon(t) \equiv 0$. Moreover, for all nonzero $\upsilon(t) \in \mathcal{L}_2[0, \infty)$ and zero initial condition, the following equality satisfies: $\mathbf{E}\left\{\int_0^\infty e^{-\alpha t} e_f^T(t) e_f(t) dt\right\} < \gamma^2 \int_0^\infty \upsilon^T(t)\upsilon(t)dt$.

Consequently, the concerned fault detection problem in this chapter can be denoted by two steps:

1. Generate a residual signal: for the system (6.2), propose a $\mathcal{H}_\infty$ fuzzy-rule-dependent filter (6.4) to produce a residual signal. In the mean time, the filter is given to make sure that the overall fault detection system (6.6) is mean-square exponentially stable under a weighted $\mathcal{H}_\infty$ performance level $\gamma > 0$.
2. Establish a fault detection scheme: choose a threshold and an evaluation function. In this chapter, the residual evaluation function $\mathcal{J}(\chi_f)$ and threshold $\mathcal{J}_{th}^{\star}$ are given by

$$\mathcal{J}(\chi_f) \triangleq \sqrt{\int_{t_0}^{t_0+t^{\star}} \chi_f^T(t)\chi_f(t)dt}, \tag{6.7}$$

$$\mathcal{J}_{th}^{\star} \triangleq \sup_{0\neq\omega\in\mathcal{L}_2, 0\neq u\in\mathcal{L}_2, f=0} \mathcal{J}(\chi_f), \tag{6.8}$$

where t_0 is the initial evaluation instant, $t^{\star}$ denotes the evaluation time. Furthermore, the appearance of faults can be detected by making a comparison between $\mathcal{J}(\chi_f)$ and $\mathcal{J}_{th}$ on account of the following standard:

$$\begin{aligned}\mathcal{J}(\chi_f) > \mathcal{J}_{th}^{\star} \quad &\Rightarrow \quad \text{Faults} \quad \Rightarrow \quad \text{Alarm},\\ \mathcal{J}(\chi_f) \leq \mathcal{J}_{th}^{\star} \quad &\Rightarrow \quad \text{No Faults}.\end{aligned}$$

6.3 Main Results

6.3.1 System Performance Analysis

Theorem 6.4 *For the given scalars $\alpha > 0, \gamma > 0$, suppose there are matrices $P^{[j]} > 0$ such that for $j \in \mathcal{N}$,*

$$\Pi_{ii}^{[j]} < 0, \quad i = 1, 2, \ldots, r, \tag{6.9a}$$

$$\frac{1}{r-1}\Pi_{ii}^{[j]} + \frac{1}{2}\left(\Pi_{il}^{[j]} + \Pi_{li}^{[j]}\right) < 0, \quad 1 \leqslant i < l \leqslant r, \tag{6.9b}$$

where

$$\Pi_{il}^{[j]} \triangleq \begin{bmatrix} \Pi_{11il}^{[j]} & P^{[j]}\tilde{B}_{il}^{[j]} & K^T\left(\tilde{E}_{il}^{[j]}\right)^T P^{[j]} & \left(\tilde{C}_{fl}^{[j]}\right)^T \\ \star & -\gamma^2 I & 0 & 0 \\ \star & \star & -P^{[j]} & 0 \\ \star & \star & \star & -I \end{bmatrix}, \tag{6.10}$$

with $\Pi_{11il}^{[j]} \triangleq P^{[j]}\tilde{A}_{il}^{[j]} + \left(\tilde{A}_{il}^{[j]}\right)^T P^{[j]} + \alpha P^{[j]}$. The resulting dynamic system in (6.6) is mean-square exponentially stable with a weighted $\mathcal{H}_\infty$ error performance (γ, α) under any switching signal with $T_a > T_a^* = \frac{\ln \mu}{\alpha}$ and for $\mu \geq 1$,

$$P^{[j]} \leq \mu P^{[s]}, \quad \forall j, s \in \mathcal{N}. \tag{6.11}$$

Furthermore, an estimate of the state decay satisfies

$$\mathbf{E}\left\{\|\zeta(t)\|^2\right\} \leq \eta e^{-\lambda t}\|\zeta(0)\|^2, \tag{6.12}$$

where

$$\begin{cases} \lambda = \alpha - \ln \mu / T_a > 0, \quad b = \max_{\forall j \in \mathscr{I}} \lambda_{\max}\left(P^{[j]}\right), \\ \eta = b/a \geq 1, \quad a = \min_{\forall j \in \mathscr{I}} \lambda_{\min}\left(P^{[j]}\right). \end{cases} \tag{6.13}$$

Proof On the basis of the switching signal and fuzzy basis functions, it yields from (6.9a)–(6.9b) that

$$\sum_{j=1}^{N}\sigma_j\sum_{i=1}^{r}h_i^{[j]}\left(\theta^{[j]}\right)\sum_{l=1}^{r}h_l^{[j]}\left(\theta^{[j]}\right)$$
$$\begin{bmatrix} \Pi_{11il}^{[j]} & P^{[j]}\tilde{B}_{il}^{[j]} & K^T\left(\tilde{E}_{il}^{[j]}\right)^T P^{[j]} & \left(\tilde{C}_{il}^{[j]}\right)^T \\ \star & -\gamma^2 I & 0 & 0 \\ \star & \star & -P^{[j]} & 0 \\ \star & \star & \star & -I \end{bmatrix} < 0. \tag{6.14}$$

Introduce the following Lyapunov functional:

$$V(\zeta_t, t) \triangleq \zeta^T(t)\left(\sum_{j=1}^{N}\sigma_j P^{[j]}\right)\zeta(t), \tag{6.15}$$

where $P^{[j]} > 0$, $j \in \mathcal{N}$ are to be decided. For a fixed $\sigma_j(t)$ and $\upsilon(t) = 0$, along the trajectory of (6.6) and employing the Itô formula, we have

$$dV(\zeta_t, t) = \mathscr{L}V(\tilde{x}_t, t)dt + 2\sum_{j=1}^{N}\sigma_j\sum_{i=1}^{r}h_i^{[j]}\left(\theta^{[j]}\right)$$
$$\sum_{l=1}^{r}h_l^{[j]}\left(\theta^{[j]}\right)\zeta^T(t)P^{[j]}\tilde{E}_{il}^{[j]}K\zeta(t)d\varpi(t), \tag{6.16a}$$
$$\mathscr{L}V(\zeta_t, t) = \sum_{j=1}^{N}\sigma_j\sum_{i=1}^{r}h_i^{[j]}\left(\theta^{[j]}\right)\sum_{l=1}^{r}h_l^{[j]}\left(\theta^{[j]}\right)\zeta^T(t)$$
$$\left\{2P^{[j]}\tilde{A}_{il}^{[j]} + \left(K\tilde{E}_{il}^{[j]}\right)^T P^{[j]}\tilde{E}_{il}^{[j]}K\right\}\zeta(t). \tag{6.16b}$$

Using the Schur complement approach and considering (6.14) and (6.16b), it yields

$$\mathscr{L}V(\zeta_t, t) < -\alpha\zeta^T(t)P^{[j]}\zeta(t) = -\alpha V(\zeta_t, t). \tag{6.17}$$

Notice that

$$d[e^{\alpha t}V(\zeta_t, t)] < 2\sum_{j=1}^{N}\sigma_j\sum_{i=1}^{r}h_i^{[j]}\left(\theta^{[j]}\right)\sum_{l=1}^{r}h_l^{[j]}\left(\theta^{[j]}\right)$$
$$e^{\alpha t}\zeta^T(t)P^{[j]}\tilde{E}_{il}^{[j]}K\zeta^T(t)d\varpi(t). \tag{6.18}$$

Integrate two sides of (6.18) from $t^\star > 0$ to t and make expectations. It is not hard to get

$$\mathbf{E}\{V(\zeta_t, t)\} < e^{-\alpha(t-t^\star)}\mathbf{E}\left\{V(\zeta_{t^\star}, t^\star)\right\}. \tag{6.19}$$

For any $t > 0$ and an arbitrary piecewise switching signal $\sigma_j(t)$, denote $0 = t_0 < t_1 < \cdots < t_k < \cdots (k = 0, 1, \ldots)$ and the switching points of $\sigma_j(t)$ over the interval $(0, t)$. As we mentioned before, the j_kth subsystem is activated during $t \in \left[t_k, t_{k+1}\right)$. Denoting $t^\star = t_k$ in (6.19) yields

$$\mathbf{E}\{V(\zeta_t,t)\} < e^{-\alpha(t-t_k)}\mathbf{E}\left\{V(\zeta_{t_k},t_k)\right\}. \tag{6.20}$$

At the switching moment t_k, it follows from (6.11) and (6.15) that

$$\mathbf{E}\left\{V(\zeta_{t_k},t_k)\right\} \leq \mu\mathbf{E}\left\{V(\zeta_{t_k^-},t_k^-)\right\}. \tag{6.21}$$

Hence, based on (6.20)–(6.21) and $\vartheta = N_{\sigma_j}(0,t) \leq (t-0)/T_a$, we can obtain

$$\mathbf{E}\{V(\zeta_t,t)\} \leq e^{-(\alpha-\ln\mu/T_a)t}V(\zeta_0,0). \tag{6.22}$$

It yields from (6.15) that

$$\mathbf{E}\{V(\zeta_t,t)\} \geq a\mathbf{E}\left\{\|\zeta(t)\|^2\right\}, \quad V(\zeta_0,0) \leq b\|\zeta(0)\|^2, \tag{6.23}$$

where a and b are denoted in (6.13). Combining (6.22)–(6.23), then we get

$$\mathbf{E}\left\{\|\zeta(t)\|^2\right\} \leq \frac{1}{a}\mathbf{E}\{V(\zeta_t,t)\} \leq \frac{b}{a}e^{-(\alpha-\ln\mu/T_a)t}\|\zeta(0)\|^2. \tag{6.24}$$

Consider Definition 6.1 when $t_0 = 0$, the overall dynamics in (6.6) with $\upsilon(t) = 0$ is mean-square exponentially stable.

Next, the weighted $\mathcal{H}_\infty$ performance for (6.6) defined in Definition 6.3 is introduced. For any nonzero $\upsilon(t) \in \mathcal{L}_2[0,\infty)$, along the solution of dynamic error system with a fixed β, we can get

$$\begin{aligned}
&\mathscr{L}V(\zeta_t,t) + \alpha V(\zeta_t,t) + e_f^T(t)e_f(t) - \gamma^2 u^T(t)u(t)\\
&\leq \sum\nolimits_{j=1}^{N}\sigma_j\sum\nolimits_{i=1}^{r}h_i^{[j]}\left(\theta^{[j]}\right)\sum\nolimits_{l=1}^{r}h_l^{[j]}\left(\theta^{[j]}\right)\\
&\begin{bmatrix}\zeta(t)\\ \upsilon(t)\end{bmatrix}^T\begin{bmatrix}\tilde{\Pi}_{11il}^{[j]} & P^{[j]}\tilde{B}_{il}^{[j]}\\ \star & -\gamma^2 I\end{bmatrix}\begin{bmatrix}\zeta(t)\\ \upsilon(t)\end{bmatrix},
\end{aligned} \tag{6.25}$$

where $\tilde{\Pi}_{11il}^{[j]} \triangleq P^{[j]}\tilde{A}_{il}^{[j]} + \left(P^{[j]}\tilde{A}_{il}^{[j]}\right)^T + \alpha P^{[j]} + \left(\tilde{E}_{il}^{[j]}K\right)^T P^{[j]}\tilde{E}_{il}^{[j]}K + \left(\tilde{C}_{il}^{[j]}\right)^T\tilde{C}_{il}^{[j]}$. Using the Schur's complement, it can be seen that (6.14) equals to

$$\sum_{j=1}^{N}\sigma_j\sum_{i=1}^{r}h_i^{[j]}\left(\theta^{[j]}\right)\sum_{l=1}^{r}h_l^{[j]}\left(\theta^{[j]}\right)\begin{bmatrix}\tilde{\Pi}_{11il}^{[j]} & P^{[j]}\tilde{B}_{il}^{[j]}\\ \star & -\gamma^2 I\end{bmatrix} < 0,$$

and

$$\mathscr{L}V(\zeta_t,t) + \alpha V(\zeta_t,t) + e_f^T(t)e_f(t) - \gamma^2 u^T(t)u(t) < 0. \tag{6.26}$$

Denote $\Gamma(t) \triangleq e_f^T(t)e_f(t) - \gamma^2 u^T(t)u(t)$, then (6.26) can be converted to

$$\mathscr{L}V(\zeta_t, t) < -\alpha V(\zeta_t, t) - \Gamma(t), \tag{6.27}$$

and

$$dV(\zeta_t, t) < 2\sum_{j=1}^{N} \sigma_j \sum_{i=1}^{r} h_i^{[j]}\left(\theta^{[j]}\right) \sum_{l=1}^{r} h_l^{[j]}\left(\theta^{[j]}\right)\zeta^T(t)$$
$$P^{[j]}\tilde{E}_{il}^{[j]} K\zeta^T(t)d\varpi(t) - \alpha V(\zeta_t, \beta, t)dt - \Gamma(t)dt.$$

Notice that

$$d[e^{\alpha t}V(\zeta_t, t)] < e^{\alpha t}\Bigg[- \Gamma(t)dt + 2\sum_{j=1}^{N} \sigma_j \sum_{i=1}^{r} h_i^{[j]}\left(\theta^{[j]}\right)$$
$$\sum_{l=1}^{r} h_l^{[j]}\left(\theta^{[j]}\right)\zeta^T(t)P^{[j]}\tilde{E}_{il}^{[j]}K\zeta^T(t)d\varpi(t)\Bigg]. \tag{6.28}$$

Utilizing the similar strategy as the proof of the previous exponential stability, we have

$$\mathbf{E}\left\{V(\zeta_t, t)\right\} < e^{-\alpha(t-t_k)}\mathbf{E}\left\{V(\zeta_{t_k}, t_k)\right\} - \mathbf{E}\left\{\int_{t_k}^{t} e^{-\alpha(t-s)}\Gamma(s)ds\right\}. \tag{6.29}$$

Under the zero initial condition, based on (6.21), (6.29) and $\vartheta = N_{\sigma_j}(0, t) \le (t - 0)/T_a$, it gives

$$\int_0^t e^{-\alpha(t-s)-N_{\sigma_j}(0,s)\ln\mu} e_f^T(s)e_f(s)ds \le \gamma^2 \int_0^t e^{-\alpha(t-s)}u^T(s)u(s)ds. \tag{6.30}$$

Due to $N_{\sigma_j}(0, t) \le t/T_a$ and $T_a > T_a^* = \ln\mu/\alpha$, we can get $N_{\sigma_j}(0, t)\ln\mu \le \alpha t$. Therefore, (6.30) signifies

$$\int_0^t e^{-\alpha(t-s)-\alpha s} e_f^T(s)e_f(s)ds \le \gamma^2 \int_0^t e^{-\alpha(t-s)}u^T(s)u(s)ds. \tag{6.31}$$

Integrate the above inequality from $t = 0$ to ∞ and it's easy to obtain the weighted $\mathcal{H}_\infty$ performance level (γ, α). Thus this completes the proof.

6.3.2 Fault Detection Filter Design

On the basis of proposed weighted $\mathcal{H}_\infty$ performance conditions and the convex linearization technique, we are going to deal with the fault detection issue for fuzzy switched stochastic systems.

Theorem 6.5 *Given scalars $\alpha > 0$ and $\gamma > 0$, assume there are matrices $0 < \mathcal{U}^{[j]} \in \mathbb{R}^{n\times n}$, $0 < \mathcal{V}^{[j]} \in \mathbb{R}^{n\times n}$, $0 < V^{[j]} \in \mathbb{R}^{k} \times \mathbb{R}^{k}$, $\mathcal{A}_{fi}^{[j]} \in \mathbb{R}^{n\times n}$, $\mathcal{B}_{fi}^{[j]} \in \mathbb{R}^{n\times p}$, and $\mathcal{C}_{fi}^{[j]} \in \mathbb{R}^{l\times n}$ such that for $j \in \mathcal{N}$, $i = 1, 2, \dots, r$, $l = 1, 2, \dots, r$,*

$$\Psi_{ii}^{[j]} < 0, \quad i = 1, 2, \dots, r, \tag{6.32a}$$

$$\frac{1}{r-1}\Psi_{ii}^{[j]} + \frac{1}{2}\left(\Psi_{il}^{[j]} + \Psi_{li}^{[j]}\right) < 0, \quad 1 \leqslant i < l \leqslant r, \tag{6.32b}$$

$$\begin{bmatrix} \mathcal{U}^{[j]} & \mathcal{V}^{[j]} \\ \star & \mathcal{V}^{[j]} \end{bmatrix} > 0, \tag{6.32c}$$

where

$$\Psi_{il}^{[j]} \triangleq \begin{bmatrix} \Psi_{1il}^{[j]} & \Psi_{2il}^{[j]} & \Psi_{3il}^{[j]} \\ \star & -\gamma^2 I & 0 \\ \star & \star & \Psi_{4il}^{[j]} \end{bmatrix}, \Psi_{1il}^{[j]} \triangleq \begin{bmatrix} \Psi_{11il}^{[j]} & \Psi_{12il}^{[j]} & 0 \\ \star & \Psi_{22l}^{[j]} & 0 \\ \star & \star & \Psi_{33}^{[j]} \end{bmatrix}, \Psi_{2il}^{[j]} \triangleq \begin{bmatrix} \Psi_{14il}^{[j]} & \Psi_{15il}^{[j]} & \Psi_{16il}^{[j]} \\ \Psi_{24il}^{[j]} & \Psi_{25il}^{[j]} & \Psi_{26il}^{[j]} \\ 0 & 0 & V^{[j]}B_w \end{bmatrix},$$

$$\Psi_{3il}^{[j]} \triangleq \begin{bmatrix} \left(\Psi_{17il}^{[j]}\right)^T & \left(\Psi_{18il}^{[j]}\right)^T & 0 & 0 \\ 0 & 0 & 0 & \left(\mathcal{C}_{fl}^{[j]}\right)^T \\ 0 & 0 & 0 & -C_w^T \end{bmatrix},$$

$$\Psi_{4il}^{[j]} \triangleq diag\left\{\begin{bmatrix} -\mathcal{U}^{[j]} & -\mathcal{V}^{[j]} \\ \star & -\mathcal{V}^{[j]} \end{bmatrix}, -V^{[j]}, -I\right\},$$

with

$$\begin{aligned}
\Psi_{11il}^{[j]} &\triangleq \mathcal{U}^{[j]}A_i^{[j]} + \mathcal{B}_{fl}^{[j]}C_i^{[j]} + \left(A_i^{[j]}\right)^T\left(\mathcal{U}^{[j]}\right)^T + \left(C_i^{[j]}\right)^T\left(\mathcal{B}_{fl}^{[j]}\right)^T + \alpha\mathcal{U}^{[j]},\\
\Psi_{12il}^{[j]} &\triangleq \mathcal{A}_{fl}^{[j]} + \left(A_i^{[j]}\right)^T\mathcal{V}^{[j]} + \left(C_i^{[j]}\right)^T\left(\mathcal{B}_{fl}^{[j]}\right)^T + \alpha\mathcal{V}^{[j]},\\
\Psi_{22l}^{[j]} &\triangleq \mathcal{A}_{fl}^{[j]} + \left(\mathcal{A}_{fl}^{[j]}\right)^T + \alpha\mathcal{V}^{[j]}, \ \Psi_{17il}^{[j]} \triangleq \mathcal{U}^{[j]}E_i^{[j]} + \mathcal{B}_{fl}^{[j]}F_i^{[j]},\\
\Psi_{14il}^{[j]} &\triangleq \mathcal{U}^{[j]}B_{0i}^{[j]} + \mathcal{B}_{fl}^{[j]}D_{0i}^{[j]}, \ \Psi_{15il}^{[j]} \triangleq \mathcal{U}^{[j]}B_i^{[j]} + \mathcal{B}_{fl}^{[j]}D_i^{[j]},\\
\Psi_{18il}^{[j]} &\triangleq \mathcal{V}^{[j]}E_i^{[j]} + \mathcal{B}_{fl}^{[j]}F_i^{[j]}, \ \Psi_{33}^{[j]} \triangleq V^{[j]}A_w + A_w^TV^{[j]} + \alpha V^{[j]},\\
\Psi_{16il}^{[j]} &\triangleq \mathcal{U}^{[j]}B_{1i}^{[j]} + \mathcal{B}_{fl}^{[j]}D_{1i}^{[j]}, \ \Psi_{24il}^{[j]} \triangleq \mathcal{V}^{[j]}B_{0i}^{[j]} + \mathcal{B}_{fl}^{[j]}D_{0i}^{[j]},\\
\Psi_{25il}^{[j]} &\triangleq \mathcal{V}^{[j]}B_i^{[j]} + \mathcal{B}_{fl}^{[j]}D_i^{[j]}, \ \Psi_{26il}^{[j]} \triangleq \mathcal{V}^{[j]}B_{1i}^{[j]} + \mathcal{B}_{fl}^{[j]}D_{1i}^{[j]}.
\end{aligned}$$

There exists a fuzzy-basis-dependent filter as (6.3) to assure the overall error system in (6.6) is mean-square exponentially stable with a weighted $\mathcal{H}_\infty$ performance (γ, α) under any switching signal over the ADT satisfying $T_a > T_a^* = \frac{\ln\mu}{\alpha}$. In addition, if the above conditions have feasible solutions $\left\{\mathcal{U}^{[j]}, \mathcal{V}^{[j]}, V^{[j]}, \mathcal{A}_{fi}^{[j]}, \mathcal{B}_{fi}^{[j]}, \mathcal{C}_{fi}^{[j]}\right\}$, the desired fault detection filter matrices in (6.3) can be described as

$$\begin{bmatrix} A_{fi}^{[j]} & B_{fi}^{[j]} \\ C_{fi}^{[j]} & 0 \end{bmatrix} = \begin{bmatrix} \left(\mathcal{V}^{[j]}\right)^{-1} & 0 \\ 0 & I \end{bmatrix} \begin{bmatrix} \mathcal{A}_{fi}^{[j]} & \mathcal{B}_{fi}^{[j]} \\ \mathcal{C}_{fi}^{[j]} & 0 \end{bmatrix}. \tag{6.33}$$

Proof Based on Theorem 6.4, define $P^{[j]} \triangleq \operatorname{diag}\{U^{[j]}, V^{[j]}\}$ in (6.9), with $U^{[j]} \in \mathbb{R}^{2n} \times \mathbb{R}^{2n}$ and $V^{[j]} \in \mathbb{R}^{k} \times \mathbb{R}^{k}$. Particularly, for a set of (γ, α), the resulting system (6.6) is mean-square exponentially stable subject to a weighted $\mathcal{H}_\infty$ performance (γ, α) if there are some suitable matrices and the following equations satisfy:

$$\Phi_{ii}^{[j]} < 0, \quad i = 1, 2, \ldots, r, \tag{6.34a}$$

$$\frac{1}{r-1}\Phi_{ii}^{[j]} + \frac{1}{2}\left(\Phi_{il}^{[j]} + \Phi_{li}^{[j]}\right) < 0, \quad 1 \leqslant i < l \leqslant r, \tag{6.34b}$$

where

$$\Phi_{il}^{[j]} \triangleq \begin{bmatrix} \Phi_{11il}^{[j]} & 0 & U^{[j]}\hat{B}_{il}^{[j]} & \left(\hat{E}_{il}^{[j]}\right)^T U^{[j]} & 0 & \left(\hat{C}_{fl}^{[j]}\right)^T \\ \star & \Phi_{22}^{[j]} & V^{[j]}\hat{B}_w & 0 & 0 & -C_w^T \\ \star & \star & -\gamma^2 I & 0 & 0 & 0 \\ \star & \star & \star & -U^{[j]} & 0 & 0 \\ \star & \star & \star & \star & -V^{[j]} & 0 \\ \star & \star & \star & \star & \star & -I \end{bmatrix},$$

with

$$\Phi_{11il}^{[j]} \triangleq U^{[j]}\hat{A}_{il}^{[j]} + \left(\hat{A}_{il}^{[j]}\right)^T U^{[j]} + \alpha U^{[j]}, \quad \hat{C}_{fl} \triangleq \left[0 \; C_{fl}^{[j]}\right],$$

$$\Phi_{22}^{[j]} \triangleq V^{[j]}A_w + A_w^T V^{[j]} + \alpha V^{[j]}, \quad \hat{B}_w \triangleq \left[0 \; 0 \; B_w\right],$$

$$\hat{A}_{il}^{[j]} \triangleq \begin{bmatrix} A_i^{[j]} & 0 \\ B_{fl}^{[j]}C_i^{[j]} & A_{fl}^{[j]} \end{bmatrix}, \quad \hat{E}_{il}^{[j]} \triangleq \begin{bmatrix} E_i^{[j]} & 0 \\ B_{fl}^{[j]}F_i^{[j]} & 0 \end{bmatrix},$$

$$\hat{B}_{il}^{[j]} \triangleq \begin{bmatrix} B_{0i}^{[j]} & B_i^{[j]} & B_{1i}^{[j]} \\ B_{fl}^{[j]}D_{0i}^{[j]} & B_{fl}^{[j]}D_i^{[j]} & B_{fl}^{[j]}D_{1i}^{[j]} \end{bmatrix}.$$

Then $U^{[j]}$ is partitioned as $U^{[j]} \triangleq \begin{bmatrix} U_1^{[j]} & U_2^{[j]} \\ \star & U_3^{[j]} \end{bmatrix} > 0$, where $U_1^{[j]} \in \mathbb{R}^n \times \mathbb{R}^n$, $U_2^{[j]} \in \mathbb{R}^n \times \mathbb{R}^n$, and $U_3^{[j]} \in \mathbb{R}^n \times \mathbb{R}^n$. Owing to a full-order filter is proposed, $U_2^{[j]}$ is a square matrix. Without loss of generality, provided that $U_2^{[j]}$ is nonsingular (If not, $U_2^{[j]}$ may be perturbed by $\Delta U_2^{[j]}$ with sufficiently small norm such that $U_2^{[j]} + \Delta U_2^{[j]}$) and satisfies (6.34).

Introduce several nonsingular matrices as

$$\mathcal{F}^{[j]} \triangleq \begin{bmatrix} I & 0 \\ 0 & \left(U_3^{[j]}\right)^{-1}\left(U_2^{[j]}\right)^T \end{bmatrix}, \begin{array}{l} \mathcal{V}^{[j]} \triangleq U_2^{[j]}\left(U_3^{[j]}\right)^{-1}\left(U_2^{[j]}\right)^T, \\ \mathcal{U}^{[j]} \triangleq U_1^{[j]}, \end{array} \tag{6.35}$$

and

$$\begin{bmatrix} \mathcal{A}_{fi}^{[j]} & \mathcal{B}_{fi}^{[j]} \\ \mathcal{C}_{fi}^{[j]} & 0 \end{bmatrix} \triangleq \begin{bmatrix} U_2^{[j]} & 0 \\ 0 & I \end{bmatrix} \begin{bmatrix} A_{fi}^{[j]} & B_{fi}^{[j]} \\ C_{fi}^{[j]} & 0 \end{bmatrix} \begin{bmatrix} \left(U_3^{[j]}\right)^{-1}\left(U_2^{[j]}\right)^T & 0 \\ 0 & I \end{bmatrix}. \tag{6.36}$$

Perform a congruence transformation to (6.34) with $\Gamma \triangleq \{\mathcal{F}^{[j]}, I, I, \mathcal{F}^{[j]}, I, I\}$ and we can get

$$\Gamma^T \Phi_{ii}^{[j]} \Gamma < 0, \tag{6.37a}$$

$$\Gamma^T \left[\frac{1}{r-1}\Phi_{ii}^{[j]} + \frac{1}{2}\left(\Phi_{il}^{[j]} + \Phi_{li}^{[j]}\right) \right] \Gamma < 0. \tag{6.37b}$$

On account of (6.35)–(6.36), it yields

$$\left.\begin{array}{l}
\left(\mathcal{F}^{[j]}\right)^T U^{[j]} \hat{A}_{il}^{[j]} \mathcal{F}^{[j]} = \begin{bmatrix} \mathcal{U}^{[j]} A_i^{[j]} + \mathcal{B}_{fl}^{[j]} C_i^{[j]} & \mathcal{A}_{fl}^{[j]} \\ \mathcal{V}^{[j]} A_i^{[j]} + \mathcal{B}_{fl}^{[j]} C_i^{[j]} & \mathcal{A}_{fl}^{[j]} \end{bmatrix}, \\
\left(\mathcal{F}^{[j]}\right)^T U^{[j]} \hat{E}_{il}^{[j]} \mathcal{F}^{[j]} = \begin{bmatrix} \mathcal{U}^{[j]} E_i^{[j]} + \mathcal{B}_{fl}^{[j]} F_i^{[j]} & 0 \\ \mathcal{V}^{[j]} E_i^{[j]} + \mathcal{B}_{fl}^{[j]} F_i^{[j]} & 0 \end{bmatrix}, \\
\left(\mathcal{F}^{[j]}\right)^T U^{[j]} \hat{B}_{il}^{[j]} = \left[\begin{array}{lll} \mathcal{U}^{[j]} B_{0i}^{[j]} + \mathcal{B}_{fl}^{[j]} D_{0i}^{[j]} & & \\ \mathcal{V}^{[j]} B_{0i}^{[j]} + \mathcal{B}_{fl}^{[j]} D_{0i}^{[j]} & & \end{array} \right. \\
\qquad \left. \begin{array}{ll} \mathcal{U}^{[j]} B_i^{[j]} + \mathcal{B}_{fl}^{[j]} D_i^{[j]} & \mathcal{U}^{[j]} B_{1i}^{[j]} + \mathcal{B}_{fl}^{[j]} D_{1i}^{[j]} \\ \mathcal{V}^{[j]} B_i^{[j]} + \mathcal{B}_{fl}^{[j]} D_i^{[j]} & \mathcal{V}^{[j]} B_{1i}^{[j]} + \mathcal{B}_{fl}^{[j]} D_{1i}^{[j]} \end{array} \right], \\
\left(\mathcal{F}^{[j]}\right)^T U^{[j]} \mathcal{F}^{[j]} = \begin{bmatrix} \mathcal{U}^{[j]} & \mathcal{V}^{[j]} \\ \mathcal{V}^{[j]} & \mathcal{V}^{[j]} \end{bmatrix}, \hat{C}_{fl}^{[j]} \mathcal{F}^{[j]} = \begin{bmatrix} 0 & \mathcal{C}_{fl}^{[j]} \end{bmatrix}.
\end{array}\right\} \tag{6.38}$$

On the basis of (6.38), it follows from (6.32) that (6.38). In addition, (6.36) is equal to

$$\begin{aligned}
\begin{bmatrix} A_{fi}^{[j]} & B_{fi}^{[j]} \\ C_{fi}^{[j]} & 0 \end{bmatrix} &= \begin{bmatrix} \left(U_2^{[j]}\right)^{-1} & 0 \\ 0 & I \end{bmatrix} \begin{bmatrix} \mathcal{A}_{fi}^{[j]} & \mathcal{B}_{fi}^{[j]} \\ \mathcal{C}_{fi}^{[j]} & 0 \end{bmatrix} \times \begin{bmatrix} \left(U_2^{[j]}\right)^{-T} U_3^{[j]} & 0 \\ 0 & I \end{bmatrix} \\
&= \begin{bmatrix} \left(\left(U_2^{[j]}\right)^{-T} U_3^{[j]}\right)^{-1} \left(\mathcal{V}^{[j]}\right)^{-1} & 0 \\ 0 & I \end{bmatrix} \begin{bmatrix} \mathcal{A}_{fi}^{[j]} & \mathcal{B}_{fi}^{[j]} \\ \mathcal{C}_{fi}^{[j]} & 0 \end{bmatrix} \\
&\quad \times \begin{bmatrix} \left(U_2^{[j]}\right)^{-T} U_3^{[j]} & 0 \\ 0 & I \end{bmatrix}.
\end{aligned} \tag{6.39}$$

It can be observed that the filter matrices $A_{fi}^{[j]}$, $B_{fi}^{[j]}$ and $C_{fi}^{[j]}$ in (6.3) can be rewritten as (6.39), which means $\left(U_2^{[j]}\right)^{-T} U_3^{[j]}$ can be regarded as the similarity transformation on the state-space realization for the designed filter, and has no influence in the filter mapping from y to χ_f. Without loss of generality, denote $\left(U_2^{[j]}\right)^{-T} U_3^{[j]} = I$, thus it gives (6.33). As a consequence, the fuzzy-basis-dependent fault detection filter in (6.3) is established via (6.33). The proof is then completed.

6.4 Illustrative Example

Consider the nonlinear switched stochastic system in (6.1) with $\mathcal{N} = 2$, which can be formulated as the T-S fuzzy model and the relevant matrices are given as follows:

Subsystem 1.

$$
A_1^{[1]} = \begin{bmatrix} -3.0 & 0.2 & 0.4 \\ 0.3 & -1.7 & 0.5 \\ 0.2 & 0.5 & -2.5 \end{bmatrix},\ B_1^{[1]} = \begin{bmatrix} 0.5 \\ 0.8 \\ 0.6 \end{bmatrix},\ B_{01}^{[1]} = \begin{bmatrix} 0.3 \\ 0.6 \\ 0.5 \end{bmatrix},
$$

$$
A_2^{[1]} = \begin{bmatrix} -2.7 & 0.3 & 0.6 \\ 0.2 & -1.5 & 0.8 \\ 0.3 & 0.4 & -2.4 \end{bmatrix},\ B_2^{[1]} = \begin{bmatrix} 0.5 \\ 0.6 \\ 0.3 \end{bmatrix},\ B_{02}^{[1]} = \begin{bmatrix} 0.4 \\ 0.6 \\ 0.7 \end{bmatrix},
$$

$$
B_{11}^{[1]} = \begin{bmatrix} 0.4 \\ 0.5 \\ 0.4 \end{bmatrix},\ B_{12}^{[1]} = \begin{bmatrix} 0.6 \\ 0.4 \\ 0.3 \end{bmatrix},\ \begin{array}{l} F_1^{[1]} = \begin{bmatrix} 0.1 & 0.2 & 0.4 \end{bmatrix}, \\ F_2^{[1]} = \begin{bmatrix} 0.1 & 0.2 & 0.2 \end{bmatrix}, \end{array}
$$

$$
E_1^{[1]} = \begin{bmatrix} 0.2 & 0 & 0.1 \\ 0.3 & 0.1 & 0.2 \\ 0.0 & 0.1 & 0.2 \end{bmatrix},\ E_2^{[1]} = \begin{bmatrix} 0.3 & 0.1 & 0.2 \\ 0.1 & 0.2 & 0.2 \\ 0.0 & 0.3 & 0.1 \end{bmatrix},
$$

$$
C_1^{[1]} = \begin{bmatrix} 1.5 & 0.6 & 1.3 \end{bmatrix},\ D_{01}^{[1]} = 0.6,\ D_1^{[1]} = 0.3,\ D_{11}^{[1]} = 0.4,
$$

$$
C_2^{[1]} = \begin{bmatrix} 1.0 & 1.3 & 0.7 \end{bmatrix},\ D_{02}^{[1]} = 0.4,\ D_2^{[1]} = 0.3,\ D_{12}^{[1]} = 0.4.
$$

Subsystem 2.

$$
A_1^{[2]} = \begin{bmatrix} -3.2 & 0.1 & 0.3 \\ 0.3 & -1.8 & 0.4 \\ 0.3 & 0.4 & -2.6 \end{bmatrix},\ B_1^{[2]} = \begin{bmatrix} 0.4 \\ 0.9 \\ 0.5 \end{bmatrix},\ B_{01}^{[2]} = \begin{bmatrix} 0.2 \\ 0.5 \\ 0.4 \end{bmatrix},
$$

$$
A_2^{[2]} = \begin{bmatrix} -2.9 & 0.4 & 0.7 \\ 0.3 & -1.3 & 0.6 \\ 0.4 & 0.3 & -2.2 \end{bmatrix},\ B_2^{[2]} = \begin{bmatrix} 0.4 \\ 0.5 \\ 0.4 \end{bmatrix},\ B_{02}^{[2]} = \begin{bmatrix} 0.5 \\ 0.3 \\ 0.8 \end{bmatrix},
$$

$$B_{11}^{[2]} = \begin{bmatrix} 0.3 \\ 0.4 \\ 0.5 \end{bmatrix}, \quad B_{12}^{[2]} = \begin{bmatrix} 0.2 \\ 0.1 \\ 0.5 \end{bmatrix}, \quad \begin{matrix} F_1^{[2]} = \begin{bmatrix} 0.2 \; 0.3 \; 0.5 \end{bmatrix}, \\ F_2^{[2]} = \begin{bmatrix} 0.2 \; 0.1 \; 0.2 \end{bmatrix}, \end{matrix}$$

$$E_1^{[2]} = \begin{bmatrix} 0.1 & 0 & 0.3 \\ 0.4 & 0.2 & 0.1 \\ 0.0 & 0.2 & 0.1 \end{bmatrix}, \quad E_2^{[2]} = \begin{bmatrix} 0.1 & 0.2 & 0.1 \\ 0.4 & 0.1 & 0.3 \\ 0.0 & 0.4 & 0.2 \end{bmatrix},$$

$$C_1^{[2]} = \begin{bmatrix} 1.6 \; 0.4 \; 1.4 \end{bmatrix}, \; D_{01}^{[2]} = 0.5, \; D_1^{[2]} = 0.2, \; D_{11}^{[2]} = 0.5,$$
$$C_2^{[2]} = \begin{bmatrix} 0.9 \; 1.1 \; 0.8 \end{bmatrix}, \; D_{02}^{[2]} = 0.2, \; D_2^{[2]} = 0.4, \; D_{12}^{[2]} = 0.5.$$

The weighting matrix $W(s)$ is assumed as $W(s) = 5/(s+5)$. And the corresponding parameters in (6.5) are set as $A_w = -5$, $B_w = 5$ and $C_w = 1$.

Solving the LMIs in Theorem 6.5, the minimized feasible γ can be computed as $\gamma^* = 1.0876$, and

$$A_{f1}^{[1]} = \begin{bmatrix} -5.9191 & -0.3430 & -0.1660 \\ -1.9168 & -2.0822 & -1.7281 \\ -1.2755 & 0.6872 & -5.6867 \end{bmatrix}, \; B_{f1}^{[1]} = \begin{bmatrix} -1.0772 \\ -1.3602 \\ -1.1693 \end{bmatrix},$$

$$A_{f2}^{[1]} = \begin{bmatrix} -5.8700 & -1.8943 & 0.0918 \\ -1.0856 & -2.5724 & -0.5889 \\ -0.2762 & -0.0162 & -4.6259 \end{bmatrix}, \; B_{f2}^{[1]} = \begin{bmatrix} -1.7928 \\ -1.1643 \\ -0.9090 \end{bmatrix},$$

$$C_{f1}^{[1]} = \begin{bmatrix} -0.0323 \\ -0.0497 \\ -0.3092 \end{bmatrix}^T, \quad C_{f2}^{[1]} = \begin{bmatrix} -0.2410 \\ -0.0741 \\ 0.0016 \end{bmatrix}^T,$$

$$A_{f1}^{[2]} = \begin{bmatrix} -4.4854 & -0.2213 & -0.7589 \\ -1.8817 & -2.6268 & -0.3243 \\ -0.8301 & 0.1435 & -5.2591 \end{bmatrix}, \quad B_{f1}^{[2]} = \begin{bmatrix} -0.7733 \\ -1.1491 \\ -1.2526 \end{bmatrix},$$

$$A_{f2}^{[2]} = \begin{bmatrix} -3.3415 & -0.2332 & 0.3802 \\ -0.2229 & -2.1478 & 1.1747 \\ 0.1982 & -1.1113 & -4.3050 \end{bmatrix}, \quad B_{f2}^{[2]} = \begin{bmatrix} -0.5826 \\ -0.3967 \\ -1.4342 \end{bmatrix},$$

$$C_{f1}^{[2]} = \begin{bmatrix} -0.1938 \\ -0.0701 \\ -0.2003 \end{bmatrix}^T, \quad C_{f2}^{[2]} = \begin{bmatrix} -0.0016 \\ 0.0452 \\ -0.3281 \end{bmatrix}^T.$$

In this example, the fuzzy basis functions are chosen as $h_1(x_1(t)) \triangleq \exp\left[-\frac{(x_1(t)-\vartheta)^2}{2\sigma^2}\right]$, $h_2(x_1(t)) \triangleq 1 - \exp\left[-\frac{(x_1(t)-\vartheta)^2}{2\sigma^2}\right]$, with $\vartheta = 5$ and $\sigma = 1$. The disturbance signal $\omega(t)$ is assumed to be random noise and the known input is set to $u(t) = \sin(t),\ 0 \leq t \leq 10$; and the fault signal is given by $f(t) = \begin{cases} 1, & 2.5 \leq t \leq 5, \\ 0, & \text{otherwise.} \end{cases}$. Figure 6.1 plots the weighting fault signal $f_w(t)$.

The evaluation function and the threshold are chosen as (6.7)–(6.8). And we simulate the standard Brownian motion via employing the discretization technique

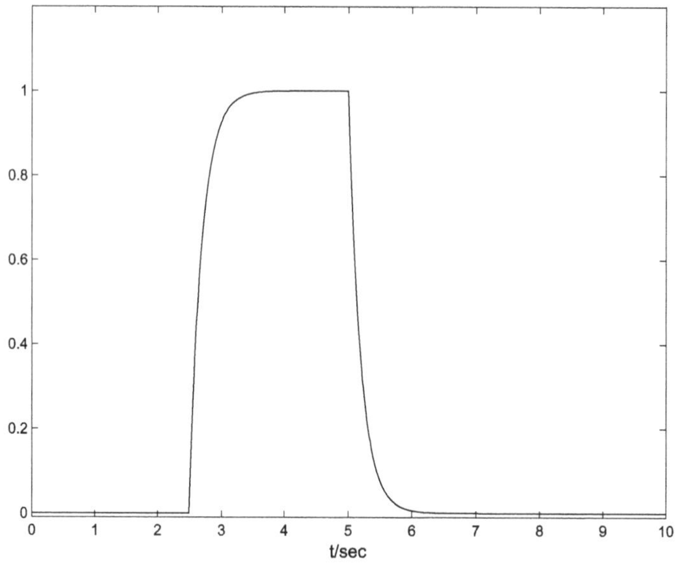

Fig. 6.1 Weighting fault signal $f_w(t)$

in [89]. In addition, the simulation time $t \in [0, T^*]$ with $T^* = 10$, the normally distributed variance $\delta t = \frac{T^*}{N^*}$ with $N^* = 2^{11}$, step size $\Delta t = \rho\delta t$ with $\rho = 2$. For the proposed filter with fuzzy-fuzzy-dependent situation, simulation results are shown in Figs. 6.2–6.3. Among them, Fig. 6.2 draws the residual signal $\chi_f(t)$ and Fig. 6.3 plots the evaluation function of $\mathcal{J}(\chi_f)$ for the fault-free case (dash-dot line) and fault case (solid line). When the residual signal is produced, then establish the fault detection measure. For the fuzzy-rule-dependent situation, with a threshold given as $\mathcal{J}_{th}^{\star} = 0.1136$, Fig. 6.2 displays $\mathcal{J}(\chi_f) > \mathcal{J}_{th}^{\star}$ for $t = 2.7$, which implies the fault signal $f(t)$ can be detected $0.2s$ after its appearance.

6.5 Conclusion

Weighted $\mathcal{H}_\infty$ fault detection filtering has been a key focus for nonlinear switched systems with stochastic disturbances, and the corresponding issue was addressed using the fuzzy-rule-dependent approach. Based on the piecewise Lyapunov functions and ADT method, sufficient conditions were established such that the mean-square exponential stability with weighted $\mathcal{H}_\infty$ performance for the dynamic error system could be ensured. Moreover, the solvable conditions of the proposed filter were derived based on the linearization procedure. In the last, to demonstrate the availability of our proposed scheme, a simulation has been provided.

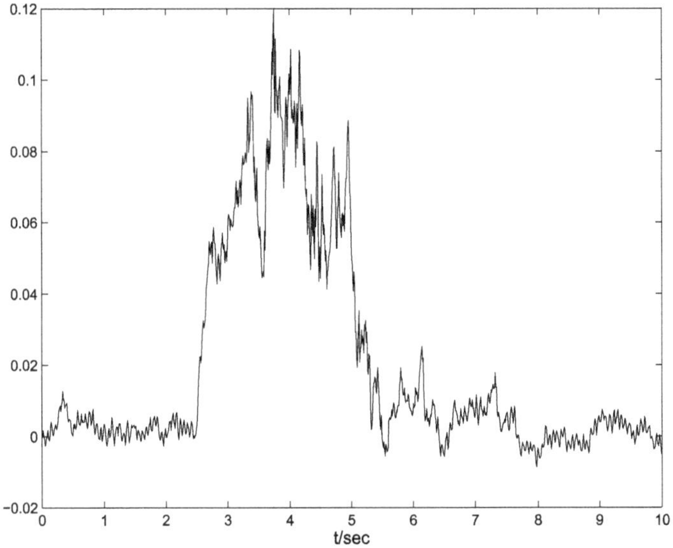

Fig. 6.2 Residual signal $\chi_f(t)$

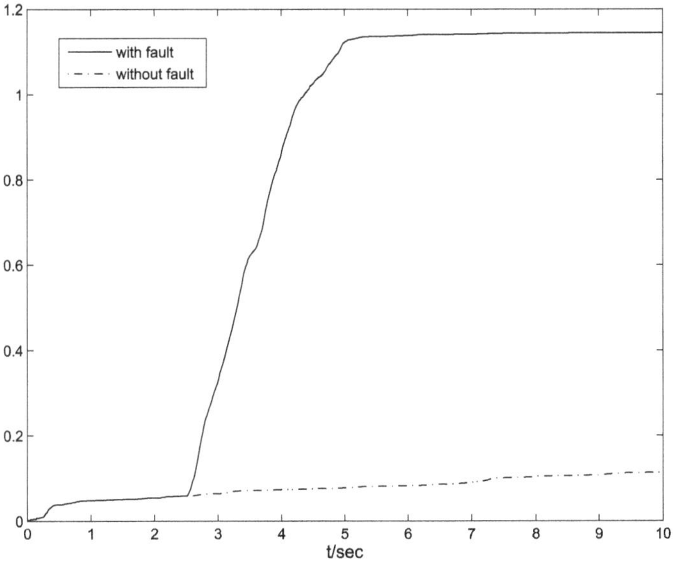

Fig. 6.3 Evaluation function of $\mathcal{J}(\chi_f)$

Chapter 7
Reliable Filtering for T-S Fuzzy Time-Delay Systems

7.1 Introduction

In this chapter, the reliable filtering problem with strict dissipative performance is considered for discrete-time T-S fuzzy systems with time-delays. A reliable filter design is prepared to ensure strict dissipativity for dynamic filtering systems. First, sufficient conditions for the reliable dissipativity analysis are established using a reciprocally convex method for T-S fuzzy systems subject to sensor failures. In addition, a reliable filter is established based on the convex optimization technique, which can be readily resolved through the standard numerical toolbox.

7.2 System Description and Preliminaries

7.2.1 System Description

Consider the nonlinear system, which can be approximated with the T-S fuzzy model with the time-varying delay as follows:
♦ **Plant Form**:

Rule i: IF $\theta_1(k)$ is M_{i1} and … and $\theta_p(k)$ is M_{ip}, THEN

$$\begin{cases} x(k+1) = A_i x(k) + A_{di} x(k - d(k)) + B_i \omega(k), \\ \quad y(k) = C_i x(k) + C_{di} x(k - d(k)) + D_i \omega(k), \\ \quad z(k) = L_i x(k) + L_{di} x(k - d(k)) + F_i \omega(k), \\ \quad x(\iota) = \psi(\iota), \quad \iota = -d_2, -d_2 + 1, \ldots, 0, \\ \qquad\qquad i = 1, 2, \ldots, r, \end{cases} \tag{7.1}$$

X. Su et al., *Intelligent Control, Filtering and Model Reduction Analysis for Fuzzy-Model-Based Systems*, Studies in Systems, Decision and Control 385,
https://doi.org/10.1007/978-3-030-81214-0_7

where $x(k) \in \mathbb{R}^n$ denotes the state vector; $y(k) \in \mathbb{R}^l$ denotes the measured output signal; $\omega(k) \in \mathbb{R}^p$ represents the disturbance input which is belong to $\ell_2[0, \infty)$; $z(k) \in \mathbb{R}^q$ denotes the signal to be estimated; the system delay $d(k)$ is a positive integer, which is supposed to be time-varying and satisfies $1 \leqslant d_1 \leqslant d(k) \leqslant d_2$, where d_2 and d_1 are positive constant scalars denoting the maximum and minimum delay, respectively. M_{ij} represents the fuzzy set; r is the quantity of IF-THEN fuzzy rules; $\theta(k) = \left[\theta_1(k), \theta_2(k), \ldots, \theta_p(k)\right]$ denotes the premise variables. A_i, A_{di}, B_i, $C_i, C_{di}, D_i, L_i, L_{di}$ and F_i are constant matrices; $\psi(k)$ stands for the initial condition. The fuzzy membership functions are expressed as $h_i\left(\theta(k)\right) \triangleq \frac{\prod_{j=1}^{p} M_{ij}\left(\theta_j(k)\right)}{\sum_{i=1}^{r}\prod_{j=1}^{p} M_{ij}\left(\theta_j(k)\right)}$, where $M_{ij}\left(\theta_j(k)\right)$ denotes the grade of membership of $\theta_j(k)$ in M_{ij}. Consequently, we can obtain $h_i\left(\theta(k)\right) \geq 0$, $i = 1, 2, \ldots, r$ and $\sum_{i=1}^{r} h_i\left(\theta(k)\right) = 1$ for all k.

A more complete expression of the T-S fuzzy model with time-varying delay can be formulated as (Σ):

$$\begin{cases} x(k+1) = \sum\limits_{i=1}^{r} h_i(\theta(k))\left[A_i x(k) + A_{di} x(k-d(k)) + B_i \omega(k)\right], \\ y(k) = \sum\limits_{i=1}^{r} h_i(\theta(k))\left[C_i x(k) + C_{di} x(k-d(k)) + D_i \omega(k)\right], \\ z(k) = \sum\limits_{i=1}^{r} h_i(\theta(k))\left[L_i x(k) + L_{di} x(k-d(k)) + F_i \omega(k)\right], \\ x(\iota) = \psi(\iota), \quad \iota = -d_2, -d_2+1, \ldots, 0. \end{cases} \tag{7.2}$$

In this note, an efficient filter with sensor failures is proposed for the estimation of $z(k)$:

$$(\Sigma_f): \quad \begin{cases} \hat{x}(k+1) = A_f \hat{x}(k) + B_f \hat{y}(k), \\ \hat{z}(k) = C_f \hat{x}(k) + D_f \hat{y}(k), \end{cases} \tag{7.3}$$

where $\hat{x}(k) \in \mathbb{R}^k$ denotes the state vector of our designed filter (7.3) with $k \leq n$; $\hat{z}(k) \in \mathbb{R}^q$ denotes the estimation of $z(k)$; A_f, B_f, C_f and D_f stand for some suitably dimensioned matrices to be decided, and $\hat{y}(k) \triangleq \left[\, \hat{y}_1(k) \ \ldots \ \hat{y}_m(k) \,\right]^T$ represents the signal from the sensor which might be faulty.

Remark 7.1 On account of the fact that the fuzzy rule is taken into account, thus the fuzzy-rule-dependent filter has less conservativeness than a fuzzy-rule-independent one. In this chapter, we design the fuzzy-rule-independent filters as (7.3) and moreover, the developed techniques can be broaden into the design for fuzzy-rule-dependent filters.

The failure form in [73] is utilized as

$$\hat{y}_j(k) = \beta_{\varepsilon j} y_j(k), \quad j = 1, 2, \ldots, m,$$

where

$$0 \leq \underline{\beta}_{\varepsilon j} \leq \beta_{\varepsilon j} \leq \bar{\beta}_{\varepsilon j}, \quad j = 1, 2, \ldots, m,$$

with $0 \leq \beta_{\varepsilon j} \leq 1$ in which the variables $\beta_{\varepsilon j}$ quantify the faults of sensors. And it follows that

$$\hat{y}(k) = B_\varepsilon y(k), \quad B_\varepsilon = \text{diag}\{\beta_{\varepsilon 1}, \beta_{\varepsilon 2}, \ldots, \beta_{\varepsilon \text{m}}\}.$$

Remark 7.2 Considering the above model, when $\underline{\beta}_{\varepsilon j} = \bar{\beta}_{\varepsilon j}$, it is a normal fully operating case, $y_i^F(k) = y_i(k)$; when $\underline{\beta}_{\varepsilon j} = 0$, it includes the outage case in [279]; when $\underline{\beta}_{\varepsilon j} \neq 0$ and $\bar{\beta}_{\varepsilon j} \neq 1$, it becomes the case that the intensity of the feedback signal from actuator may variate.

Define

$$\begin{aligned}
\bar{B}_\varepsilon &\triangleq \text{diag}\{\bar{\beta}_{\varepsilon 1}, \bar{\beta}_{\varepsilon 2}, \ldots, \bar{\beta}_{\varepsilon j}, \ldots, \bar{\beta}_{\varepsilon \text{m}}\}, \\
\underline{B}_\varepsilon &\triangleq \text{diag}\{\underline{\beta}_{\varepsilon 1}, \underline{\beta}_{\varepsilon 2}, \ldots, \underline{\beta}_{\varepsilon j}, \ldots, \underline{\beta}_{\varepsilon \text{m}}\}, \\
B_{\varepsilon 0} &\triangleq \text{diag}\{\beta_{\varepsilon 01}, \beta_{\varepsilon 02}, \ldots, \beta_{\varepsilon 0j}, \ldots, \beta_{\varepsilon 0\text{m}}\}, \\
\Lambda &\triangleq \text{diag}\{\alpha_1, \alpha_2, \ldots, \alpha_j, \ldots, \alpha_\text{m}\}, \\
E_\varepsilon &\triangleq \text{diag}\{\epsilon_{\varepsilon 1}, \epsilon_{\varepsilon 2}, \ldots, \epsilon_{\varepsilon j}, \ldots, \epsilon_{\varepsilon \text{m}}\},
\end{aligned}$$

where $\beta_{\varepsilon 0j} \triangleq \frac{\bar{\beta}_{\varepsilon j} + \underline{\beta}_{\varepsilon j}}{2}$ and $\alpha_j \triangleq \frac{\bar{\beta}_{\varepsilon j} - \underline{\beta}_{\varepsilon j}}{2}$. It yields

$$B_\varepsilon = B_{\varepsilon 0} + E_\varepsilon, \quad |\epsilon_{\varepsilon j}| \leq \alpha_j. \tag{7.4}$$

Extending the model (Σ) to contain our proposed filter (Σ_f), the following dynamic filtering system can be obtained, which is denoted as (Σ_e):

$$\begin{cases}
\xi(k+1) = \sum_{i=1}^{r} h_i(\theta(k)) \left[\bar{A}_i \xi(k) + \bar{A}_{di} \xi(k - d(k)) + \bar{B}_i \omega(k) \right], \\
e(k) = \sum_{i=1}^{r} h_i(\theta(k)) \left[\bar{L}_i \xi(k) + \bar{L}_{di} \xi(k - d(k)) + \bar{F}_i \omega(k) \right], \\
\xi(\iota) = \varphi(\iota), \quad \iota = -d_2, -d_2 + 1, \ldots, 0,
\end{cases} \tag{7.5}$$

where $\xi(k) \triangleq \left[x^T(k) \; \hat{x}^T(k) \right]^T$, $e(k) \triangleq z(k) - \hat{z}(k)$ and

$$\left.\begin{aligned}
\bar{A}_i &\triangleq \begin{bmatrix} A_i & 0 \\ B_f B_\varepsilon C_i & A_f \end{bmatrix}, \quad \bar{A}_{di} \triangleq \begin{bmatrix} A_{di} & 0 \\ B_f B_\varepsilon C_{di} & 0 \end{bmatrix}, \\
\bar{B}_i &\triangleq \begin{bmatrix} B_i \\ B_f B_\varepsilon D_i \end{bmatrix}, \quad \bar{L}_{di} \triangleq \left[L_{di} - D_f B_\varepsilon C_{di} \; 0 \right], \\
\bar{L}_i &\triangleq \left[L_i - D_f B_\varepsilon C_i \; -C_f \right], \quad \bar{F}_i \triangleq F_i - D_f B_\varepsilon D_i.
\end{aligned}\right\} \tag{7.6}$$

Furthermore, define

$$\left.\begin{aligned} \bar{A}(k) &\triangleq \sum_{i=1}^{r} h_i(\theta(k))\bar{A}_i, & \bar{A}_d(k) &\triangleq \sum_{i=1}^{r} h_i(\theta(k))\bar{A}_{di}, \\ \bar{B}(k) &\triangleq \sum_{i=1}^{r} h_i(\theta(k))\bar{B}_i, & \bar{L}(k) &\triangleq \sum_{i=1}^{r} h_i(\theta(k))\bar{L}_i, \\ \bar{F}(k) &\triangleq \sum_{i=1}^{r} h_i(\theta(k))\bar{F}_i, & \bar{L}_d(k) &\triangleq \sum_{i=1}^{r} h_i(\theta(k))\bar{L}_{di}. \end{aligned}\right\} \tag{7.7}$$

Definition 7.3 The dynamic filtering system in (7.5) is asymptotically stable if $\omega(k) = 0$,

$$\lim_{k\to\infty} |\xi(k)| = 0.$$

For a system (Σ_e), which is asymptotically stable, it follows that $e = \{e(k)\} \in \ell_2[0, \infty)$ when $\omega = \{\omega(k)\} \in \ell_2[0, \infty)$.

7.2.2 Dissipativity Definition

In this section, discussions on dissipative systems are introduced. Dissipative systems can be regarded as a generalization of passive systems with more general internal and supplied energies [341]. A system is called "dissipative" if "power dissipation" exists in the system. Dissipative systems are those that cannot store more energy than that supplied by the environment and/or by other systems connected to them, i.e., dissipative systems can only dissipate but not generate energy [220]. Based on [99], associated with the discrete-time T-S fuzzy time-varying delay system in (7.5) is a real valued function $\mathcal{G}(\omega(k), e(k))$ called the *supply rate* which is formally defined as follows.

The classical form of dissipativity in [99] is obviously applicable to the discrete-time T-S fuzzy time-varying delay system in (7.5) in the following.

Remark 7.4 Passive systems are a kind of particular dissipative systems, which owes a bilinear supply rate, i.e. $\mathcal{G}(\omega, e) = e^T\omega$. If a system with a constant *positive* feed forward of $\mathcal{X}$ is passive, then the procedure is dissipative in regard to the supply rate $\mathcal{G}(\omega, e) = e^T\omega + \omega^T\mathcal{X}\omega$, where $\mathcal{X} = \mathcal{X}^T \in \mathbb{R}^{p\times p}$. In a similar way, if a system with a constant *negative* feedback of $\mathcal{Z}$ is passive, then the procedure is dissipative in regard to the supply rate $\mathcal{G}(\omega, e) = e^T\mathcal{Z}e + e^T\omega$, where $\mathcal{Z} = \mathcal{Z}^T \in \mathbb{R}^{p\times p}$.

Definition 7.5 For the given matrices $\mathcal{Z} \in \mathbb{R}^{q\times q}$, $\mathcal{X} \in \mathbb{R}^{p\times p}$, $\mathcal{Y} \in \mathbb{R}^{q\times p}$, where $\mathcal{Z}$ and $\mathcal{X}$ are symmetric, for the real function $\varrho(\cdot)$ with $\varrho(0) = 0$, the discrete T-S fuzzy system with time-varying delay as (7.5) is dissipative if the following condition satisfies:

$$\sum_{k=0}^{T^*} \begin{bmatrix} e(k) \\ \omega(k) \end{bmatrix}^T \begin{bmatrix} \mathcal{Z} & \mathcal{Y} \\ \star & \mathcal{X} \end{bmatrix} \begin{bmatrix} e(k) \\ \omega(k) \end{bmatrix} + \varrho(\psi(0)) \geq 0, \ \forall T^* \geq 0. \tag{7.8}$$

Moreover, for $\delta > 0$, if the following equality holds:

$$\sum_{k=0}^{T^*} \begin{bmatrix} e(k) \\ \omega(k) \end{bmatrix}^T \begin{bmatrix} \mathcal{Z} & \mathcal{Y} \\ \star & \mathcal{X} \end{bmatrix} \begin{bmatrix} e(k) \\ \omega(k) \end{bmatrix} + \varrho(\psi(0)) \geq \delta \sum_{k=0}^{T^*} \omega^T(k)\omega(k), \forall T^* \geq 0, \tag{7.9}$$

then the resulting system (7.5) is called as strictly dissipative.

Provided that $\mathcal{Z} \leq 0$, then it follows $-\mathcal{Z} = (\mathcal{Z}_-^{\frac{1}{2}})^2$, for $\mathcal{Z}_-^{\frac{1}{2}} \geq 0$.

Remark 7.6 The dissipative theory generalizes some different system theories, containing small gain theorem, passivity theorem, circle criterion, and so forth. Some special circumstances can be obtained by choosing different $(\mathcal{Z}, \mathcal{Y}, \mathcal{X})$, which are shown as follows:

- If $\mathcal{Z} = -I$, $\mathcal{Y} = 0$ and $\mathcal{X} = \gamma^2 I$ $(\gamma > 0)$, the strict dissipative performance becomes $\mathcal{H}_\infty$ performance.
- If $\mathcal{Z} = 0$, $\mathcal{Y} = I$ and $\mathcal{X} = 0$, the strict dissipative performance changes to the positive real performance.
- If $\mathcal{Z} = -\theta I, \mathcal{Y} = 1 - \theta$ and $\mathcal{X} = \theta\gamma^2 I$ $(\gamma > 0, \theta \in [0, 1])$, it reduces to the mixed performance.
- $\mathcal{Z} = -I, \mathcal{Y} = \frac{1}{2}(\mathcal{K}_1 + \mathcal{K}_2)^T$ and $\mathcal{X} = -\frac{1}{2}(\mathcal{K}_1^T\mathcal{K}_2 + \mathcal{K}_2^T\mathcal{K}_1)$ $(\gamma > 0$, for constant matrices $\mathcal{K}_1, \mathcal{K}_2)$, it converts into the sector bounded performance.

7.2.3 Reciprocally Convex Approach

This part discusses a specialized class of function compositions, which can be utilizing to deal with the double summation terms in Lyapunov functions for T-S fuzzy systems with time-varying delay.

Definition 7.7 Set $\Psi_1, \Psi_2, \ldots,$ and $\Psi_N : \mathbb{R}^m \to \mathbb{R}^n$ as a limited number of functions such that they owe positive values among an open subset **D** of $\mathbb{R}^m$. And the reciprocally convex combination over **D** is of the form as

$$\frac{1}{\vartheta_1}\Psi_1 + \frac{1}{\vartheta_2}\Psi_2 + \cdots + \frac{1}{\vartheta_N}\Psi_N : \mathbb{D} \to \mathbb{R}^n, \tag{7.10}$$

where $\vartheta_i > 0$ and $\sum_i \vartheta_i = 1$.

A lower bound for the reciprocally convex method of scalar positive functions $\Psi_i = f_i$ is introduced in the following lemma.

Lemma 7.8 *[206] Let $f_1, f_2, \ldots,$ and $f_N : \mathbb{R}^m \to \mathbb{R}$ have positive values in an open subset* **D** *of $\mathbb{R}^m$. The reciprocally convex combination of f_i over* **D** *satisfies the following condition:*

$$\min_{\{\vartheta_i|\vartheta_i>0,\sum_i\vartheta_i=1\}}\sum_i\frac{1}{\vartheta_i}f_i(\theta)=\sum_i f_i(\theta)+\max_{g_{i,j}(\theta)}\sum_{i\neq j}g_{i,j}(\theta),$$

where

$$\left\{g_{i,j}:\mathbb{R}^m\to\mathbb{R},\,g_{j,i}(\theta)=g_{i,j}(\theta),\begin{bmatrix}f_i(\theta) & g_{i,j}(\theta)\\ g_{j,i}(\theta) & f_j(\theta)\end{bmatrix}\geq 0\right\}.$$

We focus on designing a reliable filter (Σ_f) for system (7.2) and achieve the following goals simultaneously:

- The dynamic filtering system in (7.5) with $\omega(k)=0$ is asymptotically stable.
- The dynamic filtering system in (7.5) is strictly dissipative.

Assumption 7.1 System (Σ) in (7.2) is asymptotically stable.

Remark 7.9 In fact, we have omitted the control input in system (Σ), then the original system to be estimated must be asymptotically stable, which is a precondition for the dynamic filtering system (Σ_e) to be asymptotically stable.

7.3 Main Results

7.3.1 *Reliable Dissipativity Analysis*

In this part, the Lyapunov functions and reciprocally convex approach are utilized to investigate the asymptotic stability with strict dissipativity for the augmented filtering system in (7.5). Denote $d=d_2-d_1$ and

$$\left.\begin{aligned}Q_1(k)&\triangleq\sum_{i=1}^{r}h_i(\theta(k))Q_{1i},\\ Q_2(k)&\triangleq\sum_{i=1}^{r}h_i(\theta(k))Q_{2i},\\ Q_3(k)&\triangleq\sum_{i=1}^{r}h_i(\theta(k))Q_{3i},\end{aligned}\right\}\tag{7.11}$$

where $Q_{1i}>0$, $Q_{2i}>0$, $Q_{3i}>0$, $i=1,2,\ldots,r$, are all $(n+k)\times(n+k)$ matrices.

Theorem 7.10 *For matrices* $0\geq\mathcal{Z}\in\mathbb{R}^{q\times q}$, $\mathcal{X}\in\mathbb{R}^{p\times p}$, $\mathcal{Y}\in\mathbb{R}^{q\times p}$ *with* $\mathcal{Z}$ *and* $\mathcal{X}$ *being symmetric, and scalar* $\delta>0$, *if there are matrices* $0<P\in\mathbb{R}^{(n+k)\times(n+k)}$, $0<Q_{1i}\in\mathbb{R}^{(n+k)\times(n+k)}$, $0<Q_{2i}\in\mathbb{R}^{(n+k)\times(n+k)}$, $0<Q_{3i}\in\mathbb{R}^{(n+k)\times(n+k)}$, $0<S_1\in\mathbb{R}^{(n+k)\times(n+k)}$, $0<S_2\in\mathbb{R}^{(n+k)\times(n+k)}$ *and* $M\in\mathbb{R}^{(n+k)\times(n+k)}$ *such that for* $i,j,s,t=1,\ldots,r$,

$$\Xi_{ijst} < 0, \tag{7.12}$$

$$\begin{bmatrix} S_2 & M^T \\ \star & S_2 \end{bmatrix} \geq 0, \tag{7.13}$$

where

$$\Xi_{ijst} \triangleq \begin{bmatrix} \Xi_{11i} & \Xi_{12} & 0 & 0 & \Xi_{15i} & \Xi_{16i} \\ \star & \Xi_{22j} & \Xi_{23} & \Xi_{24} & 0 & 0 \\ \star & \star & \Xi_{33s} & \Xi_{34} & \Xi_{35i} & \Xi_{36i} \\ \star & \star & \star & \Xi_{44t} & 0 & 0 \\ \star & \star & \star & \star & \Xi_{55i} & \Xi_{56i} \\ \star & \star & \star & \star & \star & \Xi_{66} \end{bmatrix},$$

$$\begin{aligned}
&\Xi_{11i} \triangleq -P + Q_{1i} + Q_{2i} + (d+1)Q_{3i} - S_1,\ \Xi_{12} \triangleq S_1,\ \Xi_{15i} \triangleq -\bar{L}_i^T \mathcal{Y}, \\
&\Xi_{35i} \triangleq -\bar{L}_{di}^T \mathcal{Y},\ \Xi_{22j} \triangleq -Q_{1j} - S_1 - S_2,\ \Xi_{23} \triangleq -M + S_2, \\
&\Xi_{33s} \triangleq -Q_{3s} - 2S_2 + M + M^T,\ \Xi_{34} \triangleq -M + S_2,\ \Xi_{44t} \triangleq -Q_{2t} - S_2, \\
&\Xi_{55i} \triangleq -\bar{F}_i^T \mathcal{Y} - \mathcal{Y}^T \bar{F}_i - \mathcal{X} + \delta I, \quad \Xi_{24} \triangleq M, \\
&\Xi_{16i} \triangleq \left[\bar{A}_i^T \ \ d_1(\bar{A}_i^T - I) \ \ d(\bar{A}_i^T - I) \ \ \bar{L}_i^T \mathcal{Z}_-^{\frac{1}{2}} \right], \\
&\Xi_{36i} \triangleq \left[\bar{A}_{di}^T \ \ d_1 \bar{A}_{di}^T \ \ d\bar{A}_{di}^T \ \ \bar{L}_{di}^T \mathcal{Z}_-^{\frac{1}{2}} \right],\ \Xi_{56i} \triangleq \left[\bar{B}_i^T \ \ d_1 \bar{B}_i^T \ \ d\bar{B}_i^T \ \ \bar{F}_i^T \mathcal{Z}_-^{\frac{1}{2}} \right], \\
&\Xi_{66} \triangleq \mathrm{diag}\{-P^{-1}, -S_1^{-1}, -S_2^{-1}, -I\},
\end{aligned}$$

then the dynamic filtering system in (7.5) with sensor failure is asymptotically stable with strict dissipativity based on Definition 7.5.

Proof Considering the fuzzy membership functions, it follows from (7.12) that

$$\sum_{i=1}^{r} h_i(\theta(k)) \sum_{j=1}^{r} h_j(\theta(k-d_1)) \sum_{s=1}^{r} h_s(\theta(k-d(k))) \sum_{t=1}^{r} h_t(\theta(k-d_2)) \Xi_{ijst} < 0.$$

A more complete expression can be presented as

$$\Xi(k) < 0, \tag{7.14}$$

where

$$\Xi(k) \triangleq \begin{bmatrix} \Xi_{11}(k) & \Xi_{12} & 0 & 0 & \Xi_{15}(k) & \Xi_{16}(k) \\ \star & \Xi_{22}(k) & \Xi_{23} & \Xi_{24} & 0 & 0 \\ \star & \star & \Xi_{33}(k) & \Xi_{34} & \Xi_{35}(k) & \Xi_{36}(k) \\ \star & \star & \star & \Xi_{44}(k) & 0 & 0 \\ \star & \star & \star & \star & \Xi_{55}(k) & \Xi_{56}(k) \\ \star & \star & \star & \star & \star & \Xi_{66} \end{bmatrix},$$

$$\begin{aligned}
\Xi_{11}(k) &\triangleq -P + Q_1(k) + Q_2(k) + (d+1)Q_3(k) - S_1,\ \Xi_{15}(k) \triangleq -\bar{L}^T(k)\mathcal{Y}, \\
\Xi_{22}(k) &\triangleq -Q_1(k-d_1) - S_1 - S_2,\ \Xi_{44}(k) \triangleq -Q_2(k-d_2) - S_2, \\
\Xi_{33}(k) &\triangleq -Q_3(k-d(k)) - 2S_2 + M + M^T,\ \Xi_{35}(k) \triangleq -\bar{L}_d^T(k)\mathcal{Y}, \\
\Xi_{55}(k) &\triangleq -\bar{F}^T(k)\mathcal{Y} - \mathcal{Y}^T\bar{F}(k) - \mathcal{X} + \delta I, \\
\Xi_{16}(k) &\triangleq \left[\, \bar{A}^T(k)\ d_1(\bar{A}^T(k)-I)\ d(\bar{A}^T(k)-I)\ \bar{L}^T(k)\mathcal{Z}_-^{\frac{1}{2}} \,\right], \\
\Xi_{36}(k) &\triangleq \left[\, \bar{A}_d^T(k)\ d_1\bar{A}_d^T(k)\ d\bar{A}_d^T(k)\ \bar{L}_d^T(k)\mathcal{Z}_-^{\frac{1}{2}} \,\right], \\
\Xi_{56}(k) &\triangleq \left[\, \bar{B}^T(k)\ d_1\bar{B}^T(k)\ d\bar{B}^T(k)\ \bar{F}^T(k)\mathcal{Z}_-^{\frac{1}{2}} \,\right].
\end{aligned}$$

Employing the Schur complement method, (7.14) signifies

$$\bar{\Xi}(k) < 0, \tag{7.15}$$

where

$$\bar{\Xi}(k) \triangleq \begin{bmatrix} \bar{\Xi}_{11}(k) & \Xi_{12} & \bar{\Xi}_{13}(k) & 0 & \bar{\Xi}_{15}(k) \\ \star & \Xi_{22}(k) & \Xi_{23} & \Xi_{24} & 0 \\ \star & \star & \bar{\Xi}_{33}(k) & \Xi_{34} & \bar{\Xi}_{35}(k) \\ \star & \star & \star & \Xi_{44}(k) & 0 \\ \star & \star & \star & \star & \bar{\Xi}_{55}(k) \end{bmatrix},$$

$$\begin{aligned}
\bar{\Xi}_{11}(k) &\triangleq \bar{A}^T(k)P\bar{A}(k) - P + Q_1(k) - \bar{L}^T(k)\mathcal{Z}\bar{L}(k) + (d+1)Q_3(k) \\
&\quad + d_1^2[\bar{A}(k)-I]^T S_1[\bar{A}(k)-I] + d^2[\bar{A}(k)-I]^T S_2[\bar{A}(k)-I] \\
&\quad + Q_2(k) - S_1, \\
\bar{\Xi}_{13}(k) &\triangleq d_1^2[\bar{A}(k)-I]^T S_1\bar{A}_d(k) - \bar{L}^T(k)\mathcal{Z}\bar{L}_d(k) + \bar{A}^T(k)P\bar{A}_d(k) \\
&\quad + d^2[\bar{A}(k)-I]^T S_2\bar{A}_d(k), \\
\bar{\Xi}_{15}(k) &\triangleq d_1^2[\bar{A}(k)-I]^T S_1\bar{B}(k) + d^2[\bar{A}(k)-I]^T S_2 B(k) + \bar{A}^T(k)P\bar{B}(k) \\
&\quad - \bar{L}^T(k)\mathcal{Z}\bar{F}(k) - \bar{L}^T(k)\mathcal{Y}, \\
\bar{\Xi}_{33}(k) &\triangleq \bar{A}_d^T(k)P\bar{A}_d(k) - Q_3(k-d(k)) - 2S_2 + M^T + d^2\bar{A}_d^T(k)S_2\bar{A}_d(k) \\
&\quad + M - \bar{L}_d^T(k)\mathcal{Z}\bar{L}_d(k) + d_1^2\bar{A}_d^T(k)S_1\bar{A}_d(k), \\
\bar{\Xi}_{35}(k) &\triangleq \bar{A}_d^T(k)P\bar{B}(k) + d_1^2\bar{A}_d^T(k)S_1\bar{B}(k) - \bar{L}_d^T(k)\mathcal{Y} + d^2\bar{A}_d^T(k)S_2 B(k) \\
&\quad - \bar{L}_d^T(k)\mathcal{Z}\bar{F}(k), \\
\bar{\Xi}_{55}(k) &\triangleq \bar{B}^T(k)P\bar{B}(k) + d_1^2\bar{B}^T(k)S_1\bar{B}(k) + \delta I - \bar{F}^T(k)\mathcal{Z}\bar{F}(k) \\
&\quad - \bar{F}^T(k)\mathcal{Y} - \mathcal{Y}^T\bar{F}(k) - \mathcal{X} + d^2\bar{B}^T(k)S_2 B(k).
\end{aligned}$$

Next, the fuzzy Lyapunov functions are considered as follows:

$$V(k) \triangleq \sum_{i=1}^{5} V_i(k), \tag{7.16}$$

where

$$\begin{cases} V_1(k) \triangleq \xi^T(k)P\xi(k), \\ V_2(k) \triangleq \sum_{j=1}^{2} \sum_{i=k-d_j}^{k-1} \xi^T(i)Q_j(i)\xi(i), \\ V_3(k) \triangleq \sum_{i=k-d(k)}^{k-1} \xi^T(i)Q_3(i)\xi(i) + \sum_{j=-d_2+1}^{-d_1} \sum_{i=k+j}^{k-1} \xi^T(i)Q_3(i)\xi(i), \\ V_4(k) \triangleq \sum_{j=-d_1}^{-1} \sum_{i=k+j}^{k-1} d_1\zeta^T(i)S_1\zeta(i), \\ V_5(k) \triangleq \sum_{j=-d_2}^{-d_1-1} \sum_{i=k+j}^{k-1} d\zeta^T(i)S_2\zeta(i), \\ \zeta(k) \triangleq \xi(k+1) - \xi(k). \end{cases}$$

Along the trajectories of the dynamic filtering system (Σ_e) in (7.5), and taking the difference of the Lyapunov functions in (7.16), it yields that

$$\Delta V(k) \triangleq V(k+1) - V(k) = \sum_{i=1}^{5} \Delta V_i(k), \tag{7.17}$$

where

$$\begin{aligned}
\Delta V_1(k) &= \xi^T(k+1)P\xi(k+1) - \xi^T(k)P\xi(k), \\
\Delta V_2(k) &= \xi^T(k)Q_1(k)\xi(k) + \xi^T(k)Q_2(k)\xi(k) \\
&\quad -\xi^T(k-d_1)Q_1(k-d_1)\xi(k-d_1) \\
&\quad -\xi^T(k-d_2)Q_2(k-d_2)\xi(k-d_2), \\
\Delta V_3(k) &= (d+1)\xi^T(k)Q_3(k)\xi(k) + \sum_{i=k-d(k+1)+1}^{k-1} \xi^T(i)Q_3(i)\xi(i) \\
&\quad - \sum_{i=k-d(k)+1}^{k-1} \xi^T(i)Q_3(i)\xi(i) - \sum_{i=k-d_2+1}^{k-d_1} \xi^T(i)Q_3(i)\xi(i) \\
&\quad -\xi^T(k-d(k))Q_3(k-d(k))\xi(k-d(k)) \\
&\le -\xi^T(k-d(k))Q_3(k-d(k))\xi(k-d(k)) \\
&\quad +(d+1)\xi^T(k)Q_3(k)\xi(k), \\
\Delta V_4(k) &= d_1^2\zeta^T(k)S_1\zeta(k) - d_1 \sum_{i=k-d_1}^{k-1} \zeta^T(i)S_1\zeta(i) \\
&\le -\left[\sum_{i=k-d_1}^{k-1} \zeta(i)\right]^T S_1 \left[\sum_{i=k-d_1}^{k-1} \zeta^T(i)\right] + d_1^2\zeta^T(k)S_1\zeta(k) \\
&= -\left[\xi(k) - \xi(k-d_1)\right] S_1 \left[\xi(k) - \xi(k-d_1)\right] + d_1^2\zeta^T(k)S_1\zeta(k).
\end{aligned}$$

Due to $\begin{bmatrix} S_2 & M^T \\ \star & S_2 \end{bmatrix} \geq 0$, the following condition satisfies:

$$\begin{bmatrix} \sqrt{\frac{\vartheta_1}{\vartheta_2}}\zeta_1(k) \\ -\sqrt{\frac{\vartheta_2}{\vartheta_1}}\zeta_2(k) \end{bmatrix}^T \begin{bmatrix} S_2 & M^T \\ \star & S_2 \end{bmatrix} \begin{bmatrix} \sqrt{\frac{\vartheta_1}{\vartheta_2}}\zeta_1(k) \\ -\sqrt{\frac{\vartheta_2}{\vartheta_1}}\zeta_2(k) \end{bmatrix} \geq 0,$$

where

$$\zeta_1(k) \triangleq \xi(k-d(k)) - \xi(k-d_2), \quad \vartheta_1 \triangleq \frac{d_2 - d(k)}{d},$$
$$\zeta_2(k) \triangleq \xi(k-d_1) - \xi(k-d(k)), \quad \vartheta_2 \triangleq \frac{d(k) - d_1}{d}.$$

Applying Lemma 7.8, it follows for $d_1 \leq d(k) \leq d_2$ that

$$\begin{aligned}
\Delta V_5(k) &= d^2\zeta^T(k)S_2\zeta(k) - d\sum_{i=k-d_2}^{k-d(k)-1} \zeta^T(i)S_2\eta(i) \\
&\quad - d\sum_{i=k-d(k)}^{k-d_1-1} \zeta^T(i)S_2\zeta(i) \\
&\leq -\frac{d}{d_2-d(k)}\left[\sum_{i=k-d_2}^{k-d(k)-1} \zeta(i)\right]^T S_2 \left[\sum_{i=k-d_2}^{k-d(k)-1} \zeta(i)\right] \\
&\quad - \frac{d}{d(k)-d_1}\left[\sum_{i=k-d(k)}^{k-d_1-1} \zeta(i)\right]^T S_2 \left[\sum_{i=k-d(k)}^{k-d_1-1} \zeta(i)\right] + d^2\zeta^T(k)S_2\zeta(k) \\
&\leq -\begin{bmatrix} \zeta_1(k) \\ \zeta_2(k) \end{bmatrix}^T \begin{bmatrix} S_2 & M^T \\ \star & S_2 \end{bmatrix} \begin{bmatrix} \zeta_1(k) \\ \zeta_2(k) \end{bmatrix} + d^2\zeta^T(k)S_2\zeta(k) \\
&= -\begin{bmatrix} \xi(k-d_1) \\ \xi(k-d(k)) \end{bmatrix}^T \begin{bmatrix} S_2 & -S_2 \\ \star & S_2 \end{bmatrix} \begin{bmatrix} \xi(k-d_1) \\ \xi(k-d(k)) \end{bmatrix} \\
&\quad - \begin{bmatrix} \xi(k-d(k)) \\ \xi(k-d_2) \end{bmatrix}^T \begin{bmatrix} S_2 & -S_2 \\ \star & S_2 \end{bmatrix} \begin{bmatrix} \xi(k-d(k)) \\ \xi(k-d_2) \end{bmatrix} \\
&\quad - \begin{bmatrix} \xi(k-d_1) \\ \xi(k-d(k)) \\ \xi(k-d_2) \end{bmatrix}^T \begin{bmatrix} 0 & M & -M \\ \star & -M-M^T & M \\ \star & \star & 0 \end{bmatrix} \times \begin{bmatrix} \xi(k-d_1) \\ \xi(k-d(k)) \\ \xi(k-d_2) \end{bmatrix} \\
&\quad + d^2\zeta^T(k)S_2\zeta(k).
\end{aligned}$$

When $d(k) = d_1$ or $d(k) = d_2$, it has $\zeta_1(k) = 0$ or $\zeta_2(k) = 0$. Thus, the inequality in $\Delta V_5(k)$ still satisfies. And we can get the following condition:

$$\Delta V(k) = \bar{\zeta}^T(k)\hat{\Xi}(k)\bar{\zeta}(k), \tag{7.18}$$

where

$$\begin{aligned}
\bar{\zeta}(k) &\triangleq \left[\xi^T(k)\ \ \xi^T(k-d_1)\ \ \xi^T(k-d(k))\ \xi^T(k-d_2)\ \ \omega^T(k)\right]^T, \\
\hat{\Xi}(k) &\triangleq \begin{bmatrix} \hat{\Xi}_{11}(k) & \Xi_{12} & \hat{\Xi}_{13}0 & \hat{\Xi}_{15}(k) & \\ \star & \Xi_{22}(k) & \Xi_{23} & \Xi_{24} & 0 \\ \star & \star & \hat{\Xi}_{33}(k) & \Xi_{34} & \hat{\Xi}_{35}(k) \\ \star & \star & \star & \Xi_{44}(k) & 0 \\ \star & \star & \star & \star & \hat{\Xi}_{55}(k) \end{bmatrix}, \\
\hat{\Xi}_{11}(k) &\triangleq \bar{A}^T(k)P\bar{A}(k) - P + Q_1(k) + Q_2(k) - S_1 + (d+1)Q_3(k) \\
&\quad + d_1^2[\bar{A}(k)-I]^T S_1[\bar{A}(k)-I] + d^2[\bar{A}(k)-I]^T S_2[\bar{A}(k)-I], \\
\hat{\Xi}_{13}(k) &\triangleq \bar{A}^T(k)P\bar{A}_d(k) + d_1^2[\bar{A}(k)-I]^T S_1\bar{A}_d(k) + d^2[\bar{A}(k)-I]^T S_2\bar{A}_d(k), \\
\hat{\Xi}_{15}(k) &\triangleq \bar{A}^T(k)P\bar{B}(k) + d_1^2[\bar{A}(k)-I]^T S_1\bar{B}(k) + d^2[\bar{A}(k)-I]^T S_2 B(k), \\
\hat{\Xi}_{33}(k) &\triangleq \bar{A}_d^T(k)P\bar{A}_d(k) + d_1^2\bar{A}_d^T(k)S_1\bar{A}_d(k) - 2S_2 + M - Q_3(k-d(k)) \\
&\quad + d^2\bar{A}_d^T(k)S_2\bar{A}_d(k) + M^T, \\
\hat{\Xi}_{35}(k) &\triangleq \bar{A}_d^T(k)P\bar{B}(k) + d_1^2\bar{A}_d^T(k)S_1\bar{B}(k) + d^2\bar{A}_d^T(k)S_2 B(k), \\
\hat{\Xi}_{55}(k) &\triangleq \bar{B}^T(k)P\bar{B}(k) + d_1^2\bar{B}^T(k)S_1\bar{B}(k) + d^2\bar{B}^T(k)S_2 B(k).
\end{aligned}$$

On the basis of (7.13), (7.15), (7.18) and the zero input signal $\omega(k) = 0$, we have $\Delta V(k) < 0$, hence the dynamic filtering system (7.5) is asymptotically stable.

In the following, the strict dissipativity of system (7.5) is discussed. Define

$$\mathcal{J}(T^\star) \triangleq \sum_{k=0}^{T^\star} \begin{bmatrix} e(k) \\ \omega(k) \end{bmatrix}^T \begin{bmatrix} \mathcal{Z} & \mathcal{Y} \\ \star & \mathcal{X} \end{bmatrix} \begin{bmatrix} e(k) \\ \omega(k) \end{bmatrix} - \delta \sum_{k=0}^{T^\star} \omega^T(k)\omega(k), \ \ \forall T^* \geq 0.$$

When $\xi(k) = 0$, that is the zero initial condition, for $k = -d_2, -d_2+1, \ldots, 0$ and any non-zero $\omega(k) \in \ell_2[0, \infty)$, it is observed that

$$\begin{aligned}
& V(T^\star + 1) - V(0) - \mathcal{J}(T^\star) \\
&= \sum_{k=0}^{T^\star} \left\{ \Delta V(k) - \begin{bmatrix} e(k) \\ \omega(k) \end{bmatrix}^T \begin{bmatrix} \mathcal{Z} & \mathcal{Y} \\ \star & \mathcal{X} - \delta I \end{bmatrix} \begin{bmatrix} e(k) \\ \omega(k) \end{bmatrix} \right\} \\
&= \bar{\zeta}^T(k)\bar{\Xi}(k)\bar{\zeta}(k) < 0.
\end{aligned} \tag{7.19}$$

Considering (7.19) and $V(T^\star + 1) > 0$, we can get

$$\mathcal{J}(T^\star) > 0. \tag{7.20}$$

On the basis of Definition 7.5, the system in (7.5) is asymptotically stable with strict dissipativity. Then this completes the proof.

Notice that when $\omega(k) = 0$ in the system (Σ), it has

$$x(k+1) = A(k)x(k) + A_d(k)x(k - d(k)), \tag{7.21}$$

where

$$A(k) \triangleq \sum_{i=1}^{r} h_i(\theta(k))A_i, \quad A_d(k) \triangleq \sum_{i=1}^{r} h_i(\theta(k))A_{di}.$$

Then we will provide a stability condition for the concerned system. Consider the system in (7.21) and let

$$\left.\begin{aligned} \mathcal{P}(k) &\triangleq \sum_{i=1}^{r} h_i(\theta(k))\mathcal{P}_i,\ \mathcal{Q}_1(k) \triangleq \textstyle\sum_{i=1}^{r} h_i(\theta(k))\mathcal{Q}_{1i}, \\ \mathcal{S}_1(k) &\triangleq \sum_{i=1}^{r} h_i(\theta(k))\mathcal{S}_{1i},\ \mathcal{Q}_2(k) \triangleq \textstyle\sum_{i=1}^{r} h_i(\theta(k))\mathcal{Q}_{2i}, \\ \mathcal{S}_2(k) &\triangleq \sum_{i=1}^{r} h_i(\theta(k))\mathcal{S}_{2i},\ \mathcal{Q}_3(k) \triangleq \sum_{i=1}^{r} h_i(\theta(k))\mathcal{Q}_{3i}, \\ \bar{\mathcal{S}}_1(k) &\triangleq \sum_{i=1}^{r} h_i(\theta(k))\bar{\mathcal{S}}_{1i},\ \bar{\mathcal{S}}_2(k) \triangleq \sum_{i=1}^{r} h_i(\theta(k))\bar{\mathcal{M}}_2, \\ \bar{\mathcal{M}}(k) &\triangleq \sum_{i=1}^{r} h_i(\theta(k))\bar{\mathcal{M}}_i, \end{aligned}\right\} \tag{7.22}$$

where $\mathcal{P}_i > 0$, $\mathcal{S}_{1i} > 0$, $\mathcal{S}_{2i} > 0$, $\mathcal{Q}_{1i} > 0$, $\mathcal{Q}_{2i} > 0$, $\mathcal{Q}_{3i} > 0$, $\bar{\mathcal{S}}_{1i} > 0$, $\bar{\mathcal{S}}_{2i} > 0$ and $\mathcal{M}_i$ $(i = 1, 2, \ldots, r)$ are all $n \times n$ matrices.

Based on Theorem 7.10, the additional conditions for (7.21) are proposed.

Corollary 7.11 *For $1 \le d_1 \le d_2$, the system in (7.21) with time-varying delay $d(k)$ satisfying $d_1 \le d(k) \le d_2$ is asymptotically stable provided that there are matrices $\mathcal{P}_i > 0$, $\mathcal{S}_{1i} > 0$, $\mathcal{S}_{2i} > 0$, $\mathcal{Q}_{1i} > 0$, $\mathcal{Q}_{2i} > 0$, $\mathcal{Q}_{3i} > 0$, $\bar{\mathcal{S}}_{1i} > 0$, $\bar{\mathcal{S}}_{2i} > 0$ and $\mathcal{M}_i$ such that the following conditions satisfy:*

$$\Gamma_{ostlii} < 0, \quad o, s, t, l, i \in (1, 2, \ldots, r), \tag{7.23}$$

$$\left.\begin{aligned} \Gamma_{ostlij} + \Gamma_{ostlji} < 0, \quad & 1 \le i \ne j \le r, \\ & o, s, t, l, i, j \in (1, 2, \ldots, r), \end{aligned}\right\} \tag{7.24}$$

$$\begin{bmatrix} \bar{\mathcal{S}}_{2i} & \mathcal{M}_i^T \\ \star & \bar{\mathcal{S}}_{2i} \end{bmatrix} \ge 0, \quad i \in (1, 2, \ldots, r), \tag{7.25}$$

$$\mathcal{S}_{1i} - \bar{\mathcal{S}}_{1j} \ge 0, \quad i, j \in (1, 2, \ldots, r), \tag{7.26}$$

$$\mathcal{S}_{2i} - \bar{\mathcal{S}}_{2j} \ge 0, \quad i, j \in (1, 2, \ldots, r), \tag{7.27}$$

where

$$\Gamma_{ostlji} \triangleq \begin{bmatrix} \Gamma_{11tlji} & \Gamma_{12t} & \Gamma_{13lji} & 0 \\ \star & \Gamma_{22ti} & \Gamma_{23i} & \Gamma_{24i} \\ \star & \star & \Gamma_{33slji} & \Gamma_{34i} \\ \star & \star & \star & \Gamma_{44oi} \end{bmatrix},$$

with

$$\begin{aligned}
\Gamma_{11tlji} &\triangleq A_j^T \mathcal{P}_l A_j + \mathcal{Q}_{1i} + \mathcal{Q}_{2i} + d_1^2 [A_j - I]^T \mathcal{S}_{1i} [A_j - I] + (d+1)\mathcal{Q}_{3i} \\
&\quad - \bar{\mathcal{S}}_{1t} + d^2 [A_j - I]^T \mathcal{S}_{2i} [A_j - I] - \mathcal{P}_i, \\
\Gamma_{13lji} &\triangleq A_j^T \mathcal{P}_l A_{dj} + d_1^2 [A_j - I]^T \mathcal{S}_{1i} A_{dj} + d^2 [A_j - I]^T \mathcal{S}_{2i} A_{dj}, \\
\Gamma_{22ti} &\triangleq -\mathcal{Q}_{1t} - \bar{\mathcal{S}}_{1t} - \bar{\mathcal{S}}_{2i}, \ \Gamma_{12t} \triangleq \bar{\mathcal{S}}_{1t}, \ \Gamma_{23i} \triangleq -\mathcal{M}_i + \bar{\mathcal{S}}_{2i}, \\
\Gamma_{33slji} &\triangleq A_{dj}^T \mathcal{P}_l A_{dj} - \mathcal{Q}_{3s} + d_1^2 A_{dj}^T \mathcal{S}_{1i} A_{dj} - 2\bar{\mathcal{S}}_{2i} + \mathcal{M}_i + \mathcal{M}_i^T + d^2 A_{dj}^T \mathcal{S}_{2i} A_{dj}, \\
\Xi_{44oi} &\triangleq -\mathcal{Q}_{2o} - \bar{\mathcal{S}}_{2i}, \ \Gamma_{34i} \triangleq -\mathcal{M}_i + \bar{\mathcal{S}}_{2i}, \ \Gamma_{24i} \triangleq \mathcal{M}_i.
\end{aligned}$$

Proof The fuzzy Lyapunov functions are selected as

$$\mathcal{V}(k) \triangleq \sum_{i=1}^{5} \mathcal{V}_i(k),$$

where

$$\begin{cases}
\mathcal{V}_1(k) \triangleq x^T(k)\mathcal{P}(k)x(k), \ \ \eta(k) \triangleq x(k+1) - x(k), \\
\mathcal{V}_2(k) \triangleq \sum\limits_{j=1}^{2} \sum\limits_{i=k-d_j}^{k-1} x^T(i)\mathcal{Q}_j(i)x(i), \\
\mathcal{V}_3(k) \triangleq \sum\limits_{i=k-d(k)}^{k-1} x^T(i)\mathcal{Q}_3(i)x(i) + \sum\limits_{j=-d_2+1}^{-d_1} \sum\limits_{i=k+j}^{k-1} x^T(i)\mathcal{Q}_3(i)x(i), \\
\mathcal{V}_4(k) \triangleq \sum\limits_{j=-d_1}^{-1} \sum\limits_{i=k+j}^{k-1} d_1 \eta^T(i)\mathcal{S}_1(i)\eta(i), \\
\mathcal{V}_5(k) \triangleq \sum\limits_{j=-d_2}^{-d_1-1} \sum\limits_{i=k+j}^{k-1} d\eta^T(i)\mathcal{S}_2(i)\eta(i).
\end{cases}$$

Using the similar ways as the proof in Theorem 7.10, we can get the corresponding results readily.

Remark 7.12 In Corollary 7.11, sufficient stability conditions for (7.21) are established via reciprocally convex method. There exist some techniques to handle the T-S fuzzy systems with time-delays, such as the delay partition method [299], circumventing the bounding inequalities approach [82], the input-output method [251]. Compared with these methods, our proposed results have advantages as twofold. Firstly, our proposed stability conditions don't include the interval delay item in Lyapunov functions and reduce the computational complexity in simulations. Secondly, the conservativeness is further reduced via utilizing the reciprocally convex approach. Furthermore, compare Corollary 7.11 with the existing results [82, 251, 299], see Table 7.1 in Example 7.17, it is observable that the results we obtained are more efficient than the results in [82, 251, 299].

7.3.2 Reliable Filter Design with Dissipativity

On the basis of Theorem 7.10, the following theorems present the reliable filter with strict dissipativity for T-S fuzzy system with time-varying delays in (7.2). Firstly, consider the dynamic filtering system in (7.5) owes known sensor failure parameters.

Theorem 7.13 *For the matrices* $0 \geq \mathcal{Z} \in \mathbb{R}^{q\times q}$, $\mathcal{X} \in \mathbb{R}^{p\times p}$, $\mathcal{Y} \in \mathbb{R}^{q\times p}$ *with* $\mathcal{Z}$ *and* $\mathcal{X}$ *being symmetric, and scalars* $\delta > 0$, *provided that there are matrices* $0 < \mathcal{O} \in \mathbb{R}^{n\times n}$, $0 < \mathcal{L} \in \mathbb{R}^{k\times k}$, $0 < \bar{Q}_{1i} \in \mathbb{R}^{(n+k)\times(n+k)}$, $0 < \bar{Q}_{2i} \in \mathbb{R}^{(n+k)\times(n+k)}$, $0 < \bar{Q}_{3i} \in \mathbb{R}^{(n+k)\times(n+k)}$, $0 < \bar{S}_1 \in \mathbb{R}^{(n+k)\times(n+k)}$, $0 < \bar{S}_2 \in \mathbb{R}^{(n+k)\times(n+k)}$, $\bar{M} \in \mathbb{R}^{(n+k)\times(n+k)}$, $\mathcal{W}_1 \in \mathbb{R}^{n\times n}$, $\mathcal{W}_2 \in \mathbb{R}^{n\times k}$, $\mathcal{A}_f \in \mathbb{R}^{k\times k}$, $\mathcal{B}_f \in \mathbb{R}^{k\times p}$, $\mathcal{C}_f \in \mathbb{R}^{q\times k}$ *and* $\mathcal{D}_f \in \mathbb{R}^{q\times p}$ *such that for* $i, j, s, t = 1, \ldots, r$,

$$\Upsilon_{ijst} < 0, \tag{7.28}$$

$$\Pi \geq 0, \tag{7.29}$$

where

$$\Pi \triangleq \begin{bmatrix} S_{21} & S_{22} & M_1^T & M_3^T \\ \star & S_{24} & M_2^T & M_4^T \\ \star & \star & S_{21} & S_{22} \\ \star & \star & \star & S_{24} \end{bmatrix},$$

$$\Upsilon_{ijst} \triangleq \begin{bmatrix} \Upsilon_{11i} & \Upsilon_{12} & 0 & 0 & \Upsilon_{15i} & \Upsilon_{16i} \\ \star & \Upsilon_{22j} & \Upsilon_{23} & \Upsilon_{24} & 0 & 0 \\ \star & \star & \Upsilon_{33s} & \Upsilon_{34} & \Upsilon_{35i} & \Upsilon_{36i} \\ \star & \star & \star & \Upsilon_{44t} & 0 & 0 \\ \star & \star & \star & \star & \Upsilon_{55i} & \Upsilon_{56i} \\ \star & \star & \star & \star & \star & \Upsilon_{66} \end{bmatrix},$$

with

$$\bar{Q}_{1i} \triangleq \begin{bmatrix} Q_{1i1} & Q_{2i2} \\ \star & Q_{1i4} \end{bmatrix}, \quad \bar{Q}_{2i} \triangleq \begin{bmatrix} Q_{2i1} & Q_{2i2} \\ \star & Q_{2i4} \end{bmatrix}, \quad \bar{Q}_{3i} \triangleq \begin{bmatrix} Q_{3i1} & Q_{3i2} \\ \star & Q_{3i4} \end{bmatrix},$$

$$\Upsilon_{11i} \triangleq \begin{bmatrix} \Upsilon_{11i1} & \Upsilon_{11i2} \\ \star & \Upsilon_{11i4} \end{bmatrix}, \quad \Upsilon_{22j} \triangleq \begin{bmatrix} -Q_{1j1} - S_{11} - S_{21} & -Q_{1j2} - S_{12} - S_{22} \\ \star & -Q_{1j4} - S_{14} - S_{24} \end{bmatrix},$$

$$\Upsilon_{33s} \triangleq \begin{bmatrix} \Upsilon_{33s1} & \Upsilon_{33s2} \\ \star & \Upsilon_{33s4} \end{bmatrix}, \quad \bar{M} \triangleq \begin{bmatrix} M_1 & M_2 \\ M_3 & M_4 \end{bmatrix}, \quad \Upsilon_{15i} \triangleq \begin{bmatrix} -L_i^T \mathcal{Y} + C_i^T B_\varepsilon^T \mathcal{D}_f^T \mathcal{Y} \\ \mathcal{C}_f^T \mathcal{Y} \end{bmatrix},$$

$$\Upsilon_{55i} \triangleq -(F_i^T \mathcal{Y} - D_i^T B_\varepsilon^T \mathcal{D}_f^T \mathcal{Y}) - \mathcal{X} + \delta I - (F_i^T \mathcal{Y} - D_i^T B_\varepsilon^T \mathcal{D}_f^T \mathcal{Y})^T,$$

$$\Upsilon_{35i} \triangleq \begin{bmatrix} -L_{di}^T \mathcal{Y} + C_{di}^T B_\varepsilon^T \mathcal{D}_f^T \mathcal{Y} \\ 0 \end{bmatrix}, \ \Upsilon_{12} \triangleq \begin{bmatrix} S_{11} & S_{12} \\ S_{12}^T & S_{14} \end{bmatrix}, \Upsilon_{24} \triangleq \begin{bmatrix} M_1 & M_2 \\ M_3 & M_4 \end{bmatrix},$$

$$\Upsilon_{11i1} \triangleq -\mathcal{O} + Q_{1i1} + Q_{2i1} + (d+1)Q_{3i1} - S_{11},$$

$$\Upsilon_{11i2} \triangleq -\mathcal{I}\mathcal{L} + Q_{1i2} + Q_{2i2} + (d+1)Q_{3i2} - S_{12},$$

$$\Upsilon_{11i4} \triangleq -\mathcal{L}^T + Q_{1i4} + Q_{2i4} + (d+1)Q_{3i4} - S_{14},$$

$$\Upsilon_{33s1} \triangleq -Q_{3s1} - S_{21} - S_{21}^T + M_1 + M_1^T, \Upsilon_{56i} \triangleq \begin{bmatrix} \Upsilon_{56i1} & \Upsilon_{56i2} & \Upsilon_{56i3} & \Upsilon_{56i4} \end{bmatrix},$$

$$\Upsilon_{16i} \triangleq \begin{bmatrix} \Upsilon_{16i1} & \Upsilon_{16i2} & \Upsilon_{16i3} & \Upsilon_{16i4} \end{bmatrix}, \Upsilon_{36i} \triangleq \begin{bmatrix} \Upsilon_{36i1} & \Upsilon_{36i2} & \Upsilon_{36i3} & \Upsilon_{36i4} \end{bmatrix},$$

$$\Upsilon_{33s2} \triangleq -Q_{3s2} - S_{22} - S_{22}^T + M_2 + M_3^T,$$

$$\Upsilon_{33s4} \triangleq -Q_{3s4} - S_{24} - S_{24}^T + M_4 + M_4^T,$$

$$\Upsilon_{662} \triangleq \begin{bmatrix} S_{11} - \mathcal{W}_1 - \mathcal{W}_1^T & S_{12} - \mathcal{W}_2 - (\mathcal{L}^T \mathcal{I}^T)^T \\ \star & S_{14} - \mathcal{L}^T - \mathcal{L} \end{bmatrix},$$

$$\Upsilon_{663} \triangleq \begin{bmatrix} S_{21} - \mathcal{W}_1 - \mathcal{W}_1^T & S_{22} - \mathcal{W}_2 - (\mathcal{L}^T \mathcal{I}^T)^T \\ \star & S_{24} - \mathcal{L}^T - \mathcal{L} \end{bmatrix},$$

$$\Upsilon_{16i1} \triangleq \begin{bmatrix} A_i^T \mathcal{O} + C_i^T B_\varepsilon^T \mathcal{B}_f^T \mathcal{I}^T & A_i^T \mathcal{I}\mathcal{L} + C_i^T B_\varepsilon^T \mathcal{B}_f^T \\ \mathcal{A}_f^T \mathcal{I}^T & \mathcal{A}_f^T \end{bmatrix},$$

$$\Upsilon_{36i1} \triangleq \begin{bmatrix} A_{di}^T \mathcal{O} + C_{di}^T B_\varepsilon^T \mathcal{B}_f^T \mathcal{I}^T & A_{di}^T \mathcal{I}\mathcal{L} + C_{di}^T B_\varepsilon^T \mathcal{B}_f^T \\ 0 & 0 \end{bmatrix},$$

$$\Upsilon_{56i1} \triangleq \begin{bmatrix} B_i^T \mathcal{O} + D_i^T B_\varepsilon^T \mathcal{B}_f^T \mathcal{I}^T & B_i^T \mathcal{I}\mathcal{L} + D_i^T B_\varepsilon^T \mathcal{B}_f^T \end{bmatrix},$$

$$\Upsilon_{16i2} \triangleq d_1 \begin{bmatrix} A_i^T \mathcal{W}_1 + C_i^T B_\varepsilon^T \mathcal{B}_f^T \mathcal{I}^T - \mathcal{W}_1 & A_i^T \mathcal{W}_2 + C_i^T B_\varepsilon^T \mathcal{B}_f^T - \mathcal{W}_2 \\ \mathcal{A}_f^T \mathcal{I}^T - \mathcal{L}^T \mathcal{I}^T & \mathcal{A}_f^T - \mathcal{L}^T \end{bmatrix},$$

$$\Upsilon_{36i2} \triangleq d_1 \begin{bmatrix} A_{di}^T \mathcal{W}_1 + C_{di}^T B_\varepsilon^T \mathcal{B}_f^T \mathcal{I}^T & A_{di}^T \mathcal{W}_2 + C_{di}^T B_\varepsilon^T \mathcal{B}_f^T \\ 0 & 0 \end{bmatrix},$$

$$\Upsilon_{56i2} \triangleq d_1 \begin{bmatrix} B_i^T \mathcal{W}_1 + D_i^T B_\varepsilon^T \mathcal{B}_f^T \mathcal{I}^T & B_i^T \mathcal{W}_2 + D_i^T B_\varepsilon^T \mathcal{B}_f^T \end{bmatrix},$$

$$\Upsilon_{16i3} \triangleq d \begin{bmatrix} A_i^T \mathcal{W}_1 + C_i^T B_\varepsilon^T \mathcal{B}_f^T \mathcal{I}^T - \mathcal{W}_1 & A_i^T \mathcal{W}_2 + C_i^T B_\varepsilon^T \mathcal{B}_f^T - \mathcal{W}_2 \\ \mathcal{A}_f^T \mathcal{I}^T - \mathcal{L}^T \mathcal{I}^T & \mathcal{A}_f^T - \mathcal{L}^T \end{bmatrix},$$

$$\Upsilon_{36i3} \triangleq d \begin{bmatrix} A_{di}^T \mathcal{W}_1 + C_{di}^T B_\varepsilon^T \mathcal{B}_f^T \mathcal{I}^T & A_{di}^T \mathcal{W}_2 + C_{di}^T B_\varepsilon^T \mathcal{B}_f^T \\ 0 & 0 \end{bmatrix},$$

$$\Upsilon_{56i3} \triangleq d \begin{bmatrix} B_i^T \mathcal{W}_1 + D_i^T B_\varepsilon^T \mathcal{B}_f^T \mathcal{I}^T & B_i^T \mathcal{W}_2 + D_i^T B_\varepsilon^T \mathcal{B}_f^T \end{bmatrix},$$

$$\Upsilon_{23} \triangleq \begin{bmatrix} -M_1 + S_{21} & -M_2 + S_{22} \\ -M_3 + S_{22}^T & -M_4 + S_{24} \end{bmatrix}, \ \bar{S}_2 \triangleq \begin{bmatrix} S_{21} & S_{22} \\ \star & S_{24} \end{bmatrix},$$

$$\Upsilon_{34} \triangleq \begin{bmatrix} -M_1 + S_{21} & -M_2 + S_{22} \\ -M_3 + S_{22}^T & -M_4 + S_{24} \end{bmatrix}, \ \bar{S}_1 \triangleq \begin{bmatrix} S_{11} & S_{12} \\ \star & S_{14} \end{bmatrix},$$

$$\Upsilon_{44t} \triangleq \begin{bmatrix} -Q_{2t1} - S_{21} & -Q_{2t2} - S_{22} \\ \star & -Q_{2t4} - S_{24} \end{bmatrix},$$

$$\Upsilon_{16i4} \triangleq \begin{bmatrix} L_i^T \mathcal{Z}_-^{\frac{1}{2}} - C_i^T B_\varepsilon^T \mathcal{D}_f^T \mathcal{Z}_-^{\frac{1}{2}} \\ -\mathcal{C}_f^T \mathcal{Z}_-^{\frac{1}{2}} \end{bmatrix}, \ \Upsilon_{661} \triangleq \begin{bmatrix} -\mathcal{O} & -\mathcal{I}\mathcal{L} \\ \star & -\mathcal{L}^T \end{bmatrix},$$

$$\Upsilon_{36i4} \triangleq \begin{bmatrix} L_{di}^T \mathcal{Z}_-^{\frac{1}{2}} - C_{di}^T B_\varepsilon^T \mathcal{D}_f^T \mathcal{Z}_-^{\frac{1}{2}} \\ 0 \end{bmatrix}, \ \mathcal{I} \triangleq \begin{bmatrix} I_{k\times k} \\ 0_{(n-k)\times k} \end{bmatrix},$$

$$\Upsilon_{56i4} \triangleq F_i^T \mathcal{Z}_-^{\frac{1}{2}} - D_i^T B_\varepsilon^T \mathcal{D}_f^T \mathcal{Z}_-^{\frac{1}{2}}, \ \Upsilon_{66i} \triangleq diag\{\Upsilon_{661}, \Upsilon_{662}, \Upsilon_{663}, -I\},$$

the dynamic filtering system with sensor failures in (7.5) is asymptotically stable with strict dissipativity based on Definition 7.5. In addition, if the conditions have feasible solutions ($\mathcal{O}$, $\mathcal{L}$, $\mathcal{A}_f$, $\mathcal{B}_f$, $\mathcal{C}_f$, $\mathcal{D}_f$, $\mathcal{W}_1$, $\mathcal{W}_2$, $\mathcal{Q}_{1i1}$, $\mathcal{Q}_{1i2}$, $\mathcal{Q}_{1i4}$, $\mathcal{Q}_{2i1}$, $\mathcal{Q}_{2i2}$, $\mathcal{Q}_{2i4}$, $\mathcal{Q}_{3i1}$, $\mathcal{Q}_{3i2}$, $\mathcal{Q}_{3i4}$, S_{11}, S_{12}, S_{14}, S_{21}, S_{22}, S_{24}, M_1, M_2, M_3, M_4), and the corresponding parameters of our proposed filter (Σ_f) in (7.3) are described as

$$A_f = \mathcal{L}^{-1}\mathcal{A}_f, \ B_f = \mathcal{L}^{-1}\mathcal{B}_f, \ C_f = \mathcal{C}_f, \ D_f = \mathcal{D}_f. \tag{7.30}$$

Proof Based on Theorem 7.10, it is not difficult to conclude the dynamic filtering system in (7.5) is asymptotically stable and strictly dissipative if there are matrices $0 < P \in \mathbb{R}^{(n+k)\times(n+k)}$, $0 < Q_{1i} \in \mathbb{R}^{(n+k)\times(n+k)}$, $0 < Q_{2i} \in \mathbb{R}^{(n+k)\times(n+k)}$, $0 < Q_{3i} \in \mathbb{R}^{(n+k)\times(n+k)}$, $(i = 1, \dots, r)$, $0 < S_1 \in \mathbb{R}^{(n+k)\times(n+k)}$, $0 < S_2 \in \mathbb{R}^{(n+k)\times(n+k)}$, $M \in \mathbb{R}^{(n+k)\times(n+k)}$ and $\mathcal{W} \in \mathbb{R}^{(n+k)\times(n+k)}$, which satisfy (7.13) and

$$\bar{\Xi}_{ijst} < 0, \tag{7.31}$$

where

$$\bar{\Xi}_{16i} \triangleq \begin{bmatrix} \bar{A}_i^T P & d_1(\bar{A}_i^T - I)\mathcal{W} & d(\bar{A}_i^T - I)\mathcal{W} & \bar{L}_i^T \mathcal{Z}_-^{\frac{1}{2}} \end{bmatrix},$$

$$\bar{\Xi}_{36i} \triangleq \begin{bmatrix} \bar{A}_{di}^T P & d_1 \bar{A}_{di}^T \mathcal{W} & d\bar{A}_{di}^T \mathcal{W} & \bar{L}_{di}^T \mathcal{Z}_-^{\frac{1}{2}} \end{bmatrix},$$

$$\bar{\Xi}_{56i} \triangleq \begin{bmatrix} \bar{B}_i^T P & d_1 \bar{B}_i^T \mathcal{W} & d\bar{B}_i^T \mathcal{W} & \bar{F}_i^T \mathcal{Z}_-^{\frac{1}{2}} \end{bmatrix},$$

$$\bar{\Xi}_{66} \triangleq \text{diag}\{-P, S_1 - \mathcal{W} - \mathcal{W}^T, S_2 - \mathcal{W} - \mathcal{W}^T, -I\},$$

$$\bar{\Xi}_{ijst} \triangleq \begin{bmatrix} \Xi_{11i} & \Xi_{12} & 0 & 0 & \Xi_{15i} & \bar{\Xi}_{16i} \\ \star & \Xi_{22j} & \Xi_{23} & \Xi_{24} & 0 & 0 \\ \star & \star & \Xi_{33s} & \Xi_{34} & \Xi_{35i} & \bar{\Xi}_{36i} \\ \star & \star & \star & \Xi_{44t} & 0 & 0 \\ \star & \star & \star & \star & \Xi_{55i} & \bar{\Xi}_{56i} \\ \star & \star & \star & \star & \star & \bar{\Xi}_{66} \end{bmatrix}.$$

Partition P as

$$P \triangleq \begin{bmatrix} P_1 & P_2 \\ \star & P_3 \end{bmatrix} > 0, \quad P_2 \triangleq \begin{bmatrix} P_4 \\ 0_{(n-k)\times k} \end{bmatrix}, \tag{7.32}$$

where $0 < P_1 \in \mathbb{R}^{n\times n}$, $0 < P_3 \in \mathbb{R}^{k\times k}$ and $P_4 \in \mathbb{R}^{k\times k}$. Without loss of generality, P_4 is assumed to be nonsingular. Denote $\mathcal{N} \triangleq P + \epsilon\mathcal{T}$, with ϵ being a positive scalar and

$$\mathcal{T} \triangleq \begin{bmatrix} 0_{n\times n} & \mathcal{I} \\ \star & 0_{k\times k} \end{bmatrix}, \ \mathcal{N} \triangleq \begin{bmatrix} \mathcal{N}_1 & \mathcal{N}_2 \\ \star & \mathcal{N}_3 \end{bmatrix}, \ \mathcal{N}_2 \triangleq \begin{bmatrix} \mathcal{N}_4 \\ 0_{(n-k)\times k} \end{bmatrix}.$$

Due to $P > 0$, it yields $\mathcal{N} > 0$ for $\epsilon > 0$ near the origin. Then it can be seen that there exists an arbitrarily small $\epsilon > 0$ such that $\mathcal{N}_4$ is nonsingular and (7.31) is feasible with P replaced by $\mathcal{N}$. Because of $\mathcal{N}_4$ being nonsingular, it follows that P_4 is nonsingular.

Define the following nonsingular matrices as

$$\left.\begin{aligned} \mathcal{F} &\triangleq \begin{bmatrix} I & 0 \\ 0 & P_3^{-1}P_4^T \end{bmatrix}, & \mathcal{W} &\triangleq \begin{bmatrix} \mathcal{W}_1 & \mathcal{W}_2 P_4^{-T} P_3 \\ (\mathcal{I}P_4)^T & P_3 \end{bmatrix}, \\ \mathcal{L} &\triangleq P_4 P_3^{-1} P_4^T, & M &\triangleq \mathcal{F}^{-T}\bar{M}\mathcal{F}^{-1}, \\ S_1 &\triangleq \mathcal{F}^{-T}\bar{S}_1\mathcal{F}^{-1}, & S_2 &\triangleq \mathcal{F}^{-T}\bar{S}_2\mathcal{F}^{-1}, \\ Q_{1i} &\triangleq \mathcal{F}^{-T}\bar{Q}_{1i}\mathcal{F}^{-1}, & Q_{3i} &\triangleq \mathcal{F}^{-T}\bar{Q}_{3i}\mathcal{F}^{-1}, \\ \mathcal{O} &\triangleq P_1, & Q_{2i} &\triangleq \mathcal{F}^{-T}\bar{Q}_{2i}\mathcal{F}^{-1}, \end{aligned}\right\} \tag{7.33}$$

and

$$\left.\begin{aligned} \mathcal{A}_f &\triangleq P_4 A_f P_3^{-1} P_4^T, & \mathcal{B}_f &\triangleq P_4 B_f, \\ \mathcal{C}_f &\triangleq C_f P_3^{-1} P_4^T, & \mathcal{D}_f &\triangleq D_f. \end{aligned}\right\} \tag{7.34}$$

Then it yields

$$\begin{aligned}
\mathcal{F}^T \bar{A}_i^T P\mathcal{F} &= \begin{bmatrix} A_i^T\mathcal{O} + C_i^T B_\varepsilon^T \mathcal{B}_f^T \mathcal{I}^T & A_i^T \mathcal{I}\mathcal{L} + C_i^T B_\varepsilon^T \mathcal{B}_f^T \\ \mathcal{A}_f^T \mathcal{I}^T & \mathcal{A}_f^T \end{bmatrix}, \\
\mathcal{F}^T \bar{A}_{di}^T P\mathcal{F} &= \begin{bmatrix} A_{di}^T\mathcal{O} + C_{di}^T B_\varepsilon^T \mathcal{B}_f^T \mathcal{I}^T & A_{di}^T \mathcal{I}\mathcal{L} + C_{di}^T B_s^T \mathcal{B}_f^T \\ 0 & 0 \end{bmatrix}, \\
\bar{B}_i^T P\mathcal{F} &= \begin{bmatrix} B_i^T\mathcal{O} + D_i^T B_\varepsilon^T \mathcal{B}_f^T \mathcal{I}^T & B_i^T \mathcal{I}\mathcal{L} + D_i^T B_\varepsilon^T \mathcal{B}_f^T \end{bmatrix}, \\
\mathcal{F}^T \bar{A}_i^T \mathcal{W}\mathcal{F} &= \begin{bmatrix} A_i^T\mathcal{W}_1 + C_i^T B_\varepsilon^T \mathcal{B}_f^T \mathcal{I}^T & A_i^T \mathcal{W}_2 + C_i^T B_\varepsilon^T \mathcal{B}_f^T \\ \mathcal{A}_f^T \mathcal{I}^T & \mathcal{A}_f^T \end{bmatrix}, \\
\mathcal{F}\bar{A}_{di}^T \mathcal{W}\mathcal{F} &= \begin{bmatrix} A_{di}^T\mathcal{W}_1 + C_{di}^T B_\varepsilon^T \mathcal{B}_f^T \mathcal{I}^T & A_{di}^T \mathcal{W}_2 + C_{di}^T B_\varepsilon^T \mathcal{B}_f^T \\ 0 & 0 \end{bmatrix}, \\
\bar{B}_i^T \mathcal{W}\mathcal{F} &= \begin{bmatrix} B_i^T\mathcal{W}_1 + D_i^T B_\varepsilon^T \mathcal{B}_f^T \mathcal{I}^T & B_i^T \mathcal{W}_2 + D_i^T B_\varepsilon^T \mathcal{B}_f^T \end{bmatrix}, \\
\mathcal{F}^T \bar{L}_i^T \mathcal{Z}_-^{\frac{1}{2}} &= \begin{bmatrix} L_i^T \mathcal{Z}_-^{\frac{1}{2}} - C_i^T B_\varepsilon^T \mathcal{D}_f^T \mathcal{Z}_-^{\frac{1}{2}} \\ -\mathcal{C}_f^T \mathcal{Z}_-^{\frac{1}{2}} \end{bmatrix}, \mathcal{F}^T \mathcal{W}\mathcal{F} = \begin{bmatrix} \mathcal{W}_1 & \mathcal{W}_2 \\ \mathcal{L}^T\mathcal{I}^T & \mathcal{L}^T \end{bmatrix}, \\
\mathcal{F}^T \bar{L}_{di}^T \mathcal{Z}_-^{\frac{1}{2}} &= \begin{bmatrix} L_{di}^T \mathcal{Z}_-^{\frac{1}{2}} - C_{di}^T B_\varepsilon^T \mathcal{D}_f^T \mathcal{Z}_-^{\frac{1}{2}} \\ 0 \end{bmatrix}, \mathcal{F}^T P\mathcal{F} = \begin{bmatrix} \mathcal{O} & \mathcal{I}\mathcal{L} \\ \mathcal{L}^T\mathcal{I}^T & \mathcal{L}^T \end{bmatrix}, \\
\mathcal{F}^T \bar{L}_i^T \mathcal{Y} &= \begin{bmatrix} L_i^T \mathcal{Y} - C_i^T B_\varepsilon^T \mathcal{D}_f^T \mathcal{Y} \\ -\mathcal{C}_f^T \mathcal{Y} \end{bmatrix}, \mathcal{F}^T \bar{L}_{di}^T \mathcal{Y} = \begin{bmatrix} L_{di}^T \mathcal{Y} - C_{di}^T B_\varepsilon^T \mathcal{D}_f^T \mathcal{Y} \\ 0 \end{bmatrix}, \\
\bar{F}_i^T \mathcal{Z}_-^{\frac{1}{2}} &= F_i^T \mathcal{Z}_-^{\frac{1}{2}} - D_i^T B_\varepsilon^T \mathcal{D}_f^T \mathcal{Z}_-^{\frac{1}{2}}, \ \bar{F}_i^T \mathcal{Y} = F_i^T \mathcal{Y} - D_i^T B_\varepsilon^T \mathcal{D}_f^T \mathcal{Y}.
\end{aligned} \tag{7.35}$$

Performing congruence transformations to (7.13) and (7.31) with diag$(\mathcal{F}, \mathcal{F})$ and diag $(\mathcal{F}, \mathcal{F}, \mathcal{F}, \mathcal{F}, I, \mathcal{F}, \mathcal{F}, \mathcal{F}, I)$, respectively, and based on (7.33)–(7.35), we can get (7.28)–(7.29). Furthermore, (7.30) is equal to

$$\left.\begin{array}{l} A_f \triangleq P_4^{-1}\mathcal{A}_f P_4^{-T} P_3 = (P_4^{-T} P_3)^{-1}\mathcal{L}^{-1}\mathcal{A}_f P_4^{-T} P_3, \\ B_f \triangleq P_4^{-1}\mathcal{B}_f = (P_4^{-T} P_3)^{-1}\mathcal{L}^{-1}\mathcal{B}_f, \\ C_f \triangleq \mathcal{C}_f P_4^{-T} P_3, \quad D_f \triangleq \mathcal{D}_f. \end{array}\right\} \tag{7.36}$$

Actually, A_f, B_f, C_f and D_f in (7.3) can be expressed by (7.36), which means $P_4^{-T} P_3$ can be regarded to be a similarity transformation on the state-space realization for the filter and has no influence in the filter mapping from y to $\hat{e}$. Without loss of generality, denote $P_4^{-T} P_3 = I$, then we get (7.30). Hence, the filter (Σ_f) in (7.3) can be established via (7.30). The proof is completed.

Next, on the basis of Theorem 7.13, we design a reliable filter with strict dissipativity when the sensor failure parameter is unknown but satisfies the condition in (7.4).

Theorem 7.14 *For the given matrices $0 \geq \mathcal{Z} \in \mathbb{R}^{q\times q}$, $\mathcal{X} \in \mathbb{R}^{p\times p}$, $\mathcal{Y} \in \mathbb{R}^{q\times p}$ with $\mathcal{Z}$ and $\mathcal{X}$ being symmetric, and scalar $\delta > 0$, if there exist matrices $0 < \mathcal{O} \in \mathbb{R}^{n\times n}$, $0 < \mathcal{L} \in \mathbb{R}^{k\times k}$, $0 < \bar{Q}_{1i} \in \mathbb{R}^{(n+k)\times(n+k)}$, $0 < \bar{Q}_{2i} \in \mathbb{R}^{(n+k)\times(n+k)}$, $0 < \bar{Q}_{3i} \in \mathbb{R}^{(n+k)\times(n+k)}$, $0 < \bar{S}_1 \in \mathbb{R}^{(n+k)\times(n+k)}$, $0 < \bar{S}_2 \in \mathbb{R}^{(n+k)\times(n+k)}$, $\bar{M} \in \mathbb{R}^{(n+k)\times(n+k)}$, $\mathcal{W}_1 \in \mathbb{R}^{n\times n}$, $\mathcal{W}_2 \in \mathbb{R}^{n\times k}$, $\mathcal{A}_f \in \mathbb{R}^{k\times k}$, $\mathcal{B}_f \in \mathbb{R}^{k\times p}$, $\mathcal{C}_f \in \mathbb{R}^{q\times k}$ $\mathcal{D}_f \in \mathbb{R}^{q\times p}$, and $\pi > 0$ such that for $i, j, s, t = 1, \ldots, r$, (7.29) and*

$$\begin{bmatrix} \hat{\varUpsilon}_{ijst} & \hat{\varUpsilon}_{1i} & \hat{\varUpsilon}_2 \\ \star & -\pi\Lambda^{-2} & 0 \\ \star & \star & -\pi I \end{bmatrix} < 0, \tag{7.37}$$

where

$$\hat{\varUpsilon}_{15i} \triangleq \begin{bmatrix} -L_i^T\mathcal{Y} + C_i^T B_{\varepsilon 0}^T\mathcal{D}_f^T\mathcal{Y} \\ \mathcal{C}_f^T\mathcal{Y} \end{bmatrix}, \hat{\varUpsilon}_{35i} \triangleq \begin{bmatrix} -L_{di}^T\mathcal{Y} + C_{di}^T B_{\varepsilon 0}^T\mathcal{D}_f^T\mathcal{Y} \\ 0 \end{bmatrix},$$

$$\hat{\varUpsilon}_{55i} \triangleq -(F_i^T\mathcal{Y} - D_i^T B_{\varepsilon 0}^T\mathcal{D}_f^T\mathcal{Y}) - (F_i^T\mathcal{Y} - D_i^T B_{\varepsilon 0}^T\mathcal{D}_f^T\mathcal{Y})^T - \mathcal{X} + \delta I,$$

$$\hat{\varUpsilon}_3^T \triangleq \begin{bmatrix} [\,B_f^T\mathcal{I}^T \;\; B_f^T\,] & [\,B_f^T\mathcal{I}^T \;\; B_f^T\,] & [\,B_f^T\mathcal{I}^T \;\; B_f^T\,] & -D_f^T\mathcal{Z}_-^{\frac{1}{2}} \end{bmatrix},$$

$$\hat{\varUpsilon}_{16i} \triangleq \begin{bmatrix} \hat{\varUpsilon}_{16i1} & \hat{\varUpsilon}_{16i2} & \hat{\varUpsilon}_{16i3} & \hat{\varUpsilon}_{16i4} \end{bmatrix}, \hat{\varUpsilon}_{36i} \triangleq \begin{bmatrix} \hat{\varUpsilon}_{36i1} & \hat{\varUpsilon}_{36i2} & \hat{\varUpsilon}_{36i3} & \hat{\varUpsilon}_{36i4} \end{bmatrix},$$

$$\hat{\varUpsilon}_{56i} \triangleq \begin{bmatrix} \hat{\varUpsilon}_{56i1} & \hat{\varUpsilon}_{56i2} & \hat{\varUpsilon}_{56i3} & \hat{\varUpsilon}_{56i4} \end{bmatrix}, \hat{\varUpsilon}_{56i4} \triangleq F_i^T\mathcal{Z}_-^{\frac{1}{2}} - D_i^T B_{\varepsilon 0}^T\mathcal{D}_f^T\mathcal{Z}_-^{\frac{1}{2}},$$

$$\hat{\varUpsilon}_{16i1} \triangleq \begin{bmatrix} A_i^T\mathcal{O} + C_i^T B_{\varepsilon 0}^T\mathcal{B}_f^T\mathcal{I}^T & A_i^T\mathcal{I}\mathcal{L} + C_i^T B_{\varepsilon 0}^T\mathcal{B}_f^T \\ \mathcal{A}_f^T\mathcal{I}^T & \mathcal{A}_f^T \end{bmatrix},$$

$$\hat{\varUpsilon}_{36i1} \triangleq \begin{bmatrix} A_{di}^T\mathcal{O} + C_{di}^T B_{\varepsilon 0}^T\mathcal{B}_f^T\mathcal{I}^T & A_{di}^T\mathcal{I}\mathcal{L} + C_{di}^T B_{\varepsilon 0}^T\mathcal{B}_f^T \\ 0 & 0 \end{bmatrix},$$

$$\hat{\Upsilon}_{56i1} \triangleq \left[B_i^T \mathcal{O} + D_i^T B_{\varepsilon 0}^T \mathcal{B}_f^T \mathcal{I}^T \;\; B_i^T \mathcal{I}\mathcal{L} + D_i^T B_{\varepsilon 0}^T \mathcal{B}_f^T \right],$$

$$\hat{\Upsilon}_{16i2} \triangleq d_1 \begin{bmatrix} A_i^T \mathcal{W}_1 + C_i^T B_{\varepsilon 0}^T \mathcal{B}_f^T \mathcal{I}^T - \mathcal{W}_1 & A_i^T \mathcal{W}_2 + C_i^T B_{\varepsilon 0}^T \mathcal{B}_f^T - \mathcal{W}_2 \\ \mathcal{A}_f^T \mathcal{I}^T - \mathcal{L}^T \mathcal{I}^T & \mathcal{A}_f^T - \mathcal{L}^T \end{bmatrix},$$

$$\hat{\Upsilon}_{36i2} \triangleq d_1 \begin{bmatrix} A_{di}^T \mathcal{W}_1 + C_{di}^T B_{\varepsilon 0}^T \mathcal{B}_f^T \mathcal{I}^T & A_{di}^T \mathcal{W}_2 + C_{di}^T B_{\varepsilon 0}^T \mathcal{B}_f^T \\ 0 & 0 \end{bmatrix},$$

$$\hat{\Upsilon}_{56i2} \triangleq d_1 \left[B_i^T \mathcal{W}_1 + D_i^T B_{\varepsilon 0}^T \mathcal{B}_f^T \mathcal{I}^T \;\; B_i^T \mathcal{W}_2 + D_i^T B_{\varepsilon 0}^T \mathcal{B}_f^T \right],$$

$$\hat{\Upsilon}_{16i3} \triangleq d \begin{bmatrix} A_i^T \mathcal{W}_1 + C_i^T B_{\varepsilon 0}^T \mathcal{B}_f^T \mathcal{I}^T - \mathcal{W}_1 & A_i^T \mathcal{W}_2 + C_i^T B_{\varepsilon 0}^T \mathcal{B}_f^T - \mathcal{W}_2 \\ \mathcal{A}_f^T - \mathcal{L}^T & \mathcal{A}_f^T - \mathcal{L}^T \end{bmatrix},$$

$$\hat{\Upsilon}_{36i3} \triangleq d \begin{bmatrix} A_{di}^T \mathcal{W}_1 + C_{di}^T B_{\varepsilon 0}^T \mathcal{B}_f^T \mathcal{I}^T & A_{di}^T \mathcal{W}_2 + C_{di}^T B_{\varepsilon 0}^T \mathcal{B}_f^T \\ 0 & 0 \end{bmatrix},$$

$$\hat{\Upsilon}_{56i3} \triangleq d \left[B_i^T \mathcal{W}_1 + D_i^T B_{\varepsilon 0}^T \mathcal{B}_f^T \mathcal{I}^T \;\; B_i^T \mathcal{W}_2 + D_i^T B_{\varepsilon 0}^T \mathcal{B}_f^T \right],$$

$$\hat{\Upsilon}_{16i4} \triangleq \begin{bmatrix} L_i^T \mathcal{Z}_-^{\frac{1}{2}} - C_i^T B_{\varepsilon 0}^T \mathcal{D}_f^T \mathcal{Z}_-^{\frac{1}{2}} \\ -\mathcal{C}_f^T \mathcal{Z}_-^{\frac{1}{2}} \end{bmatrix}, \hat{\Upsilon}_{36i4} \triangleq \begin{bmatrix} L_{di}^T \mathcal{Z}_-^{\frac{1}{2}} - C_{di}^T B_{\varepsilon 0}^T \mathcal{D}_f^T \mathcal{Z}_-^{\frac{1}{2}} \\ 0 \end{bmatrix},$$

$$\hat{\Upsilon}_{ijst} \triangleq \begin{bmatrix} \Upsilon_{11i} & \Upsilon_{12} & 0 & 0 & \hat{\Upsilon}_{15i} & \hat{\Upsilon}_{16i} \\ \star & \Upsilon_{22j} & \Upsilon_{23} & \Upsilon_{24} & \Upsilon_{25} & \Upsilon_{26} \\ \star & \star & \Upsilon_{33s} & \Upsilon_{34} & \hat{\Upsilon}_{35i} & \hat{\Upsilon}_{36i} \\ \star & \star & \star & \Upsilon_{44t} & \Upsilon_{45} & \Upsilon_{46} \\ \star & \star & \star & \star & \hat{\Upsilon}_{55i} & \hat{\Upsilon}_{56i} \\ \star & \star & \star & \star & \star & \Upsilon_{66} \end{bmatrix}, \hat{\Upsilon}_{1i} \triangleq \begin{bmatrix} \begin{bmatrix} C_i^T \\ 0 \end{bmatrix} \\ 0 \\ \begin{bmatrix} C_i^T \\ 0 \end{bmatrix} \\ 0 \\ D_i^T \\ 0 \end{bmatrix},$$

$$\hat{\Upsilon}_2 \triangleq \left[0\; 0\; 0\; 0\; \pi (D_f^T \mathcal{Y})^T \;\; \pi \hat{\Upsilon}_3^T \right]^T,$$

and Υ_{11i}, Υ_{12}, Υ_{22j}, Υ_{23}, Υ_{24}, Υ_{33s}, Υ_{34}, Υ_{44t}, Υ_{66} *are noted in Theorem 7.13.*

Then the dynamic filtering system in (7.5) with sensor failure is asymptotically stable with strict dissipativity on account of Definition 7.5. And the filter matrices in the form of (7.3) can be obtained by (7.30).

Proof Substituting B_ε by $B_{\varepsilon 0} + E_\varepsilon$ in (7.28), then

$$\bar{\Upsilon}_{ijst} = \hat{\Upsilon}_{ijst} + \hat{\Upsilon}_{1i} E_\varepsilon \hat{\Upsilon}_2^T + \hat{\Upsilon}_2 E_\varepsilon \hat{\Upsilon}_{1i}^T < 0. \tag{7.38}$$

Employing $x^T y + y^T x \leq \pi x^T x + \pi^{-1} y^T y$ for $\pi > 0$, consider (7.4) and we have

$$\bar{\Upsilon}_{ijst} \leq \hat{\Upsilon}_{ijst} + \pi \hat{\Upsilon}_{1i} \Lambda^2 \hat{\Upsilon}_{1i}^T + \pi^{-1} \hat{\Upsilon}_2 \hat{\Upsilon}_2^T. \tag{7.39}$$

Therefore, (7.39) is true if (7.37) holds via using the Schur complement method. Thus it completes the proof.

Remark 7.15 It can be observed that the conditions in Theorem 7.14 are strict LMIs. Consequently, the reliable filter design issue can be settled via convex optimization algorithms and the filter matrices can be readily computed with the standard software in MATLAB.

Remark 7.16 Many practical systems can be modeled as T-S fuzzy systems with time delay, such as automotive systems, robotics systems, chemical procedures and so on. In some engineering fields, to estimate system state of the corresponding system with noise inputs and prevent the occurrence of contingent failures and uninterrupted signal measurements, it is necessary to design an efficient filter. Such as the Henon mapping system in [276] can be approximated with the T-S fuzzy system with time-varying delay, and system states are estimated via filter design. In the next section, the reliable filter method based on reciprocally convex strategy can be applied into the Henon mapping system in Example 7.18 to estimate system states.

7.4 Illustrative Example

In this section, Example 7.17 is given to illustrate the validity and advantage of our proposed methods. Example 7.18 is presented to apply the reliable filter design scheme to the Henon mapping system in [276], which is a representative system exhibiting chaotic behavior and usually used as the proving ground in the theory of dynamic systems, such as the Ising model in statistical mechanics.

Example 7.17 Consider the T-S fuzzy system with time-varying delay in (7.21), and the relevant system parameters are set as

$$\begin{aligned} A_1 &= \begin{bmatrix} -0.291 & 1 \\ 0 & 0.95 \end{bmatrix}, \quad A_{d1} = \begin{bmatrix} 0.012 & 0.014 \\ 0 & 0.015 \end{bmatrix}, \\ A_2 &= \begin{bmatrix} -0.1 & 0 \\ 1 & -0.2 \end{bmatrix}, \quad A_{d2} = \begin{bmatrix} 0.01 & 0 \\ 0.01 & 0.015 \end{bmatrix}, \end{aligned} \tag{7.40}$$

which has been introduced in [82, 299].

In this example, $d(k)$ stands for the time-varying delay, and the upper delay bound can be obtained via employing proposed approaches in Corollary 7.11. Table 7.1 shows the specific comparison, where the obtained upper bounds of time delay are given as for their corresponding lower bounds. It can be observed that the developed techniques in this chapter are better than the results shown in [82, 299].

Example 7.18 The Henon mapping system with time-varying delay is considered:

$$\begin{cases} x_1(k+1) = -[\mu x_1(k) + (1-\mu)x_1(k-d(k))]^2 \\ \qquad\qquad\quad +0.3x_2(k) + \omega(k), \\ x_2(k+1) = \mu x_1(k) + (1-\mu)x_1(k-d(k)), \\ \quad\;\; y(k) \;= \mu x_1(k) + (1-\mu)x_1(k-d(k)) + \omega(k), \\ \quad\;\; z(k) \;= x_1(k), \end{cases} \tag{7.41}$$

Table 7.1 Allowable upper bound of d_2 for different values of d_1

For different d_1	$d_1 = 3$	$d_1 = 5$	$d_1 = 10$	$d_1 = 12$
Corollary 7.11 of [251]	$d_2 = 13$	$d_2 = 14$	$d_2 = 19$	$d_2 = 22$
Theorem 7.10 of [82]	$d_2 = 14$	$d_2 = 16$	$d_2 = 20$	$d_2 = 21$
Theorem 7.14 of [299]	$d_2 = 23$	$d_2 = 25$	$d_2 = 29$	$d_2 = 32$
Corollary 7.11	$d_2 = 26$	$d_2 = 28$	$d_2 = 33$	$d_2 = 35$

where $\omega(k)$ denotes the disturbance input and $\mu \in [0, 1]$ denotes the retarded coefficient.

Set $\theta(k) = \mu x_1(k) + (1 - \mu)x_1(k - d)$. It is assumed that $\theta(k) \in [-\nu, \nu],\ \nu > 0$. Utilizing the similar process in [276], the nonlinear term $\theta^2(k)$ can be completely described as

$$\theta^2(k) = h_1(\theta(k))(-\nu)\theta(k) + h_2(\theta(k))\nu\theta(k),$$

where $h_1(\theta(k)), h_2(\theta(k)) \in [0, 1]$, and $h_1(\theta(k)) + h_2(\theta(k)) = 1$. It is easy to get the fuzzy basis functions as

$$h_1(\theta(k)) = \frac{1}{2}\left(1 - \frac{\theta(k)}{\nu}\right), \quad h_2(\theta(k)) = \frac{1}{2}\left(1 + \frac{\theta(k)}{\nu}\right).$$

From the above descriptions, we have $h_1(\theta(k)) = 1$ and $h_2(\theta(k)) = 0$ when $\theta(k)$ is $-\nu$, $h_1(\theta(k)) = 0$ and $h_2(\theta(k)) = 1$ when $\theta(k)$ is ν. The nonlinear system in (7.41) can be approximated by the T-S fuzzy model as follows:

♦ **Plant Form**:

Rule 1: IF $\theta(k)$ is $-\nu$, THEN

$$\begin{cases} x(k+1) = A_1x(k) + A_{d1}x(k - d(k)) + B_1\omega(k), \\ \quad y(k) = C_1x(k) + C_{d1}x(k - d(k)) + D_1\omega(k), \\ \quad z(k) = L_1x(k). \end{cases}$$

Rule 2: IF $\theta(k)$ is ν, THEN

$$\begin{cases} x(k+1) = A_2x(k) + A_{d2}x(k - d(k)) + B_2\omega(k), \\ \quad y(k) = C_2x(k) + C_{d2}x(k - d(k)) + D_2\omega(k), \\ \quad z(k) = L_2x(k), \end{cases}$$

where

$$A_1 = \begin{bmatrix} \mu\nu & 0.3 \\ \mu & 0 \end{bmatrix}, \ A_{d1} = \begin{bmatrix} (1-\mu)\nu & 0 \\ 1-\mu & 0 \end{bmatrix}, \ B_1 = \begin{bmatrix} 1 \\ 0 \end{bmatrix},$$
$$A_2 = \begin{bmatrix} -\mu\nu & 0.3 \\ C & 0 \end{bmatrix}, \ A_{d2} = \begin{bmatrix} -(1-\mu)\nu & 0 \\ 1-\mu & 0 \end{bmatrix}, \ B_2 = \begin{bmatrix} 1 \\ 0 \end{bmatrix},$$
$$C_1 = \begin{bmatrix} \mu & 0 \end{bmatrix}, \ C_{d1} = \begin{bmatrix} 1-\mu & 0 \end{bmatrix}, \ D_1 = 1, \ L_1 = \begin{bmatrix} 1 & 0 \end{bmatrix},$$
$$C_2 = \begin{bmatrix} \mu & 0 \end{bmatrix}, \ C_{d2} = \begin{bmatrix} 1-\mu & 0 \end{bmatrix}, \ D_2 = 0.5, \ L_2 = \begin{bmatrix} 1 & 0 \end{bmatrix}.$$

In this example, $x(k) = \begin{bmatrix} x_1^T(k) & x_2^T(k) \end{bmatrix}^T$, $\mu = 0.8$, $\nu = 0.2$, $1 \le d(k) \le 3$. Resolving the conditions in Theorems 7.13 and 7.14, we can obtain following results for different filtering cases:

- $\mathcal{H}_\infty$ performance case: $\mathcal{Z} = -I$, $\mathcal{Y} = 0$, $\mathcal{X} = \gamma^2 I$, $B_\varepsilon = 1$. Resolving the conditions in Theorem 7.13, it yields $\gamma_{\min} = 1.3774$, and the relevant filter parameters are given by

$$\left.\begin{aligned} A_f &= \begin{bmatrix} 1.1457 & 0.3504 \\ -0.3086 & 0.4303 \end{bmatrix}, \quad B_f = \begin{bmatrix} 0.7705 \\ -1.0511 \end{bmatrix}, \\ C_f &= 10^{-3} \times \begin{bmatrix} -0.7191 & -0.2939 \end{bmatrix}, \quad D_f = 0.5592. \end{aligned}\right\} \tag{7.42}$$

- Strictly dissipative case: $\mathcal{Z} = -0.25$, $\mathcal{Y} = -0.2$, $\mathcal{X} = 1$, $\underline{B}_\varepsilon = 0.8$, $\bar{B}_\varepsilon = 0.9$. Resolving the conditions in Theorem 7.14, the relevant filter parameters are obtained as

$$\left.\begin{aligned} A_f &= \begin{bmatrix} 0.8147 & 0.0378 \\ -0.0847 & 0.6813 \end{bmatrix}, \quad B_f = \begin{bmatrix} 0.6635 \\ -0.9081 \end{bmatrix}, \\ C_f &= \begin{bmatrix} -0.0258 & -0.0031 \end{bmatrix}, \quad D_f = 0.7645. \end{aligned}\right\} \tag{7.43}$$

Set the zero initial condition, that is $x(0) = 0$ $(\hat{x}(0) = 0)$, and the disturbance signal $\omega(k)$ is assumed as $\omega(k) = \frac{3\sin(0.9k)}{(0.75k)^2+3.5}$. The corresponding simulation results for the designed filters are shown in Figs. 7.2, 7.3, 7.4 and 7.5. Figure 7.1 plots the time-varying delay $d(k)$ changing randomly between $d_1 = 1$ and $d_2 = 3$. Figures 7.2 and 7.4 draw the signal $z(k)$ (solid line) and its estimations $\hat{z}(k)$ (dash-dot line), separately. The estimation errors $e(k)$ are displayed in Figs. 7.3 and 7.5. It can be seen that the estimation error of the reliable dissipative filter as (7.43) is smaller than the $\mathcal{H}_\infty$ filter as (7.42). In fact, this is obvious due to the fact that the $\mathcal{H}_\infty$ performance is a particular situation in strict dissipative performance and thus owes more limiting conditions.

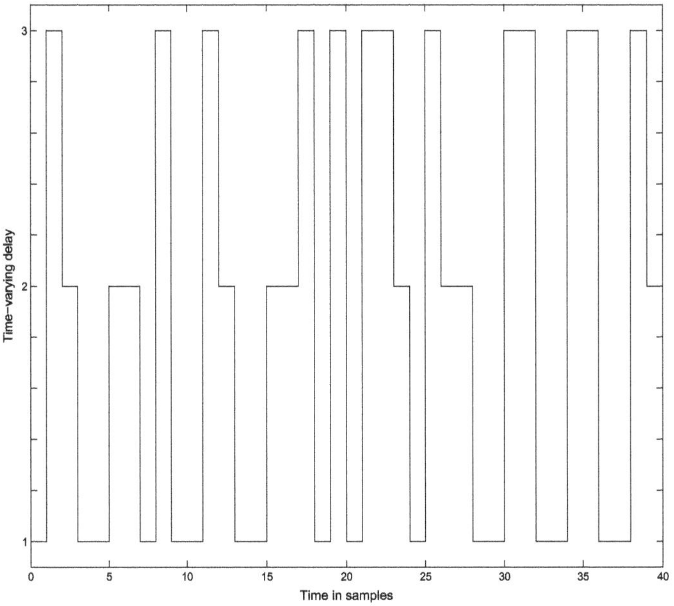

Fig. 7.1 Time-varying delays $d(k)$

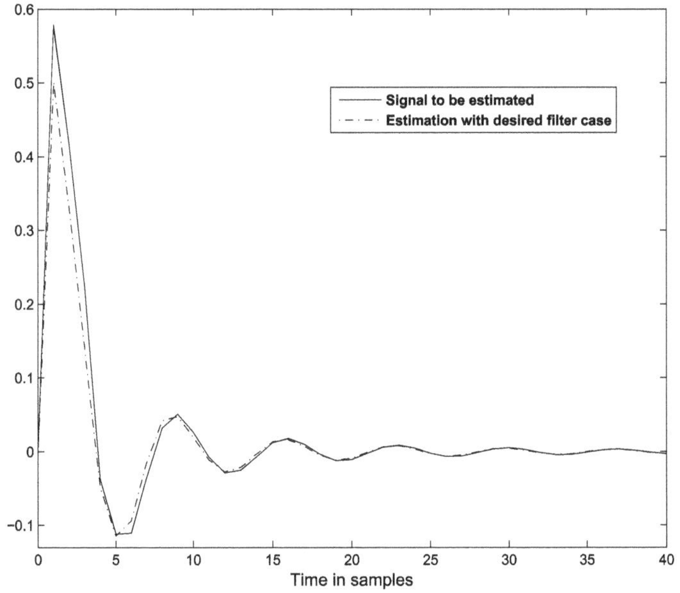

Fig. 7.2 Signal $z(k)$ and its estimation $\hat{z}(k)$ of the $\mathcal{H}_\infty$ filter

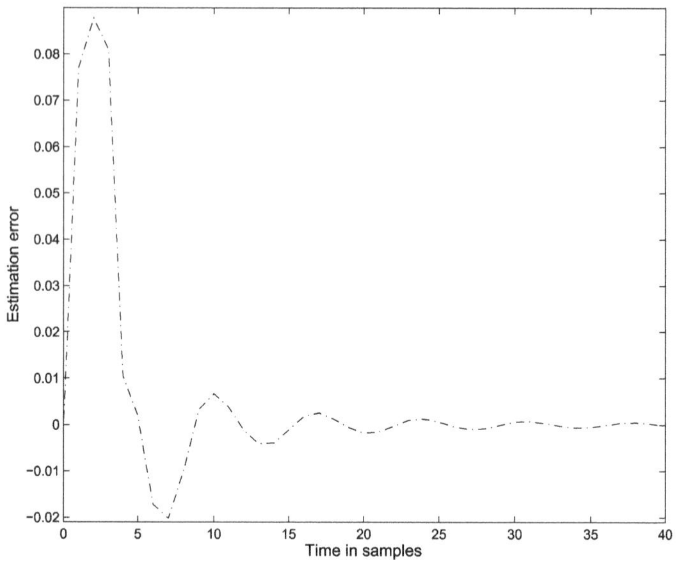

Fig. 7.3 Estimation error $e(k)$ for the $\mathcal{H}_\infty$ performance case

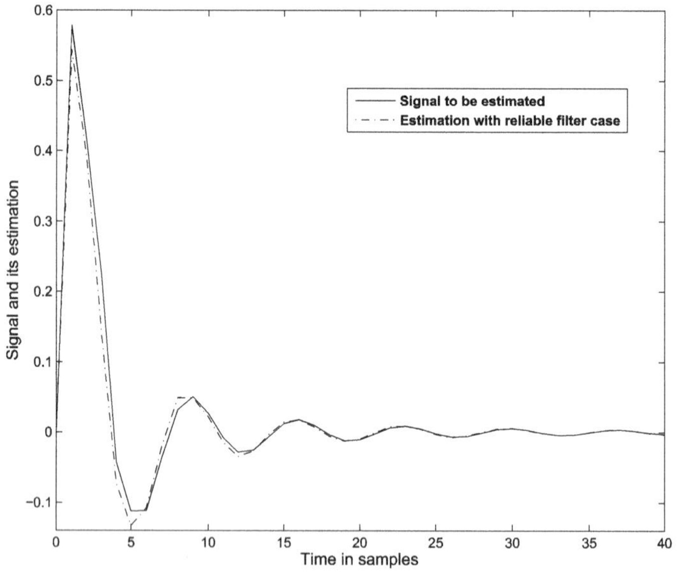

Fig. 7.4 Signal $z(k)$ and its estimation $\hat{z}(k)$ of the dissipative reliable filter

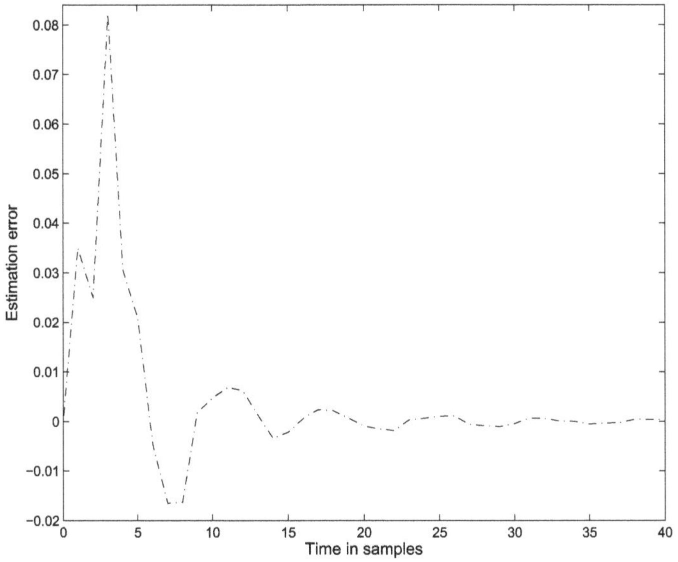

Fig. 7.5 Estimation error $e(k)$ for the dissipative case

7.5 Conclusion

A reliable filtering technique with a dissipative performance for discrete-time T-S fuzzy delayed systems was designed. First, the sufficient conditions were formulated to ensure the asymptotical stability and strict dissipativity for the dynamic filtering system by using the reciprocally convex method. The reliable filter design issue could be transformed to a convex optimization issue. Finally, some illustrative examples were put forward to show the availability of the developed techniques.

Part III
Model Reduction and Reduced-Order Synthesis

Chapter 8
Reduced-Order Model Approximation of Switched Systems

8.1 Introduction

This chapter considers the reduced-order model approximation issue for discrete hybrid switched nonlinear systems through T-S fuzzy modelling. We attempt to establish a reduced-order model for a high-dimension hybrid switched system, which can approximate the original high-order model subject to the prescribed system performance index. First, the mean-square exponential stability conditions are formulated to ensure the specific $\mathcal{H}_\infty$ performance for the resulting dynamic error system via the efficient Lyapunov stability method and ADT approach. The solutions of the relevant model reduction problems are derived using the projection technique, through which the algorithms of the reduced-order model parameters are established using a CCL method.

8.2 System Description and Preliminaries

Consider the following discrete-time high-order nonlinear hybrid stochastic switched systems:

$$x(k+1) = \sum_{j=1}^{N} \rho_j(k)\Big[\mathscr{C}_j\big(x(k), \omega(k)\big) + \mathscr{G}_j\big(x(k), \omega(k)\big)\varpi(k)\Big], \tag{8.1a}$$

$$y(k) = \sum_{j=1}^{N} \rho_j(k)\Big[\mathscr{H}_j\big(x(k), \omega(k)\big) + \mathscr{J}_j\big(x(k), \omega(t)\big)\varpi(k)\Big], \tag{8.1b}$$

where the state variable $x(\bullet) \in \mathbb{R}^n$ denotes the state vector; $\omega(\bullet) \in \mathbb{R}^m$ denotes the disturbance input belonging to $\ell_2[0, \infty)$; $\omega(\bullet)$ is supposed as energy bounded

X. Su et al., *Intelligent Control, Filtering and Model Reduction Analysis for Fuzzy-Model-Based Systems*, Studies in Systems, Decision and Control 385,
https://doi.org/10.1007/978-3-030-81214-0_8

and $\|\omega(\bullet)\|_2 \triangleq \sqrt{\sum_{k=0}^{\infty}\omega^T(\bullet)\omega(\bullet)}$; $y(\bullet) \in \mathbb{R}^p$ represents the measure output; $\varpi(\bullet)$ denotes a stochastic process on the probability space $(\Omega, \mathcal{F}, \mathcal{P})$ related to an increasing family $(\mathcal{F}_k)_{k\in\mathbb{N}}$ of σ-algebras $\mathcal{F}_k \subset \mathcal{F}$ produced by $\big(\varpi(k)\big)_{k\in\mathbb{N}}$. The stochastic process $\varpi(\bullet)$ is independent and $\mathbf{E}\{\varpi(k)\} = 0$, $\mathbf{E}\{\varpi(k)^2\} = \mu$; N is a positive integer representing the quantity of subsystems.

$$\rho_j(k) : [0, \infty) \to \{0, 1\}, \text{ and } \sum_{j=1}^{\mathcal{N}} \rho_j(k) = 1, k \in [1, \infty), j \in \mathcal{N} = \{1, 2, \cdots, N\},$$

stands for the stochastic switching signal implying which subsystem is attainable at the switching instant, ρ_j is introduced for simplicity; $\mathscr{C}_j(\bullet)$, $\mathscr{G}_j(\bullet)$, $\mathscr{H}_j(\bullet)$ and $\mathscr{J}_j(\bullet)$ represent a group of nonlinear regular functions.

At the discrete sampling time k, $\rho_j(k)$ may be established by $x(\bullet)$ or k, or both, or other hybrid schemes. Based on [186], the real-time value of ρ_j is assumed to be accessible. For switching signal ρ_j, the switching sequence is

$$\Big\{(j_0, k_0), (j_1, k_1), \ldots, (j_\kappa, k_\kappa), \ldots, \mid j_\kappa \in \mathcal{N}, \kappa = 0, 1, \ldots\Big\} \text{ with } k_0 = 0,$$

which means the j_κth subsystem is activated during $k \in \big[k_\kappa, k_{\kappa+1}\big)$.

The T-S fuzzy modelling method is utilized to deal with the reduced-order model approximation issue for nonlinear hybrid stochastic switched systems.

Fuzzy Rule $\mathcal{R}_i^{[j]}$: IF $\vartheta_1^{[j]}(k)$ is $\mathcal{M}_{i1}^{[j]}$ and $\vartheta_2^{[j]}(k)$ is $\mathcal{M}_{i2}^{[j]}$ and $\cdots$ and $\vartheta_p^{[j]}(k)$ is $\mathcal{M}_{ip}^{[j]}$, THEN

$$\begin{aligned} x(k+1) &= A_i^{[j]}x(k) + B_i^{[j]}\omega(k) + E_i^{[j]}x(k)\varpi(k), \\ y(k) &= C_i^{[j]}x(k) + D_i^{[j]}\omega(k), \end{aligned}$$

where $i = 1, 2, \ldots, r$, and r stands for the quantity of fuzzy rules; $\mathcal{M}_{i1}^{[j]}, \ldots, \mathcal{M}_{ip}^{[j]}$ represent the fuzzy sets; $\vartheta_1^{[j]}(\bullet), \vartheta_2^{[j]}(\bullet), \ldots, \vartheta_p^{[j]}(\bullet)$ represent the premise variables, denoted as $\vartheta_p^{[j]}$; $\Big\{\Big(A_i^{[j]}, B_i^{[j]}, C_i^{[j]}, D_i^{[j]}, E_i^{[j]}\Big) : j \in \mathcal{N}\Big\}$ are a group of matrices parameterized by $\mathcal{N} = \{1, 2, \ldots, N\}$, and $A_i^{[j]}$, $B_i^{[j]}$, $C_i^{[j]}$, $D_i^{[j]}$ and $E_i^{[j]}$ are given system matrices.

It is assumed that the premise variables are independent on $\omega(\bullet)$. As for a set of $\big(x(\bullet), \omega(\bullet)\big)$, the final output of hybrid switched fuzzy models can be expressed as

$$x(k+1) = \sum_{j=1}^{N}\rho_j\sum_{i=1}^{r}h_i^{[j]}\Big(\vartheta^{[j]}\Big)\Big[A_i^{[j]}x(k)+B_i^{[j]}\omega(k)+E_i^{[j]}x(k)\varpi(k)\Big], \tag{8.2a}$$

$$y(k) = \sum_{j=1}^{N}\rho_j\sum_{i=1}^{r}h_i^{[j]}\Big(\vartheta^{[j]}\Big)\Big[C_i^{[j]}x(k)+D_i^{[j]}\omega(k)\Big], \tag{8.2b}$$

where $h_i^{[j]}\left(\vartheta^{[j]}\right) = \mathcal{M}_i^{[j]}\left(\vartheta^{[j]}\right) / \sum_{i=1}^{r} \mathcal{M}_i^{[j]}\left(\vartheta^{[j]}\right), \mathcal{M}_i^{[j]}\left(\vartheta^{[j]}\right) = \prod_{l=1}^{p} \mathcal{M}_{il}^{[j]}\left(\vartheta_l^{[j]}\right)$, with $\mathcal{M}_{il}^{[j]}\left(\vartheta_l^{[j]}\right)$ being the grade of membership of $\vartheta_l^{[j]}$ in $\mathcal{M}_{il}^{[j]}$. It is assumed that $\mathcal{M}_i^{[j]}\left(\vartheta^{[j]}\right) \geqslant 0$, $i = 1, 2, \ldots, r$, $\sum_{i=1}^{r} \mathcal{M}_i^{[j]}\left(\vartheta^{[j]}\right) > 0$ for all k, then $h_i^{[j]}\left(\vartheta^{[j]}\right) \geqslant 0$ for $i = 1, 2, \ldots, r$ and $\sum_{i=1}^{r} h_i^{[j]}\left(\vartheta^{[j]}\right) = 1$ for all k.

In this chapter, for the hybrid switched system in (8.2), the original high-order system is approximated by the reduced-order model as follows:

$$\tilde{x}(k+1) = \sum_{j=1}^{N} \rho_j \left[\tilde{A}_r^{[j]} \tilde{x}(k) + \tilde{B}_r^{[j]} \omega(k) + \tilde{E}_r^{[j]} x(k) \varpi(k) \right], \tag{8.3a}$$

$$\tilde{y}(k) = \sum_{j=1}^{N} \rho_j \left[\tilde{C}_r^{[j]} \tilde{x}(k) + \tilde{D}_r^{[j]} \omega(k) \right], \tag{8.3b}$$

where $\tilde{x}(k) \in \mathbb{R}^k$ denotes the state vector of designed reduced-order model, and $k < n$; $\tilde{y}(k) \in \mathbb{R}^p$ represents the output of designed reduced-order model; $\tilde{A}_r^{[j]}$, $\tilde{B}_r^{[j]}$, $\tilde{C}_r^{[j]}$, $\tilde{D}_r^{[j]}$ and $\tilde{E}_r^{[j]}$ are matrices with appropriate dimensions to be decided lately.

Introduce $\breve{x}(k) = \begin{bmatrix} x(k) \\ \tilde{x}(k) \end{bmatrix}$, $e_r(k) \triangleq y(k) - \tilde{y}(k)$, the resulting dynamic error system can be reformulated as

$$\breve{x}(k+1) = \sum_{j=1}^{N} \rho_j \sum_{i=1}^{r} h_i^{[j]}\left(\vartheta^{[j]}\right) \left[\breve{A}_i^{[j]} \breve{x}(k) + \breve{B}_i^{[j]} \omega(k) + \breve{E}_i^{[j]} \breve{x}(k) \varpi(k) \right], \tag{8.4a}$$

$$e_r(k) = \sum_{j=1}^{N} \rho_j \sum_{i=1}^{r} h_i^{[j]}\left(\vartheta^{[j]}\right) \left[\breve{C}_i^{[j]} \breve{x}(k) + \breve{D}_i^{[j]} \omega(k) \right], \tag{8.4b}$$

where

$$\begin{cases} \breve{A}_i^{[j]} \triangleq \begin{bmatrix} A_i^{[j]} & 0 \\ 0 & \tilde{A}_r^{[j]} \end{bmatrix}, \ \breve{B}_i^{[j]} \triangleq \begin{bmatrix} B_i^{[j]} \\ \tilde{B}_r^{[j]} \end{bmatrix}, \ \breve{E}_i^{[j]} \triangleq \begin{bmatrix} E_i^{[j]} & 0 \\ 0 & \tilde{E}_r^{[j]} \end{bmatrix}, \\ \breve{C}_i^{[j]} \triangleq \begin{bmatrix} C_i^{[j]} & -\tilde{C}_r^{[j]} \end{bmatrix}, \ \breve{D}_i^{[j]} \triangleq D_i^{[j]} - \tilde{D}_r^{[j]}. \end{cases} \tag{8.5}$$

Definition 8.1 The overall dynamic error system in (8.4) with $\omega(k) = 0$ is called as mean-square exponentially stable under $\rho_j(k)$ if $\breve{x}(k)$ satisfies the following condition:

$$\mathbf{E}\Big\{ \|\breve{x}(k)\| \Big\} \leqslant \eta \, \|\breve{x}(k_0)\| \, \varrho^{(k-k_0)}, \quad \forall k \geqslant k_0,$$

for $\eta \geqslant 1$ and $0 < \varrho < 1$.

Definition 8.2 For $\gamma > 0$ and $0 < \beta < 1$, the dynamic error system in (8.4) is referred to as owing a given $\mathcal{H}_\infty$ performance index (γ, β) if it is mean-square exponentially stable with $\omega(k) = 0$, and when $x(k) = 0$, the following condition holds for all nonzero $\omega(k) \in \ell_2[0, \infty)$:

$$\mathbf{E}\left\{\sum_{s=k_0}^{\infty} \beta^s e_r^T(s) e_r(s)\right\} < \gamma^2 \sum_{s=k_0}^{\infty} \omega^T(s)\omega(s). \tag{8.6}$$

8.3 Main Results

8.3.1 Pre-specified Performance Analysis

Firstly, sufficient conditions of the prescribed $\mathcal{H}_\infty$ performance for the concerned error system (8.4) are given.

Theorem 8.3 *Given scalars $0 < \beta < 1$, $\gamma > 0$ and $\sigma \geqslant 1$, if there exists matrix $P^{[j]} \in \mathbb{R}^{(n+k)\times(n+k)}$ and $P^{[j]} > 0$ such that for $j \in \mathcal{N}$, $i = 1, 2, \ldots, r$,*

$$\Xi_i^{[j]} \triangleq \begin{bmatrix} -\beta P^{[j]} & 0 & \left(\breve{A}_i^{[j]}\right)^T & \left(\breve{C}_i^{[j]}\right)^T & \left(\breve{E}_i^{[j]}\right)^T \\ \star & -\gamma^2 I & \left(\breve{B}_i^{[j]}\right)^T & \left(\breve{D}_i^{[j]}\right)^T & 0 \\ \star & \star & -\left(P^{[j]}\right)^{-1} & 0 & 0 \\ \star & \star & \star & -I & 0 \\ \star & \star & \star & \star & -\left(\mu P^{[j]}\right)^{-1} \end{bmatrix} < 0, \tag{8.7}$$

the dynamic error system in (8.4) owns the $\mathcal{H}_\infty$ performance index (γ, β) with mean-square exponential stability for any stochastic switching signal under the ADT satisfying $T_a > T_a^\star = \frac{\ln\sigma}{\beta}$, where $\sigma \geqslant 1$ and

$$P^{[j]} \leqslant \sigma P^{[s]}, \quad \forall j, s \in \mathcal{N}. \tag{8.8}$$

Moreover, an upper bound of the state decay estimate function is given by

$$\mathbf{E}\left\{\|\breve{x}(k)\|\right\} \leqslant \eta \left\|\breve{x}(k_0)\right\| \varrho^{(k-k_0)}, \tag{8.9}$$

where

$$\varrho \triangleq \sqrt{\beta\sigma^{\frac{1}{T_a}}}, \eta \triangleq \sqrt{\frac{b}{a}}, a \triangleq \min_{\forall j\in\mathcal{N}} \lambda_{\min}\left(P^{[j]}\right), b \triangleq \max_{\forall j\in\mathcal{N}} \lambda_{\max}\left(P^{[j]}\right). \tag{8.10}$$

Proof Consider the stochastic switching signal $\rho_j(k)$ and fuzzy basis functions, it follows from (8.7) that

$$\sum_{j=1}^{N} \rho_j \sum_{i=1}^{r} h_i^{[j]}\left(\vartheta^{[j]}\right) \Xi_i^{[j]} < 0. \tag{8.11}$$

The piecewise smooth Lyapunov function is constructed as

$$V\left(\breve{x}(k), \rho_j\right) \triangleq \breve{x}^T(k)\left(\sum_{j=1}^{N} \rho_j P^{[j]}\right)\breve{x}(k), \tag{8.12}$$

where $\mathbb{R}^{(n+k)\times(n+k)} \ni P^{[j]} > 0,\ j \in \mathcal{N}$ are to be determined. For $k \in [k_l, k_{l+1})$, define

$$\begin{aligned}\mathbf{E}\left\{\triangle V\left(\breve{x}(k), \rho_j\right)\right\} &\triangleq \mathbf{E}\left\{V\left(\breve{x}(k+1), \rho_j\right) - V\left(\breve{x}(k), \rho_j\right)\right\} \\ &= \sum_{j=1}^{N} \rho_j \sum_{i=1}^{r} h_i^{[j]}\left(\vartheta^{[j]}\right) \breve{x}^T(k) \\ &\quad \left[\left(\breve{A}_i^{[j]}\right)^T P^{[j]} \breve{A}_i^{[j]} - P^{[j]} + \mu\left(\breve{E}_i^{[j]}\right)^T P^{[j]} \breve{E}_i^{[j]}\right] \breve{x}(k).\end{aligned}$$

Then we have

$$\begin{aligned}&\mathbf{E}\left\{\triangle V\left(\breve{x}(k), \rho_j\right)\right\} + \mathbf{E}\left\{(1-\beta) V\left(\breve{x}(k), \rho_j\right)\right\} \\ &= \sum_{j=1}^{N} \rho_j \sum_{i=1}^{r} h_i^{[j]}\left(\vartheta^{[j]}\right) \breve{x}^T(k) \\ &\quad \left[\left(\breve{A}_i^{[j]}\right)^T P^{[j]} \breve{A}_i^{[j]} - \beta P^{[j]} + \mu\left(\breve{E}_i^{[j]}\right)^T P^{[j]} \breve{E}_i^{[j]}\right] \breve{x}(k).\end{aligned} \tag{8.13}$$

Based on (8.11), it yields

$$\mathbf{E}\left\{\triangle V\left(\breve{x}(k), \rho_j\right) + (1-\beta) V\left(\breve{x}(k), \rho_j\right)\right\} < 0, \forall k \in [k_l, k_{l+1}), \forall j \in \mathcal{N}. \tag{8.14}$$

As for any switching signal and $k > 0$, set

$$k_0 < k_1 < \cdots < k_l < \cdots < k_N, \quad (l = 1, \ldots, N)$$

which denote the piecewise transition points of ρ_j during the space of time $(0, k)$. Due to the j_lth subsystem is activated over $k \in [k_l, k_{l+1})$. Hence, for $k \in [k_l, k_{l+1})$, it follows from (8.14) that

$$\mathbf{E}\Big\{V\Big(\breve{x}(k),\rho_j\Big)\Big\} < \beta^{k-k_l}\mathbf{E}\Big\{V\Big(\breve{x}(k_l),\rho_j(k_l)\Big)\Big\}. \tag{8.15}$$

Considering (8.8) and (8.12), we can get

$$\mathbf{E}\Big\{V\Big(\breve{x}(k_l),\rho_j(k_l)\Big)\Big\} < \sigma\mathbf{E}\Big\{V\Big(\breve{x}(k_l),\rho_j(k_{l-1})\Big)\Big\}. \tag{8.16}$$

Then it can be concluded from (8.15)–(8.16) and $N_a(k_0,k) \leqslant \frac{k-k_0}{T_a}$ that

$$\mathbf{E}\Big\{V\Big(\breve{x}(k),\rho_j\Big)\Big\} \leqslant (\beta\sigma^{\frac{1}{T_a}})^{k-k_0}\mathbf{E}\Big\{V\Big(\breve{x}(k_0),\rho_j(k_0)\Big)\Big\}. \tag{8.17}$$

From (8.12), it is workable to seek positive scalars a and b, with $a \leqslant b$, which are shown as (8.10), then

$$\mathbf{E}\Big\{V\Big(\breve{x}(k),\rho_j\Big)\Big\} \geqslant a\mathbf{E}\Big\{\|\breve{x}(k)\|^2\Big\}, \tag{8.18}$$

$$\mathbf{E}\Big\{V\Big(\breve{x}(k_0),\rho_j(k_0)\Big)\Big\} \leqslant b\|\breve{x}(k_0)\|^2. \tag{8.19}$$

Combine (8.17) and (8.19), then we have

$$\mathbf{E}\Big\{\|\breve{x}(k)\|^2\Big\} \leqslant \frac{1}{a}\mathbf{E}\Big\{V\Big(\breve{x}(k),\rho_j\Big)\Big\} \leqslant \frac{b}{a}(\beta\sigma^{\frac{1}{T_a}})^{k-k_0}\|\breve{x}(k_0)\|^2.$$

Define $\varrho \triangleq \sqrt{\beta\sigma^{\frac{1}{T_a}}}$ and we can get

$$\mathbf{E}\Big\{\|\breve{x}(k)\|\Big\} \leqslant \sqrt{\frac{b}{a}}\varrho^{k-k_0}\|\breve{x}(k_0)\|.$$

It yields from Definition 8.1 that if $0 < \varrho < 1$, that is $T_a > T_a^{\star} = ceil(-\frac{\ln\sigma}{\ln\beta})$, the resulting dynamic system in (8.4) with $\omega(k) = 0$ is mean-square exponentially stable, where the function $ceil(f)$ represents the rounding real scalar f to the nearest integer, which is equal to or greater than f.

Next, the $\mathcal{H}_\infty$ performance index (γ, β) for concerned system is analyzed then. An index is introduced as

$$\begin{aligned}\mathcal{J}(k) \triangleq \mathbf{E}\Big\{&\Delta V\Big(\breve{x}(k),\rho_j\Big) + (1-\beta)V\Big(\breve{x}(k),\rho_j\Big) + e_r^T(k)e_r(k)\\ &-\gamma^2\omega^T(k)\omega(k)\Big\}.\end{aligned}$$

Then it yields

$$\mathcal{J}(k)=\sum_{j=1}^{N}\rho_j\sum_{i=1}^{r}h_i^{[j]}\left(\vartheta^{[j]}\right)\begin{bmatrix}\breve{x}(k)\\ \omega(k)\end{bmatrix}^T\begin{bmatrix}\Xi_{11i}^{[j]} & \Xi_{12i}^{[j]}\\ \star & \Xi_{22i}^{[j]}\end{bmatrix}\begin{bmatrix}\breve{x}(k)\\ \omega(k)\end{bmatrix},$$

where

$$\Xi_{11i}^{[j]}\triangleq\left(\breve{A}_i^{[j]}\right)^T P^{[j]}\breve{A}_i^{[j]}+\left(\breve{C}_i^{[j]}\right)^T\breve{C}_i^{[j]}+\mu\left(\breve{E}_i^{[j]}\right)^T P^{[j]}\breve{E}_i^{[j]}-\beta P^{[j]},$$
$$\Xi_{12i}^{[j]}\triangleq\left(\breve{A}_i^{[j]}\right)^T P^{[j]}\breve{B}_i^{[j]}+\left(\breve{C}_i^{[j]}\right)^T\breve{D}_i^{[j]},$$
$$\Xi_{22i}^{[j]}\triangleq\left(\breve{B}_i^{[j]}\right)^T P^{[j]}\breve{B}_i^{[j]}+\left(\breve{D}_i^{[j]}\right)^T\breve{D}_i^{[j]}-\gamma^2 I.$$

On account of (8.11) and Schur's complement method, for $k\in[k_l,k_{l+1})$, it is easy to obtain $\mathcal{J}(k)<0$. Set $\Omega(k)\triangleq e_r^T(k)e_r(k)-\gamma^2\omega^T(k)\omega(k)$, then

$$\mathbf{E}\Big\{\triangle V\Big(\breve{x}(k),\rho_j\Big)\Big\}<\mathbf{E}\Big\{-(1-\beta)V\Big(\breve{x}(k),\rho_j\Big)-\Omega(k)\Big\}. \tag{8.20}$$

Hence, for $k\in[k_l,k_{l+1})$, we can obtain from (8.20) that

$$\begin{aligned}\mathbf{E}\Big\{V\Big(\breve{x}(k),\rho_j\Big)\Big\}&<\beta^{k-k_l}\mathbf{E}\Big\{V\Big(\breve{x}(k_l),\rho_j(k_l)\Big)\Big\}\\&\quad-\mathbf{E}\Big\{\sum_{s=k_l}^{k-1}\beta^{k-1-s}\Omega(s)\Big\}.\end{aligned} \tag{8.21}$$

Considering (8.16) and (8.21), it yields

$$\mathbf{E}\Big\{V\Big(\breve{x}(k),\rho_j\Big)\Big\}<\beta^{k-k_l}\mathbf{E}\Big\{V\Big(\breve{x}(k_l),\rho_j(k_l)\Big)\Big\}-\mathbf{E}\Big\{\sum_{s=k_l}^{k-1}\beta^{k-1-s}\Omega(s)\Big\},$$
$$\vdots$$
$$\begin{aligned}\mathbf{E}\Big\{V\Big(\breve{x}(k_1),\rho_j(k_1)\Big)\Big\}&<\beta^{k_1-k_0}\sigma\mathbf{E}\Big\{V\Big(\breve{x}(k_0),\rho_j(k_0)\Big)\Big\}\\&\quad-\sigma\mathbf{E}\Big\{\sum_{s=k_0}^{k_1-1}\beta^{k_1-1-s}\Omega(s)\Big\}.\end{aligned}$$

Then on the basis of the above conditions and $N_a(k_0,k)\leqslant\frac{k-k_0}{T_a}$, it has

$$\mathbf{E}\Big\{V\Big(\breve{x}(k),\rho_j\Big)\Big\} < \beta^{k-k_0}\sigma^{N_a(k_0,k)}\mathbf{E}\Big\{V\Big(\breve{x}(k_0),\rho_j(k_0)\Big)\Big\}$$
$$-\mathbf{E}\Big\{\sum_{s=k_0}^{k-1}\beta^{k-1-s}\sigma^{N_a(s,k)}\Omega(s)\Big\}. \tag{8.22}$$

It can be seen from the zero initial condition and (8.22) that

$$\mathbf{E}\Big\{\sum_{s=k_0}^{k-1}\beta^{k-1-s}\sigma^{N_a(s,k)}\Big[e_r^T(s)e_r(s)-\gamma^2\omega^T(s)\omega(s)\Big]\Big\} < 0.$$

Then we pre- and post-multiply the above inequality by $\sigma^{-N_a(0,k)}$, and we can obtain

$$\mathbf{E}\Big\{\sum_{s=k_0}^{k-1}\beta^{k-1-s}\sigma^{-N_a(0,s)}\Big[e_r^T(s)e_r(s)-\gamma^2\omega^T(s)\omega(s)\Big]\Big\} < 0. \tag{8.23}$$

Owing to $N_a(0,s)\leqslant\frac{s}{T_a}$ and $T_a > -\frac{\ln\sigma}{\ln\beta}$, it has $N_a(0,s)\leqslant -s\frac{\ln\beta}{\ln\sigma}$. Therefore, (8.23) signifies

$$\mathbf{E}\Big\{\sum_{s=k_0}^{k-1}\beta^{k-1-s}\sigma^{s\frac{\ln\beta}{\ln\sigma}}e_r^T(s)e_r(s)\Big\} < \gamma^2\mathbf{E}\Big\{\sum_{s=k_0}^{k-1}\beta^{k-1-s}\omega^T(s)\omega(s)\Big\},$$

which follows that

$$\mathbf{E}\Big\{\sum_{s=k_0}^{\infty}\beta^s e_r^T(s)e_r(s)\Big\} < \mathbf{E}\Big\{\sum_{s=k_0}^{\infty}\gamma^2\omega^T(s)\omega(s)\Big\}.$$

On the basis of Definition 8.2, it can be concluded that the overall hybrid switched system in (8.4) is mean-square exponentially stable with a weighted $\mathcal{H}_\infty$ performance level (γ,β). Then it completes the proof.

8.3.2 *Model Approximation by Projection Technique*

The solution of the reduced-order approximation issue is proposed via an efficient projection approach.

Theorem 8.4 *Given scalars* $0<\beta<1$, $\gamma>0$ *and* $\sigma\geqslant 1$, *if there are matrices* $0<P^{[j]}\in\mathbb{R}^{(n+k)\times(n+k)}$ *and* $0<\mathscr{P}^{[j]}\in\mathbb{R}^{(n+k)\times(n+k)}$ *such that for* $j,s\in\mathcal{N}$, $i=1,2,\ldots,r$,

$$\bar{\Xi}_i^{[j]} \triangleq \begin{bmatrix} -\beta P^{[j]} & 0 & \left(\breve{A}_{i0}^{[j]}\right)^T \mathscr{H}^T & \left(\breve{E}_{i0}^{[j]}\right)^T \mathscr{H}^T \\ \star & -\gamma^2 I & \left(\breve{B}_{i0}^{[j]}\right)^T \mathscr{H}^T & 0 \\ \star & \star & -\mathscr{H}\mathscr{P}^{[j]}\mathscr{H}^T & 0 \\ \star & \star & \star & -\frac{1}{\mu}\mathscr{H}\mathscr{P}\mathscr{H}^T \end{bmatrix} < 0, \tag{8.24a}$$

$$\hat{\Xi}_i^{[j]} \triangleq \begin{bmatrix} -\beta\mathscr{H}P^{[j]}\mathscr{H}^T & \mathscr{H}\left(\breve{A}_{i0}^{[j]}\right)^T & \mathscr{H}\left(\breve{C}_{i0}^{[j]}\right)^T & \mathscr{H}\left(\breve{E}_{i0}^{[j]}\right)^T \\ \star & -\mathscr{P}^{[j]} & 0 & 0 \\ \star & \star & -I & 0 \\ \star & \star & \star & -\frac{1}{\mu}\mathscr{P}^{[j]} \end{bmatrix} < 0, \tag{8.24b}$$

$$\sigma P^{[s]} \geq P^{[j]}, \tag{8.24c}$$

$$P^{[j]}\mathscr{P}^{[j]} = I, \tag{8.24d}$$

then the dynamic error system in (8.4) owns the $\mathcal{H}_\infty$ performance index (γ, β) with mean-square exponential stability. Furthermore, the parameters of designed reduced-order model in (8.3) are described as

$$\begin{cases} \mathscr{G}^{[j]} \triangleq \left[\begin{array}{c|c} \tilde{D}_r^{[j]} & \tilde{C}_r^{[j]} \\ \hline \tilde{B}_r^{[j]} & \tilde{A}_r^{[j]} \\ \hline & \tilde{E}_r^{[j]} \end{array}\right] = -\Pi^{-1}U^T\Lambda^{[j]}V^T\left(V\Lambda^{[j]}V^T\right)^{-1} \\ \qquad +\Pi^{-1}\left(\Xi^{[j]}\right)^{\frac{1}{2}} L\left(V\Lambda^{[j]}V^T\right)^{-1/2}, \\ \Lambda^{[j]} = \left(U\Pi^{-1}U^T - W^{[j]}\right)^{-1} > 0, \\ \Xi^{[j]} = \Pi - U^T\left[\Lambda^{[j]} - \Lambda^{[j]}V^T\left(V\Lambda^{[j]}V^T\right)^{-1}V\Lambda^{[j]}\right]U > 0, \end{cases} \tag{8.25}$$

where $\Pi > 0$ and $\|L\| < 1$ are any matrices with proper dimensions, and

$$\begin{cases} W^{[j]} \triangleq \begin{bmatrix} -\beta P^{[j]} & 0 & \left(\breve{A}_{i0}^{[j]}\right)^T & \left(\breve{C}_{i0}^{[j]}\right)^T & \left(\breve{E}_{i0}^{[j]}\right)^T \\ \star & -\gamma^2 I & \left(\breve{B}_{i0}^{[j]}\right)^T & \left(\breve{D}_{i0}^{[j]}\right)^T & 0 \\ \star & \star & -\left(P^{[j]}\right)^{-1} & 0 & 0 \\ \star & \star & \star & -I & 0 \\ \star & \star & \star & \star & -\left(\mu P^{[j]}\right)^{-1} \end{bmatrix}, \\ U \triangleq \begin{bmatrix} 0_{(n+k)\times(p+2k)} \\ 0_{m\times(p+2k)} \\ \mathscr{X}_1 \\ \mathscr{X}_2 \\ \mathscr{X}_3 \end{bmatrix}, \quad \begin{matrix} \mathscr{X}_1 \triangleq \begin{bmatrix} 0_{n\times p} & 0_{n\times k} & 0_{n\times k} \\ 0_{k\times p} & I_{k\times k} & 0_{k\times k} \end{bmatrix}, \\ \mathscr{X}_3 \triangleq \begin{bmatrix} 0_{n\times p} & 0_{n\times k} & 0_{n\times k} \\ 0_{k\times p} & 0_{k\times k} & I_{k\times k} \end{bmatrix}, \end{matrix} \\ V \triangleq \begin{bmatrix} \mathscr{Y}_1 & \mathscr{Y}_2 & 0_{(m+k)\times(n+k)} & 0_{(m+k)\times p} & 0_{(m+k)\times(n+k)} \end{bmatrix}, \\ \mathscr{Y}_1 \triangleq \begin{bmatrix} 0_{m\times n} & 0_{m\times k} \\ 0_{k\times n} & I_{k\times k} \end{bmatrix}, \mathscr{Y}_2 \triangleq \begin{bmatrix} I_{m\times m} \\ 0_{k\times m} \end{bmatrix}, \breve{C}_{i0}^{[j]} \triangleq \begin{bmatrix} C_i^{[j]} & 0_{p\times k} \end{bmatrix}, \\ \mathscr{X}_2 \triangleq \begin{bmatrix} -I_{p\times p} & 0_{p\times k} & 0_{p\times k} \end{bmatrix}, \mathscr{H} \triangleq \begin{bmatrix} I_{n\times n} & 0_{n\times k} \end{bmatrix}, \breve{D}_{i0}^{[j]} \triangleq D_i^{[j]}, \\ \breve{A}_{i0}^{[j]} \triangleq \begin{bmatrix} A_i^{[j]} & 0_{n\times k} \\ 0_{k\times n} & 0_{k\times k} \end{bmatrix}, \tilde{B}_{i0}^{[j]} \triangleq \begin{bmatrix} B_i^{[j]} \\ 0_{k\times m} \end{bmatrix}, \breve{E}_{i0}^{[j]} \triangleq \begin{bmatrix} E_i^{[j]} & 0_{n\times k} \\ 0_{k\times n} & 0_{k\times k} \end{bmatrix}. \end{cases} \tag{8.26}$$

Proof $\breve{A}_i^{[j]}$, $\breve{B}_i^{[j]}$, $\breve{C}_i^{[j]}$, $\breve{D}_i^{[j]}$ and $\breve{E}_i^{[j]}$ can be rewritten as

$$\begin{cases} \breve{A}_i^{[j]} \triangleq \breve{A}_{i0}^{[j]} + \mathscr{X}_1\mathscr{G}^{[j]}\mathscr{Y}_1, & \breve{B}_i^{[j]} \triangleq \breve{B}_{i0}^{[j]} + \mathscr{X}_1\mathscr{G}^{[j]}\mathscr{Y}_2, \\ \breve{C}_i^{[j]} \triangleq \breve{C}_{i0}^{[j]} + \mathscr{X}_2\mathscr{G}^{[j]}\mathscr{Y}_1, & \breve{D}_i^{[j]} \triangleq \breve{D}_{i0}^{[j]} + \mathscr{X}_2\mathscr{G}^{[j]}\mathscr{Y}_2, \\ \breve{E}_i^{[j]} \triangleq \breve{E}_{i0}^{[j]} + \mathscr{X}_3\mathscr{G}^{[j]}\mathscr{Y}_1, \end{cases} \tag{8.27}$$

where $\mathscr{G}^{[j]}$, $\breve{A}_{i0}^{[j]}$, $\breve{B}_{i0}^{[j]}$, $\breve{C}_{i0}^{[j]}$, $\breve{D}_{i0}^{[j]}$, $\breve{E}_{i0}^{[j]}$, $\mathscr{X}_1$, $\mathscr{X}_2$, $\mathscr{X}_3$, $\mathscr{Y}_1$ and $\mathscr{Y}_2$ are given in (8.25) and (8.26). Based on (8.27), the condition (8.7) in Theorem 8.3 can be converted into

$$W^{[j]} + U\mathscr{G}^{[j]}V + \left(U\mathscr{G}^{[j]}V\right)^T < 0, \tag{8.28}$$

where the notations of $W^{[j]}$, U and V are shown in (8.26). Define

$$V^{T\perp} \triangleq \begin{bmatrix} \mathscr{H} & 0 & 0 & 0 & 0 \\ 0 & 0 & I & 0 & 0 \\ 0 & 0 & 0 & I & 0 \\ 0 & 0 & 0 & 0 & I \end{bmatrix}, \quad U^{\perp} \triangleq \begin{bmatrix} I & 0 & 0 & 0 & 0 \\ 0 & I & 0 & 0 & 0 \\ 0 & 0 & \mathscr{H} & 0 & 0 \\ 0 & 0 & 0 & 0 & \mathscr{H} \end{bmatrix},$$

where $\mathscr{H}$ is given in (8.26). And employing the projection method, the condition (8.28) is feasible for $\mathscr{G}^{[j]}$ if and only if

$$U^{\perp}WU^{T\perp} < 0, V^{T\perp}WV^{\perp} < 0,$$

and it can be reformulated as

$$\begin{bmatrix} -\beta P^{[j]} & 0 & \left(\breve{A}_{i0}^{[j]}\right)^T \mathscr{H}^T & \left(\breve{E}_{i0}^{[j]}\right)^T \mathscr{H}^T \\ \star & -\gamma^2 I & \left(\breve{B}_{i0}^{[j]}\right)^T \mathscr{H}^T & 0 \\ \star & \star & -\mathscr{H}\left(P^{[j]}\right)^{-1}\mathscr{H}^T & 0 \\ \star & \star & \star & -\mathscr{H}\left(\mu P^{[j]}\right)^{-1}\mathscr{H}^T \end{bmatrix} < 0, \tag{8.29a}$$

$$\begin{bmatrix} -\beta \mathscr{H} P^{[j]} \mathscr{H}^T & \mathscr{H}\left(\breve{A}_{i0}^{[j]}\right)^T & \mathscr{H}\left(\tilde{C}_{i0}^{[j]}\right)^T & \mathscr{H}\left(\breve{E}_{i0}^{[j]}\right)^T \\ \star & -\left(P^{[j]}\right)^{-1} & 0 & 0 \\ \star & \star & -I & 0 \\ \star & \star & \star & -\left(\mu P^{[j]}\right)^{-1} \end{bmatrix} < 0. \tag{8.29b}$$

Denoting $\mathscr{P}^{[j]} \triangleq \left(P^{[j]}\right)^{-1}$, it follows from (8.29a)–(8.29b) that (8.24a)–(8.24b) hold. Besides, when the conditions in (8.24) satisfy, the model parametrization in (8.25) has a feasible solution, which can be solved with the projection strategy. Hence, the proof is completed.

Remark 8.5 Sufficient conditions in Theorem 8.4 are not all LMIs due to (8.24d). And to settle this issue, the CCL method is utilized.

In addition, solutions of the relevant non-convex feasible issue can be obtained by solving the following sequential optimization issue.

Reduced-Order Model Approximation Problem:

$$\begin{aligned} \min \quad & \text{trace}\left(\sum_{j\in\mathcal{N}} P^{[j]}\mathscr{P}^{[j]}\right) \\ \text{subject to} \quad & (8.24\text{a})-(8.24\text{c}), \quad \text{and for } j \in \mathcal{N}, \\ & \begin{bmatrix} P^{[j]} & I \\ I & \mathscr{P}^{[j]} \end{bmatrix} \geqslant 0. \end{aligned} \tag{8.30}$$

Remark 8.6 It is noticed that the simulation results in Theorem 8.4 are feasible, provided that the problem of $\min trace\ (\sum_{j\in\mathcal{N}} P^{[j]}\mathscr{P}^{[j]}) = 2Nk$ can be solved accordingly. The iteration method is applied in the subsequent example to resolve the concerned reduced-order model approximation problem. The iterations of the proposed solutions will be terminated, when the conditions in (8.24a)–(8.24c) are satisfied with a specified performance index, or it achieves the maximum quantity of iterations.

8.4 Illustrative Example

Example 8.7 Introduce the hybrid stochastic switched systems in (8.1) with $\mathcal{N} = 2$, and the relevant matrices are chosen as follows:

Subsystem 1.

$$
A_1^{[1]} = \begin{bmatrix} 0.57 & 0.45 & -0.42 & 0.86 \\ 0.48 & -0.32 & 0.54 & -0.26 \\ -0.34 & -0.22 & -0.38 & -0.40 \\ -0.42 & 0.60 & 0.36 & 0.46 \end{bmatrix}, \quad B_1^{[1]} = \begin{bmatrix} 0.70 \\ -0.42 \\ -0.42 \\ 0.36 \end{bmatrix},
$$

$$
A_2^{[1]} = \begin{bmatrix} 0.76 & 0.30 & -0.42 & 0.82 \\ 0.44 & -0.32 & 0.56 & -0.48 \\ -0.32 & -0.28 & -0.28 & -0.36 \\ -0.42 & 0.60 & 0.34 & 0.48 \end{bmatrix}, \quad B_2^{[1]} = \begin{bmatrix} 0.62 \\ -0.76 \\ -0.42 \\ 0.34 \end{bmatrix},
$$

$$
E_1^{[1]} = \begin{bmatrix} 0.07 & 0.04 & 0.02 & 0.05 \\ 0.02 & 0.06 & 0.08 & 0.10 \\ 0.04 & 0.02 & 0.06 & 0.02 \\ 0.02 & 0.08 & 0.02 & 0.04 \end{bmatrix}, \quad C_1^{[1]} = \begin{bmatrix} 1.25 \\ 0.62 \\ 1.32 \\ 0.62 \end{bmatrix}^T,
$$

$$
E_2^{[1]} = \begin{bmatrix} 0.07 & 0.04 & 0.02 & 0.05 \\ 0.08 & 0.06 & 0.04 & 0.10 \\ 0.04 & 0.02 & 0.06 & 0.08 \\ 0.01 & 0.02 & 0.01 & 0.02 \end{bmatrix}, \quad C_2^{[1]} = \begin{bmatrix} 1.15 \\ 0.66 \\ 1.36 \\ 0.64 \end{bmatrix}^T,
$$

$$
D_1^{[1]} = 1.90, \quad D_2^{[1]} = 1.50. \tag{8.31}
$$

Subsystem 2.

$$
A_1^{[2]} = \begin{bmatrix} 0.56 & 0.30 & -0.42 & 0.91 \\ 0.46 & -0.32 & 0.58 & -0.57 \\ -0.32 & -0.22 & -0.26 & -0.40 \\ -0.42 & 0.68 & 0.30 & 0.46 \end{bmatrix}, \quad B_1^{[2]} = \begin{bmatrix} 0.64 \\ -0.46 \\ -0.48 \\ 0.32 \end{bmatrix},
$$

$$
A_2^{[2]} = \begin{bmatrix} 0.52 & 0.12 & -0.08 & 0.76 \\ 0.48 & -0.32 & 0.54 & -0.24 \\ -0.30 & -0.22 & -0.24 & -0.28 \\ -0.42 & 0.60 & 0.36 & 0.42 \end{bmatrix}, \quad B_2^{[2]} = \begin{bmatrix} 0.70 \\ -0.68 \\ -0.40 \\ 0.38 \end{bmatrix},
$$

$$
E_1^{[2]} = \begin{bmatrix} 0.07 & 0.04 & 0.02 & 0.05 \\ 0.08 & 0.06 & 0.04 & 0.10 \\ 0.04 & 0.02 & 0.06 & 0.05 \\ 0.02 & 0.04 & 0.02 & 0.05 \end{bmatrix}, \quad C_1^{[2]} = \begin{bmatrix} 1.15 \\ 0.52 \\ 1.34 \\ 0.64 \end{bmatrix}^T,
$$

$$
E_2^{[2]} = \begin{bmatrix} 0.07 & 0.04 & 0.02 & 0.06 \\ 0.02 & 0.06 & 0.08 & 0.10 \\ 0.04 & 0.08 & 0.06 & 0.02 \\ 0.02 & 0.04 & 0.02 & 0.08 \end{bmatrix}, \quad C_2^{[2]} = \begin{bmatrix} 1.16 \\ 0.54 \\ 1.30 \\ 0.68 \end{bmatrix}^T,
$$

$$
D_1^{[2]} = 1.90, \quad D_2^{[2]} = 1.50, \tag{8.32}
$$

and $\beta = 0.8$. Let $\sigma = 1.02$, it follows from Theorem 8.3 that the overall system is mean-square exponentially stable.

In this note, we concentrate on designing reduced-order models in (8.3), that is, Case 1: $k = 1$; Case 2: $k = 2$; Case 3: $k = 3$, which can facilitate the original high-order system with prescribed $\mathcal{H}_\infty$ performance. Based on CCL approach, resolving the sequential optimization issue in Theorem 8.4, then the parameters of designed reduced-order models are calculated as below.

Case 1. When $k = 1$, the minimized system performance index is obtained as 0.1391 and

$$\mathscr{G}^{[1]} = \left[\begin{array}{c|c} 1.5974 & -0.6204 \\ \hline -1.0057 & 0.4770 \\ \hline & 0.0556 \end{array}\right], \tag{8.33}$$

$$\mathscr{G}^{[2]} = \left[\begin{array}{c|c} 1.4117 & -0.6318 \\ \hline -1.2455 & -0.2945 \\ \hline & 0.0015 \end{array}\right]. \tag{8.34}$$

Case 2. When $k = 2$, the minimized system performance index is obtained as 0.1166 and

$$\mathscr{G}^{[1]} = \left[\begin{array}{c|cc} 1.4902 & -0.5756 & -0.4607 \\ \hline -0.5435 & 0.2649 & 0.0866 \\ -0.6248 & 0.3817 & 0.3742 \\ \hline & 0.0987 & 0.0549 \\ & 0.0313 & 0.0270 \end{array}\right], \tag{8.35}$$

$$\mathscr{G}^{[2]} = \left[\begin{array}{c|cc} 1.5044 & -0.5850 & -0.5215 \\ \hline -0.7510 & 0.0602 & -0.5025 \\ -0.4039 & 0.1989 & 0.1270 \\ \hline & 0.0576 & -0.0283 \\ & 0.0234 & 0.0202 \end{array}\right]. \tag{8.36}$$

Case 3. When $k = 3$, the minimized system performance index is obtained as 0.0836 and

$$\left\{\begin{array}{l} \mathscr{G}^{[1]}=\left[\begin{array}{c|ccc} 1.4711 & -0.5484 & -0.4326 & -0.1501 \\ \hline -0.5445 & 0.2717 & 0.0053 & -0.0262 \\ -0.5441 & 0.4005 & 0.3705 & 0.1318 \\ -0.2047 & 0.1658 & 0.1751 & 0.0758 \\ \hline & 0.1070 & 0.0493 & 0.0122 \\ & 0.0236 & 0.0229 & 0.0085 \\ & -0.0003 & 0.0057 & 0.0028 \end{array}\right], \\ \mathscr{G}^{[2]}=\left[\begin{array}{c|ccc} 1.5189 & -0.5718 & -0.4762 & -0.1687 \\ \hline -0.7023 & 0.1416 & -0.5328 & -0.2735 \\ -0.3596 & 0.2357 & 0.1545 & 0.0449 \\ -0.0977 & 0.1000 & 0.1292 & 0.0607 \\ \hline & 0.0708 & -0.0287 & -0.0217 \\ & 0.0193 & 0.0207 & 0.0080 \\ & 0.0016 & 0.0132 & 0.0062 \end{array}\right]. \end{array}\right. \tag{8.37}$$

To demonstrate the validity of the $\mathcal{H}_\infty$ performance of concerned hybrid switched systems, let $\breve{x}(0) = 0$ ($x(0) = 0$, $\tilde{x}(0) = 0$), and the fuzzy basis functions are given by

$$\left\{\begin{array}{l} h_1^{[j]}\left(x_2^{[j]}(k)\right) \triangleq \frac{\left[1-\sin\left(x_2(k)\right)\right]}{2}, \\ h_2^{[j]}\left(x_2^{[j]}(k)\right) \triangleq \frac{\left[1+\sin\left(x_2(k)\right)\right]}{2}. \end{array}\right.$$

The input $\omega(\bullet)$ is set to $\omega(k) = \exp(-0.1k)\sin(0.9k)$, $k > 0$.

Figure 8.1 plots the hybrid stochastic switching signal, which is activated arbitrarily under various modes. The values of vertical axis, '1' and '2', denote the first and second stochastic submode, separately. It is observed from Fig. 8.1 that $T_a \geqslant \frac{\ln\sigma}{\beta} = \frac{\ln 1.02}{0.8} = 0.0248$. Figure 8.2 depicts the outputs of original system (8.31)–(8.32), the third-dimension model (8.37), the second-dimension model (8.36) and the first-dimension model (8.34). Moreover, the output errors between the original system and the concerned reduced-order models are shown in Fig. 8.3.

8.5 Conclusion

This chapter aimed on the reduced-order model approximation issue subject to a prescribed system performance sense for nonlinear hybrid stochastic switched systems through T-S fuzzy modelling strategy by using the following steps: (1) Designed a reduced-order model to approximate the given high-order hybrid switched system; (2) Obtained the resulting dynamic error system by using the piecewise blending Lyapunov function and ADT method; the system was ensured to be mean-square exponentially stable with a particular performance index. In addition, the feasible

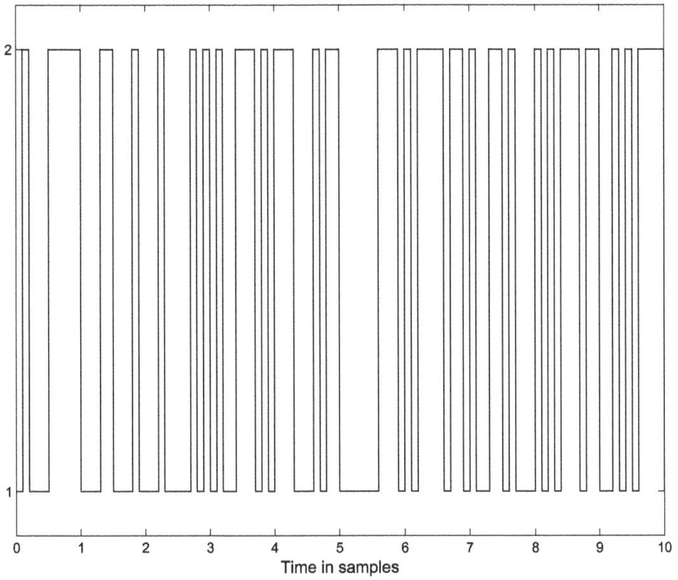

Fig. 8.1 Stochastic switching signal with the average dwell time $T_a \geqslant 0.1$

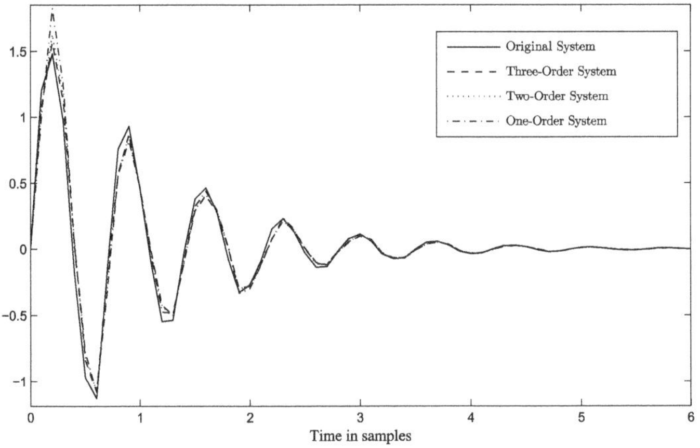

Fig. 8.2 Outputs of the original nonlinear hybrid switched system and the reduced-order hybrid switched models

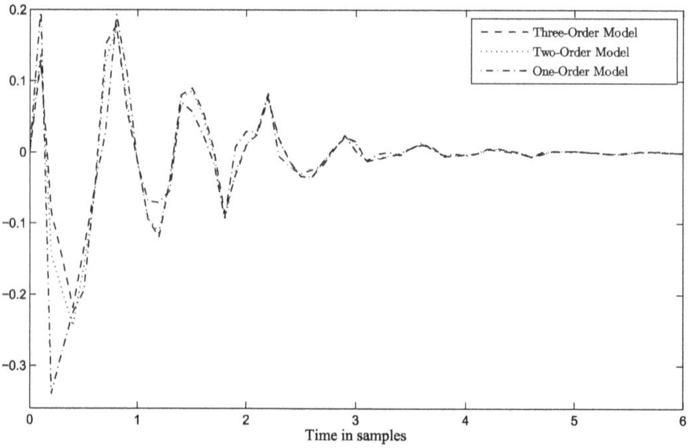

Fig. 8.3 Output errors between the original nonlinear hybrid switched system and the reduced-order hybrid switched models

conditions and corresponding parameters of designed reduced-order models were set based on the projection approach and CCL algorithm by solving the sequential minimization problem. In the end, the simulation results have verified the validity of our proposed model reduction scheme.

Chapter 9
Model Reduction of Time-Varying Delay Fuzzy Systems

9.1 Introduction

This chapter presents a novel solution for the model approximation problem for dynamic time-varying delay systems based on fuzzy modelling. We attempt to establish a reduced-order model, which can approximate the original high-order model under a specific system performance index and ensure the asymptotic stability for the overall system. A less conservative stability condition for the overall error system with a specific performance level is derived using the reciprocally convex strategy. The reduced-order model can be ultimately established via the projection strategy, which transfers the model approximation work to a sequential minimization issue under LMI constraints based on the CCL algorithms.

9.2 System Description and Preliminaries

Consider the nonlinear system, which is described as the following continuous-time T-S fuzzy model:
♦ **Plant Form**:

Rule i: IF $\theta_1(t)$ is μ_{i1} and $\theta_2(t)$ is μ_{i2} and $\cdots$ and $\theta_p(t)$ is μ_{ip}, THEN

$$\begin{cases} \dot{x}(t) = A_i x(t) + A_{di} x(t - d(t)) + B_i \omega(t), \\ y(t) = C_i x(t) + C_{di} x(t - d(t)) + D_i \omega(t), \\ \quad i = 1, 2, \ldots, r, \end{cases} \tag{9.1}$$

where $x(t) \in \mathbb{R}^n$ denotes the state vector, $\omega(t) \in \mathbb{R}^m$ denotes the disturbance input belonging to $\mathcal{L}_2[0, \infty)$, $y(t) \in \mathbb{R}^p$ represents the output, $d(t)$ represents the time-varying delay with $0 < d_1 \leqslant d(t) \leqslant d_2 < \infty, \dot{d}(t) \leqslant \tau$, where d_1, d_2 and τ stand for

X. Su et al., *Intelligent Control, Filtering and Model Reduction Analysis for Fuzzy-Model-Based Systems*, Studies in Systems, Decision and Control 385,
https://doi.org/10.1007/978-3-030-81214-0_9

real constant scalars. A_i, A_{di}, B_i, C_i, C_{di}, D_i are given matrices of suitable dimensions, r denotes the quantity of IF-THEN fuzzy rules, $\theta_1(t), \theta_2(t), \ldots, \theta_p(t)$ represent the premise variables, and $\mu_{i1}, \ldots, \mu_{ip}$ denote the fuzzy sets.

The complete T-S fuzzy systems with time delay in (9.1) can be expressed as

$$\begin{cases} \dot{x}(t) = \sum_{i=1}^{r} h_i(\theta(t)) \left[A_i x(t) + A_{di} x(t-d(t)) + B_i \omega(t)\right], \\ y(t) = \sum_{i=1}^{r} h_i(\theta(t)) [C_i x(t) + C_{di} x(t-d(t)) + D_i \omega(t)], \end{cases} \tag{9.2}$$

where

$$h_i(\theta(t)) = \frac{\nu_i(\theta(t))}{\sum_{i=1}^{r} \nu_i(\theta(t))}, \quad \nu_i(\theta(t)) = \prod_{j=1}^{p} \mu_{ij}(\theta_j(t)),$$

where $\mu_{ij}(\theta_j(t))$ denotes the grade of membership of $\theta_j(t)$ in μ_{ij}. For all t, it is easy to obtain $\nu_i(\theta(t)) \geqslant 0$, i = 1, 2, ..., r, $\sum_{i=1}^{r} h_i(\theta(t)) = 1$.

Owing to the difficulty on analysis and synthesis of high-order mathematic models with common methods. In this work, we focus on the approximation of the original system (9.2) by a reduced-order model, which is described as

$$\begin{cases} \dot{\hat{x}}(t) = \hat{A}\hat{x}(t) + \hat{A}_d \hat{x}(t-d(t)) + \hat{B}\omega(t), \\ \hat{y}(t) = \hat{C}\hat{x}(t) + \hat{C}_d \hat{x}(t-d(t)) + \hat{D}\omega(t), \end{cases} \tag{9.3}$$

where $\hat{y}(t) \in \mathbb{R}^p$ stands for the output, $\hat{x}(t) \in \mathbb{R}^k$ stands for the state of the reduced-order model (9.3) with $k < n$, $\hat{A}$, $\hat{A}_d$, $\hat{B}$, $\hat{C}$, $\hat{C}_d$ and $\hat{D}$ are properly dimensioned system matrices which are to be decided.

Combining the system (9.2) and proposed reduced-order model (9.3), we can get the overall dynamic system as below:

$$\begin{cases} \dot{\xi}(t) = \sum_{i=1}^{r} h_i(\theta(t)) [\tilde{A}_i \xi(t) + \tilde{A}_{di} \xi(t-d(t)) + \tilde{B}_i \omega(t)], \\ e(t) = \sum_{i=1}^{r} h_i(\theta(t)) [\tilde{C}_i \xi(t) + \tilde{C}_{di} \xi(t-d(t)) + \tilde{D}_i \omega(t)], \end{cases} \tag{9.4}$$

where $\xi(t) \triangleq \begin{bmatrix} x(t) \\ \hat{x}(t) \end{bmatrix}$, $e(t) \triangleq y(t) - \hat{y}(t)$ and

$$\begin{cases} \tilde{A}_i \triangleq \begin{bmatrix} A_i & 0 \\ 0 & \hat{A} \end{bmatrix}, \ \tilde{A}_{di} \triangleq \begin{bmatrix} A_{di} & 0 \\ 0 & \hat{A}_d \end{bmatrix}, \ \tilde{B}_i \triangleq \begin{bmatrix} B_i \\ \hat{B} \end{bmatrix}, \\ \tilde{C}_i \triangleq [\, C_i \ -\hat{C} \,], \ \tilde{C}_{di} \triangleq [\, C_{di} \ -\hat{C}_d \,], \ \tilde{D}_i \triangleq (D_i - \hat{D}). \end{cases} \tag{9.5}$$

The block diagram of model approximation for T-S fuzzy system with time-varying delay is shown in Fig. 9.1.

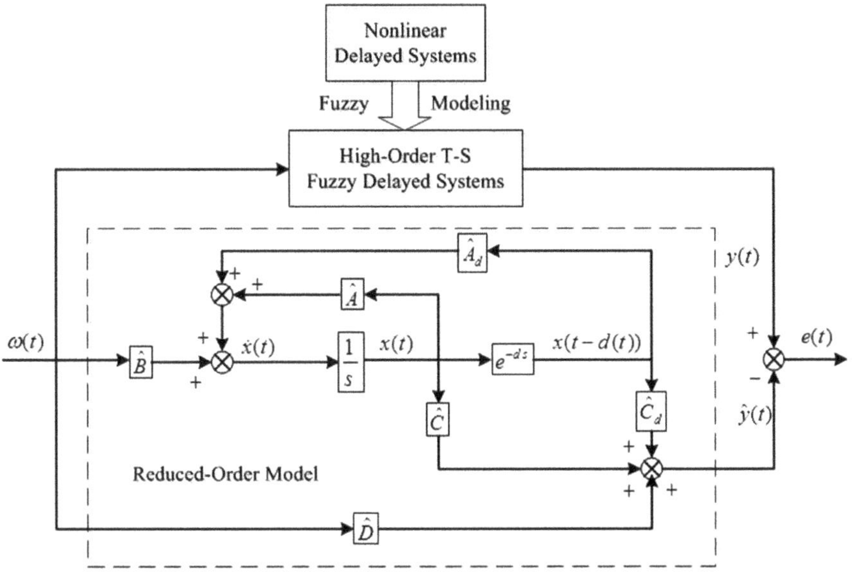

Fig. 9.1 Block diagram of model approximation for T-S fuzzy delayed systems

For the sake of simplicity, define

$$\begin{cases} \tilde{A}(t) \triangleq \sum_{i=1}^{r} h_i(\theta(t))\tilde{A}_i, & \tilde{A}_d(t) \triangleq \sum_{i=1}^{r} h_i(\theta(t))\tilde{A}_{di}, \\ \tilde{C}(t) \triangleq \sum_{i=1}^{r} h_i(\theta(t))\tilde{C}_i, & \tilde{C}_d(t) \triangleq \sum_{i=1}^{r} h_i(\theta(t))\tilde{C}_{di}, \\ \tilde{B}(t) \triangleq \sum_{i=1}^{r} h_i(\theta(t))\tilde{B}_i, & \tilde{D}(t) \triangleq \sum_{i=1}^{r} h_i(\theta(t))\tilde{D}_i. \end{cases} \tag{9.6}$$

In this chapter, we concentrate on getting $(\hat{A}, \hat{A}_d, \hat{B}, \hat{C}, \hat{C}_d, \hat{D})$ for proposed reduced-order model, which can ensure the system (9.4) is asymptotically stable with a specific $\mathcal{H}_\infty$ performance sense. The whole conditions are given under this circumstance that (9.2) is asymptotically stable when $\omega(t) = 0$.

Then some lemmas are introduced, which play a part in the derivation processes of our subsequent work.

Lemma 9.1 *[76] Set $W = W^T \in \mathbb{R}^{n\times n}$, $U \in \mathbb{R}^{n\times m}$ and $V \in \mathbb{R}^{k\times n}$ as given matrices, and assume* $\text{rank}(U) < n$, $\text{rank}(V) < n$. *Consider finding suitable matrix $\mathcal{G}$ which satisfies*

$$W + U\mathcal{G}V + (U\mathcal{G}V)^T < 0. \tag{9.7}$$

And the above condition is feasible for $\mathcal{G}$ if and only if

$$U^{\perp}WU^{\perp T} < 0, \quad V^{T\perp}WV^{T\perp T} < 0. \tag{9.8}$$

In addition, if (9.8) satisfies, the solutions of $\mathcal{G}$ can be obtained by

$$\mathcal{G} = U_R^+ \Psi V_L^+ + \Phi - U_R^+ U_R \Phi V_L V_L^+,$$

with

$$\begin{cases} \Psi = -\Pi^{-1}U_L^T \Lambda V_R^T (V_R \Lambda V_R^T)^{-1} + \Pi^{-1}\Xi^{\frac{1}{2}} L (V_R \Lambda V_R^T)^{-\frac{1}{2}}, \\ \Lambda = (U_L \Pi^{-1} U_L^T - W)^{-1} > 0, \\ \Xi = \Pi - U_L^T (\Lambda - \Lambda V_R^T (V_R \Lambda V_R^T)^{-1} V_R \Lambda) U_L > 0, \end{cases}$$

where Φ, Π, L stand for any suitably dimensioned matrices, and $\Pi > 0$, $\| L \| < 1$.

Lemma 9.2 *[206] Let $f_1, f_2, \ldots, f_N$: $\mathbb{R}^m \rightarrow \mathbb{R}$ have positive values in an open subset D of $\mathbb{R}^m$. The reciprocally convex combination of f_i in D satisfies the following condition:*

$$\min_{\{\beta_i | \beta_i > 0, \sum_i \beta_i = 1\}} \sum_i \frac{1}{\beta_i} f_i(t) = \sum_i f_i(t) + \max_{g_{i,j}(t)} \sum_{i \neq j} g_{i,j}(t),$$

with

$$\left\{ g_{i,j} : \mathbb{R}^m \rightarrow \mathbb{R}, \, g_{j,i}(t) = g_{i,j}(t), \begin{bmatrix} f_i(t) & g_{i,j}(t) \\ g_{j,i}(t) & f_j(t) \end{bmatrix} \geqslant 0 \right\}.$$

9.3 Main Results

9.3.1 Performance Analysis via Reciprocally Convex Technique

In this section, based on the reciprocally convex method, a less conservative stability criteria for the corresponding error system (9.4) is presented.

Theorem 9.3 *For the given positive scalars γ, d_1, d_2 and τ, the overall system (9.4) is asymptotically stable with a prescriptive $\mathcal{H}_\infty$ performance sense γ, if there are matrices $P > 0$, $Y > 0$, $Q_1 > 0$, $Q_2 > 0$, $S_0 > 0$, $S_1 > 0$, $R_1 > 0$, $R_2 > 0$ and Z_1, such that the following requirements are satisfied for $i = 1, 2 \ldots, r$:*

$$\Phi_i < 0, \tag{9.9}$$

$$\begin{bmatrix} S_1 & Z_1 \\ \star & S_1 \end{bmatrix} \geqslant 0, \tag{9.10}$$

where

$$\Phi_i \triangleq \begin{bmatrix} \Phi_{11i} & S_0 & 0 & P\tilde{A}_{di} & d_1R_1 & dR_2 & P\tilde{B}_i & \Phi_{18i} \\ \star & \Phi_{22} & -Z_1 & \Phi_{24} & 0 & 0 & 0 & 0 \\ \star & \star & \Phi_{33} & \Phi_{34} & 0 & 0 & 0 & 0 \\ \star & \star & \star & \Phi_{44} & 0 & 0 & 0 & \Phi_{48i} \\ \star & \star & \star & \star & -R_1 & 0 & 0 & 0 \\ \star & \star & \star & \star & \star & -R_2 & 0 & 0 \\ \star & \star & \star & \star & \star & \star & -\gamma^2 I & \Phi_{78i} \\ \star & \star & \star & \star & \star & \star & \star & \Phi_{88} \end{bmatrix},$$

with

$$\begin{aligned}
&\Phi_{11i} \triangleq P\tilde{A}_i + \tilde{A}_i^T P + Q_1 + Y, \quad \Phi_{22} \triangleq Q_2 - Q_1 - S_0 - S_1, \\
&\Phi_{24} \triangleq S_1 + Z_1, \quad \Phi_{34} \triangleq S_1 + Z_1^T, \quad \Phi_{33} \triangleq -Q_2 - S_1, \\
&\Phi_{44} \triangleq (\tau - 1)Y - 2S_1 - Z_1 - Z_1^T, \quad \Phi_{18i} \triangleq \left[\tilde{A}_i^T \ \tilde{A}_i^T \ \tilde{A}_i^T \ \tilde{A}_i^T \ \tilde{C}_i^T \right], \\
&\Phi_{48i} \triangleq \left[\tilde{A}_{di}^T \ \tilde{A}_{di}^T \ \tilde{A}_{di}^T \ \tilde{A}_{di}^T \ \tilde{C}_{di}^T \right], \quad \Phi_{78i} \triangleq \left[\tilde{B}_i^T \ \tilde{B}_i^T \ \tilde{B}_i^T \ \tilde{B}_i^T \ \tilde{D}_i^T \right], \\
&\Phi_{88} \triangleq diag(-\frac{1}{d_1^2}S_0^{-1}, \ -\frac{1}{d^2}S_1^{-1}, \ -\frac{4}{d_1^4}R_1^{-1}, \ -\frac{4}{(d_2^2 - d_1^2)^2}R_2^{-1}, \ -I).
\end{aligned}$$

Proof On the basis of fuzzy membership functions and (9.9), we can get

$$\Phi(t) \triangleq \sum_{i=1}^{r} h_i(\theta(t))\Phi_i < 0.$$

A more complete expression of the above inequality can be constructed as

$$\Phi(t) \triangleq \begin{bmatrix} \Phi_{11}(t) & S_0 & 0 & P\tilde{A}_d(t) & d_1R_1 & dR_2 & P\tilde{B}(t) & \Phi_{18}(t) \\ \star & \Phi_{22} & -Z_1 & \Phi_{24} & 0 & 0 & 0 & 0 \\ \star & \star & \Phi_{33} & \Phi_{34} & 0 & 0 & 0 & 0 \\ \star & \star & \star & \Phi_{44} & 0 & 0 & 0 & \Phi_{48}(t) \\ \star & \star & \star & \star & -R_1 & 0 & 0 & 0 \\ \star & \star & \star & \star & \star & -R_2 & 0 & 0 \\ \star & \star & \star & \star & \star & \star & -\gamma^2 I & \Phi_{78}(t) \\ \star & \star & \star & \star & \star & \star & \star & \Phi_{88} \end{bmatrix} < 0, \quad (9.11)$$

where

$$\begin{aligned}
&\Phi_{11}(t) \triangleq P\tilde{A}(t) + \tilde{A}^T(t)P + Q_1 + Y, \\
&\Phi_{18}(t) \triangleq \left[\tilde{A}^T(t) \ \tilde{A}^T(t) \ \tilde{A}^T(t) \ \tilde{A}^T(t) \ \tilde{C}^T(t) \right], \\
&\Phi_{48}(t) \triangleq \left[\tilde{A}_d^T(t) \ \tilde{A}_d^T(t) \ \tilde{A}_d^T(t) \ \tilde{A}_d^T(t) \ \tilde{C}_d^T(t) \right], \\
&\Phi_{78}(t) \triangleq \left[\tilde{B}^T(t) \ \tilde{B}^T(t) \ \tilde{B}^T(t) \ \tilde{B}^T(t) \ \tilde{D}^T(t) \right].
\end{aligned}$$

Consider the Schur complement theory [294], it is simple to conclude that (9.11) is equal to

$$\tilde{\Phi}(t) \triangleq \begin{bmatrix} \tilde{\Phi}_{11}(t) & S_0 & 0 & \tilde{\Phi}_{14}(t) & d_1 R_1 & d R_2 & \tilde{\Phi}_{17}(t) \\ \star & \Phi_{22} & -Z_1 & \Phi_{24} & 0 & 0 & 0 \\ \star & \star & \Phi_{33} & \Phi_{34} & 0 & 0 & 0 \\ \star & \star & \star & \tilde{\Phi}_{44}(t) & 0 & 0 & \tilde{\Phi}_{47}(t) \\ \star & \star & \star & \star & -R_1 & 0 & 0 \\ \star & \star & \star & \star & \star & -R_2 & 0 \\ \star & \star & \star & \star & \star & \star & \tilde{\Phi}_{77}(t) \end{bmatrix} < 0, \tag{9.12}$$

where

$$\begin{aligned}
\tilde{\Phi}_{11}(t) &\triangleq d_1^2 \tilde{A}^T(t) S_0 \tilde{A}(t) + d^2 \tilde{A}^T(t) S_1 \tilde{A}(t) + \frac{d_1^4}{4} \tilde{A}^T(t) R_1 \tilde{A}(t) \\
&\quad + \frac{(d_2^2 - d_1^2)^2}{4} \tilde{A}^T(t) R_2 \tilde{A}(t) + \tilde{C}^T(t) \tilde{C}(t) + \Phi_{11}(t), \\
\tilde{\Phi}_{14}(t) &\triangleq d_1^2 \tilde{A}^T(t) S_0 \tilde{A}_d(t) + d^2 \tilde{A}^T(t) S_1 \tilde{A}_d(t) + \frac{d_1^4}{4} \tilde{A}^T(t) R_1 \tilde{A}_d(t) \\
&\quad + \frac{(d_2^2 - d_1^2)^2}{4} \tilde{A}^T(t) R_2 \tilde{A}_d(t) + \tilde{C}^T(t) \tilde{C}_d(t) + P \tilde{A}_d(t), \\
\tilde{\Phi}_{17}(t) &\triangleq d_1^2 \tilde{A}^T(t) S_0 \tilde{B}(t) + d^2 \tilde{A}^T(t) S_1 \tilde{B}(t) + \frac{d_1^4}{4} \tilde{A}^T(t) R_1 \tilde{B}(t) \\
&\quad + \frac{(d_2^2 - d_1^2)^2}{4} \tilde{A}^T(t) R_2 \tilde{B}(t) + \tilde{C}^T(t) \tilde{D}(t) + P \tilde{B}(t), \\
\tilde{\Phi}_{44}(t) &\triangleq d_1^2 \tilde{A}_d^T(t) S_0 \tilde{A}_d(t) + d^2 \tilde{A}_d^T(t) S_1 \tilde{A}_d(t) + \frac{d_1^4}{4} \tilde{A}_d^T(t) R_1 \tilde{A}_d(t) \\
&\quad + \frac{(d_2^2 - d_1^2)^2}{4} \tilde{A}_d^T(t) R_2 \tilde{A}_d(t) + \tilde{C}_d^T(t) \tilde{C}_d(t) + \Phi_{44}, \\
\tilde{\Phi}_{47}(t) &\triangleq d_1^2 \tilde{A}_d^T(t) S_0 \tilde{B}(t) + d^2 \tilde{A}_d^T(t) S_1 \tilde{B}(t) + \frac{d_1^4}{4} \tilde{A}_d^T(t) R_1 \tilde{B}(t) \\
&\quad + \frac{(d_2^2 - d_1^2)^2}{4} \tilde{A}_d^T(t) R_2 \tilde{B}(t) + \tilde{C}_d^T(t) \tilde{D}(t), \\
\tilde{\Phi}_{77}(t) &\triangleq d_1^2 \tilde{B}^T(t) S_0 \tilde{B}(t) + d^2 \tilde{B}^T(t) S_1 \tilde{B}(t) + \frac{d_1^4}{4} \tilde{B}^T(t) R_1 \tilde{B}(t) \\
&\quad + \frac{(d_2^2 - d_1^2)^2}{4} \tilde{B}^T(t) R_2 \tilde{B}(t) + \tilde{D}^T(t) \tilde{D}(t) - \gamma^2 I.
\end{aligned}$$

The LKF is selected as follows:

$$V(t) \triangleq \sum_{i=1}^{5} V_i(t),$$

where

$$
\begin{aligned}
V_1(t) &\triangleq \xi^T(t)P\xi(t),\\
V_2(t) &\triangleq \sum_{i=1}^{2}\int_{t-d_i}^{t-d_{i-1}} \xi^T(s)Q_i\xi(s)ds,\\
V_3(t) &\triangleq \int_{t-d(t)}^{t} \xi^T(s)Y\xi(s)ds + d_1\int_{-d_1}^{0}\int_{t+\theta}^{t} \dot{\xi}^T(s)S_0\dot{\xi}(s)dsd\theta,\\
V_4(t) &\triangleq d\int_{-d_2}^{-d_1}\int_{t+\theta}^{t} \dot{\xi}^T(s)S_1\dot{\xi}(s)dsd\theta,\\
V_5(t) &\triangleq \sum_{i=1}^{2}\frac{d_i^2-d_{i-1}^2}{2}\int_{-d_i}^{-d_{i-1}}\int_{\theta}^{0}\int_{t+\lambda}^{t} \dot{\xi}^T(s)R_i\dot{\xi}(s)dsd\lambda d\theta.
\end{aligned}
$$

Denote $d_0 = 0$ and $d = d_2 - d_1$. The time derivative of $V(t)$ along the system (9.4) can be obtained as

$$
\begin{aligned}
\dot{V}_1(t) &= \xi^T(t)P\dot{\xi}(t) + \dot{\xi}^T(t)P\xi(t),\\
\dot{V}_2(t) &= \sum_{i=1}^{2}\left\{\xi^T(t-d_{i-1})Q_i\xi(t-d_{i-1}) - \xi^T(t-d_i)Q_i\xi(t-d_i)\right\},\\
\dot{V}_3(t) &= \xi^T(t)Y\xi(t) - (1-\dot{d}(t))\xi^T(t-d(t))Y\xi(t-d(t)) + d_1^2\dot{\xi}^T(t)S_0\dot{\xi}(t)\\
&\quad -d_1\int_{t-d_1}^{t} \dot{\xi}^T(s)S_0\dot{\xi}(s)ds,\\
\dot{V}_4(t) &= d^2\dot{\xi}^T(t)S_1\dot{\xi}(t) - d\int_{t-d_2}^{t-d_1} \dot{\xi}^T(s)S_1\dot{\xi}(s)ds,\\
\dot{V}_5(t) &= \sum_{i=1}^{2}\frac{(d_i^2-d_{i-1}^2)^2}{4}\dot{\xi}^T(t)R_i\dot{\xi}(t)\\
&\quad -\sum_{i=1}^{2}\frac{d_i^2-d_{i-1}^2}{2}\int_{-d_i}^{-d_{i-1}}\int_{t+\theta}^{t} \dot{\xi}^T(s)R_i\dot{\xi}(s)dsd\theta. \qquad (9.13)
\end{aligned}
$$

It is not difficult to obtain $\dot{V}_3(t)$ and $\dot{V}_5(t)$ from Jensen inequality method [18] as below:

$$\begin{aligned}\dot{V}_3(t) &\leqslant \xi^T(t)Y\xi(t) - (1-\tau)\xi^T(t-d(t))Y\xi(t-d(t)) \\ &\quad + d_1^2\dot{\xi}^T(t)S_0\dot{\xi}(t) - \big(\xi(t) - \xi(t-d_1)\big)^T S_0\big(\xi(t) - \xi(t-d_1)\big) \\ &\leqslant \xi^T(t)Y\xi(t) - (1-\tau)\xi^T(t-d(t))Y\xi(t-d(t)) + d_1^2\dot{\xi}^T(t)S_0\dot{\xi}(t) \\ &\quad + \xi^T(t)S_0\xi(t-d_1) + \xi^T(t-d_1)S_0\xi(t) - \xi^T(t-d_1)S_0\xi(t-d_1),\end{aligned} \tag{9.14}$$

$$\begin{aligned}\dot{V}_5(t) &\leqslant \sum_{i=1}^{2}\frac{(d_i^2 - d_{i-1}^2)^2}{4}\dot{\xi}^T(t)R_i\dot{\xi}(t) \\ &\quad - \left(d_1\xi(t) - \int_{t-d_1}^{t}\xi(s)ds\right)^T R_1\left(d_1\xi(t) - \int_{t-d_1}^{t}\xi(s)ds\right) \\ &\quad - \left(d\xi(t) - \int_{t-d_2}^{t-d_1}\xi(s)ds\right)^T R_2\left(d\xi(t) - \int_{t-d_2}^{t-d_1}\xi(s)ds\right) \\ &\leqslant \sum_{i=1}^{2}\frac{(d_i^2 - d_{i-1}^2)^2}{4}\dot{\xi}^T(t)R_i\dot{\xi}(t) + d_1\xi^T(t)R_1\int_{t-d_1}^{t}\xi(s)ds \\ &\quad + d_1\int_{t-d_1}^{t}\xi^T(s)dsR_1\xi(t) - \int_{t-d_1}^{t}\xi^T(s)dsR_1\int_{t-d_1}^{t}\xi(s)ds \\ &\quad + d\xi^T(t)R_2\int_{t-d_2}^{t-d_1}\xi(s)ds + d\int_{t-d_2}^{t-d_1}\xi^T(s)dsR_2\xi(t) \\ &\quad - \int_{t-d_2}^{t-d_1}\xi^T(s)dsR_2\int_{t-d_2}^{t-d_1}\xi(s)ds.\end{aligned} \tag{9.15}$$

Due to $d_1 < d(t) < d_2$, we can easily get

$$\begin{aligned}\dot{V}_4(t) &= d^2\dot{\xi}^T(t)S_1\dot{\xi}(t) - d\int_{t-d(t)}^{t-d_1}\dot{\xi}^T(s)S_1\dot{\xi}(s)ds - d\int_{t-d_2}^{t-d(t)}\dot{\xi}^T(s)S_1\dot{\xi}(s)ds \\ &\leqslant d^2\dot{\xi}^T(t)S_1\dot{\xi}(t) - \frac{d}{d(t)-d_1}(\xi(t-d(t)) - \xi(t-d_1))^T S_1(\xi(t-d(t)) \\ &\quad - \frac{d}{d_2 - d(t)}(\xi(t-d(t)) - \xi(t-d_2))^T S_1(\xi(t-d(t)) - \xi(t-d_1)) \\ &\quad - \xi(t-d_2)).\end{aligned}$$

It yields from the reciprocally convex method in Lemma 9.2 that

$$\begin{aligned}\dot{V}_4(t) &\leqslant d^2\,\dot{\xi}^T(t)S_1\dot{\xi}(t) - \begin{bmatrix}\xi(t-d(t)) - \xi(t-d_1)\\ \xi(t-d(t)) - \xi(t-d_2)\end{bmatrix}^T \begin{bmatrix}S_1 & Z_1\\ \star & S_1\end{bmatrix} \\ &\quad \begin{bmatrix}\xi(t-d(t)) - \xi(t-d_1)\\ \xi(t-d(t)) - \xi(t-d_2)\end{bmatrix}.\end{aligned} \tag{9.16}$$

When $d(t) = d_1$ or $d(t) = d_2$, (9.16) still meets because $\xi(t-d(t)) - \xi(t-d_1) = 0, \xi(t-d(t)) - \xi(t-d_2) = 0$. Thus, considering (9.13)–(9.16) when $\omega(t) = 0$, we have

$$\dot{V}(t) \leqslant \zeta^T(t)\hat{\Phi}(t)\zeta(t),$$

where

$$\zeta(t) \triangleq \Big[\xi^T(t)\ \xi^T(t-d_1)\ \xi^T(t-d_2)\ \xi^T(t-d(t))\ \int_{t-d_1}^{t} \xi^T(s)ds \\ \int_{t-d_2}^{t-d_1} \xi^T(s)ds\Big]^T,$$

and

$$\hat{\Phi}(t) \triangleq \begin{bmatrix} \hat{\Phi}_{11}(t) & S_0 & 0 & \hat{\Phi}_{14}(t) & d_1 R_1 & dR_2 \\ \star & \Phi_{22} & -Z_1 & \Phi_{24} & 0 & 0 \\ \star & \star & \Phi_{33} & \Phi_{34} & 0 & 0 \\ \star & \star & \star & \hat{\Phi}_{44}(t) & 0 & 0 \\ \star & \star & \star & \star & -R_1 & 0 \\ \star & \star & \star & \star & \star & -R_2 \end{bmatrix}, \tag{9.17}$$

with

$$\begin{aligned}
\hat{\Phi}_{11}(t) &\triangleq d_1^2 \tilde{A}^T(t) S_0 \tilde{A}(t) + d^2 \tilde{A}^T(t) S_1 \tilde{A}(t) + \frac{d_1^4}{4} \tilde{A}^T(t) R_1 \tilde{A}(t) \\
&\quad + \frac{(d_2^2 - d_1^2)^2}{4} \tilde{A}^T(t) R_2 \tilde{A}(t) + \Phi_{11}(t), \\
\hat{\Phi}_{14}(t) &\triangleq d_1^2 \tilde{A}^T(t) S_0 \tilde{A}_d(t) + d^2 \tilde{A}^T(t) S_1 \tilde{A}_d(t) + \frac{d_1^4}{4} \tilde{A}^T(t) R_1 \tilde{A}_d(t) \\
&\quad + \frac{(d_2^2 - d_1^2)^2}{4} \tilde{A}^T(t) R_2 \tilde{A}_d(t) + P\tilde{A}_d(t), \\
\hat{\Phi}_{44}(t) &\triangleq d_1^2 \tilde{A}_d^T(t) S_0 \tilde{A}_d(t) + d^2 \tilde{A}_d^T(t) S_1 \tilde{A}_d(t) + \frac{d_1^4}{4} \tilde{A}_d^T(t) R_1 \tilde{A}_d(t) \\
&\quad + \frac{(d_2^2 - d_1^2)^2}{4} \tilde{A}_d^T(t) R_2 \tilde{A}_d(t) + \Phi_{44}.
\end{aligned}$$

Employing the Schur complement approach, we can obtain $\hat{\Phi}(t) < 0$. Hence, when $\zeta(t) \neq 0$, $\dot{V}(t) \leqslant \zeta^T(t)\hat{\Phi}(t)\zeta(t) < 0$ is always true, which means the concerned system (9.4) is asymptotically stable.

In the following, we are going to establish the $\mathcal{H}_\infty$ performance index for (9.4). As for nonzero $\omega(t) \in \mathcal{L}_2[0, \infty)$, the following index is proposed:

$$\mathcal{J} \triangleq \int_0^\infty e^T(t)e(t)dt - \gamma^2 \int_0^\infty \omega^T(t)\omega(t)dt. \tag{9.18}$$

Based on the zero initial condition $V(t)\mid_{t=0}=0$, we have

$$\begin{aligned}\mathcal{J} &\leqslant \int_0^\infty e^T(t)e(t)dt - \gamma^2\int_0^\infty \omega^T(t)\omega(t)dt + V(\infty) - V(0)\\ &= \int_0^\infty \left[e^T(t)e(t) - \gamma^2\omega^T(t)\omega(t) + \dot{V}(t)\right]dt\\ &= \int_0^\infty \eta^T(t)\tilde{\Phi}(t)\eta(t)dt. \end{aligned} \tag{9.19}$$

Therefore, when $\eta(t) \triangleq [\zeta^T(t)\ \omega^T(t)]^T \neq 0$, for all nonzero $\omega(t) \in \mathcal{L}_2[0, \infty)$, it can be concluded that $\mathcal{J} < 0$ on account of (9.12). That is to say, the resulting fuzzy error system in (9.4) satisfies the prescribed $\mathcal{H}_\infty$ performance. Then the proof is completed.

Remark 9.4 Theorem 9.3 presents a stability criteria for the dynamic system (9.4) subject to a specific $\mathcal{H}_\infty$ performance. On the basis of obtained results, the proposed reduced-order model can be cast via a projection method, which converts the original model approximation issue to a sequential minimization issue under LMI constraints by utilizing the CCL approach.

When $\omega(t) = 0$, the T-S fuzzy system in (9.2) is expressed as

$$\dot{x}(t) = \sum_{i=1}^{r} h_i(\theta(t))\left[A_i x(t) + A_{di}x(t-d(t))\right]. \tag{9.20}$$

Then we will propose a stability criterion for the above time-delay system. In the first place, we give the LKF as follows:

$$V(t) \triangleq \sum_{i=1}^{5} V_i(t), \tag{9.21}$$

where

$$\begin{aligned} V_1(t) &\triangleq x^T(t)\mathscr{P}x(t),\\ V_2(t) &\triangleq \sum_{i=1}^{2}\int_{t-d_i}^{t-d_{i-1}} x^T(s)\mathscr{Q}_i x(s)ds,\\ V_3(t) &\triangleq \int_{t-d(t)}^{t} x^T(s)\mathscr{Y}x(s)ds + d_1\int_{-d_1}^{0}\int_{t+\theta}^{t}\dot{x}^T(s)\mathscr{S}_0\dot{x}(s)dsd\theta,\\ V_4(t) &\triangleq d\int_{-d_2}^{-d_1}\int_{t+\theta}^{t}\dot{x}^T(s)\mathscr{S}_1\dot{x}(s)dsd\theta,\\ V_5(t) &\triangleq \sum_{i=1}^{2}\frac{d_i^2 - d_{i-1}^2}{2}\int_{-d_i}^{-d_{i-1}}\int_{\theta}^{0}\int_{t+\lambda}^{t}\dot{x}^T(s)\mathscr{R}_i\dot{x}(s)dsd\lambda d\theta. \end{aligned}$$

Then we can get the following results on account of (9.21) and the reciprocally convex method.

Corollary 9.5 *For the given positive scalars d_1, d_2 and τ, the system in (9.20) is asymptotically stable, if there are matrices $\mathscr{P} > 0$, $\mathscr{Y} > 0$, $\mathscr{Q}_1 > 0$, $\mathscr{Q}_2 > 0$, $\mathscr{S}_0 > 0$, $\mathscr{S}_1 > 0$, $\mathscr{R}_1 > 0$, $\mathscr{R}_2 > 0$ and $\mathscr{Z}_1$, such that the following conditions satisfy for $i = 1, 2 \ldots, r$,*

$$\Xi_i < 0,$$
$$\begin{bmatrix} \mathscr{S}_1 & \mathscr{Z}_1 \\ \star & \mathscr{S}_1 \end{bmatrix} \geqslant 0,$$

where

$$\Xi_i \triangleq \begin{bmatrix} \Xi_{11i} & \mathscr{S}_0 & 0 & \Xi_{14i} & d_1\mathscr{R}_1 & d\mathscr{R}_2 \\ \star & \Xi_{22} & -\mathscr{Z}_1 & \Xi_{24} & 0 & 0 \\ \star & \star & \Xi_{33} & \Xi_{34} & 0 & 0 \\ \star & \star & \star & \Xi_{44i} & 0 & 0 \\ \star & \star & \star & \star & -\mathscr{R}_1 & 0 \\ \star & \star & \star & \star & \star & -\mathscr{R}_2 \end{bmatrix},$$

with

$$\begin{aligned}
\Xi_{22} &\triangleq \mathscr{Q}_2 - \mathscr{Q}_1 - \mathscr{S}_0 - \mathscr{S}_1, \ \Xi_{24} \triangleq \mathscr{S}_1 + \mathscr{Z}_1, \ \Xi_{33} \triangleq -\mathscr{Q}_2 - \mathscr{S}_1, \\
\Xi_{11i} &\triangleq d_1^2 A_i^T \mathscr{S}_0 A_i + d^2 A_i^T \mathscr{S}_1 A_i + \frac{d_1^4}{4} A_i^T \mathscr{R}_1 A_i + \frac{(d_2^2 - d_1^2)^2}{4} A_i^T \mathscr{R}_2 A_i \\
&\quad + \mathscr{P} A_i + A_i^T \mathscr{P} + \mathscr{Q}_1 + \mathscr{Y} - \mathscr{S}_0 - d_1^2 \mathscr{R}_1 - d^2 \mathscr{R}_2, \\
\Xi_{14i} &\triangleq d_1^2 A_i^T \mathscr{S}_0 A_{di} + d^2 A_i^T \mathscr{S}_1 A_{di} + \frac{d_1^4}{4} A_i^T \mathscr{R}_1 A_{di} + \frac{(d_2^2 - d_1^2)^2}{4} A_i^T \mathscr{R}_2 A_{di} \\
&\quad + \mathscr{P} A_{di}, \ \Xi_{34} \triangleq \mathscr{S}_1 + \mathscr{Z}_1^T, \\
\Xi_{44i} &\triangleq d_1^2 A_{di}^T \mathscr{S}_0 A_{di} + d^2 A_{di}^T \mathscr{S}_1 A_{di} + \frac{d_1^4}{4} A_{di}^T \mathscr{R}_1 A_{di} + \frac{(d_2^2 - d_1^2)^2}{4} A_{di}^T \mathscr{R}_2 A_{di} \\
&\quad (\tau - 1)\mathscr{Y} - 2\mathscr{S}_1 - \mathscr{Z}_1 - \mathscr{Z}_1^T.
\end{aligned}$$

Proof It is easy to obtain the results by considering the Lyapunov function in (9.21) and using similar ways as the proof in Theorem 9.3. And the specific proof is ignored here.

Remark 9.6 In Corollary 9.5, the sufficient stability conditions are developed for continuous T-S fuzzy systems with time-varying delays in (9.20) based on the reciprocally convex method. During the derivation of the Lyapunov function, we magnify the negative integral terms of quadratic quantities via employing an efficient Jensen inequality method and reciprocally convex approach, instead of neglecting these terms. Our proposed scheme reduces the system conservatism and complexity in the calculation process, and a detailed comparison is shown in Tables 9.1 and 9.2 of Example 9.8.

9.3.2 Model Approximation via Projection Technique

In this part, the scheme of $\mathcal{H}_\infty$ model approximation issue is investigated for the T-S fuzzy system with time-delay (9.2) by employing the projection strategy.

Theorem 9.7 *There exists an admissible reduced-order model as (9.3), which ensures the error system in (9.4) is asymptotically stable with a particular $\mathcal{H}_\infty$ performance index γ, if there are matrices $P > 0$, $Y > 0$, $Q_1 > 0$, $Q_2 > 0$, $S_0 > 0$, $S_1 > 0$, $R_1 > 0$, $R_2 > 0$, $X_m > 0$, $m = 1, 2, 3$, $M > 0$ and Z_1 meeting the requirements in (9.10) and the following conditions for $i = 1, 2, \ldots, r$:*

$$\begin{bmatrix} \Pi_{11} & \Pi_{12} \\ \star & \Pi_{22} \end{bmatrix} < 0, \tag{9.22}$$

$$\begin{bmatrix} -X_1 & S_0 \\ \star & -M \end{bmatrix} \leqslant 0, \tag{9.23}$$

$$\begin{bmatrix} -X_2 & d_1 R_1 \\ \star & -M \end{bmatrix} \leqslant 0, \tag{9.24}$$

$$\begin{bmatrix} -X_3 & d R_2 \\ \star & -M \end{bmatrix} \leqslant 0, \tag{9.25}$$

$$\begin{bmatrix} \Omega_{11} & \Omega_{12} & \Omega_{13} \\ \star & \Omega_{22} & 0 \\ \star & \star & \Omega_{33} \end{bmatrix} < 0, \tag{9.26}$$

$$\left.\begin{array}{r} P\mathcal{P} = I,\ Q_1\mathcal{Q}_1 = I,\ Y\mathcal{Y} = I,\ S_0\mathcal{S}_0 = I, \\ S_1\mathcal{S}_1 = I,\ R_1\mathcal{R}_1 = I,\ R_2\mathcal{R}_2 = I,\ M\mathcal{M} = I, \end{array}\right\} \tag{9.27}$$

where

$$\Pi_{12} \triangleq \begin{bmatrix} H\tilde{A}_{0i}^T & H\tilde{A}_{0i}^T & H\tilde{A}_{0i}^T & H\tilde{A}_{0i}^T & H\tilde{C}_{0i}^T \\ 0 & 0 & 0 & 0 & 0 \\ 0 & 0 & 0 & 0 & 0 \\ H\tilde{A}_{d0i}^T & H\tilde{A}_{d0i}^T & H\tilde{A}_{d0i}^T & H\tilde{A}_{d0i}^T & H\tilde{C}_{d0i}^T \\ 0 & 0 & 0 & 0 & 0 \\ 0 & 0 & 0 & 0 & 0 \end{bmatrix}, \quad \Omega_{13} \triangleq \begin{bmatrix} H\mathcal{P} & H\mathcal{P} & H\mathcal{P} \\ 0 & 0 & 0 \\ 0 & 0 & 0 \\ 0 & 0 & 0 \\ 0 & 0 & 0 \\ 0 & 0 & 0 \\ 0 & 0 & 0 \end{bmatrix},$$

$$\Pi_{11} \triangleq \begin{bmatrix} H\hat{\Phi}_{11i}H^T & HS_0 & 0 & HP\tilde{A}_{d0i}H^T & d_1HR_1 & dHR_2 \\ \star & \Phi_{22} & -Z_1 & \Phi_{24}H^T & 0 & 0 \\ \star & \star & \Phi_{33} & \Phi_{34}H^T & 0 & 0 \\ \star & \star & \star & H\Phi_{44}H^T & 0 & 0 \\ \star & \star & \star & \star & -R_1 & 0 \\ \star & \star & \star & \star & \star & -R_2 \end{bmatrix},$$

$$\Pi_{22} \triangleq diag\left(-\frac{1}{d_1^2}\mathcal{S}_0,\ -\frac{1}{d^2}\mathcal{S}_1,\ -\frac{4}{d_1^4}\mathcal{R}_1,\ -\frac{4}{(d_2^2-d_1^2)^2}\mathcal{R}_2,\ -I\right),$$

$$\Omega_{11} \triangleq \begin{bmatrix} \Omega_{11-1} & 0 & 0 & H\tilde{A}_{d0i} & 0 & 0 & H\tilde{B}_{0i} \\ \star & \Omega_{11-2} & -Z_1 & \Phi_{24} & 0 & 0 & 0 \\ \star & \star & \Phi_{33} & \Phi_{34} & 0 & 0 & 0 \\ \star & \star & \star & \Phi_{44} & 0 & 0 & 0 \\ \star & \star & \star & \star & \Omega_{11-3} & 0 & 0 \\ \star & \star & \star & \star & \star & \Omega_{11-4} & 0 \\ \star & \star & \star & \star & \star & \star & -\gamma^2 I \end{bmatrix},$$

$$\Omega_{12} \triangleq \begin{bmatrix} H\mathcal{P}\tilde{A}_{0i}^T H^T & H\mathcal{P}\tilde{A}_{0i}^T H^T & H\mathcal{P}\tilde{A}_{0i}^T H^T & H\mathcal{P}\tilde{A}_{0i}^T H^T \\ 0 & 0 & 0 & 0 \\ 0 & 0 & 0 & 0 \\ \tilde{A}_{d0i}^T H^T & \tilde{A}_{d0i}^T H^T & \tilde{A}_{d0i}^T H^T & \tilde{A}_{d0i}^T H^T \\ 0 & 0 & 0 & 0 \\ 0 & 0 & 0 & 0 \\ \tilde{B}_{0i}^T H^T & \tilde{B}_{0i}^T H^T & \tilde{B}_{0i}^T H^T & \tilde{B}_{0i}^T H^T \end{bmatrix},$$

$$\Omega_{22} \triangleq diag\left(-\frac{1}{d_1^2}H\mathcal{S}_0H^T,\ -\frac{1}{d^2}H\mathcal{S}_1H^T,\ -\frac{4}{d_1^4}H\mathcal{R}_1H^T,\ -\frac{4}{(d_2^2-d_1^2)^2}H\mathcal{R}_2H^T\right),$$

$$\Omega_{33} \triangleq diag\left(-\mathcal{Q}_1,\ -\mathcal{Y},\ -\frac{1}{3}\mathcal{M}\right),$$

with

$$\hat{\Phi}_{11i} \triangleq P\tilde{A}_{0i} + \tilde{A}_{0i}^T P + Q_1 + Y,\quad \Omega_{11-1} \triangleq H(\tilde{A}_{0i}\mathcal{P} + \mathcal{P}\tilde{A}_{0i}^T)H^T,$$
$$\Omega_{11-2} \triangleq \Phi_{22} + X_1,\quad \Omega_{11-3} \triangleq -R_1 + X_2,\quad \Omega_{11-4} \triangleq -R_2 + X_3.$$

If the above conditions have feasible solutions, the parameters of the designed reduced-order model can be described by

$$\mathcal{G} \triangleq \begin{bmatrix} \hat{A} & \hat{A}_d & \hat{B} \\ \hat{C} & \hat{C}_d & \hat{D} \end{bmatrix}, \tag{9.28}$$

where

$$\begin{cases} \mathcal{G} \triangleq -\Pi^{-1}U^T\Lambda V^T(V\Lambda V^T)^{-1} + \Pi^{-1}\Xi^{\frac{1}{2}}L(V\Lambda V^T)^{-\frac{1}{2}}, \\ \Lambda \triangleq (U\Pi^{-1}U^T - W)^{-1} > 0, \\ \Xi \triangleq \Pi - U^T(\Lambda - \Lambda V^T(V\Lambda V^T)^{-1}V\Lambda)U > 0, \end{cases}$$

where Π and L are properly-dimensioned matrices satisfying $\Pi > 0$, $\| L \| < 1$ and the related matrices are given by

$$\begin{aligned}
\tilde{A}_{0i} &\triangleq \begin{bmatrix} A_i & 0 \\ 0 & 0 \end{bmatrix}, \quad \tilde{A}_{d0i} \triangleq \begin{bmatrix} A_{di} & 0 \\ 0 & 0 \end{bmatrix}, \quad \tilde{B}_{0i} \triangleq \begin{bmatrix} B_i \\ 0 \end{bmatrix}, \tilde{C}_{0i} \triangleq \begin{bmatrix} C_i & 0 \end{bmatrix}, \\
\tilde{C}_{d0i} &\triangleq \begin{bmatrix} C_{di} & 0 \end{bmatrix}, \quad \tilde{D}_{0i} \triangleq D_i, \quad F \triangleq \begin{bmatrix} 0 & -I \end{bmatrix}, \quad H \triangleq \begin{bmatrix} I & 0 \end{bmatrix}, \\
E &\triangleq \begin{bmatrix} 0 & 0 \\ I & 0 \end{bmatrix}, \quad R \triangleq \begin{bmatrix} 0 & I \\ 0 & 0 \\ 0 & 0 \end{bmatrix}, \quad S \triangleq \begin{bmatrix} 0 & 0 \\ 0 & I \\ 0 & 0 \end{bmatrix}, \quad T \triangleq \begin{bmatrix} 0 \\ 0 \\ I \end{bmatrix}, \\
V &\triangleq \begin{bmatrix} R & 0 & 0 & S & 0 & 0 & T & 0 & 0 & 0 & 0 & 0 \end{bmatrix}, \\
U &\triangleq \begin{bmatrix} E^T P & 0 & 0 & 0 & 0 & 0 & 0 & E^T & E^T & E^T & E^T & F^T \end{bmatrix}^T, \\
W &\triangleq \begin{bmatrix} \hat{\Phi}_{11i} & S_0 & 0 & P\tilde{A}_{d0i} & d_1R_1 & dR_2 & P\tilde{B}_{0i} & W_{18i} \\ \star & \Phi_{22} & -Z_1 & \Phi_{24} & 0 & 0 & 0 & 0 \\ \star & \star & \Phi_{33} & \Phi_{34} & 0 & 0 & 0 & 0 \\ \star & \star & \star & \Phi_{44} & 0 & 0 & 0 & W_{48i} \\ \star & \star & \star & \star & -R_1 & 0 & 0 & 0 \\ \star & \star & \star & \star & \star & -R_2 & 0 & 0 \\ \star & \star & \star & \star & \star & \star & -\gamma^2 I & W_{78i} \\ \star & \star & \star & \star & \star & \star & \star & W_{88} \end{bmatrix},
\end{aligned} \tag{9.29}$$

with

$$\begin{aligned}
W_{18i} &\triangleq \left[\tilde{A}_{0i}^T \ \tilde{A}_{0i}^T \ \tilde{A}_{0i}^T \ \tilde{A}_{0i}^T \ \tilde{C}_{0i}^T \right], W_{48i} \triangleq \left[\tilde{A}_{d0i}^T \ \tilde{A}_{d0i}^T \ \tilde{A}_{d0i}^T \ \tilde{A}_{d0i}^T \ \tilde{C}_{d0i}^T \right], \\
W_{78i} &\triangleq \left[\tilde{B}_{0i}^T \ \tilde{B}_{0i}^T \ \tilde{B}_{0i}^T \ \tilde{B}_{0i}^T \ \tilde{D}_{0i}^T \right], \\
W_{88} &\triangleq diag\left(-\frac{1}{d_1^2}S_0^{-1}, \ -\frac{1}{d^2}S_1^{-1}, \ -\frac{4}{d_1^4}R_1^{-1}, \ -\frac{4}{(d_2^2 - d_1^2)^2}R_2^{-1}, \ -I \right).
\end{aligned}$$

Proof Firstly, rewrite $\tilde{A}_i$, $\tilde{A}_{di}$, $\tilde{B}_i$, $\tilde{C}_i$, $\tilde{C}_{di}$ and $\tilde{D}_i$ as

$$\begin{cases} \tilde{A}_i \triangleq \tilde{A}_{0i} + E\mathcal{G}R, & \tilde{A}_{di} \triangleq \tilde{A}_{d0i} + E\mathcal{G}S, & \tilde{B}_i \triangleq \tilde{B}_{0i} + E\mathcal{G}T, \\ \tilde{C}_i \triangleq \tilde{C}_{0i} + F\mathcal{G}R, & \tilde{C}_{di} \triangleq \tilde{C}_{d0i} + F\mathcal{G}S, & \tilde{D}_i \triangleq \tilde{D}_{0i} + F\mathcal{G}T. \end{cases} \tag{9.30}$$

where $\mathcal{G}$, $\tilde{A}_{0i}$, $\tilde{A}_{d0i}$, $\tilde{B}_{0i}$, $\tilde{C}_{0i}$, $\tilde{C}_{d0i}$, $\tilde{D}_{0i}$, E, F, R, S and T are denoted in (9.28) and (9.29). According to (9.30), the condition in (9.9) can be converted to

$$W + U\mathcal{G}V + (U\mathcal{G}V)^T < 0, \tag{9.31}$$

where W, U and V are given in (9.29). Select

$$U^{\perp} \triangleq \begin{bmatrix} U_{11}^{\perp} & 0 \end{bmatrix}, \quad V^{T\perp} \triangleq \begin{bmatrix} V_{11}^{T\perp} & 0 & 0 \\ 0 & 0 & V_{23}^{T\perp} \end{bmatrix},$$

with

$$\begin{aligned} U_{11}^{\perp} &\triangleq diag(HP^{-1}, I, I, I, I, I, I, H, H, H, H), \\ V_{11}^{T\perp} &\triangleq diag(H, I, I, H, I, I), \quad V_{23}^{T\perp} \triangleq diag(I, I, I, I, I). \end{aligned}$$

Based on Lemma 9.1, (9.31) is feasible for $\mathcal{G}$ if and only if (9.22) and the following condition satisfies:

$$\begin{bmatrix} \Gamma_{11} & \Gamma_{12} & \Gamma_{13} \\ \star & \Gamma_{22} & \Gamma_{23} \\ \star & \star & \Gamma_{33} \end{bmatrix} < 0, \tag{9.32}$$

where

$$\begin{aligned} \Gamma_{11} &\triangleq HP^{-1}\hat{\Phi}_{11i}P^{-1}H^T, \\ \Gamma_{12} &\triangleq \begin{bmatrix} HP^{-1}S_0 & 0 & H\tilde{A}_{d0i} & d_1HP^{-1}R_1 & dHP^{-1}R_2 & H\tilde{B}_{0i} \end{bmatrix}, \\ \Gamma_{13} &\triangleq \begin{bmatrix} HP^{-1}\tilde{A}_{0i}^TH^T & HP^{-1}\tilde{A}_{0i}^TH^T & HP^{-1}\tilde{A}_{0i}^TH^T & HP^{-1}\tilde{A}_{0i}^TH^T \end{bmatrix}, \\ \Gamma_{22} &\triangleq \begin{bmatrix} \Phi_{22} & -Z_1 & \Phi_{24} & 0 & 0 & 0 \\ \star & \Phi_{33} & \Phi_{34} & 0 & 0 & 0 \\ \star & \star & \Phi_{44} & 0 & 0 & 0 \\ \star & \star & \star & -R_1 & 0 & 0 \\ \star & \star & \star & \star & -R_2 & 0 \\ \star & \star & \star & \star & \star & -\gamma^2 I \end{bmatrix}, \\ \Gamma_{23} &\triangleq \begin{bmatrix} 0 & 0 & 0 & 0 \\ 0 & 0 & 0 & 0 \\ \tilde{A}_{d0i}^TH^T & \tilde{A}_{d0i}^TH^T & \tilde{A}_{d0i}^TH^T & \tilde{A}_{d0i}^TH^T \\ 0 & 0 & 0 & 0 \\ 0 & 0 & 0 & 0 \\ \tilde{B}_{0i}^TH^T & \tilde{B}_{0i}^TH^T & \tilde{B}_{0i}^TH^T & \tilde{B}_{0i}^TH^T \end{bmatrix}, \\ \Gamma_{33} &\triangleq diag(-\frac{1}{d_1^2}HS_0^{-1}H^T, \ -\frac{1}{d^2}HS_1^{-1}H^T, \ -\frac{4}{d_1^4}HR_1^{-1}H^T, \\ &\quad -\frac{4}{(d_2^2-d_1^2)^2}HR_2^{-1}H^T). \end{aligned}$$

Therefore, (9.32) can be reformulated as

$$\Sigma_1 + \Sigma_2 S_0 \Sigma_3 + \Sigma_3^T S_0 \Sigma_2^T + d_1 \Sigma_2 R_1 \Sigma_4 + d_1 \Sigma_4^T R_1 \Sigma_2^T \\ + d \Sigma_2 R_2 \Sigma_5 + d \Sigma_5^T R_2 \Sigma_2^T < 0, \quad (9.33)$$

where

$$\Sigma_1 \triangleq \begin{bmatrix} \Gamma_{11} & \hat{\Gamma}_{12} & \Gamma_{13} \\ \star & \Gamma_{22} & \Gamma_{23} \\ \star & \star & \Gamma_{33} \end{bmatrix},$$

$$\Sigma_2 \triangleq \left[P^{-1}H^T \; 0\,0\,0\,0\,0\,0\,0\,0\,0\,0 \right]^T, \Sigma_3 \triangleq \left[0\; I\; 0\,0\,0\,0\,0\,0\,0\,0\,0 \right],$$
$$\Sigma_4 \triangleq \left[0\,0\,0\,0\; I\; 0\,0\,0\,0\,0\,0 \right], \Sigma_5 \triangleq \left[0\,0\,0\,0\,0\; I\; 0\,0\,0\,0\,0 \right],$$

with

$$\hat{\Gamma}_{12} \triangleq \left[0\; 0\; H\tilde{A}_{d0i}\; 0\; 0\; H\tilde{B}_{0i} \right].$$

For $\Sigma_i, i = 2, 3, 4, 5, M > 0, X_1 > 0, X_2 > 0, X_3 > 0, S_0, d_1 R_1$ and $d R_2$, which satisfy $S_0 M^{-1} S_0 \leqslant X_1, d_1^2 R_1 M^{-1} R_1 \leqslant X_2$ and $d^2 R_2 M^{-1} R_2 \leqslant X_3$, hence, it yields

$$\Sigma_2 S_0 \Sigma_3 + \Sigma_3^T S_0 \Sigma_2^T \leqslant \Sigma_2 M \Sigma_2^T + \Sigma_3^T X_1 \Sigma_3,$$
$$d_1 \Sigma_2 R_1 \Sigma_4 + d_1 \Sigma_4^T R_1 \Sigma_2^T \leqslant \Sigma_2 M \Sigma_2^T + \Sigma_4^T X_2 \Sigma_4,$$
$$d \Sigma_2 R_2 \Sigma_5 + d \Sigma_5^T R_2 \Sigma_2^T \leqslant \Sigma_2 M \Sigma_2^T + \Sigma_5^T X_3 \Sigma_5.$$

And it is easy to see that (9.33) holds if (9.23)–(9.25) and the following condition (9.34) satisfies:

$$\begin{bmatrix} \tilde{\Gamma}_{11} & \hat{\Gamma}_{12} & \Gamma_{13} \\ \star & \tilde{\Gamma}_{22} & \Gamma_{23} \\ \star & \star & \Gamma_{33} \end{bmatrix} < 0, \quad (9.34)$$

where

$$\tilde{\Gamma}_{11} \triangleq HP^{-1}(\hat{\Phi}_{11i} + 3M)P^{-1}H^T,$$
$$\tilde{\Gamma}_{22} \triangleq \begin{bmatrix} \Phi_{22} + X_1 & -Z_1 & \Phi_{24} & 0 & 0 & 0 \\ \star & \Phi_{33} & \Phi_{34} & 0 & 0 & 0 \\ \star & \star & \Phi_{44} & 0 & 0 & 0 \\ \star & \star & \star & -R_1 + X_2 & 0 & 0 \\ \star & \star & \star & \star & -R_2 + X_3 & 0 \\ \star & \star & \star & \star & \star & -\gamma^2 I \end{bmatrix},$$

$\hat{\Gamma}_{12}$, Γ_{13}, Γ_{23} and Γ_{33} are denoted in (9.33) and (9.32).

On the basis of Schur complement method and (9.27), it yields (9.26) from (9.34). Consequently, employing the projection strategy, the proof can be completed.

Table 9.1 Comparisons of maximum allowable upper bound d_2: the fast varying delay case

Method	$d_1 = 0.4$	$d_1 = 0.8$	$d_1 = 1.2$
Lien et al. [170]	0.8829	1.0677	1.3181
Li et al. [146]	1.0380	1.1580	1.3590
Corollary 1	1.2736	1.3027	1.4191

Table 9.2 Comparisons of maximum allowable upper bound d_2: the slow varying delay case with $\tau = 0.1$

Method	$d_1 = 0.4$
Lien et al. [170]	1.4841
Li et al. [146]	1.4849
Zhao et al. [340]	1.5465
Corollary 1	1.6321

9.4 Illustrative Example

Two illustrative examples are given to verify the validity of our proposed schemes. The first example demonstrates the superiority compared with previous studies. Another one illustrates the availability of the reduced-order model design approach.

Example 9.8 Consider the T-S fuzzy system with time-delay in (9.20), and the relevant parameters are given by

$$A_1 = \begin{bmatrix} -2 & 0 \\ 0 & -0.9 \end{bmatrix}, \quad A_{d1} = \begin{bmatrix} -1 & 0 \\ -1 & -1 \end{bmatrix}, \quad A_2 = \begin{bmatrix} -1 & 0.5 \\ 0 & -1 \end{bmatrix}, \quad A_{d2} = \begin{bmatrix} -1 & 0 \\ 0.1 & -1 \end{bmatrix}.$$

Table 9.1 depicts the maximum allowable upper bound with diverse lower bound d_1. Under the consideration of the fast varying delay ($\tau \geqslant 1$), the maximum allowable upper bound of Corollary 9.5 is shown as the last line in Table 9.1. Compared with the studies in [146, 170], our proposed scheme reduces the conservatism and calculation complexity to some extent. Furthermore, it is observed from Table 9.1 that the smaller d_1 is, the more conservatism of proposed stability criteria reduces.

Table 9.2 plots the achieved maximum allowable upper bound under the situation of slow varying delay when $d_1 = 0.4$ and $\tau = 0.1$. It can be seen from Table 9.2 that the results of our study have the least conservatism compared with existing methods in [146, 170, 340].

Example 9.9 Consider the T-S fuzzy system with time-delay in (9.2), and the relevant parameters are chosen as

$$A_1 = \begin{bmatrix} -3.0 & 0.5 & 0.6 & 0.2 \\ 0.0 & -2.5 & 0.1 & 0.3 \\ 0.4 & 0.0 & -3.4 & 0.3 \\ 0.5 & 0.3 & 0.2 & -1.8 \end{bmatrix}, \quad A_{d1} = \begin{bmatrix} 0.2 & 0.1 & 0.2 & 0.0 \\ 0.0 & 0.4 & 0.1 & 0.2 \\ 0.2 & 0.1 & 0.6 & 0.0 \\ 0.1 & 0.0 & 0.1 & 0.3 \end{bmatrix}, \quad B_1 = \begin{bmatrix} 0.2 \\ 0.1 \\ 0.3 \\ 0.5 \end{bmatrix},$$

$$A_2 = \begin{bmatrix} -2.1 & 0.2 & 0.0 & 0.2 \\ 0.4 & -3.8 & 0.1 & 0.6 \\ 0.1 & 0.0 & -2.0 & 0.4 \\ 0.3 & 0.2 & 0.0 & -1.5 \end{bmatrix}, \quad A_{d2} = \begin{bmatrix} 0.3 & 0.1 & 0.2 & 0.0 \\ 0.1 & 0.2 & 0.0 & 0.2 \\ 0.1 & 0.0 & 0.4 & 0.1 \\ 0.0 & 0.2 & 0.0 & 0.5 \end{bmatrix}, \quad B_2 = \begin{bmatrix} 0.4 \\ 0.2 \\ 0.1 \\ 0.2 \end{bmatrix},$$

$$C_1 = \begin{bmatrix} 1.0 & 1.2 & 0.8 & 0.6 \end{bmatrix}, \quad C_{d1} = \begin{bmatrix} 0.2 & 0.1 & 0.3 & 0.2 \end{bmatrix}, \quad D_1 = 0.3,$$
$$C_2 = \begin{bmatrix} 0.6 & 1.0 & 0.6 & 0.8 \end{bmatrix}, \quad C_{d2} = \begin{bmatrix} 0.1 & 0.1 & 0.2 & 0.4 \end{bmatrix}, \quad D_2 = 0.2. \tag{9.35}$$

In this work, we concentrate on finding appropriate reduced-order models, including one-order, two-order, three-order models as (9.3) to approximate concerned systems, which guarantee the overall dynamic system is asymptotically stable and owns a specific performance level. Solving the non-convex feasible issue in Theorem 9.7, we can get the following results under different cases.
Case 1: (When $k = 3$, we have $\gamma_{min} = 0.40$)

$$\hat{A} = \begin{bmatrix} -11.1438 & 1.2601 & 1.5499 \\ 0.5764 & -2.3911 & 0.9546 \\ 0.5846 & 2.5267 & -5.0508 \end{bmatrix}, \quad \hat{A}_d = \begin{bmatrix} 0.0679 & 0.0613 & 0.0932 \\ 0.0608 & 0.1819 & 0.1928 \\ 0.0884 & 0.1869 & 0.2720 \end{bmatrix},$$

$$\hat{B} = \begin{bmatrix} 0.2952 \\ 0.2778 \\ 0.3420 \end{bmatrix}, \quad \hat{C} = \begin{bmatrix} 0.0978 & 2.3698 & 0.7115 \end{bmatrix},$$

$$\hat{C}_d = \begin{bmatrix} 0.1295 & 0.3275 & 0.3002 \end{bmatrix}, \quad \hat{D} = 0.3212. \tag{9.36}$$

Case 2: (When $k = 2$, we have $\gamma_{min} = 0.55$)

$$\hat{A} = \begin{bmatrix} -1.9885 & 1.5085 \\ 1.0086 & -5.8827 \end{bmatrix}, \quad \hat{A}_d = \begin{bmatrix} 0.2462 & 0.1024 \\ 0.0151 & -0.2189 \end{bmatrix}, \quad \hat{B} = \begin{bmatrix} -0.2463 \\ -0.2984 \end{bmatrix},$$
$$\hat{C} = \begin{bmatrix} -3.4715 & -0.7786 \end{bmatrix}, \quad \hat{C}_d = \begin{bmatrix} -0.3327 & -0.0321 \end{bmatrix}, \quad \hat{D} = 0.3135. \tag{9.37}$$

Case 3: (When $k = 1$, we have $\gamma_{min} = 0.80$)

$$\hat{A} = -4.0407, \quad \hat{A}_d = 0.2659, \quad \hat{B} = 0.4774,$$
$$\hat{C} = 4.8774, \quad \hat{C}_d = 0.3373, \quad \hat{D} = 0.4559. \tag{9.38}$$

In addition, the initial condition is assumed to be zero, that is $x(0) = 0$ ($\hat{x}(0) = 0$), and the membership functions are selected as

$$h_1(x_1(t)) = \frac{1 - \sin^2(x_1(t))}{2}, \quad h_2(x_1(t)) = \frac{1 + \sin^2(x_1(t))}{2},$$

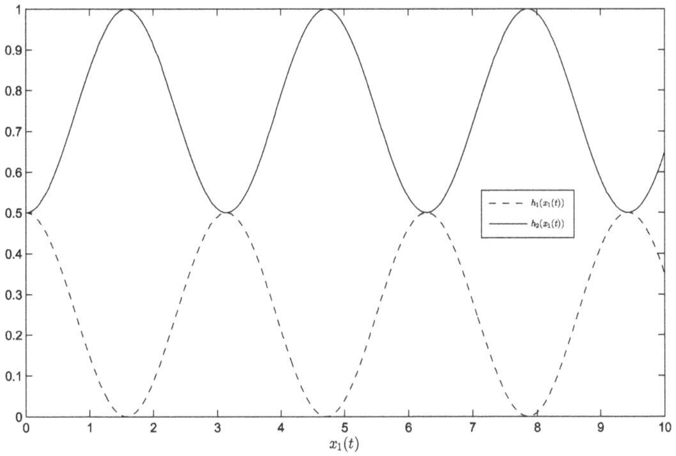

Fig. 9.2 Membership functions for the two fuzzy sets

which are plotted in Fig. 9.2, and the exogenous disturbance is set to

$$\omega(t) = \begin{cases} e^{(-t+1)}\sin(t), & t \leqslant 25, \\ 0.3\sin(t), & 25 < t \leqslant 75, \\ 0, & otherwise. \end{cases}$$

Figure 9.3 draws the output trajectories of the original system (9.35) (solid line), three-order model (9.36) (dashed line), two-order model (9.37) (dotted line), and one-order model (9.38) (dashed-dotted line). $\hat{y}_3(t)$, $\hat{y}_2(t)$ and $\hat{y}_1(t)$ denote the output of the three-order, two-order, and one-order reduced model, separately. Figure 9.4 depicts the relevant output errors between the original system and reduced-order models. In this chapter, γ represents the ratio among the output error energy $\mathbb{E}(t)$ and input energy $\mathbb{W}(t)$, that is $\gamma = \mathbb{E}(t)/\mathbb{W}(t)$. $\mathbb{E}(t)$ denotes the output error energy, which means the extraction of the summation of $e^T(t)e(t)$, that is $\mathbb{E}(t) \triangleq \sqrt{\int_0^\infty e^T(t)e(t)dt}$. $\mathbb{W}(t)$ denotes the input energy, which means the extraction of the summation of $\omega^T(t)\omega(t)$, that is $\mathbb{W}(t) \triangleq \sqrt{\int_0^\infty \omega^T(t)\omega(t)dt}$.

It is easy to conclude the more the order is reduced, the more obvious errors produce. Moreover, the minimum $\mathcal{H}_\infty$ performance sense γ can be seen to be smaller with the rising reduced order. In other words, the more information that reduced-order models demand, the better performance can be obtained. All the simulation results illustrate the feasibility of the presented scheme.

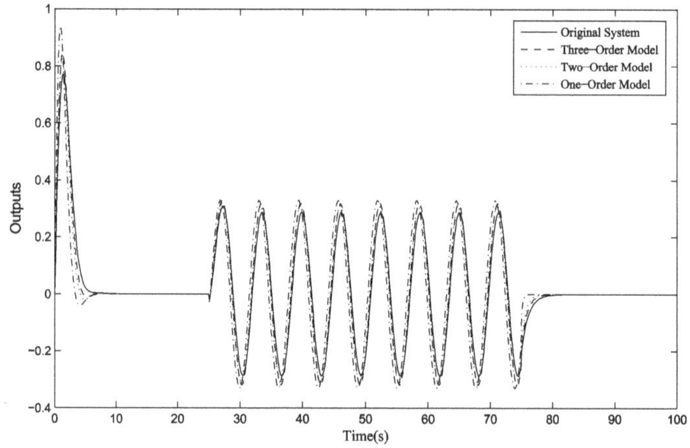

Fig. 9.3 Outputs of the original system and the reduced-order models

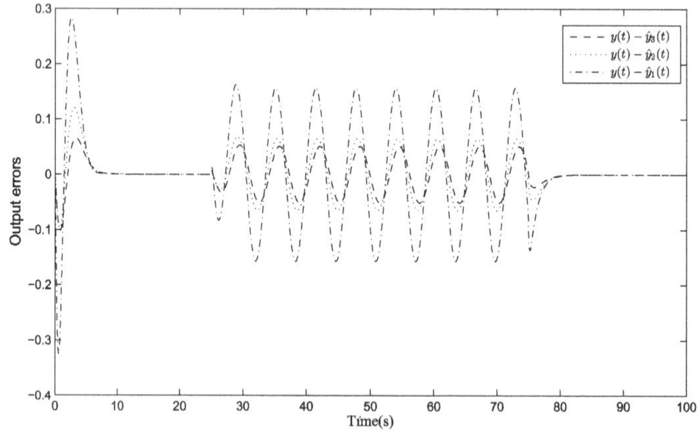

Fig. 9.4 Errors of the original system and the reduced-order models

9.5 Conclusion

In this chapter, the model approximation issue for T-S fuzzy systems with time-varying delay was examined. Asymptotic stability conditions with less conservativeness were established for the resulting error system by using the reciprocally convex method. Moreover, the reduced-order model was developed using the projection strategy, which could not only approximate the original system with a specific $\mathcal{H}_\infty$ performance level, but also facilitate the analysis and synthesis of complex high-order systems. In the final, two simulations have been given to demonstrate the availability of our presented approach.

Chapter 10
Model Approximation of Fuzzy Switched Systems

10.1 Introduction

In this chapter, the model approximation issue of a T-S fuzzy switched system with stochastic disturbances is considered. Considering a high-order system, we constructing a reduced-order model, which can approximate the original system with a Hankel-norm performance and convert it to a lower-order switched system. Sufficient conditions are derived using the ADT method and piecewise Lyapunov function to ensure the resulting error system is mean-square exponentially stable and satisfies the Hankel-norm performance sense. Based on a linearization process, the preceding model approximation can be transformed to a convex optimization issue.

10.2 System Description and Preliminaries

In this chapter, we introduce the nonlinear switched system with stochastic disturbance, which is expressed as the following T-S fuzzy switched model:

♦ **Plant Form**:

Rule $\mathcal{R}_i^{\beta}$: IF $\theta_1(\beta, t)$ is $\mu_{(i1,\beta)}$ and $\theta_2(\beta, t)$ is $\mu_{(i2,\beta)}$ and $\cdots$ and $\theta_p(\beta, t)$ is $\mu_{(ip,\beta)}$, THEN

$$\begin{aligned} dx(t) &= [A_i(\beta)x(t) + B_i(\beta)u(t)]\,dt + E_i(\beta)x(t)d\varpi(t), \\ y(t) &= C_i(\beta)x(t), \quad i = 1, 2, \ldots, r, \end{aligned}$$

where $x(t) \in \mathbb{R}^n$ denotes the system state; $u(t) \in \mathbb{R}^l$ denotes the input which belongs to $\mathcal{L}_2[0, \infty]$; $y(t) \in \mathbb{R}^m$ represents the output; $\varpi(t)$ represents the scalar Brownian motion under a probability space $(\Omega, \mathcal{F}, \{\mathcal{F}_t\}_{t\geq 0}, \mathcal{P})$ satisfying $\mathbf{E}\{d\varpi(t)\} = 0$ and

X. Su et al., *Intelligent Control, Filtering and Model Reduction Analysis for Fuzzy-Model-Based Systems*, Studies in Systems, Decision and Control 385,
https://doi.org/10.1007/978-3-030-81214-0_10

$\mathbf{E}\left\{d\varpi^2(t)\right\} = dt$; r is the quantity of fuzzy rules, $\mu_{i1}(\beta), \ldots, \mu_{ip}(\beta)$ denote the fuzzy sets; $\theta_1(\beta, t), \theta_2(\beta, t), \ldots, \theta_p(\beta, t)$ denote premise variables; $\{(A_i(\beta), B_i(\beta), E_i(\beta), C_i(\beta)) : \beta \in \mathscr{I}\}$ are some matrices parameterized by $\mathscr{I} = \{1, 2, \ldots, S\}$ and $\beta : \mathbb{R} \to \mathscr{I}$ denotes a piecewise constant function of t referred to as a switching signal. For the given time t, $\beta(t)$, simplified as β, may associate with t or $x(t)$, or both, or other hybrid combination. From [58], $\beta(t)$ is assumed to be unknown, but its momentary value can be accessible in real time. For every probable value $\beta(t) = j$, $j \in \mathscr{I}$, the system matrices related to the mode j are given by

$$\begin{aligned} A_i(j) &= A_i(\beta), \quad B_i(j) = B_i(\beta), \\ C_i(j) &= C_i(\beta), \quad E_i(j) = E_i(\beta), \end{aligned}$$

where $A_i(j)$, $B_i(j)$, $C_i(j)$ and $E_i(j)$ denote constant matrices. Considering the switching signal $\beta(t)$, we set the switching sequence as

$$\{(j_0, t_0), (j_1, t_1), \ldots, (j_k, t_k), \ldots, \mid j_k \in \mathscr{I},\ k = 0, 1, \ldots\}$$

with $t_0 = 0$, which signifies the j_kth subsystem can be activated during $t \in [t_k, t_{k+1})$.

Under the assumption that the premise variables don't rely on $u(t)$. Then for a couple of $(x(t), u(t))$, the fuzzy switched systems with stochastic disturbance can be described by the following form:

$$\begin{aligned} dx(t) &= \sum_{i=1}^{r} h_{(i,\beta)}(\theta(\beta, t)) \Big\{ \Big[A_i(\beta)x(t) + B_i(\beta)u(t) \Big] dt \\ &\quad + E_i(\beta)x(t)d\varpi(t) \Big\}, \qquad (10.1a) \\ y(t) &= \sum_{i=1}^{r} h_{(i,\beta)}(\theta(\beta, t)) C_i(\beta)x(t),\ i = 1, 2, \ldots, r, \qquad (10.1b) \end{aligned}$$

where

$$\begin{aligned} h_{(i,\beta)}(\theta(\beta, t)) &= \nu_{(i,\beta)}(\theta(\beta, t)) / \sum_{i=1}^{r} \nu_{(i,\beta)}(\theta(\beta, t)), \\ \nu_{(i,\beta)}(\theta(\beta, t)) &= \prod_{l=1}^{p} \mu_{(il,\beta)}(\theta_l(\beta, t)), \end{aligned}$$

and $\mu_{(il,\beta)}(\theta_l(\beta, t))$ represents the grade of membership of $\theta_l(\beta, t)$ in $\mu_{(il,\beta)}$. Provided that $\nu_{(i,\beta)}(\theta(\beta, t)) \geq 0$, $i = 1, 2, \ldots, r$, $\sum_{i=1}^{r} \nu_{(i,\beta)}(\theta(\beta, t)) > 0$ for all t. Hence, $h_{(i,\beta)}(\theta(\beta, t)) \geq 0$ for $i = 1, 2, \ldots, r$ and $\sum_{i=1}^{r} h_{(i,\beta)}(\theta(\beta, t)) = 1$ for all t.

Let us suppose the premise variable of the original model $\theta(\beta, t)$ is accessible for the reduced-order model, which means $h_{(i,\beta)}(\theta(\beta, t))$ is accessible as well. The previous system (10.1) can be approximated with the reduced-order fuzzy switched model as follows:

♦ **Reduced-Order Plant Form**:

Rule $\mathcal{R}_i^{\beta}$: IF $\theta_1(\beta, t)$ is $\mu_{(i1,\beta)}$ and $\theta_2(\beta, t)$ is $\mu_{(i2,\beta)}$ and $\cdots$ and $\theta_p(\beta, t)$ is $\mu_{(ip,\beta)}$, THEN

$$\begin{aligned} d\hat{x}(t) &= \left[\hat{A}_i(\beta)\hat{x}(t) + \hat{B}_i(\beta)u(t)\right]dt + \hat{E}_i(\beta)\hat{x}(t)d\varpi(t), \\ \hat{y}(t) &= \hat{C}_i(\beta)\hat{x}(t), \end{aligned}$$

where $\hat{x}(t) \in \mathbb{R}^k$ denotes the state of the reduced-order model and $k < n$; $\hat{A}_i(\beta)$, $\hat{B}_i(\beta)$, $\hat{C}_i(\beta)$ and $\hat{E}_i(\beta)$ represent some suitably dimensioned matrices to be computed later.

The reduced-order model can be formulated with a more complete expression as below:

$$\begin{aligned} d\hat{x}(t) &= \sum_{i=1}^{r} h_{(i,\beta)}(\theta(\beta, t))\Big\{\left[\hat{A}_i(\beta)\hat{x}(t) + \hat{B}_i(\beta)u(t)\right]dt \\ &\quad + \hat{E}_i(\beta)\hat{x}(t)d\varpi(t)\Big\}, \end{aligned} \tag{10.2a}$$

$$\hat{y}(t) = \sum_{i=1}^{r} h_{(i,\beta)}(\theta(\beta, t))\hat{C}_i(\beta)\hat{x}(t), \; i = 1, 2, \ldots, r. \tag{10.2b}$$

Let the original fuzzy switched system (10.1) contain the information in reduced-order model (10.2), we can get the overall dynamic error system:

$$\begin{aligned} d\tilde{x}(t) &= \sum_{i=1}^{r} h_{(i,\beta)}(\theta(\beta, t)) \sum_{l=1}^{r} h_{(l,\beta)}(\theta(\beta, t))\Big\{\tilde{E}_{il}(\beta)\tilde{x}(t)d\varpi(t) \\ &\quad + \left[\tilde{A}_{il}(\beta)\tilde{x}(t) + \tilde{B}_{il}(\beta)u(t)\right]dt\Big\}, \end{aligned} \tag{10.3a}$$

$$e(t) = \sum_{i=1}^{r} h_{(i,\beta)}(\theta(\beta, t)) \sum_{l=1}^{r} h_{(l,\beta)}(\theta(\beta, t))\tilde{C}_{il}(\beta)\tilde{x}(t), \tag{10.3b}$$

where $\tilde{x}(t) \triangleq \begin{bmatrix} x(t) \\ \hat{x}(t) \end{bmatrix}$, $e(t) \triangleq y(t) - \hat{y}(t)$ and

$$\begin{cases} \tilde{A}_{il}(\beta) \triangleq \begin{bmatrix} A_i(\beta) & 0 \\ 0 & \hat{A}_l(\beta) \end{bmatrix}, & \tilde{E}_{il}(\beta) \triangleq \begin{bmatrix} E_i(\beta) & 0 \\ 0 & \hat{E}_l(\beta) \end{bmatrix}, \\ \tilde{C}_{il}(\beta) \triangleq \begin{bmatrix} C_i(\beta) & -\hat{C}_l(\beta) \end{bmatrix}, & \tilde{B}_{il}(\beta) \triangleq \begin{bmatrix} B_i(\beta) \\ \hat{B}_l(\beta) \end{bmatrix}. \end{cases}$$

Some relevant definitions are given as follows.

Definition 10.1 For the system in (10.3), the equilibrium $\tilde{x}^{\star}(t) = 0$ under $u(t) = 0$ is termed as to have the mean-square exponential stability with $\beta(t)$ if the following inequality holds:

$$\mathbf{E}\left\{\|\tilde{x}(t)\|^2\right\} \leq \eta \, \|\tilde{x}(t_0)\|^2 \, e^{-\lambda(t-t_0)}, \quad \forall t \geq t_0,$$

where $\eta \geq 1$ and $\lambda > 0$.

Definition 10.2 *[125]* For any $T_2 > T_1 \geq 0$, $N_\beta(T_1, T_2)$ represents the number of switchings of $\beta(t)$ over (T_1, T_2). If $N_\beta(T_1, T_2) \leq N_0 + (T_2 - T_1)/T_a$ is satisfied for $T_a > 0$, $N_0 \geq 0$, then T_a is referred to the ADT.

Definition 10.3 *[25]* For $\alpha > 0$, $\gamma > 0$, the system in (10.3) is termed as to have the mean-square exponential stability subject to a Hankel-norm error performance (γ, α) under $\beta(t)$, if it is mean-square exponentially stable with $u(t) = 0$ and the following condition satisfies:

$$\mathbf{E}\left\{\int_T^\infty e^{-\alpha t} e^T(t) e(t) dt\right\} < \gamma^2 \int_0^T u^T(t) u(t) dt,$$

for all $u(t) \in \mathcal{L}_2[0, \infty)$ with $u(t) = 0$, $\forall t \geq T$.

Assumption 10.1 The original system in (10.1) is mean-square exponentially stable.

Remark 10.4 Assumption 10.1 is on the basis of there being no control in original system (10.1). Thus, in order to let the obtained dynamic error system (10.3) be mean-square exponentially stable, the original system we want to approximate must be mean-square exponentially stable, which is a precondition.

10.3 Main Results

10.3.1 Hankel-Norm Performance Analysis

In this part, we propose parameter-dependent sufficient conditions, such that the overall system in (10.3) is mean-square exponentially stable and satisfies the Hankel-norm error performance.

Theorem 10.5 *For the given scalars $\alpha > 0$ and $\gamma > 0$, provided that there are matrices $0 < P_1(j) \in \mathbb{R}^{(n+k)\times(n+k)}$ and $0 < P_2(j) \in \mathbb{R}^{(n+k)\times(n+k)}$, such that for $j \in \mathscr{I}$, $i = 1, 2, \ldots, r$, $l = 1, 2, \ldots, r$,*

$$\Pi_{il}(j) \triangleq \begin{bmatrix} \Pi_{11il}(j) & P_1(j)\tilde{B}_{il}(j) & \tilde{E}_{il}^T(j)P_1(j) \\ \star & -\gamma^2 I & 0 \\ \star & \star & -P_1(j) \end{bmatrix} < 0, \tag{10.4a}$$

$$\Phi_{il}(j) \triangleq \begin{bmatrix} \Phi_{11il}(j) & \tilde{E}_{il}^T(j)P_2(j) & \tilde{C}_{il}^T(j) \\ \star & -P_2(j) & 0 \\ \star & \star & -I \end{bmatrix} < 0, \tag{10.4b}$$

$$P_2(j) - P_1(j) < 0, \tag{10.4c}$$

where

$$\begin{cases} \Pi_{11il}(j) \triangleq P_1(j)\tilde{A}_{il}(j) + \tilde{A}_{il}^T(j)P_1(j) + \alpha P_1(j), \\ \Phi_{11il}(j) \triangleq P_2(j)\tilde{A}_{il}(j) + \tilde{A}_{il}^T(j)P_2(j) + \alpha P_2(j). \end{cases}$$

Then the derivative system in (10.3) has the mean-square exponential stability and Hankel-norm error performance (γ, α) during any switching signal under $T_a > T_a^\star = \frac{\ln \mu}{\alpha}$, where $\mu \geq 1$ and

$$P_1(j) \leq \mu P_1(s), \;\; P_2(j) \leq \mu P_2(s), \;\; \forall j, s \in \mathcal{I}. \tag{10.5}$$

Besides, a state decay estimate is described as

$$\mathbf{E}\left\{\|\tilde{x}(t)\|^2\right\} \leq \eta e^{-\lambda t} \|\tilde{x}(0)\|^2, \tag{10.6}$$

where

$$\begin{cases} \lambda = \alpha - \frac{\ln \mu}{T_a} > 0, & b = \max\limits_{\forall j \in \mathcal{I}} \lambda_{\max}(P_1(j)), \\ \eta = \frac{b}{a} \geq 1, & a = \min\limits_{\forall j \in \mathcal{I}} \lambda_{\min}(P_1(j)). \end{cases} \tag{10.7}$$

Proof Considering the fuzzy membership functions and (10.4a)–(10.4b), we have

$$\sum_{i=1}^{r} h_{(i,\beta)}(\theta(\beta, t)) \sum_{l=1}^{r} h_{(l,\beta)}(\theta(\beta, t))\Pi_{il}(\beta) < 0, \tag{10.8a}$$

$$\sum_{i=1}^{r} h_{(i,\beta)}(\theta(\beta, t)) \sum_{l=1}^{r} h_{(l,\beta)}(\theta(\beta, t))\Phi_{il}(\beta) < 0, \tag{10.8b}$$

where $\Pi_{il}(\beta)$ and $\Phi_{il}(\beta)$ are given in (10.4).

Construct the following piecewise Lyapunov function:

$$V(\tilde{x}_t, \beta, t) \triangleq \tilde{x}^T(t)P_1(\beta)\tilde{x}(t), \tag{10.9}$$

where $P_1(\beta) > 0, \beta \in \mathcal{I}$ are still to be decided. Based on the Itô formula, and along the solution of overall system for a settled β and $u(t) = 0$, the stochastic differential can be obtained as follows:

$$
\begin{aligned}
dV(\tilde{x}_t, \beta, t) &= \mathscr{L}V(\tilde{x}_t, \beta, t)dt + 2\sum_{i=1}^{r} h_{(i,\beta)}(\theta(\beta, t)) \\
&\quad \sum_{l=1}^{r} h_{(l,\beta)}(\theta(\beta, t))\tilde{x}^T(t)P_1(\beta)\tilde{E}_{il}(\beta)\tilde{x}^T(t)d\varpi(t), \\
\mathscr{L}V(\tilde{x}_t, \beta, t) &= \sum_{i=1}^{r} h_{(i,\beta)}(\theta(\beta, t)) \sum_{l=1}^{r} h_{(l,\beta)}(\theta(\beta, t)) \\
&\quad 2\tilde{x}^T(t)P_1(\beta)\tilde{A}_{il}(\beta)\tilde{x}(t) + \sum_{i=1}^{r} h_{(i,\beta)}(\theta(\beta, t)) \\
&\quad \sum_{l=1}^{r} h_{(l,\beta)}(\theta(\beta, t)) \sum_{o=1}^{r} h_{(o,\beta)}(\theta(\beta, t)) \\
&\quad \sum_{s=1}^{r} h_{(s,\beta)}(\theta(\beta, t))\tilde{x}^T(t)\tilde{E}_{il}^T(\beta)P_1(\beta)\tilde{E}_{os}(\beta)\tilde{x}(t) \\
&\leqslant \sum_{i=1}^{r} h_{(i,\beta)}(\theta(\beta, t)) \sum_{l=1}^{r} h_{(l,\beta)}(\theta(\beta, t))\tilde{x}^T(t)\Big\{P_1(\beta) \\
&\quad \tilde{A}_{il}(\beta) + \tilde{A}_{il}^T(\beta)P_1(\beta)\Big\}\tilde{x}(t) + \sum_{i=1}^{r} h_{(i,\beta)}(\theta(\beta, t)) \\
&\quad \sum_{l=1}^{r} h_{(l,\beta)}(\theta(\beta, t))\tilde{x}^T(t)\tilde{E}_{il}^T(\beta)P_1(\beta)\tilde{E}_{il}(\beta)\tilde{x}(t). \qquad (10.10)
\end{aligned}
$$

On account of (10.8a) and (10.10), employing Schur complement method, we have

$$
\mathscr{L}V(\tilde{x}_t, \beta, t) < -\alpha\tilde{x}^T(t)P_1(\beta)\tilde{x}(t) = -\alpha V(\tilde{x}_t, \beta, t),
$$

$$
\begin{aligned}
dV(\tilde{x}_t, \beta, t) &< -\alpha V(\tilde{x}_t, \beta, t)dt + 2\sum_{i=1}^{r} h_{(i,\beta)}(\theta(\beta, t)) \\
&\quad \sum_{l=1}^{r} h_{(l,\beta)}(\theta(\beta, t))\tilde{x}^T(t)P_1(\beta)\tilde{E}_{il}(\beta)\tilde{x}^T(t)d\varpi(t).
\end{aligned}
$$

Then

$$
\begin{aligned}
d[e^{\alpha t}V(\tilde{x}_t,\beta,t)] &= \alpha e^{\alpha t}V(\tilde{x}_t,\beta,t)dt + e^{\alpha t}dV(\tilde{x}_t,\beta,t)\\
&< 2\sum_{i=1}^{r} h_{(i,\beta)}(\theta(\beta,t))\sum_{l=1}^{r} h_{(l,\beta)}(\theta(\beta,t))\\
&\quad e^{\alpha t}\tilde{x}^T(t)P_1(\beta)\tilde{E}_{il}(\beta)\tilde{x}^T(t)d\varpi(t).
\end{aligned}
\tag{10.11}
$$

Integrating two sides in (10.11) from $t^\star > 0$ to t, then make expectations. By using some efficient mathematical calculations, it yields

$$
\mathbf{E}\{V(\tilde{x}_t,\beta,t)\} < e^{-\alpha(t-t^\star)}\mathbf{E}\left\{V(\tilde{x}_{t^\star},\beta,t^\star)\right\}. \tag{10.12}
$$

Given any piecewise constant switching signal β, for $t > 0$, set $0 = t_0 < t_1 < \cdots < t_k < \cdots$, $k = 0, 1, \ldots$, which stand for the switching points of β during the time interval $(0, t)$. The j_kth subsystem is activated during $t \in [t_k, t_{k+1})$. Denote $t^\star = t_k$ in (10.12), which yields

$$
\mathbf{E}\{V(\tilde{x}_t,\beta,t)\} < e^{-\alpha(t-t_k)}\mathbf{E}\left\{V(\tilde{x}_{t_k},\beta,t_k)\right\}. \tag{10.13}
$$

Considering (10.5) and (10.9), at the switching moment t_k, we can get

$$
\mathbf{E}\left\{V(\tilde{x}_{t_k},\beta,t_k)\right\} \le \mu\mathbf{E}\left\{V(\tilde{x}_{t_k^-},\beta,t_k^-)\right\}. \tag{10.14}
$$

Consequently, based on (10.13)–(10.14) and $\vartheta = N_\beta(0,t) \le (t-0)/T_a$, we have

$$
\begin{aligned}
\mathbf{E}\{V(\tilde{x}_t,\beta,t)\} &\le e^{-\alpha(t-t_k)}\mu\mathbf{E}\left\{V(\tilde{x}_{t_k^-},\beta,t_k^-)\right\}\\
&\le \cdots \le e^{-\alpha(t-0)}\mu^{\vartheta}\mathbf{E}\{V(\tilde{x}_0,\beta,0)\}\\
&\le e^{-(\alpha-\ln\mu/T_a)t}\mathbf{E}\{V(\tilde{x}_0,\beta,0)\}\\
&= e^{-(\alpha-\ln\mu/T_a)t}V(\tilde{x}_0,\beta,0).
\end{aligned}
\tag{10.15}
$$

From (10.9), it yields

$$
\mathbf{E}\{V(\tilde{x}_t,\beta,t)\} \ge a\mathbf{E}\left\{\|\tilde{x}(t)\|^2\right\},\ V(\tilde{x}_0,\beta,0) \le b\|\tilde{x}(0)\|^2, \tag{10.16}
$$

where a and b have been denoted in (10.7). Combining (10.15)–(10.16), we can obtain

$$
\mathbf{E}\left\{\|\tilde{x}(t)\|^2\right\} \le \frac{1}{a}\mathbf{E}\{V(\tilde{x}_t,\beta,t)\} \le \frac{b}{a}e^{-(\alpha-\ln\mu/T_a)t}\|\tilde{x}(0)\|^2,
$$

which signifies (10.6). When $t_0 = 0$, in consideration of Definition 1, the dynamic error system in (10.3) with $u(t) = 0$ satisfies the mean-square exponential stability.

Next, we are going to verify the Hankel-norm performance for the dynamic error system in (10.3). As for any nonzero $u(t) \in \mathcal{L}_2[0,\infty)$, we can get

$$
\begin{aligned}
&\mathscr{L}V(\tilde{x}_t,\beta,t)+\alpha V(\tilde{x}_t,\beta,t)-\gamma^2u^T(t)u(t)\\
&=\sum_{i=1}^{r}h_{(i,\beta)}(\theta(\beta,t))\sum_{l=1}^{r}h_{(l,\beta)}(\theta(\beta,t))2\tilde{x}^T(t)P_1(\beta)\left[\tilde{A}_{il}(\beta)\tilde{x}(t)+\tilde{B}_{il}(\beta)u(t)\right]\\
&\quad+\alpha\tilde{x}^T(t)P_1(\beta)\tilde{x}(t)-\gamma^2u^T(t)u(t)+\sum_{i=1}^{r}h_{(i,\beta)}(\theta(\beta,t))\sum_{l=1}^{r}h_{(l,\beta)}(\theta(\beta,t))\\
&\quad\sum_{o=1}^{r}h_{(o,\beta)}(\theta(\beta,t))\sum_{s=1}^{r}h_{(s,\beta)}(\theta(\beta,t))\tilde{x}^T(t)\tilde{E}_{il}^T(\beta)P_1(\beta)\tilde{E}_{os}(\beta)\tilde{x}(t)\\
&\leqslant\sum_{i=1}^{r}h_{(i,\beta)}(\theta(\beta,t))\sum_{l=1}^{r}h_{(l,\beta)}(\theta(\beta,t))2\tilde{x}^T(t)P_1(\beta)\left[\tilde{A}_{il}(\beta)\tilde{x}(t)+\tilde{B}_{il}(\beta)u(t)\right]\\
&\quad+\alpha\tilde{x}^T(t)P_1(\beta)\tilde{x}(t)-\gamma^2u^T(t)u(t)+\sum_{i=1}^{r}h_{(i,\beta)}(\theta(\beta,t))\sum_{l=1}^{r}h_{(l,\beta)}(\theta(\beta,t))\\
&\quad\tilde{x}^T(t)\tilde{E}_{il}^T(\beta)P_1(\beta)\tilde{E}_{il}(\beta)\tilde{x}(t)\\
&=\tilde{\xi}^T(t)\left[\sum_{i=1}^{r}h_{(i,\beta)}(\theta(\beta,t))\sum_{l=1}^{r}h_{(l,\beta)}(\theta(\beta,t))\tilde{\Pi}_{il}(\beta)\right]\tilde{\xi}(t),
\end{aligned}
$$

where $\tilde{\xi}(t)\triangleq\begin{bmatrix}\tilde{x}(t)\\u(t)\end{bmatrix}$ and $\tilde{\Pi}_{il}(\beta)\triangleq\begin{bmatrix}\tilde{\Pi}_{11il}(\beta) & \tilde{\Pi}_{12il}(\beta)\\ \star & \tilde{\Pi}_{22}(\beta)\end{bmatrix}$, with

$$
\begin{cases}
\tilde{\Pi}_{11il}(\beta)\triangleq\alpha P_1(\beta)+\tilde{E}_{il}^T(\beta)P_1(\beta)\tilde{E}_{il}(\beta)+P_1(\beta)\tilde{A}_{il}(\beta)+\tilde{A}_{il}^T(\beta)P_1(\beta),\\
\tilde{\Pi}_{12il}(\beta)\triangleq P_1(\beta)\tilde{B}_{il}(\beta),\quad \tilde{\Pi}_{22}(\beta)\triangleq-\gamma^2I.
\end{cases}
$$

Employing the Schur's complement method, (10.8a) is equivalent to

$$
\sum_{i=1}^{r}h_{(i,\beta)}(\theta(\beta,t))\sum_{l=1}^{r}h_{(l,\beta)}(\theta(\beta,t))\tilde{\Pi}_{il}(\beta)<0,
$$

hence,

$$
\mathscr{L}V(\tilde{x}_t,\beta,t)+\alpha V(\tilde{x}_t,\beta,t)-\gamma^2u^T(t)u(t)<0. \tag{10.17}
$$

Set $\Gamma(t)\triangleq-\gamma^2u^T(t)u(t)$, then

$$
\mathscr{L}V(\tilde{x}_t,\beta,t)<-\alpha V(\tilde{x}_t,\beta,t)-\Gamma(t),
$$

and

$$\begin{aligned} dV(\tilde{x}_t,\beta,t) &= \mathscr{L}V(\tilde{x}_t,\beta,t)dt + 2\sum_{i=1}^{r} h_{(i,\beta)}(\theta(\beta,t)) \\ &\quad \sum_{l=1}^{r} h_{(l,\beta)}(\theta(\beta,t))\tilde{x}^T(t)P_1(\beta)\tilde{E}_{il}(\beta)\tilde{x}^T(t)d\varpi(t) \\ &< -\alpha V(\tilde{x}_t,\beta,t)dt + 2\sum_{i=1}^{r} h_{(i,\beta)}(\theta(\beta,t)) \\ &\quad \sum_{i=1}^{r} h_{(l,\beta)}(\theta(\beta,t))\tilde{x}^T(t)P_1(\beta)\tilde{E}_{il}(\beta)\tilde{x}^T(t)d\varpi(t) - \Gamma(t)dt. \end{aligned}$$

Notice that

$$\begin{aligned} d[e^{\alpha t}V(\tilde{x}_t,\beta,t)] &= \alpha e^{\alpha t}V(\tilde{x}_t,\beta,t)dt + e^{\alpha t}dV(\tilde{x}_t,\beta,t) \\ &< e^{\alpha t}\Bigg[-\Gamma(t)dt + 2\sum_{i=1}^{r} h_{(i,\beta)}(\theta(\beta,t))\sum_{l=1}^{r} h_{(l,\beta)}(\theta(\beta,t)) \\ &\quad \tilde{x}^T(t)P_1(\beta)\tilde{E}_{il}(\beta)\tilde{x}^T(t)d\varpi(t)\Bigg]. \end{aligned} \tag{10.18}$$

Integrating two sides of (10.18) from $t^\star > 0$ to t, then make expectations as

$$\mathbf{E}\left\{e^{\alpha t}V(\tilde{x}_t,\beta,t)\right\} - \mathbf{E}\left\{e^{\alpha t^\star}V(\tilde{x}_{t^\star},\beta,t^\star)\right\} < -\mathbf{E}\left\{\int_{t^\star}^{t} e^{\alpha s}\Gamma(s)ds\right\},$$

that is,

$$\mathbf{E}\{V(\tilde{x}_t,\beta,t)\} < e^{-\alpha(t-t^\star)}\mathbf{E}\left\{V(\tilde{x}_{t^\star},\beta,t^\star)\right\} - \mathbf{E}\left\{\int_{t^\star}^{t} e^{-\alpha(t-s)}\Gamma(s)ds\right\}. \tag{10.19}$$

Utilizing the similar methods as the proof of stability analysis, it yields

$$\mathbf{E}\{V(\tilde{x}_t,\beta,t)\} < e^{-\alpha(t-t_k)}\mathbf{E}\left\{V(\tilde{x}_{t_k},\beta,t_k)\right\} - \mathbf{E}\left\{\int_{t_k}^{t} e^{-\alpha(t-s)}\Gamma(s)ds\right\}. \tag{10.20}$$

Thus, on the basis of (10.14), (10.20) and $\vartheta = N_\beta(0,t) \le (t-0)/T_a$, we have

$$\begin{aligned} &\mathbf{E}\{V(\tilde{x}_t,\beta,t)\} \\ &\le e^{-\alpha(t-t_k)}\mu\mathbf{E}\left\{V(\tilde{x}_{t_k^-},\beta,t_k^-)\right\} - \mathbf{E}\left\{\int_{t_k}^{t} e^{-\alpha(t-s)}\Gamma(s)ds\right\} \end{aligned}$$

$$
\begin{aligned}
&\leq \mu^{\vartheta} e^{-\alpha t}\mathbf{E}\{V(\tilde{x}_0,\beta,0)\} - \mu^{\vartheta}\mathbf{E}\left\{\int_0^{t_1} e^{-\alpha(t-s)}\Gamma(s)ds\right\} \\
&\quad -\mu^{\vartheta-1}\mathbf{E}\left\{\int_{t_1}^{t_2} e^{-\alpha(t-s)}\Gamma(s)ds\right\} - \cdots - \mu^{0}\mathbf{E}\left\{\int_{t_k}^{t} e^{-\alpha(t-s)}\Gamma(s)ds\right\} \\
&= -\mathbf{E}\left\{\int_0^{t} e^{-\alpha(t-s)+N_\beta(t,s)\ln\mu}\Gamma(s)ds\right\} + e^{-\alpha t+N_\beta(0,t)\ln\mu}V(\tilde{x}_0,\beta,0). \quad (10.21)
\end{aligned}
$$

Considering the zero initial condition $\tilde{x}(0)=0$, (10.21) means

$$
\mathbf{E}\{V(\tilde{x}_t,\beta,t)\} \leq \gamma^2\int_0^t e^{-\alpha(t-s)+N_\beta(s,t)\ln\mu}u^T(s)u(s)ds. \quad (10.22)
$$

Multiply two sides of (10.22) by $e^{-N_\beta(0,t)\ln\mu}$ and it gets

$$
\begin{aligned}
&e^{-N_\beta(0,t)\ln\mu}\mathbf{E}\{V(\tilde{x}_t,\beta,t)\} \\
&\leq \gamma^2\int_0^t e^{-\alpha(t-s)-N_\beta(0,s)\ln\mu}u^T(s)u(s)ds \\
&\leq \gamma^2\int_0^t e^{-\alpha(t-s)}u^T(s)u(s)ds \leq \gamma^2\int_0^t u^T(s)u(s)ds. \quad (10.23)
\end{aligned}
$$

Due to $N_\beta(0,t) \leq t/T_a$ and $T_a > T_a^\star = \ln\mu/\alpha$, we have $N_\beta(0,t)\ln\mu \leq \alpha t$. Then (10.23) signifies

$$
e^{-\alpha t}\mathbf{E}\{V(\tilde{x}_t,\beta,t)\} \leq \gamma^2\int_0^t u^T(s)u(s)ds.
$$

Denoting $u(t)=0, \forall t \geq T$, and it yields

$$
e^{-\alpha t}\mathbf{E}\{V(\tilde{x}_t,\beta,t)\} \leq \gamma^2\int_0^T u^T(t)u(t)dt. \quad (10.24)
$$

Give the Lyapunov function as follow:

$$
W(\tilde{x}_t,\beta,t) \triangleq \tilde{x}^T(t)P_2(\beta)\tilde{x}(t), \quad (10.25)
$$

where $P_2(\beta) > 0$, $\beta \in \mathscr{I}$ is to be decided. When $u(t)=0, \forall t \geq T$, we can get

$$
\begin{aligned}
&\mathscr{L}W(\tilde{x}_t,\beta,t) + \alpha W(\tilde{x}_t,\beta,t) + e^T(t)e(t) \\
&\leqslant \sum_{i=1}^{r} h_{(i,\beta)}(\theta(\beta,t))\sum_{l=1}^{r} h_{(l,\beta)}(\theta(\beta,t))\tilde{x}^T(t)\Big\{P_2(\beta)\tilde{A}_{il}(\beta) + \tilde{A}_{il}^T(\beta)P_2(\beta)\Big\}\tilde{x}(t) \\
&\quad + \sum_{i=1}^{r} h_{(i,\beta)}(\theta(\beta,t))\sum_{l=1}^{r} h_{(l,\beta)}(\theta(\beta,t))\tilde{x}^T(t)\tilde{E}_{il}^T(\beta)P_2(\beta)\tilde{E}_{il}(\beta)\tilde{x}(t)
\end{aligned}
$$

$$
\begin{aligned}
&+\sum_{i=1}^{r} h_{(i,\beta)}(\theta(\beta,t)) \sum_{l=1}^{r} h_{(l,\beta)}(\theta(\beta,t)) \tilde{x}^T(t) \tilde{C}_{il}^T(\beta) \tilde{C}_{il}(\beta) \tilde{x}^T(t) \\
&+\alpha \tilde{x}^T(t) P_2(\beta) \tilde{x}(t) \\
= \tilde{x}^T(t) &\left[\sum_{i=1}^{r} h_{(i,\beta)}(\theta(\beta,t)) \sum_{l=1}^{r} h_{(l,\beta)}(\theta(\beta,t)) \bar{\Pi}_{il}(\beta) \right] \tilde{x}(t),
\end{aligned}
$$

where

$$
\begin{aligned}
\bar{\Pi}_{il}(\beta) \triangleq\ & P_2(\beta)\tilde{A}_{il}(\beta) + \tilde{A}_i^T(\beta)P_2(\beta) + \tilde{E}_{il}^T(\beta)P_2(\beta)\tilde{E}_{il}(\beta) \\
& +\alpha P_2(\beta) + \tilde{C}_{il}^T(\beta)\tilde{C}_{il}(\beta).
\end{aligned}
$$

Based on Schur's complement approach, (10.8b) is equivalent to

$$
\sum_{i=1}^{r} h_{(i,\beta)}(\theta(\beta,t)) \sum_{l=1}^{r} h_{(l,\beta)}(\theta(\beta,t)) \bar{\Pi}_{il}(\beta) < 0,
$$

hence,

$$
\mathscr{L}W(\tilde{x}_t, \beta, t) + \alpha W(\tilde{x}_t, \beta, t) + e^T(t)e(t) < 0. \tag{10.26}
$$

Define $\Lambda(t) \triangleq e^T(t)e(t)$, then

$$
\mathscr{L}W(\tilde{x}_t, \beta, t) < -\alpha W(\tilde{x}_t, \beta, t) - \Lambda(t),
$$

and

$$
\begin{aligned}
dW(\tilde{x}_t, \beta, t) = &\ \mathscr{L}W(\tilde{x}_t, \beta, t) + 2\sum_{i=1}^{r} h_{(i,\beta)}(\theta(\beta,t)) \sum_{l=1}^{r} h_{(l,\beta)}(\theta(\beta,t)) \\
&\ \tilde{x}^T(t) P_2(\beta) \tilde{E}_{il}(\beta) \tilde{x}^T(t) d\varpi(t) \\
< &\ 2\sum_{i=1}^{r} h_{(i,\beta)}(\theta(\beta,t)) \sum_{l=1}^{r} h_{(l,\beta)}(\theta(\beta,t)) \tilde{x}^T(t) P_2(\beta) \tilde{E}_{il}(\beta) \\
&\ \tilde{x}^T(t) d\varpi(t) - \alpha W(\tilde{x}_t, \beta, t)dt - \Lambda(t)dt.
\end{aligned}
$$

On account of

$$
\begin{aligned}
& d[e^{\alpha t} W(\tilde{x}_t, \beta, t)] = \alpha e^{\alpha t} W(\tilde{x}_t, \beta, t)dt + e^{\alpha t} dW(\tilde{x}_t, \beta, t) \\
< e^{\alpha t} & \left[-\Lambda(t)dt + 2\sum_{i=1}^{r} h_{(i,\beta)}(\theta(\beta,t)) \sum_{l=1}^{r} h_{(l,\beta)}(\theta(\beta,t)) \right. \\
& \left. \tilde{x}^T(t) P_2(\beta) \tilde{E}_{il}(\beta) \tilde{x}^T(t) d\varpi(t) \right].
\end{aligned} \tag{10.27}
$$

Integrating two sides of (10.27) from T to t, then make expectations as

$$\mathbf{E}\left\{e^{\alpha t}W(\tilde{x}_t,\beta,t)\right\}-\mathbf{E}\left\{e^{\alpha T}W(\tilde{x}_T,\beta,T)\right\}<-\mathbf{E}\left\{\int_T^t e^{\alpha s}\Lambda(s)ds\right\},$$

in other words,

$$\mathbf{E}\{W(\tilde{x}_t,\beta,t)\}<e^{-\alpha(t-T)}\mathbf{E}\{W(\tilde{x}_T,\beta,T)\}-\mathbf{E}\left\{\int_T^t e^{-\alpha(t-s)}\Lambda(s)ds\right\}. \quad (10.28)$$

$T=t_0<t_1<\cdots<t_k<\cdots, k=0,1,\ldots$, which denote the switching points of β during (T,t). The i_kth subsystem is activated when $t\in\left[t_k,t_{k+1}\right)$, then

$$\mathbf{E}\{W(\tilde{x}_t,\beta,t)\}<e^{-\alpha(t-t_k)}\mathbf{E}\left\{V(\tilde{x}_{t_k},\beta,t_k)\right\}-\mathbf{E}\left\{\int_{t_k}^t e^{-\alpha(t-s)}\Lambda(s)ds\right\}. \quad (10.29)$$

At the switching moment t_k, based on (10.5) and (10.25), we can get

$$\mathbf{E}\left\{W(\tilde{x}_{t_k},\beta(t_k),t_k)\right\}\le\mu\mathbf{E}\left\{W(\tilde{x}_{t_k^-},\beta(t_k^-),t_k^-)\right\}. \quad (10.30)$$

Hence, in consideration of (10.29)–(10.30) and $\tilde{\vartheta}=N_\beta(T,t)\le(t-T)/T_a$, we have

$$\begin{aligned}
&\mathbf{E}\{W(\tilde{x}_t,\beta,t)\}\\
&\le\mu e^{-\alpha(t-t_k)}\mathbf{E}\left\{W(\tilde{x}_{t_k^-},\beta(t_k^-),t_k^-)\right\}-\mathbf{E}\left\{\int_{t_k}^t e^{-\alpha(t-s)}\Lambda(s)ds\right\}\\
&\le\mu^{\tilde{\vartheta}}e^{-\alpha(t-T)}\mathbf{E}\{W(\tilde{x}_T,\beta(T),T)\}-\mu^{\tilde{\vartheta}}\mathbf{E}\left\{\int_T^{t_1} e^{-\alpha(t-s)}\Lambda(s)ds\right\}\\
&\quad-\mu^{\tilde{\vartheta}-1}\mathbf{E}\left\{\int_{t_1}^{t_2} e^{-\alpha(t-s)}\Lambda(s)ds\right\}-\cdots-\mu^0\mathbf{E}\left\{\int_{t_k}^t e^{-\alpha(t-s)}\Lambda(s)ds\right\}\\
&=e^{-\alpha(t-T)+N_\beta(T,t)\ln\mu}\mathbf{E}\{W(\tilde{x}_T,\beta(T),T)\}\\
&\quad-\mathbf{E}\left\{\int_T^t e^{-\alpha(t-s)+N_\beta(s,t)\ln\mu}\Lambda(s)ds\right\},
\end{aligned}$$

that is

$$\begin{aligned}
&\mathbf{E}\left\{\int_T^t e^{-\alpha(t-s)+N_\beta(s,t)\ln\mu}e^T(s)e(s)ds\right\}\\
&\le e^{-\alpha(t-T)+N_\beta(T,t)\ln\mu}\mathbf{E}\{W(\tilde{x}_T,\beta(T),T)\}-\mathbf{E}\{W(\tilde{x}_t,\beta,t)\}\\
&\le e^{-\alpha(t-T)+N_\beta(T,t)\ln\mu}\mathbf{E}\{W(\tilde{x}_T,\beta(T),T)\}.
\end{aligned} \quad (10.31)$$

Multiply two sides of (10.31) by $e^{-N_\beta(T,t)\ln\mu}$ and it has

$$\mathbf{E}\left\{\int_T^t e^{-\alpha(t-s)+N_\beta(s,T)\ln\mu}e^T(s)e(s)ds\right\}\le e^{-\alpha(t-T)}\mathbf{E}\{W(\tilde{x}_T,\beta(T),T)\},$$

that is

$$\mathbf{E}\left\{\int_T^t e^{N_\beta(s,T)\ln\mu} e^T(s)e(s)ds\right\} \leq \mathbf{E}\{W(\tilde{x}_T, \beta(T), T)\}. \tag{10.32}$$

On account of $N_\beta(T,s) \leq (s-T)/T_a$ and $T_a > T_a^\star = \ln\mu/\alpha$, we have $N_\beta(T,s)\ln\mu \leq \alpha(s-T)$. Then

$$\mathbf{E}\left\{\int_T^t e^{-\alpha(s-T)} e^T(s)e(s)ds\right\} \leq \mathbf{E}\{W(\tilde{x}_T, \beta(T), T)\},$$

which means

$$\mathbf{E}\left\{\int_T^t e^{-\alpha s} e^T(s)e(s)ds\right\} \leq \mathbf{E}\left\{e^{-\alpha T} W(\tilde{x}_T, \beta(T), T)\right\}. \tag{10.33}$$

As for any time $t > T$, (10.33) holds, then we get

$$\mathbf{E}\left\{\int_T^\infty e^{-\alpha t} e^T(t)e(t)dt\right\} \leq e^{-\alpha T}\mathbf{E}\{W(\tilde{x}_T, \beta(T), T)\}. \tag{10.34}$$

In addition, it follows from (10.24) that

$$e^{-\alpha T}\mathbf{E}\{V(\tilde{x}_T, \beta(T), T)\} \leq \gamma^2 \int_0^T u^T(t)u(t)dt. \tag{10.35}$$

Based on (10.34)–(10.35) and (10.4c), we can obtain

$$\mathbf{E}\left\{\int_T^\infty e^{-\alpha t} e^T(t)e(t)dt\right\} < \gamma^2 \int_0^T u^T(t)u(t)dt,$$

which implies the Hankel-norm error performance as stated in Definition 10.3. Hence, the proof is completed.

Remark 10.6 When $\mu > 1$ and $\alpha \to 0$ in $T_a > T_a^\star = \frac{\ln\mu}{\alpha}$, it yields $T_a \to \infty$, in other words, there is no switching among considered fuzzy systems. The fuzzy switched stochastic system (10.3) always operates well at one of the subsystems, T-S fuzzy stochastic systems.

In the following, sufficient conditions on the mean-square stability and Hankel-norm error performance of considered T-S fuzzy systems are given.

♦ **Plant Form**:

Rule i: IF $\theta_1(t)$ is μ_{i1} and $\theta_2(t)$ is μ_{i2} and $\cdots$ and $\theta_p(t)$ is μ_{ip}, THEN

$$\begin{aligned} dx(t) &= [A_i x(t) + B_i u(t)]\,dt + E_i x(t)d\varpi, \\ y(t) &= C_i x(t), \quad i = 1, 2, \ldots, r. \end{aligned}$$

Provided that premise variables do not rely on the input $u(t)$. For a set of $(x(t), u(t))$, we can obtain the following T-S fuzzy stochastic system:

$$dx(t) = \sum_{i=1}^{r} h_i(\theta(t)) \{[A_i x(t) + B_i u(t)] dt + E_i x(t) d\varpi\}, \tag{10.36a}$$

$$y(t) = \sum_{i=1}^{r} h_i(\theta(t)) C_i x(t), \tag{10.36b}$$

where

$$h_i(\theta(t)) = \nu_i(\theta(t)) / \sum_{i=1}^{r} \nu_i(\theta(t)), \quad \nu_i(\theta(t)) = \prod_{l=1}^{p} \mu_{il}(\theta_l(t)),$$

and $\mu_{il}(\theta_l(t))$ stands for the grade of membership of $\theta_l(t)$ in μ_{il}. If $\nu_i(\theta(t)) \geq 0,\ i = 1, 2, \ldots, r,\ \sum_{i=1}^{r} \nu_i(\theta(t)) > 0$ for all t. Then $h_i(\theta(t)) \geq 0$ for $i = 1, 2, \ldots, r$ and $\sum_{i=1}^{r} h_i(\theta(t)) = 1$ for all t.

In this chapter, we concentrate on employing the following reduced-order model to approximate the system (10.36):

$$d\hat{x}(t) = \left[\hat{A}\hat{x}(t) + \hat{B}u(t)\right] dt + \hat{E}\hat{x}(t) d\varpi(t), \tag{10.37a}$$

$$\hat{y}(t) = \hat{C}\hat{x}(t), \tag{10.37b}$$

where $\hat{x}(t) \in \mathbb{R}^k$ denotes the state of the reduced-order model and $k < n$; $\hat{A}$, $\hat{B}$, $\hat{C}$ and $\hat{E}$ represent some relevant matrices to be decided later.

Combining the system (10.36) and reduced-order model (10.37), we can get the dynamic error system as

$$d\xi(t) = \sum_{i=1}^{r} h_i(\theta(t)) \left\{\left[\tilde{A}_i \xi(t) + \tilde{B}_i u(t)\right] dt + \tilde{E}_i \xi(t) d\varpi(t)\right\}, \tag{10.38a}$$

$$e(t) = \sum_{i=1}^{r} h_i(\theta(t)) \tilde{C}_i \xi(t), \tag{10.38b}$$

where

$$\xi(t) \triangleq \begin{bmatrix} x(t) \\ \hat{x}(t) \end{bmatrix}, \tilde{A}_i \triangleq \begin{bmatrix} A_i & 0 \\ 0 & \hat{A} \end{bmatrix}, \tilde{E}_i \triangleq \begin{bmatrix} E_i & 0 \\ 0 & \hat{E} \end{bmatrix}, \tilde{C}_i \triangleq \begin{bmatrix} C_i & -\hat{C} \end{bmatrix}, \tilde{B}_i \triangleq \begin{bmatrix} B_i \\ \hat{B} \end{bmatrix}.$$

Corollary 10.7 *For a given scalar $\gamma > 0$, the obtained system (10.38) has the mean-square asymptotic stability and Hankel-norm performance sense γ if there are matrices $0 < P_1 \in \mathbb{R}^{(n+k)\times(n+k)}$ and $0 < P_2 \in \mathbb{R}^{(n+k)\times(n+k)}$ such that for $i = 1, 2, \ldots, r$,*

$l = 1, 2, \ldots, r,$

$$\begin{bmatrix} P_1\tilde{A}_i + \tilde{A}_i^T P_1 & P_1\tilde{B}_i & \tilde{E}_i^T P_1 \\ \star & -\gamma^2 I & 0 \\ \star & \star & -P_1 \end{bmatrix} < 0,$$

$$\begin{bmatrix} P_2\tilde{A}_i + \tilde{A}_i^T P_2 & \tilde{E}_i^T P_2 & \tilde{C}_i^T \\ \star & -P_2 & 0 \\ \star & \star & -I \end{bmatrix} < 0.$$

Note that if there is no nonlinearity, the concerned system (10.1) turns into a common switched stochastic system in the form of

$$dx(t) = [A(\beta)x(t) + B(\beta)u(t)]dt + E(\beta)x(t)d\varpi(t), \tag{10.39a}$$

$$y(t) = C(\beta)x(t). \tag{10.39b}$$

Utilizing the following reduced-order switched model to approximate (10.39):

$$d\hat{x}(t) = \left[\hat{A}(\beta)\hat{x}(t) + \hat{B}(\beta)u(t)\right]dt + \hat{E}(\beta)\hat{x}(t)d\varpi(t), \tag{10.40a}$$

$$\hat{y}(t) = \hat{C}(\beta)\hat{x}(t), \tag{10.40b}$$

where $\hat{x}(t) \in \mathbb{R}^k$ denotes the state of reduced-order switched model and $k < n$; $\hat{A}(\beta)$, $\hat{B}(\beta)$, $\hat{C}(\beta)$ and $\hat{E}(\beta)$ denote some appropriate matrices to be decided later.

Combining the system in (10.39) and the model in (10.40), a corresponding error system can be inferred as

$$d\tilde{x}(t) = \left[\tilde{A}(\beta)\tilde{x}(t) + \tilde{B}(\beta)u(t)\right]dt + \tilde{E}(\beta)\tilde{x}(t)d\varpi(t), \tag{10.41a}$$

$$e(t) = \tilde{C}(\beta)\tilde{x}(t), \tag{10.41b}$$

where $\tilde{x}(t) \triangleq \begin{bmatrix} x(t) \\ \hat{x}(t) \end{bmatrix}$ and

$$\begin{cases} \tilde{A}(\beta) \triangleq \begin{bmatrix} A(\beta) & 0 \\ 0 & \hat{A}(\beta) \end{bmatrix}, & \tilde{E}(\beta) \triangleq \begin{bmatrix} E(\beta) & 0 \\ 0 & \hat{E}(\beta) \end{bmatrix}, \\ \tilde{C}(\beta) \triangleq \begin{bmatrix} C(\beta) & -\hat{C}(\beta) \end{bmatrix}, & \tilde{B}(\beta) \triangleq \begin{bmatrix} B(\beta) \\ \hat{B}(\beta) \end{bmatrix}. \end{cases}$$

For the switched stochastic system (10.39), we can get the following corollary.

Corollary 10.8 *For the given scalars $\alpha > 0$ and $\gamma > 0$, it is assumed that there are matrices $0 < P_1(j) \in \mathbb{R}^{(n+k)\times(n+k)}$ and $0 < P_2(j) \in \mathbb{R}^{(n+k)\times(n+k)}$ such that for $j \in \mathcal{I}$,*

$$\begin{bmatrix} \Pi_{11}(j) & P_1(j)\tilde{B}(j) & \tilde{E}^T(j)P_1(j) \\ \star & -\gamma^2 I & 0 \\ \star & \star & -P_1(j) \end{bmatrix} < 0,$$

$$\begin{bmatrix} \Phi_{11}(j) & \tilde{E}^T(j)P_2(j) & \tilde{C}^T(j) \\ \star & -P_2(j) & 0 \\ \star & \star & -I \end{bmatrix} < 0,$$

$$P_2(j) - P_1(j) < 0,$$

where

$$\begin{cases} \Pi_{11}(j) \triangleq P_1(j)\tilde{A}(j) + \tilde{A}^T(j)P_1(j) + \alpha P_1(j), \\ \Phi_{11}(j) \triangleq P_2(j)\tilde{A}(j) + \tilde{A}^T(j)P_2(j) + \alpha P_2(j). \end{cases}$$

Then, the concerned system in (10.41) is mean-square exponentially stable subject to the Hankel-norm error performance (γ, α) with any switching signal under $T_a > T_a^\star = \frac{\ln \mu}{\alpha}$, where $\mu \geq 1$ is denoted in (10.5). In addition, the state decay estimate is shown in (10.6), where λ, η, a and b are given by (10.7).

Proof The mentioned results can be derived based on Theorem 10.5, and the specific proof process is omitted here.

10.3.2 Model Approximation by the Hankel-Norm Approach

In the following, we are going to resolve the model approximation issue for fuzzy switched stochastic systems utilizing the convex linearization method and specific Hankel-norm error performance index.

Theorem 10.9 *For the scalars $\alpha > 0$, $\gamma > 0$ and $0 < \vartheta \leq 1$, provided that there are matrices $0 < P(j) \in \mathbb{R}^{(n+k)\times(n+k)}$ such that for $j \in \mathscr{I}$, $i = 1, 2, \ldots, r$, $l = 1, 2, \ldots, r$,*

$$\begin{bmatrix} \bar{\Pi}_{11il}(j) & P(j)\tilde{B}_{il}(j) & \tilde{E}_{il}^T(j)P(j) \\ \star & -\gamma^2 I & 0 \\ \star & \star & -P(j) \end{bmatrix} < 0, \tag{10.42a}$$

$$\begin{bmatrix} \bar{\Phi}_{11il}(j) & \vartheta \tilde{E}_{il}^T(j)P(j) & \tilde{C}_{il}^T(j) \\ \star & -\vartheta P(j) & 0 \\ \star & \star & -I \end{bmatrix} < 0, \tag{10.42b}$$

where

$$\bar{\Pi}_{11il}(j) \triangleq P(j)\tilde{A}_{il}(j) + \tilde{A}_{il}^T(j)P(j) + \alpha P(j),$$
$$\bar{\Phi}_{11il}(j) \triangleq \vartheta P(j)\tilde{A}_{il}(j) + \vartheta \tilde{A}_{il}^T(j)P(j) + \alpha\vartheta P(j).$$

Then the considered system (10.3) has the mean-square exponential stability and a Hankel-norm error performance sense (γ, α) *with any switching signal under* $T_a > T_a^\star = \frac{\ln\mu}{\alpha}$, *where* $\mu \geq 1$ *satisfies*

$$P(j) \leq \mu P(s), \ \forall j, s \in \mathscr{I}. \tag{10.43}$$

Besides, a state decay estimate is described as

$$\mathbf{E}\left\{\|\tilde{x}(t)\|^2\right\} \leq \eta e^{-\lambda t} \|\tilde{x}(0)\|^2, \tag{10.44}$$

where

$$\begin{aligned} a &= \min_{\forall j \in \mathscr{I}} \lambda_{\min}(P(j)), \quad \eta = \frac{b}{a} \geq 1, \\ b &= \max_{\forall j \in \mathscr{I}} \lambda_{\max}(P(j)), \quad \lambda = \alpha - \frac{\ln\mu}{T_a} > 0. \end{aligned}$$

Proof Due to $0 < \vartheta \leq 1$ and (10.4c), it gets $P_2(j) = \vartheta P_1(j)$. Utilizing the similar methods as the proof in Theorem 10.5 and $P_2(j) = \vartheta P_1(j)$, we can observe that the dynamic system (10.3) is mean-square exponentially stable subject to a Hankel-norm error performance (γ, α) with any switching signal under $T_a > T_a^\star = \frac{\ln\mu}{\alpha}$ if there exists $P(j) > 0$, which satisfies (10.42a)–(10.44). Then the proof is completed.

Next, the model reduction parameters of fuzzy switched systems can be obtained as follows.

Theorem 10.10 *Given the scalars* $\alpha > 0$, $\gamma > 0$ *and* $0 < \vartheta \leq 1$, *provided that there exist matrices* $0 < \mathcal{P}(j) \in \mathbb{R}^{n\times n}$, $0 < \mathcal{Q}(j) \in \mathbb{R}^{k\times k}$, $\mathcal{A}_i(j) \in \mathbb{R}^{k\times k}$, $\mathcal{B}_i(j) \in \mathbb{R}^{k\times l}$, $\mathcal{C}_i(j) \in \mathbb{R}^{m\times k}$ *and* $\mathcal{E}_i(j) \in \mathbb{R}^{k\times k}$ *such that for* $j \in \mathscr{I}$, $i = 1, 2, \ldots, r$, $l = 1, 2, \ldots, r$,

$$\begin{bmatrix} \hat{\Pi}_{11i}(j) & \hat{\Pi}_{12il}(j) & \hat{\Pi}_{13il}(j) & \hat{\Pi}_{14i}(j) & \hat{\Pi}_{15i}(j) \\ \star & \hat{\Pi}_{22l}(j) & \hat{\Pi}_{23il}(j) & \hat{\Pi}_{24l}(j) & \hat{\Pi}_{25l}(j) \\ \star & \star & -\gamma^2 I & 0 & 0 \\ \star & \star & \star & -\mathcal{P}(j) & -\mathcal{H}\mathcal{Q}(j) \\ \star & \star & \star & \star & -\mathcal{Q}(j) \end{bmatrix} < 0, \tag{10.45a}$$

$$\begin{bmatrix} \hat{\Phi}_{11i}(j) & \hat{\Phi}_{12il}(j) & \hat{\Phi}_{13i}(j) & \hat{\Phi}_{14i}(j) & C_i^T(j) \\ \star & \hat{\Phi}_{22l}(j) & \hat{\Phi}_{23l}(j) & \hat{\Phi}_{24l}(j) & -\mathcal{C}^T(j) \\ \star & \star & -\vartheta\mathcal{P}(j) & -\vartheta\mathcal{H}\mathcal{Q}(j) & 0 \\ \star & \star & \star & -\vartheta\mathcal{Q}(j) & 0 \\ \star & \star & \star & \star & -I \end{bmatrix} < 0, \tag{10.45b}$$

where

$$\begin{cases} \hat{\Pi}_{11i}(j) \triangleq \mathcal{P}(j)A_i(j) + A_i^T(j)\mathcal{P}(j) + \alpha\mathcal{P}(j), \\ \hat{\Pi}_{12il}(j) \triangleq \mathcal{H}\mathcal{A}_l(j) + A_i^T(j)\mathcal{H}\mathcal{Q}(j) + \alpha\mathcal{H}\mathcal{Q}(j), \\ \hat{\Pi}_{22l}(j) \triangleq \mathcal{A}_l(j) + \mathcal{A}_l^T(j) + \alpha\mathcal{Q}(j), \\ \hat{\Pi}_{13il}(j) \triangleq \mathcal{P}(j)B_i(j) + \mathcal{H}\mathcal{B}_l(j), \quad \mathcal{H} \triangleq \begin{bmatrix} I_{k\times k} \\ 0_{(n-k)\times k} \end{bmatrix}, \\ \hat{\Pi}_{23il}(j) \triangleq \mathcal{Q}(j)\mathcal{H}^T B_i(j) + \mathcal{B}_l(j), \\ \hat{\Pi}_{14i}(j) \triangleq E_i^T(j)\mathcal{P}(j), \quad \hat{\Pi}_{15i}(j) \triangleq E_i^T(j)\mathcal{H}\mathcal{Q}(j), \\ \hat{\Pi}_{24l}(j) \triangleq \mathcal{E}_l^T(j)\mathcal{H}^T, \quad \hat{\Pi}_{25l}(j) \triangleq \mathcal{E}_l^T(j), \\ \hat{\Phi}_{11i}(j) \triangleq \vartheta\mathcal{P}(j)A_i(j) + \vartheta A_i^T(j)\mathcal{P}(j) + \alpha\vartheta\mathcal{P}(j), \\ \hat{\Phi}_{12il}(j) \triangleq \vartheta\mathcal{H}\mathcal{A}_l(j) + \vartheta A_i^T(j)\mathcal{H}\mathcal{Q}(j) + \alpha\vartheta\mathcal{H}\mathcal{Q}(j), \\ \hat{\Phi}_{22l}(j) \triangleq \vartheta\mathcal{A}_l(j) + \vartheta\mathcal{A}_l^T(j) + \alpha\vartheta\mathcal{Q}(j), \\ \hat{\Phi}_{13i}(j) \triangleq \vartheta E_i^T(j)\mathcal{P}(j), \quad \hat{\Phi}_{14i}(j) \triangleq \vartheta E_i^T(j)\mathcal{H}\mathcal{Q}(j), \\ \hat{\Phi}_{23l}(j) \triangleq \vartheta\mathcal{E}_l^T(j)\mathcal{H}^T, \quad \hat{\Phi}_{24l}(j) \triangleq \vartheta\mathcal{E}_l^T(j). \end{cases}$$

Then the concerned error system (10.3) is mean-square exponentially stable subject to the Hankel-norm error performance (γ, α) for any switching signal with $T_a > T_a^\star = \frac{\ln\mu}{\alpha}$.

In addition, if the above-mentioned conditions have relevant feasible solutions $\{\mathcal{P}(j), \mathcal{Q}(j), \mathcal{A}_i(j), \mathcal{B}_i(j), \mathcal{C}_i(j), \mathcal{E}_i(j)\}$, the parameters of proposed reduced-order model as (10.2) can be described as

$$\begin{cases} \hat{A}_i(j) = \mathcal{Q}^{-1}(j)\mathcal{A}_i(j), & \hat{B}_i(j) = \mathcal{Q}^{-1}(j)\mathcal{B}_i(j), \\ \hat{C}_i(j) = \mathcal{C}_i(j), & \hat{E}_i(j) = \mathcal{Q}^{-1}(j)\mathcal{E}_i(j). \end{cases} \tag{10.46}$$

Proof On the basis of Theorem 10.9, $P(j)$ is nonsingular due to $P(j) > 0$. Partition $P(j)$ as

$$P(j) \triangleq \begin{bmatrix} \mathcal{P}_1(j) & \mathcal{P}_2(j) \\ \star & \mathcal{P}_3(j) \end{bmatrix}, \quad \mathcal{P}_2(j) \triangleq \begin{bmatrix} \mathcal{P}_4(j) \\ 0_{(n-k)\times k} \end{bmatrix},$$

where $\mathcal{P}_1(j) \in \mathbb{R}^{n\times n}$ and $\mathcal{P}_3(j) \in \mathbb{R}^{k\times k}$ are symmetric positive definite matrices; $\mathcal{P}_2(j) \in \mathbb{R}^{n\times k}$ and $\mathcal{P}_4(j) \in \mathbb{R}^{k\times k}$. Without loss of generality, suppose that $\mathcal{P}_4(j)$ is nonsingular. Set $M(j) \triangleq P(j) + \epsilon N$, where ϵ is a positive scalar and

$$M(j) \triangleq \begin{bmatrix} M_1(j) & M_2(j) \\ \star & M_3(j) \end{bmatrix}, \quad N \triangleq \left[\begin{array}{c|c} 0_{n\times n} & \mathcal{H} \\ \hline \star & 0_{k\times k} \end{array}\right], \quad M_2(j) \triangleq \begin{bmatrix} M_4(j) \\ 0_{(n-k)\times k} \end{bmatrix}.$$

It follows from $P(j) > 0$ that $M(j) > 0$ for $\epsilon > 0$ around the origin. It is easy to conclude that there is an arbitrarily small $\epsilon > 0$ such that $M_4(j)$ is nonsingular and (10.42a)–(10.42b) are feasible by replacing $P(j)$ with $M(j)$. Because of $M_4(j)$ is nonsingular, then it can be seen that $\mathcal{P}_4(j)$ is nonsingular.

Define some nonsingular matrices as

$$\begin{cases} \mathcal{J}(j) \triangleq \begin{bmatrix} I & 0 \\ 0 & \mathcal{P}_3^{-1}(j)\mathcal{P}_4^T(j) \end{bmatrix}, \quad \mathcal{P}(j) \triangleq \mathcal{P}_1(j), \\ \mathcal{Q}(j) \triangleq \mathcal{P}_4(j)\mathcal{P}_3^{-1}(j)\mathcal{P}_4^T(j), \end{cases} \tag{10.47}$$

and

$$\begin{bmatrix} \mathcal{A}_i(j) & \mathcal{B}_i(j) \\ \mathcal{C}_i(j) & 0 \\ \mathcal{E}_i(j) & 0 \end{bmatrix} \triangleq \begin{bmatrix} \mathcal{P}_4(j) & 0 & 0 \\ 0 & I & 0 \\ 0 & 0 & \mathcal{P}_4(j) \end{bmatrix} \begin{bmatrix} \hat{A}_i(j) & \hat{B}_i(j) \\ \hat{C}_i(j) & 0 \\ \hat{E}_i(j) & 0 \end{bmatrix} \begin{bmatrix} \mathcal{P}_3^{-1}(j)\mathcal{P}_4^T(j) & 0 \\ 0 & I \end{bmatrix}. \tag{10.48}$$

Operating the congruence transformation to (10.42) with diag $(\mathcal{J}(j), I, \mathcal{J}(j))$ and diag $(\mathcal{J}(j), \mathcal{J}(j), I)$ separately, it yields

$$\begin{bmatrix} \mathcal{J}^T(j)\bar{\Pi}_{11il}(j)\mathcal{J}(j) & \mathcal{J}^T(j)P(j)\tilde{B}_{il}(j) & \mathcal{J}^T(j)\tilde{E}_{il}^T(j)P(j)\mathcal{J}(j) \\ \star & -\gamma^2 I & 0 \\ \star & \star & -\mathcal{J}^T(j)P(j)\mathcal{J}(j) \end{bmatrix} < 0, \tag{10.49a}$$

$$\begin{bmatrix} \mathcal{J}^T(j)\bar{\Phi}_{11il}(j)\mathcal{J}(j) & \vartheta\mathcal{J}^T(j)\tilde{E}_{il}^T(j)P(j)\mathcal{J}(j) & \mathcal{J}^T(j)\tilde{C}_{il}^T(j) \\ \star & -\vartheta\mathcal{J}^T(j)P(j)\mathcal{J}(j) & 0 \\ \star & \star & -I \end{bmatrix} < 0. \tag{10.49b}$$

Based on (10.47)–(10.48), we can get

$$\left.\begin{aligned} \mathcal{J}^T(j)P(j)\tilde{A}_{il}(j)\mathcal{J}(j) &= \begin{bmatrix} \mathcal{P}(j)A_i(j) & \mathcal{H}\mathcal{A}_l(j) \\ \mathcal{Q}(j)\mathcal{H}^T A_i(j) & \mathcal{A}_l(j) \end{bmatrix}, \\ \mathcal{J}^T(j)P(j)\tilde{B}_{il}(j) &= \begin{bmatrix} \mathcal{P}(j)B_i(j) + \mathcal{H}\mathcal{B}_l(j) \\ \mathcal{Q}(j)\mathcal{H}^T B_i(j) + \mathcal{B}_l(j) \end{bmatrix}, \\ \mathcal{J}^T(j)\tilde{E}_{il}^T(j)P(j)\mathcal{J}(j) &= \begin{bmatrix} E_i^T(j)\mathcal{P}(j) & E_i^T(j)\mathcal{H}\mathcal{Q}(j) \\ \mathcal{E}_l^T(j)\mathcal{H}^T & \mathcal{E}_l^T(j) \end{bmatrix}, \\ \mathcal{J}^T(j)P(j)\mathcal{J}(j) &= \begin{bmatrix} \mathcal{P}(j) & \mathcal{H}\mathcal{Q}(j) \\ \star & \mathcal{Q}(j) \end{bmatrix}, \\ \mathcal{J}^T(j)\tilde{C}_{il}^T(j) &= \begin{bmatrix} C_i^T(j) \\ -\mathcal{C}_l^T(j) \end{bmatrix}. \end{aligned}\right\} \tag{10.50}$$

Considering (10.50), it follows from (10.49a)–(10.49b) that (10.45a)–(10.45b), separately. And (10.48) implies

$$\begin{bmatrix} \hat{A}_i(j) & \hat{B}_i(j) \\ \hat{C}_i(j) & 0 \\ \hat{E}_i(j) & 0 \end{bmatrix}$$
$$\triangleq \begin{bmatrix} \mathcal{P}_4^{-1}(j) & 0 & 0 \\ 0 & I & 0 \\ 0 & 0 & \mathcal{P}_4^{-1}(j) \end{bmatrix} \begin{bmatrix} \mathcal{A}_i(j) & \mathcal{B}_i(j) \\ \mathcal{C}_i(j) & 0 \\ \mathcal{E}_i(j) & 0 \end{bmatrix} \begin{bmatrix} \mathcal{P}_4^{-T}(j)\mathcal{P}_3(j) & 0 \\ 0 & I \end{bmatrix}$$
$$= \begin{bmatrix} \left(\mathcal{P}_4^{-T}(j)\mathcal{P}_3(j)\right)^{-1}\mathcal{Q}^{-1}(j) & 0 & 0 \\ 0 & I & 0 \\ 0 & 0 & \left(\mathcal{P}_4^{-T}(j)\mathcal{P}_3(j)\right)^{-1}\mathcal{Q}^{-1}(j) \end{bmatrix}$$
$$\begin{bmatrix} \mathcal{A}_i(j) & \mathcal{B}_i(j) \\ \mathcal{C}_i(j) & 0 \\ \mathcal{E}_i(j) & 0 \end{bmatrix} \begin{bmatrix} \mathcal{P}_4^{-T}(j)\mathcal{P}_3(j) & 0 \\ 0 & I \end{bmatrix}. \tag{10.51}$$

It is observed that $\hat{A}_i(j)$, $\hat{B}_i(j)$, $\hat{C}_i(j)$ and $\hat{E}_i(j)$ in (10.2) can be rewritten as (10.51), which means $\mathcal{P}_4^{-T}(j)\mathcal{P}_3(j)$ can be regarded as a similarity transformation on the state-space realization of the designed filter and makes no influence in the filter mapping from u to $\hat{y}$. Without loss of generality, set $\mathcal{P}_4^{-T}(j)\mathcal{P}_3(j) = I$, then (10.46). Thus, the reduced-order model in (10.2) can be expressed as (10.46). At this point, it accomplishes the proof.

Remark 10.11 On account of the given conditions in Theorem 10.10 are all in the form of LMIs, the model reduction issue of fuzzy switched stochastic systems can be converted into the following convex optimization issue:

$$\min_{\substack{\mathcal{P}(j)>0,\,\mathcal{Q}(j)>0,\\ \mathcal{A}_i(j),\mathcal{B}_i(j),\mathcal{C}_i(j),\mathcal{E}_i(j)}} \delta \quad \text{subject to (10.45)} \quad (where \quad \delta = \gamma^2).$$

On the basis of utilizing the convex linearization method, we present some relevant research results for T-S fuzzy stochastic system in (10.38) and switched stochastic system in (10.39), separately. The specific proof can be obtained via employing the similar process in Theorem 10.10.

Corollary 10.12 *As for the T-S fuzzy stochastic system in (10.38), given the scalars* $\alpha > 0$ *and* $\gamma > 0$*, there is a reduced-order model (10.37) to resolve the model reduction issue, if there exist some matrices* $0 < \mathcal{P} \in \mathbb{R}^{n\times n}$, $0 < \mathcal{Q} \in \mathbb{R}^{k\times k}$, $\mathcal{A} \in \mathbb{R}^{k\times k}$, $\mathcal{B} \in \mathbb{R}^{k\times l}$, $\mathcal{C} \in \mathbb{R}^{m\times k}$ *and* $\mathcal{E} \in \mathbb{R}^{k\times k}$*, which satisfy*

$$\begin{bmatrix} \bar{\Pi}_{11i} & \bar{\Pi}_{12i} & \bar{\Pi}_{13i} & \bar{\Pi}_{14i} & \bar{\Pi}_{15i} \\ \star & \bar{\Pi}_{22l} & \bar{\Pi}_{23i} & \bar{\Pi}_{24} & \bar{\Pi}_{25} \\ \star & \star & -\gamma^2 I & 0 & 0 \\ \star & \star & \star & -\mathcal{P} & -\mathcal{H}\mathcal{Q} \\ \star & \star & \star & \star & -\mathcal{Q} \end{bmatrix} < 0,$$

$$\begin{bmatrix} \bar{\Phi}_{11i} & \bar{\Phi}_{12i} & \bar{\Phi}_{13i} & \bar{\Phi}_{14i} & C_i^T \\ \star & \bar{\Phi}_{22} & \bar{\Phi}_{23} & \bar{\Phi}_{24} & -\mathcal{C}^T \\ \star & \star & -\vartheta\mathcal{P} & -\vartheta\mathcal{H}\mathcal{Q} & 0 \\ \star & \star & \star & -\vartheta\mathcal{Q} & 0 \\ \star & \star & \star & \star & -I \end{bmatrix} < 0,$$

where

$$\begin{cases} \bar{\Pi}_{11i} \triangleq \mathcal{P}A_i + A_i^T\mathcal{P}, \quad \bar{\Pi}_{12i} \triangleq \mathcal{H}\mathcal{A} + A_i^T\mathcal{H}\mathcal{Q}, \\ \bar{\Pi}_{22} \triangleq \mathcal{A} + \mathcal{A}^T, \quad \bar{\Pi}_{23i} \triangleq \mathcal{Q}\mathcal{H}^T B_i + \mathcal{B}, \\ \bar{\Pi}_{13i} \triangleq \mathcal{P}B_i + \mathcal{H}\mathcal{B}, \quad \mathcal{H} \triangleq \begin{bmatrix} I_{k\times k} \\ 0_{(n-k)\times k} \end{bmatrix}, \\ \bar{\Pi}_{14i} \triangleq E_i^T\mathcal{P}, \quad \bar{\Pi}_{15i} \triangleq E_i^T\mathcal{H}\mathcal{Q}, \quad \bar{\Pi}_{25} \triangleq \mathcal{E}^T, \\ \bar{\Phi}_{11i} \triangleq \vartheta\mathcal{P}A_i + \vartheta A_i^T\mathcal{P}, \quad \bar{\Phi}_{22} \triangleq \vartheta\mathcal{A} + \vartheta\mathcal{A}^T, \\ \bar{\Phi}_{12i} \triangleq \vartheta\mathcal{H}\mathcal{A} + \vartheta A_i^T\mathcal{H}\mathcal{Q}, \quad \bar{\Pi}_{24} \triangleq \mathcal{E}^T\mathcal{H}^T, \\ \bar{\Phi}_{13i} \triangleq \vartheta E_i^T\mathcal{P}, \quad \bar{\Phi}_{14i} \triangleq \vartheta E_i^T\mathcal{H}\mathcal{Q}, \\ \bar{\Phi}_{23} \triangleq \vartheta\mathcal{E}^T\mathcal{H}^T, \quad \bar{\Phi}_{24} \triangleq \vartheta\mathcal{E}^T. \end{cases}$$

In addition, if above-mentioned conditions have feasible solutions, then the reduced-order model matrices (10.37) can be obtained by

$$\begin{bmatrix} \hat{A} & \hat{B} \\ \hat{C} & 0 \\ \hat{E} & 0 \end{bmatrix} = \begin{bmatrix} \mathcal{Q}^{-1} & 0 & 0 \\ 0 & I & 0 \\ 0 & 0 & \mathcal{Q}^{-1} \end{bmatrix} \begin{bmatrix} \mathcal{A} & \mathcal{B} \\ \mathcal{C} & 0 \\ \mathcal{E} & 0 \end{bmatrix}.$$

Corollary 10.13 *As for the dynamic error system in (10.41). Given some scalars $\alpha > 0$, $\gamma > 0$ and $0 < \vartheta \le 1$, provided that there are matrices $0 < \mathcal{P}(j) \in \mathbb{R}^{n\times n}$, $0 < \mathcal{Q}(j) \in \mathbb{R}^{k\times k}$, $\mathcal{A}(j) \in \mathbb{R}^{k\times k}$, $\mathcal{B}(j) \in \mathbb{R}^{k\times l}$, $\mathcal{C}(j) \in \mathbb{R}^{m\times k}$ and $\mathcal{E}(j) \in \mathbb{R}^{k\times k}$ such that for $j \in \mathscr{I}$,*

$$\begin{bmatrix} \hat{\Pi}_{11}(j) & \hat{\Pi}_{12}(j) & \hat{\Pi}_{13}(j) & \hat{\Pi}_{14}(j) & \hat{\Pi}_{15}(j) \\ \star & \hat{\Pi}_{22}(j) & \hat{\Pi}_{23}(j) & \hat{\Pi}_{24}(j) & \hat{\Pi}_{25}(j) \\ \star & \star & -\gamma^2 I & 0 & 0 \\ \star & \star & \star & -\mathcal{P}(j) & -\mathcal{H}\mathcal{Q}(j) \\ \star & \star & \star & \star & -\mathcal{Q}(j) \end{bmatrix} < 0,$$

$$\begin{bmatrix} \hat{\Phi}_{11}(j) & \hat{\Phi}_{12}(j) & \hat{\Phi}_{13}(j) & \hat{\Phi}_{14}(j) & C^T(j) \\ \star & \hat{\Phi}_{22}(j) & \hat{\Phi}_{23}(j) & \hat{\Phi}_{24}(j) & -\mathcal{C}^T(j) \\ \star & \star & -\vartheta\mathcal{P}(j) & -\vartheta\mathcal{H}\mathcal{Q}(j) & 0 \\ \star & \star & \star & -\vartheta\mathcal{Q}(j) & 0 \\ \star & \star & \star & \star & -I \end{bmatrix} < 0,$$

where

$$\begin{cases} \hat{\Pi}_{11}(j) \triangleq \mathcal{P}(j)A(j) + A^T(j)\mathcal{P}(j) + \alpha\mathcal{P}(j), \\ \hat{\Pi}_{12}(j) \triangleq \mathcal{H}\mathcal{A}(j) + A^T(j)\mathcal{H}\mathcal{Q}(j) + \alpha\mathcal{H}\mathcal{Q}(j), \\ \hat{\Pi}_{22}(j) \triangleq \mathcal{A}(j) + \mathcal{A}^T(j) + \alpha\mathcal{Q}(j), \\ \hat{\Pi}_{13}(j) \triangleq \mathcal{P}(j)B(j) + \mathcal{H}\mathcal{B}(j), \quad \mathcal{H} \triangleq \begin{bmatrix} I_{k\times k} \\ 0_{(n-k)\times k} \end{bmatrix}, \\ \hat{\Pi}_{23}(j) \triangleq \mathcal{Q}(j)\mathcal{H}^T B(j) + \mathcal{B}(j), \\ \hat{\Pi}_{14}(j) \triangleq E^T(j)\mathcal{P}(j), \quad \hat{\Pi}_{15}(j) \triangleq E^T(j)\mathcal{H}\mathcal{Q}(j), \\ \hat{\Pi}_{24}(j) \triangleq \mathcal{E}^T(j)\mathcal{H}^T, \quad \hat{\Pi}_{25}(j) \triangleq \mathcal{E}^T(j), \\ \hat{\Phi}_{11}(j) \triangleq \vartheta\mathcal{P}(j)A(j) + \vartheta A^T(j)\mathcal{P}(j) + \alpha\vartheta\mathcal{P}(j), \\ \hat{\Phi}_{12}(j) \triangleq \vartheta\mathcal{H}\mathcal{A}(j) + \vartheta A^T(j)\mathcal{H}\mathcal{Q}(j) + \alpha\vartheta\mathcal{H}\mathcal{Q}(j), \\ \hat{\Phi}_{22}(j) \triangleq \vartheta\mathcal{A}(j) + \vartheta\mathcal{A}^T(j) + \alpha\vartheta\mathcal{Q}(j), \\ \hat{\Phi}_{13}(j) \triangleq \vartheta E^T(j)\mathcal{P}(j), \quad \hat{\Phi}_{14}(j) \triangleq \vartheta E^T(j)\mathcal{H}\mathcal{Q}(j), \\ \hat{\Phi}_{23}(j) \triangleq \vartheta\mathcal{E}^T(j)\mathcal{H}^T, \quad \hat{\Phi}_{24}(j) \triangleq \vartheta\mathcal{E}^T(j). \end{cases}$$

Then the corresponding error system in (10.41) satisfies the mean-square exponential stability and Hankel-norm error performance sense (γ, α) for any switching signal with $T_a > T_a^\star = \frac{\ln\mu}{\alpha}$.

If the above conditions have feasible solutions $\{\mathcal{P}(j), \mathcal{Q}(j), \mathcal{A}(j), \mathcal{B}(j), \mathcal{C}(j), \mathcal{E}(j)\}$, the reduced-order model matrices in (10.40) can be formulated as

$$\begin{bmatrix} \hat{A}(j) & \hat{B}(j) \\ \hat{C}(j) & 0 \\ \hat{E}(j) & 0 \end{bmatrix} = \begin{bmatrix} \mathcal{Q}^{-1}(j) & 0 & 0 \\ 0 & I & 0 \\ 0 & 0 & \mathcal{Q}^{-1}(j) \end{bmatrix} \begin{bmatrix} \mathcal{A}(j) & \mathcal{B}(j) \\ \mathcal{C}(j) & 0 \\ \mathcal{E}(j) & 0 \end{bmatrix}.$$

10.4 Illustrative Example

In this part, we give two examples to illustrate the validity of designed model approximation methods.

Example 10.14 The T-S fuzzy switched model in (10.1) is considered with $S = 2$.
Subsystem 1.

$$A_1(1) = \begin{bmatrix} -3.0 & 0.3 & 0.2 & 0.3 \\ 0.2 & -2.1 & 0.2 & 0.4 \\ 0.1 & 0.1 & -3.8 & 0.3 \\ 0.3 & 0.2 & 0.1 & -1.7 \end{bmatrix}, \; B_1(1) = \begin{bmatrix} 2.1 \\ 1.1 \\ 1.4 \\ 1.0 \end{bmatrix},$$
$$A_2(1) = \begin{bmatrix} -2.1 & 0.3 & 0.0 & 0.1 \\ 0.3 & -3.4 & 0.2 & 0.5 \\ 0.3 & 0.1 & -2.1 & 0.4 \\ 0.1 & 0.2 & 0.1 & -1.5 \end{bmatrix}, \; B_2(1) = \begin{bmatrix} 1.1 \\ 1.7 \\ 1.4 \\ 1.2 \end{bmatrix},$$
$$E_1(1) = \begin{bmatrix} 0.1 & 0.2 & 0.1 & 0.0 \\ 0.0 & 0.03 & 0.2 & 0.1 \\ 0.02 & 0.1 & 0.3 & 0.0 \\ 0.1 & 0.2 & 0.1 & 0.2 \end{bmatrix}, \; C_1(1) = \begin{bmatrix} 1.0 \\ 1.2 \\ 0.8 \\ 0.6 \end{bmatrix}^T,$$
$$E_2(1) = \begin{bmatrix} 0.2 & 0.3 & 0.1 & 0.0 \\ 0.0 & 0.4 & 0.0 & 0.02 \\ 0.1 & 0.0 & 0.04 & 0.2 \\ 0.0 & 0.2 & 0.3 & 0.1 \end{bmatrix}, \; C_2(1) = \begin{bmatrix} 0.7 \\ 1.1 \\ 0.5 \\ 0.7 \end{bmatrix}^T. \tag{10.52}$$

Subsystem 2.

$$A_1(2) = \begin{bmatrix} -3.3 & 0.1 & 0.5 & 0.3 \\ 0.2 & -2.1 & 0.2 & 0.4 \\ 0.3 & 0.0 & -3.5 & 0.2 \\ 0.4 & 0.1 & 0.4 & -1.9 \end{bmatrix}, \; B_1(2) = \begin{bmatrix} 2.0 \\ 1.3 \\ 1.5 \\ 1.0 \end{bmatrix},$$
$$A_2(2) = \begin{bmatrix} -2.5 & 0.1 & 0.3 & 0.1 \\ 0.4 & -3.3 & 0.2 & 0.2 \\ 0.2 & 0.1 & -2.1 & 0.2 \\ 0.0 & 0.2 & 0.0 & -1.5 \end{bmatrix}, \; B_2(2) = \begin{bmatrix} 1.1 \\ 1.2 \\ 1.0 \\ 1.4 \end{bmatrix},$$
$$E_1(2) = \begin{bmatrix} 0.3 & 0.2 & 0.4 & 0.1 \\ 0.0 & 0.05 & 0.2 & 0.3 \\ 0.02 & 0.1 & 0.2 & 0.0 \\ 0.2 & 0.0 & 0.3 & 0.4 \end{bmatrix}, \; C_1(2) = \begin{bmatrix} 1.4 \\ 1.1 \\ 0.2 \\ 0.4 \end{bmatrix}^T,$$
$$E_2(2) = \begin{bmatrix} 0.2 & 0.1 & 0.2 & 0.0 \\ 0.0 & 0.2 & 0.0 & 0.02 \\ 0.1 & 0.0 & 0.04 & 0.1 \\ 0.0 & 0.2 & 0.0 & 0.2 \end{bmatrix}, \; C_2(2) = \begin{bmatrix} 0.6 \\ 1.0 \\ 0.6 \\ 0.8 \end{bmatrix}^T, \tag{10.53}$$

and $\alpha = 0.8$. Let $\mu = 1.01$, it is not difficult to conclude the concerned switched system is mean-square exponentially stable.

In this note, we concentrate on designing reduced-order switched models (Case 1: $k = 3$; Case 2: $k = 2$; Case 3: $k = 1$) as $(\hat{\Sigma})$ in (10.2) to approximate the original system with a prescribed Hankel-norm property. As for different cases, we can get the following results via solving the convex feasibility issue in Theorem 10.10 with MATLAB:

$$\mathcal{G}_i(\beta) = \left[\begin{array}{c|c} \hat{A}_i(\beta) & \hat{E}_i(\beta) \\ \hline \hat{B}_i^T(\beta) & \hat{C}_i(\beta) \end{array}\right], \quad i = 1, 2, \quad \beta \in \mathscr{I}.$$

Case 1. When $k = 3$, the minimized feasible γ is computed as $\gamma^\star = 0.7678$ and (10.54).

$$\begin{cases} \mathcal{G}_i(1) = \left[\begin{array}{ccc|ccc} -3.3997 & -0.6819 & 2.6337 & 0.1362 & 0.1701 & 0.2748 \\ -0.7425 & -3.0803 & 4.2818 & 0.0477 & 0.1286 & 0.2186 \\ 1.6850 & 1.4228 & -6.8997 & 0.2108 & -0.1159 & 0.2698 \\ \hline -1.2594 & -0.0103 & -2.7860 & -1.2669 & -1.2525 & -1.2286 \end{array}\right], \\ \mathcal{G}_i(2) = \left[\begin{array}{ccc|ccc} -7.4084 & 7.3515 & -4.2197 & 0.2057 & 0.3767 & 0.2104 \\ 11.9893 & -20.9144 & 13.0282 & 0.0203 & 0.3590 & 0.0913 \\ -4.5321 & 8.0141 & -8.2907 & 0.0526 & 0.0697 & 0.1889 \\ \hline -2.1772 & 0.4992 & -2.0320 & -1.1115 & -1.2224 & -1.3360 \end{array}\right]. \end{cases} \tag{10.54}$$

Case 2. When $k = 2$, the minimized feasible γ is computed as $\gamma^\star = 1.2064$ and

$$\begin{cases} \mathcal{G}_i(1) = \left[\begin{array}{cc|cc} -5.8189 & 0.8670 & 0.4571 & 0.1165 \\ 5.2334 & -3.1756 & 0.1434 & 0.2670 \\ \hline -4.2996 & 2.3550 & -1.8935 & -1.3291 \end{array}\right], \\ \mathcal{G}_i(2) = \left[\begin{array}{cc|cc} -10.9223 & 7.7235 & 0.4194 & 0.2672 \\ 12.3134 & -13.1956 & 0.1962 & 0.2126 \\ \hline -2.3159 & -0.4537 & -2.0206 & -1.1942 \end{array}\right]. \end{cases} \tag{10.55}$$

Case 3. When $k = 1$, the minimized feasible γ is calculated as $\gamma^\star = 1.4006$ and

$$\begin{cases} \mathcal{G}_i(1) = \left[\begin{array}{c|c} -2.3653 & 0.5956 \\ \hline -1.6058 & -3.3012 \end{array}\right], \\ \mathcal{G}_i(2) = \left[\begin{array}{c|c} -2.3336 & 0.7461 \\ \hline -1.6256 & -3.0772 \end{array}\right]. \end{cases} \tag{10.56}$$

Set the zero initial condition as $\tilde{x}(0) = 0$ ($x(0) = 0$, $\hat{x}(0) = 0$), and the fuzzy basis functions are selected as

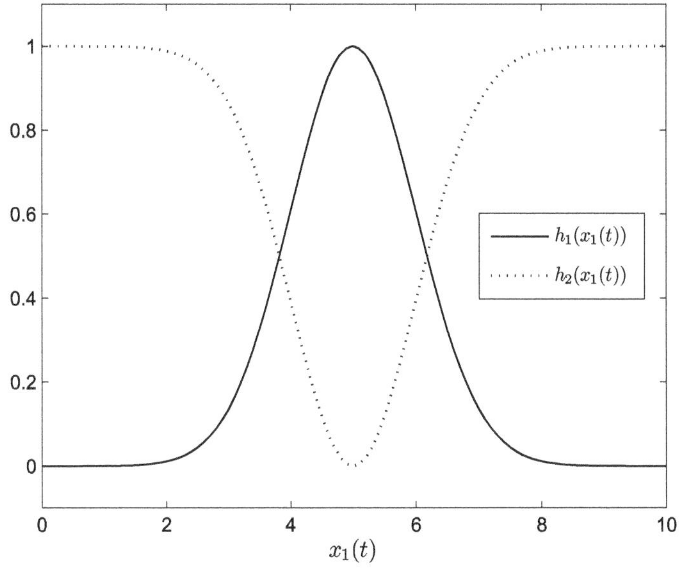

Fig. 10.1 Membership functions for the two fuzzy sets for Example 10.14

$$\begin{cases} h_1(x_1(t)) \triangleq \exp\left[-\dfrac{(x_1(t)-\vartheta)^2}{2\sigma^2}\right], \\ h_2(x_1(t)) \triangleq 1-\exp\left[-\dfrac{(x_1(t)-\vartheta)^2}{2\sigma^2}\right], \end{cases}$$

with $\vartheta = 5$ and $\sigma = 1$. And the fuzzy basis functions are plotted in Fig. 10.1. The exogenous input $u(t)$ is set to

$$u(t) = \exp(-t)\sin(t), \quad t \geq 0.$$

Figure 10.2 shows the switching signal, which is generated randomly, where '1' and '2' stand for the first and the second fuzzy stochastic subsystem, separately. From Fig. 10.2, we can see that $T_a \geq 0.1 > \frac{\ln \mu}{\alpha} = \frac{\ln 1.01}{0.8} = 0.012$. Figure 10.3 draws the outputs of the original system (10.52)–(10.53) (solid line), the third-order model (10.54) (dash line), the second-order model (10.55) (dotted line) and the first-order model (10.56) (dash-dot line) based on the above-mentioned input signal. The output errors among the original system and reduced models are depicted in Fig. 10.4.

Example 10.15 In this example, we consider the tunnel diode circuit [5] as Fig. 10.5, which is described as

$$i_D(t) = 0.002v_D(t) + 0.01v_D^3(t).$$

Set $x_1(t) = v_C(t)$, $x_2(t) = i_L(t)$ and the circuit can be reformulated as

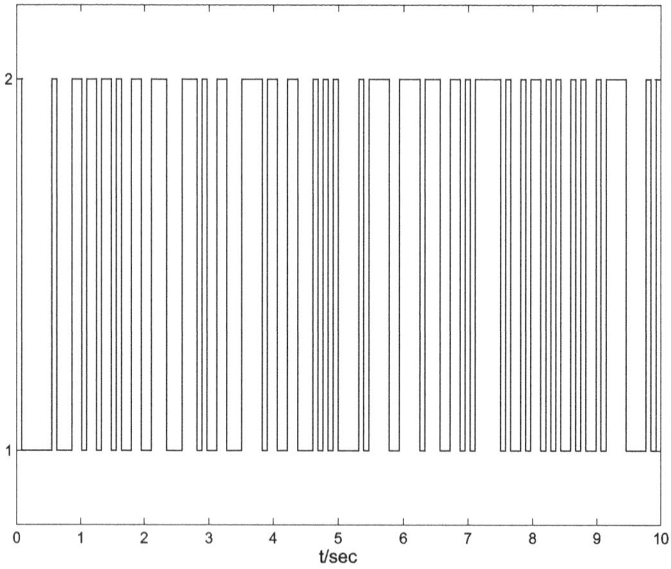

Fig. 10.2 Switching signal with the average dwell time $T_a \geq 0.1$

$$\begin{cases} C\dot{x}_1(t) = -0.002x_1(t) - 0.01x_1^3(t) + x_2(t), \\ L\dot{x}_2(t) = -x_1(t) - Rx_2(t) + u(t), \\ \quad y(t) = x_1(t), \end{cases} \tag{10.57}$$

where $u(t)$ denotes the external disturbance and $y(t)$ denotes the measurement output. The stochastic perturbation $\varpi(t)$ is generated by voltage disturbance, which is supposed to be standard scalar Brownian motion with $\mathbf{E}\{d\varpi(t)\} = 0$ and $\mathbf{E}\left\{d\varpi^2(t)\right\} = dt$. Therefore, a complete presentation of the circuit system with two rules is given as follows:

♦ **Plant Form**:

Rule 1: IF $x_1(t)$ is $M_1(x_1(t))$, THEN

$$\begin{aligned} dx(t) &= [A_1x(t) + B_1u(t)]dt + E_1x(t)d\varpi(t), \\ y(t) &= C_1x(t). \end{aligned}$$

Rule 2: IF $x_1(t)$ is $M_2(x_1(t))$, THEN

$$\begin{aligned} dx(t) &= [A_2x(t) + B_2u(t)]dt + E_2x(t)d\varpi(t), \\ y(t) &= C_2x(t), \end{aligned}$$

where

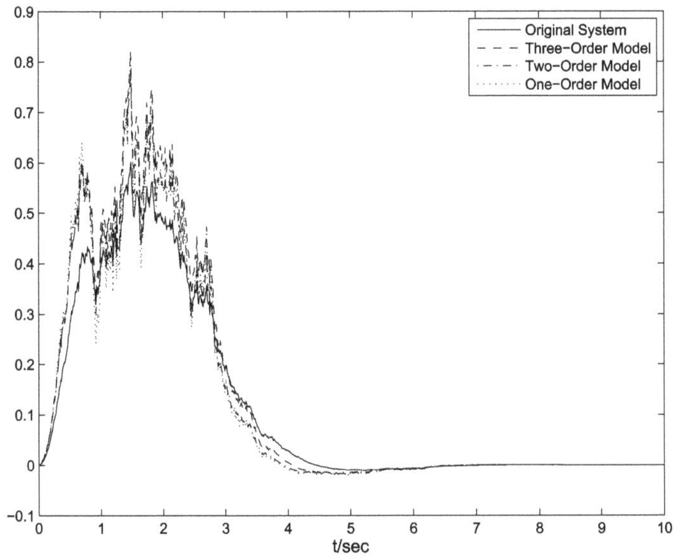

Fig. 10.3 Outputs of the original system and the reduced-order models

$$A_1 = \begin{bmatrix} -0.1 & 50 \\ -1 & -10 \end{bmatrix}, E_1 = \begin{bmatrix} -0.1 & 5 \\ -0.1 & -1 \end{bmatrix}, B_1 = \begin{bmatrix} 0 \\ 1 \end{bmatrix}, C_1 = \begin{bmatrix} 1 & 0 \end{bmatrix},$$
$$A_2 = \begin{bmatrix} -4.6 & 50 \\ -1 & -10 \end{bmatrix}, E_2 = \begin{bmatrix} -0.6 & 5 \\ -0.3 & -1 \end{bmatrix}, B_2 = \begin{bmatrix} 0 \\ 1 \end{bmatrix}, C_2 = \begin{bmatrix} 1 & 0 \end{bmatrix}.$$

The fuzzy functions with two rules are shown in Fig. 10.6.

On the basis of Corollary 10.12, the Hankel-norm performance level is set as $\gamma = 1.505$, and the reduced-order model parameters are computed as

$$\left[\begin{array}{c|c} \hat{A} & \hat{B} \\ \hline \hat{E} & \hat{C} \end{array}\right] = \left[\begin{array}{c|c} -13.5818 & -9.3129 \\ \hline -0.5347 & -1.1265 \end{array}\right].$$

Furthermore, set the zero initial condition as $\tilde{x}(0) = 0$ $(x(0) = 0, \hat{x}(0) = 0)$ and the disturbance noise as $u(t) = \exp(-0.6t)\sin(t),\ t \geq 0$. Figure 10.7 plots the outputs of the original system (black line) and reduced-order model (red line), and the error between them is drawn in Fig. 10.8.

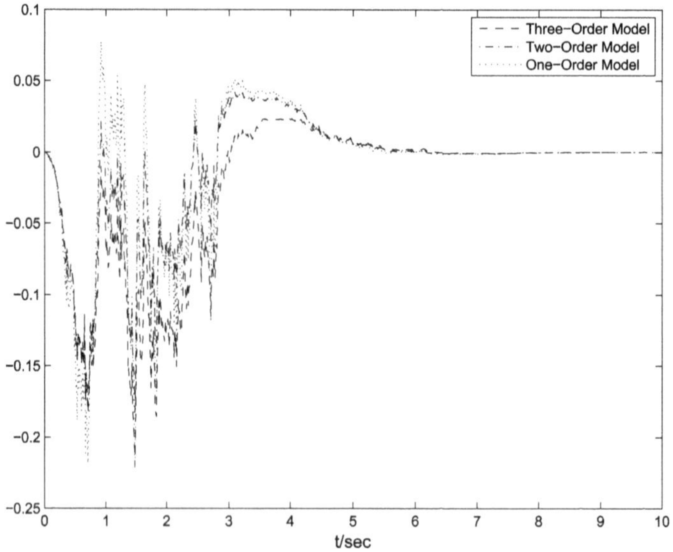

Fig. 10.4 Output errors between the original system and the reduced-order models

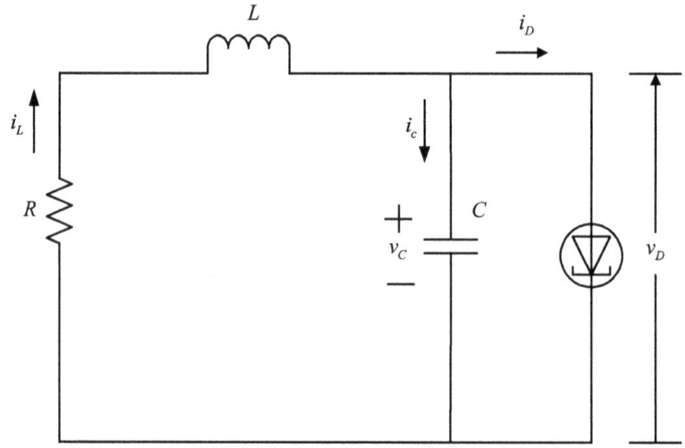

Fig. 10.5 Tunnel diode circuit

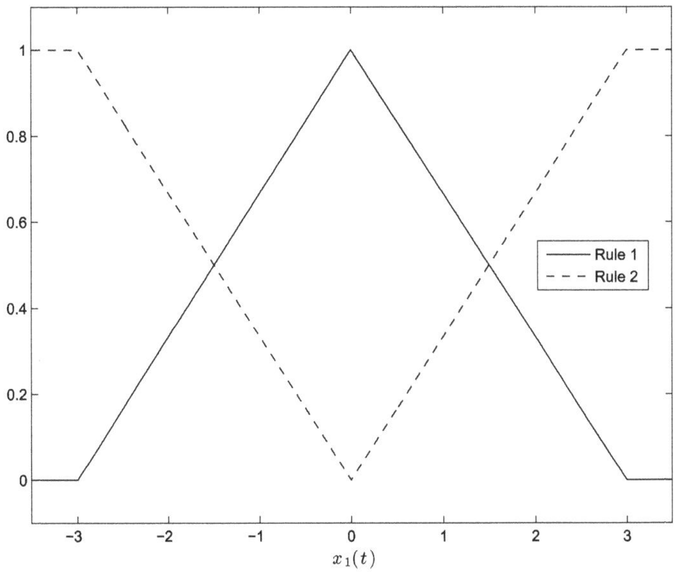

Fig. 10.6 Membership functions for the two fuzzy sets

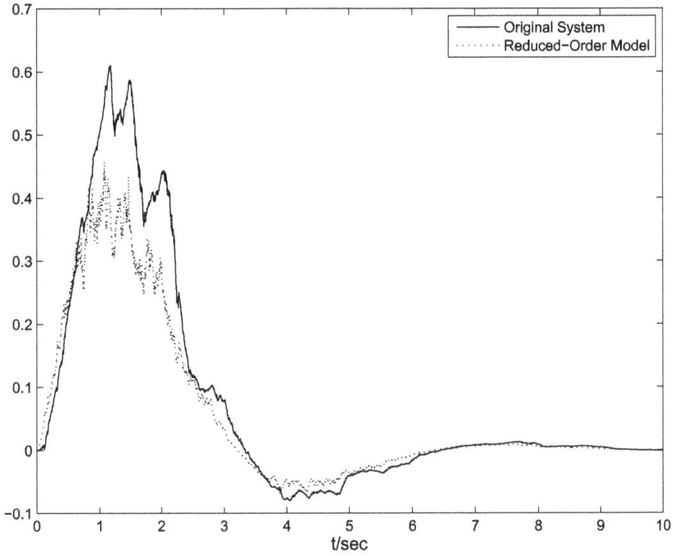

Fig. 10.7 Outputs of the original system and the reduced-order model

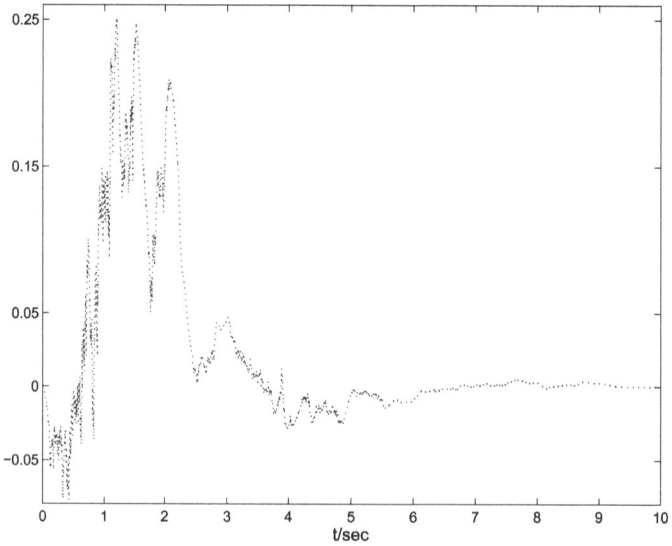

Fig. 10.8 Output error between the original system and the reduced-order model

10.5 Conclusion

The model approximation method for fuzzy switched systems with stochastic disturbances was established. By employing the piecewise Lyapunov function and ADT technique, sufficient conditions were derived to ensure the mean-square exponential stability and Hankel-norm performance for the resulting dynamic system. Furthermore, the relevant solvable conditions of the designed reduced-order models were set using an efficient linearization procedure method. In the final, simulation results have been provided to verify the validity of the presented scheme.

Chapter 11
Reduced-Order Filter Design of Fuzzy Stochastic Systems

11.1 Introduction

This chapter is interested in designing the $\mathcal{H}_\infty$ reduced-order filter for T-S fuzzy delayed systems with stochastic perturbation. Using a novel Lyapunov function and the reciprocally convex approach, the fuzzy-dependent conditions are established to ensure that the resulting filtering system is mean-square asymptotically stable and satisfies the prescribed $\mathcal{H}_\infty$ performance. Next, feasible solutions of the designed reduced-order filter are specified, which can be converted to a convex optimization problem via the convex linearization strategy. Finally, simulation examples, including that of an inverted pendulum, are given to demonstrate the validity and superiority of the presented $\mathcal{H}_\infty$ reduced-order filtering scheme.

11.2 System Description and Preliminaries

Introduce a dynamic nonlinear system, which can be approximated by the following T-S fuzzy stochastic delayed models:

♦ **Plant Form**:
Rule i: IF $\theta_1(k)$ is μ_{i1} and $\theta_2(k)$ is μ_{i2} and $\cdots$ and $\theta_p(k)$ is μ_{ip}, THEN

$$\begin{aligned} x(k+1) &= A_i x(k) + A_{di} x\big(k-d(k)\big) + B_i w(k) + B_{\bar{w}i} x(k)\bar{w}(k), \\ y(k) &= C_i x(k) + C_{di} x\big(k-d(k)\big) + D_i w(k) + D_{\bar{w}i} x(k)\bar{w}(k), \\ z(k) &= L_i x(k) + L_{di} x\big(k-d(k)\big) + E_i w(k), \\ x(k) &= \varphi(k), \ k = -d_2, -d_2+1, \ldots, 0, \end{aligned} \tag{11.1}$$

where $x(k) \in \mathbb{R}^n$ denotes the state vector; $w(k) \in \mathbb{R}^l$ denotes the noise input belonging to $\ell_2[0, \infty)$; $\bar{w}(k)$ stands for the Gaussian white noise sequence with zero mean on the probability space $(\Omega, \mathcal{F}, \mathcal{P})$ and

X. Su et al., *Intelligent Control, Filtering and Model Reduction Analysis for Fuzzy-Model-Based Systems*, Studies in Systems, Decision and Control 385,
https://doi.org/10.1007/978-3-030-81214-0_11

$$\mathbb{E}\{\bar{w}(k)\} = 0,\ \mathbb{E}\{\bar{w}^2(k)\} = 1,\ \mathbb{E}\{\bar{w}(k)\bar{w}(i)\} = 0,\ i \neq k,$$

where $\mathbb{E}$ denotes the expectation operate, and $y(k) \in \mathbb{R}^p$ denotes the measurable output; $z(k) \in \mathbb{R}^q$ represents the signal to be estimated, for $i \in \mathbb{S} \triangleq \{1, 2, \ldots, r\}$, r denotes the quantity of fuzzy rules; $\mu_{i1}, \ldots, \mu_{ip}$ stand for the fuzzy sets; $\theta(k) = \big[\theta_1(k), \theta_2(k), \ldots, \theta_p(k)\big]$ represent the premise variables, $\varphi(k)$ denotes the initial condition. A_i, A_{di}, B_i, C_i, C_{di}, D_i, L_i, L_{di}, $B_{\bar{w}i}$, $D_{\bar{w}i}$, and E_i are known constant matrices with suitable dimensions. The time-varying delay $d(k)$ satisfies

$$0 < d_1 \leqslant d(k) \leqslant d_2 < \infty,$$

where d_1, d_2 are constant positive integers representing the minimum and maximum bounds of time-varying delays, separately. The fuzzy membership functions are expressed as

$$h_i\big(\theta(k)\big) = \frac{\nu_i\big(\theta(k)\big)}{\sum_{i=1}^{r} \nu_i\big(\theta(k)\big)},\quad \nu_i\big(\theta(k)\big) = \prod_{j=1}^{p} \mu_{ij}\big(\theta_j(k)\big),$$

with $\mu_{ij}\big(\theta_j(k)\big)$ standing for the grade of membership of $\theta_j(k)$ in μ_{ij}. Thus, we can see that

$$\nu_i\big(\theta(k)\big) \geqslant 0,\ h_i\big(\theta(k)\big) \geqslant 0,\ i \in \mathbb{S},\ \sum_{i=1}^{r} h_i\big(\theta(k)\big) = 1.$$

Provided that the input variables $u(k)$ cannot influence the premise variables $\theta(k)$. For a group of $(x(k), u(k))$, the defuzzification of the discrete-time fuzzy delayed system with stochastic perturbation can be constructed as

$$\begin{aligned} x(k+1) &= \sum_{i=1}^{r} h_i\big(\theta(k)\big)\Big[A_i x(k) + A_{di} x\big(k - d(k)\big) + B_i w(k) + B_{\bar{w}i} x(k)\bar{w}(k)\Big], \\ y(k) &= \sum_{i=1}^{r} h_i\big(\theta(k)\big)\Big[C_i x(k) + C_{di} x\big(k - d(k)\big) + D_i w(k) + D_{\bar{w}i} x(k)\bar{w}(k)\Big], \\ z(k) &= \sum_{i=1}^{r} h_i\big(\theta(k)\big)\Big[L_i x(k) + L_{di} x\big(k - d(k)\big) + E_i w(k)\Big], \\ x(k) &= \varphi(k),\ k = -d_2, -d_2 + 1, \ldots, 0. \end{aligned} \tag{11.2}$$

The mentioned-above fuzzy model can be presented as the following complete form:

$$\begin{aligned} x(k+1) &= A(k)x(k) + A_d(k)x\big(k - d(k)\big) + B(k)w(k) + B_{\bar{w}}(k)x(k)\bar{w}(k), \\ y(k) &= C(k)x(k) + C_d(k)x\big(k - d(k)\big) + D(k)w(k) + D_{\bar{w}}(k)x(k)\bar{w}(k), \\ z(k) &= L(k)x(k) + L_d(k)x\big(k - d(k)\big) + E(k)w(k), \end{aligned} \tag{11.3}$$

where

$$\begin{cases} A(k) \triangleq \sum_{i=1}^{r} h_i(\theta(k))A_i, & A_d(k) \triangleq \sum_{i=1}^{r} h_i(\theta(k))A_{di}, \\ B(k) \triangleq \sum_{i=1}^{r} h_i(\theta(k))B_i, & C(k) \triangleq \sum_{i=1}^{r} h_i(\theta(k))C_i, \\ C_d(k) \triangleq \sum_{i=1}^{r} h_i(\theta(k))C_{di}, & D(k) \triangleq \sum_{i=1}^{r} h_i(\theta(k))D_i, \\ L(k) \triangleq \sum_{i=1}^{r} h_i(\theta(k))L_i, & L_d(k) \triangleq \sum_{i=1}^{r} h_i(\theta(k))L_{di}, \\ B_{\bar{w}}(k) \triangleq \sum_{i=1}^{r} h_i(\theta(k))B_{\bar{w}i}, & D_{\bar{w}}(k) \triangleq \sum_{i=1}^{r} h_i(\theta(k))D_{\bar{w}i}, \\ E(k) \triangleq \sum_{i=1}^{r} h_i(\theta(k))E_i. \end{cases}$$

An efficient reduced-order filter, which ensures $\hat{z}(k)$ to track $z(k)$ well, is proposed as follow:

$$\begin{aligned} \hat{x}(k+1) &= A_f\hat{x}(k) + B_f y(k), \\ \hat{z}(k) &= L_f\hat{x}(k) + D_f y(k), \end{aligned} \tag{11.4}$$

where $\hat{x}(k) \in \mathbb{R}^r$ represents the filter state and $r \leq n$; $\hat{z}(k) \in \mathbb{R}^q$ denotes the estimation of $z(k)$; A_f, B_f, L_f, and D_f are filter parameters to be designed.

Making a combination of the system (11.2) and filter model (11.4), we can obtain the dynamic error system with a more compact form:

$$\begin{aligned} \xi(k+1) &= \sum_{i=1}^{r} h_i(\theta(k))\Big[\hat{A}_i\xi(k) + \hat{A}_{di}\xi(k-d(k)) + \hat{B}_i w(k) + \hat{B}_{\bar{w}i}\xi(k)\bar{w}(k)\Big], \\ e(k) &= \sum_{i=1}^{r} h_i(\theta(k))\Big[\hat{L}_i\xi(k) + \hat{L}_{di}\xi(k-d(k)) + \hat{E}_i w(k) + \hat{F}_i\xi(k)\bar{w}(k)\Big], \end{aligned} \tag{11.5}$$

where $\xi(k) \triangleq \begin{bmatrix} x^T(k) \; \hat{x}^T(k) \end{bmatrix}^T$, $e(k) \triangleq z(k) - \hat{z}(k)$ and

$$\hat{A}_i \triangleq \begin{bmatrix} A_i & 0 \\ B_f C_i & A_f \end{bmatrix}, \quad \hat{A}_{di} \triangleq \begin{bmatrix} A_{di} & 0 \\ B_f C_{di} & 0 \end{bmatrix}, \quad \hat{B}_i \triangleq \begin{bmatrix} B_i \\ B_f D_i \end{bmatrix},$$

$$\hat{B}_{\bar{w}i} \triangleq \begin{bmatrix} B_{\bar{w}i} & 0 \\ B_f D_{\bar{w}i} & 0 \end{bmatrix}, \quad \hat{F}_i \triangleq \begin{bmatrix} -D_f D_{\bar{w}i} & 0 \end{bmatrix}, \quad \hat{E}_i \triangleq E_i - D_f D_i,$$

$$\hat{L}_i \triangleq \begin{bmatrix} L_i - D_f C_i & -L_f \end{bmatrix}, \quad \hat{L}_{di} \triangleq \begin{bmatrix} L_{di} - D_f C_{di} & 0 \end{bmatrix}.$$

Figure 11.1 shows the block diagram of the overall error system in (11.5), and it follows that

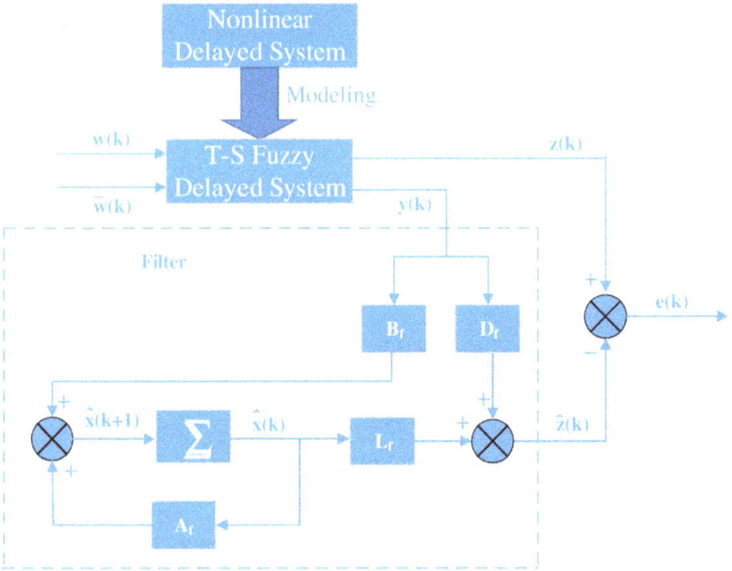

Fig. 11.1 Block diagram of the more compact presentation for the filtering error system

$$\xi(k+1) = \hat{A}(k)\xi(k) + \hat{A}_d(k)\xi\big(k - d(k)\big) + \hat{B}(k)w(k) + \hat{B}_{\bar{w}}(k)\xi(k)\bar{w}(k),$$
$$e(k) = \hat{L}(k)\xi(k) + \hat{L}_d(k)\xi\big(k - d(k)\big) + \hat{E}(k)w(k) + \hat{F}(k)\xi(k)\bar{w}(k),$$

where

$$\begin{cases} \hat{A}(k) \triangleq \sum_{i=1}^{r} h_i\big(\theta(k)\big)\hat{A}_i, & \hat{A}_d(k) \triangleq \sum_{i=1}^{r} h_i\big(\theta(k)\big)\hat{A}_{di}, \\ \hat{B}(k) \triangleq \sum_{i=1}^{r} h_i\big(\theta(k)\big)\hat{B}_i, & \hat{L}(k) \triangleq \sum_{i=1}^{r} h_i\big(\theta(k)\big)\hat{L}_i, \\ \hat{L}_d(k) \triangleq \sum_{i=1}^{r} h_i\big(\theta(k)\big)\hat{L}_{di}, & \hat{B}_{\bar{w}}(k) \triangleq \sum_{i=1}^{r} h_i\big(\theta(k)\big)\hat{B}_{\bar{w}i}, \\ \hat{E}(k) \triangleq \sum_{i=1}^{r} h_i\big(\theta(k)\big)\hat{E}_i, & \hat{F}(k) \triangleq \sum_{i=1}^{r} h_i\big(\theta(k)\big)\hat{F}_i. \end{cases}$$

Before giving the main results, some useful lemmas and definitions are introduced firstly.

Lemma 11.1 [205] *Assume that $f_1, f_2, \ldots, f_N$: $\mathbb{R}^m \to \mathbb{R}$ possess positive values in the open subset D of $\mathbb{R}^m$. The reciprocally convex combination of f_i over D satisfies the following condition:*

$$\min_{\left\{\beta_i \mid \beta_i > 0, \sum_i \beta_i = 1\right\}} \sum_i \frac{1}{\beta_i} f_i(t) = \sum_i f_i(t) + \max_{g_{i,j}(t)} \sum_{i \neq j} g_{i,j}(t),$$

with

$$\left\{g_{i,j} : \mathbb{R}^m \to \mathbb{R}, g_{j,i}(t) = g_{i,j}(t), \begin{bmatrix} f_i(t) & g_{i,j}(t) \\ g_{j,i}(t) & f_j(t) \end{bmatrix} \geqslant 0\right\}.$$

Definition 11.2 [249] When $w(k) = 0$, the resulting dynamic system in (11.5) is mean-square asymptotically stable if $\xi(k)$ satisfies the following condition:

$$\lim_{k\to\infty} \left|\xi(k)\right| = 0.$$

As for an error system (11.5) with the mean-square asymptotic stability, it follows $e = \left\{e(k)\right\} \in \ell_2[0, \infty)$ when $w = \left\{w(k)\right\} \in \ell_2[0, \infty)$.

Definition 11.3 [211] For a given scalar $\gamma > 0$, if the dynamic filtering system in (11.5) is mean-square asymptotically stable subject to the specific $\mathcal{H}_\infty$ error performance sense γ under $w(k) = 0$ and the zero initial condition, the following inequality holds:

$$\left\|e(k)\right\|_2 < \gamma\left\|w(k)\right\|_2, \ \forall 0 \neq w \in \ell_2[0, \infty),$$

where

$$\left\|e(k)\right\|_2 \triangleq \sqrt{\sum_{k=0}^{\infty} e^T(k)e(k)}.$$

Our aim in this chapter is to obtain the parameters of reduced-order filters (A_f, B_f, L_f, D_f) in the form of (11.4) and reach the requirements below:

- Utilizing some fuzzy basis-dependent existence conditions, an efficient reduced-order model is proposed such that the overall filtering system in (11.5) is mean-square asymptotically stable and satisfies a particular $\mathcal{H}_\infty$ error performance index γ.
- The presented filter design conditions are formulated as LMIs via employing the convex linearization method, which can be resolved by the standard toolbox.

Assumption 11.1 The fuzzy delayed system under stochastic perturbation (11.2) has the mean-square asymptotical stability.

11.3 Main Results

11.3.1 $\mathcal{H}_\infty$ Performance Analysis

For notational simplicity, we give the following parameters:

$$\begin{cases} P(k) \triangleq \sum_{i=1}^{r} h_i(\theta(k))P_i, & Q_1(k) \triangleq \sum_{i=1}^{r} h_i(\theta(k))Q_{1i}, \\ Q_2(k) \triangleq \sum_{i=1}^{r} h_i(\theta(k))Q_{2i}, & Q_3(k) \triangleq \sum_{i=1}^{r} h_i(\theta(k))Q_{3i}. \end{cases}$$

Theorem 11.4 *The overall filtering system in (11.5) has mean-square asymptotical stability subject to a specific $\mathcal{H}_\infty$ performance index γ, if there are positive scalars $\gamma > 0$, d_1, d_2, and suitable matrices $P_i > 0$, $Q_{1i} > 0$, $Q_{2i} > 0$, $Q_{3i} > 0$, $S_1 > 0$, $S_2 > 0$, Z_2, which satisfy the following conditions for $i \in \mathbb{S}$:*

$$\Upsilon_{ilstu} \triangleq \begin{bmatrix} \Upsilon_{11i} & \Upsilon_{12} & 0 & 0 & 0 & \Upsilon_{16i} \\ \star & \Upsilon_{22t} & \Upsilon_{23} & \Upsilon_{24} & 0 & 0 \\ \star & \star & \Upsilon_{33s} & \Upsilon_{34} & 0 & 0 \\ \star & \star & \star & \Upsilon_{44u} & 0 & \Upsilon_{46i} \\ \star & \star & \star & \star & \Upsilon_{55} & \Upsilon_{56i} \\ \star & \star & \star & \star & \star & \Upsilon_{66l} \end{bmatrix} < 0, \tag{11.6}$$

$$\begin{bmatrix} S_2 & Z_2 \\ \star & S_2 \end{bmatrix} \geqslant 0, \tag{11.7}$$

where

$$\Upsilon_{11i} \triangleq -P_i + Q_{1i} + Q_{2i} + (1+d)Q_{3i} - S_1,\ \Upsilon_{12} \triangleq S_1,\ \Upsilon_{22t} \triangleq -Q_{1t} - S_1 - S_2,$$
$$\Upsilon_{23} \triangleq Z_2,\ \Upsilon_{24} \triangleq S_2 - Z_2,\ \Upsilon_{33s} \triangleq -Q_{2s} - S_2,\ \Upsilon_{34} \triangleq -Z_2^T + S_2,$$
$$\Upsilon_{44u} \triangleq -2S_2 + Z_2^T + Z_2 - Q_{3u},\ \Upsilon_{46i} \triangleq \left[\ \hat{A}_{di}^T\ \hat{A}_{di}^T\ \hat{A}_{di}^T\ \hat{L}_{di}^T\ 0\,0\,0\,0 \right],$$
$$\Upsilon_{16i} \triangleq \left[\ \hat{A}_i^T\ \hat{A}_i^T - I\ \hat{A}_i^T - I\ \hat{L}_i^T\ B_{\bar{w}i}^T\ B_{\bar{w}i}^T\ B_{\bar{w}i}^T\ \hat{F}_i^T \right],\ \Upsilon_{55} \triangleq -\gamma^2 I,$$
$$\Upsilon_{56i} \triangleq \left[\ \hat{B}_i^T\ \hat{B}_i^T\ \hat{B}_i^T\ \hat{E}_i^T\ 0\,0\,0\,0 \right],$$
$$\Upsilon_{66l} \triangleq diag\left\{ -P_l^{-1}\ -\tfrac{1}{d_1^2}S_1^{-1}\ -\tfrac{1}{d^2}S_2^{-1}\ -I\ -P_l^{-1}\ -\tfrac{1}{d_1^2}S_1^{-1}\ -\tfrac{1}{d^2}S_2^{-1}\ -I \right\}.$$

Proof On account of the fuzzy membership functions and (11.6), we have

$$\Upsilon(k) \triangleq \sum_{i=1}^{r} h_i(\theta(k)) \sum_{l=1}^{r} h_l(\theta(k+1)) \sum_{t=1}^{r} h_t(\theta(k-d_1))$$
$$\times \sum_{s=1}^{r} h_s(\theta(k-d_2)) \sum_{u=1}^{r} h_u(\theta(k-d(k)))\Upsilon_{ilstu} < 0.$$

A more complete expression of the condition previously mentioned can be described as

$$\Upsilon(k) \triangleq \begin{bmatrix} \Upsilon_{11}(k) & \Upsilon_{12} & 0 & 0 & 0 & \Upsilon_{16}(k) \\ \star & \Upsilon_{22}(k) & \Upsilon_{23} & \Upsilon_{24} & 0 & 0 \\ \star & \star & \Upsilon_{33}(k) & \Upsilon_{34} & 0 & 0 \\ \star & \star & \star & \Upsilon_{44}(k) & 0 & \Upsilon_{46}(k) \\ \star & \star & \star & \star & \Upsilon_{55} & \Upsilon_{56}(k) \\ \star & \star & \star & \star & \star & \Upsilon_{66}(k) \end{bmatrix} < 0, \tag{11.8}$$

where

$$\begin{aligned} \Upsilon_{11}(k) &\triangleq -P(k) + Q_1(k) + Q_2(k) + (1+d)Q_3(k) - S_1, \\ \Upsilon_{22}(k) &\triangleq -Q_1(k-d_1) - S_1 - S_2, \ \Upsilon_{33}(k) \triangleq -Q_2(k-d_2) - S_2, \\ \Upsilon_{44}(k) &\triangleq -Q_3\big(k-d(k)\big) - 2S_2 + Z_2^T + Z_2, \\ \Upsilon_{16}(k) &\triangleq \Big[\hat{A}^T(k) \ \hat{A}^T(k) - I \ \hat{A}^T(k) - I \ \hat{L}^T(k) \ \hat{B}_{\bar{w}}^T(k) \ \hat{B}_{\bar{w}}^T(k) \ \hat{B}_{\bar{w}}^T(k) \ \hat{F}^T(k)\Big], \\ \Upsilon_{46}(k) &\triangleq \Big[\hat{A}_d^T(k) \ \hat{A}_d^T(k) \ \hat{A}_d^T(k) \ \hat{L}_d^T(k) \ 0\,0\,0\,0\Big], \\ \Upsilon_{56}(k) &\triangleq \Big[\hat{B}^T(k) \ \hat{B}^T(k) \ \hat{B}^T(k) \ \hat{E}^T(k) \ 0\,0\,0\,0\Big], \\ \Upsilon_{66}(k) &\triangleq diag\Big\{-P^{-1}(k+1) \ -\tfrac{1}{d_1^2}S_1^{-1} \ -\tfrac{1}{d^2}S_2^{-1} \ -I \ -P^{-1}(k+1) \ -\tfrac{1}{d_1^2}S_1^{-1} \\ &\qquad -\tfrac{1}{d^2}S_2^{-1} \ -I\Big\}. \end{aligned}$$

Utilizing the Schur complement method, (11.8) implies

$$\tilde{\Upsilon}(k) \triangleq \begin{bmatrix} \tilde{\Upsilon}_{11}(k) & \Upsilon_{12} & 0 & \tilde{\Upsilon}_{14}(k) & \tilde{\Upsilon}_{15}(k) \\ \star & \Upsilon_{22}(k) & \Upsilon_{23} & \Upsilon_{24} & 0 \\ \star & \star & \Upsilon_{33}(k) & \Upsilon_{34} & 0 \\ \star & \star & \star & \tilde{\Upsilon}_{44}(k) & \tilde{\Upsilon}_{45}(k) \\ \star & \star & \star & \star & \tilde{\Upsilon}_{55}(k) \end{bmatrix}, \tag{11.9}$$

where

$$\begin{aligned} \tilde{\Upsilon}_{11}(k) &\triangleq \hat{A}^T(k)P(k+1)\hat{A}(k) - P(k) + Q_1(k) + Q_2(k) + (1+d)Q_3(k) - S_1 \\ &\quad + d_1^2\big(\hat{A}^T(k) - I\big)S_1(\hat{A}(k) - I\big) + d^2\big(\hat{A}^T(k) - I\big)S_2(\hat{A}(k) - I\big) \\ &\quad + \hat{B}_{\bar{w}}^T(k)P(k+1)\hat{B}_{\bar{w}}(k) + d_1^2\hat{B}_{\bar{w}}^T(k)S_1\hat{B}_{\bar{w}}(k) + d^2\hat{B}_{\bar{w}}^T(k)S_2\hat{B}_{\bar{w}}(k) \\ &\quad + \hat{F}^T(k)\hat{F}(k) + \hat{L}^T(k)\hat{L}(k), \\ \tilde{\Upsilon}_{14}(k) &\triangleq \hat{A}^T(k)P(k+1)\hat{A}_d(k) + d_1^2\big(\hat{A}^T(k) - I\big)S_1\hat{A}_d(k) \\ &\quad + d^2\big(\hat{A}^T(k) - I\big)S_2\hat{A}_d(k) + \hat{L}^T(k)\hat{L}_d(k), \\ \tilde{\Upsilon}_{15}(k) &\triangleq \hat{A}^T(k)P(k+1)\hat{B}(k) + d_1^2\big(\hat{A}^T(k) - I\big)S_1\hat{B}(k) \\ &\quad + d^2\big(\hat{A}^T(k) - I\big)S_2\hat{B}(k) + \hat{L}^T(k)\hat{E}(k), \\ \tilde{\Upsilon}_{44}(k) &\triangleq \hat{A}_d^T(k)P(k+1)\hat{A}_d(k) - Q_3\big(k-d(k)\big) + d_1^2\hat{A}_d^T(k)S_1\hat{A}_d(k) \end{aligned}$$

$$
\begin{aligned}
&\quad +d^2\hat{A}_d^T(k)S_2\hat{A}_d(k)+Z_2^T+Z_2-2S_2+\hat{L}_d^T(k)\hat{L}_d(k),\\
\tilde{\Upsilon}_{45}(k) &\triangleq \hat{A}_d^T(k)P(k+1)\hat{B}(k)+d_1^2\hat{A}_d^T(k)S_1\hat{B}(k)+d^2\hat{A}_d^T(k)S_2\hat{B}(k)\\
&\quad +\hat{L}_d^T(k)\hat{E}(k),\\
\tilde{\Upsilon}_{55}(k) &\triangleq \hat{B}^T(k)P(k+1)\hat{B}(k)+d_1^2\hat{B}^T(k)S_1\hat{B}(k)+d^2\hat{B}^T(k)S_2\hat{B}(k)\\
&\quad +\hat{E}^T(k)\hat{E}(k)-\gamma^2 I.
\end{aligned}
$$

Consider the following Lyapunov function:

$$
V(k)\triangleq\sum_{i=1}^{5}V_i(k), \tag{11.10}
$$

where

$$
\begin{cases}
V_1(k)\triangleq \xi^T(k)P(k)\xi(k),\\
V_2(k)\triangleq \sum\limits_{i=k-d_1}^{k-1}\xi^T(i)Q_1(i)\xi(i)+\sum\limits_{i=k-d_2}^{k-1}\xi^T(i)Q_2(i)\xi(i),\\
V_3(k)\triangleq \sum\limits_{i=k-d(k)}^{k-1}\xi^T(i)Q_3(i)\xi(i)+\sum\limits_{j=-d_2+1}^{-d_1}\sum\limits_{i=k+j}^{k-1}\xi^T(i)Q_3(i)\xi(i),\\
V_4(k)\triangleq d_1\sum\limits_{j=-d_1}^{-d_0-1}\sum\limits_{i=k+j}^{k-1}\delta^T(i)S_1\delta(i),\\
V_5(k)\triangleq d\sum\limits_{j=-d_2}^{-d_1-1}\sum\limits_{i=k+j}^{k-1}\delta^T(i)S_2(i)\delta(i),\\
\delta(k)\triangleq \xi(k+1)-\xi(k).
\end{cases}
$$

Set $d_0=0$, $d=d_2-d_1$. Along the trajectory of system (11.5), then we can obtain

$$
\begin{aligned}
\mathbb{E}\Big\{\Delta V_1(k)\Big\} &= \mathbb{E}\Big\{\xi^T(k+1)P(k+1)\xi(k+1)-\xi^T(k)P(k)\xi(k)\Big\},\\
\mathbb{E}\Big\{\Delta V_2(k)\Big\} &= \mathbb{E}\Big\{\xi^T(k)Q_1(k)\xi(k)+\xi^T(k)Q_2(k)\xi(k)\\
&\quad -\xi^T(k-d_1)Q_1(k-d_1)\xi(k-d_1)\Big\}\\
&\quad -\mathbb{E}\Big\{\xi^T(k-d_2)Q_2(k-d_2)\xi(k-d_2)\Big\},\\
\mathbb{E}\Big\{\Delta V_3(k)\Big\} &= \mathbb{E}\Big\{(1+d)\xi^T(k)Q_3(k)\xi(k)-\xi^T\big(k-d(k)\big)Q_3\big(k-d(k)\big)\\
&\quad \xi\big(k-d(k)\big)\Big\}+\mathbb{E}\Big\{\sum_{i=k-d(k+1)+1}^{k-1}\xi^T(i)Q_3(i)\xi(i)\\
&\quad -\sum_{i=k-d(k)+1}^{k-1}\xi^T(i)Q_3(i)\xi(i)-\sum_{i=k-d_2+1}^{k-d_1}\xi^T(i)Q_3(i)\xi(i)\Big\}
\end{aligned}
$$

$$
\begin{aligned}
&\leqslant \mathbb{E}\Big\{(1+d)\xi^T(k)Q_3(k)\xi(k) - \xi^T\big(k-d(k)\big)Q_3\big(k-d(k)\big) \\
&\quad \xi\big(k-d(k)\big)\Big\},
\end{aligned}
$$

$$
\begin{aligned}
\mathbb{E}\Big\{\Delta V_4(k)\Big\} &= \mathbb{E}\Big\{d_1^2\delta^T(k)S_1\delta(k) - d_1\sum_{i=k-d_1}^{k-1}\delta^T(i)S_1\delta(i)\Big\} \\
&\leqslant \mathbb{E}\Big\{d_1^2\delta^T(k)S_1\delta(k) - \Big[\xi(k)-\xi(k-d_1)\Big]^T S_1\Big[\xi(k)-\xi(k-d_1)\Big]\Big\}, \\
\mathbb{E}\Big\{\Delta V_5(k)\Big\} &= \mathbb{E}\Big\{d^2\delta^T(k)S_2\delta(k) - d\sum_{i=k-d_2}^{k-d_1-1}\delta^T(i)S_2\delta(i)\Big\}.
\end{aligned}
\tag{11.11}
$$

For $\Delta V_5(k)$, due to $d_1 < d(k) < d_2$, considering the condition in (11.7) and Lemma 11.1, then it has

$$
\begin{aligned}
&\mathbb{E}\Big\{\Delta V_5(k)\Big\} \qquad (11.12)\\
&= \mathbb{E}\left\{d^2\delta^T(k)S_2\delta(k) - d\sum_{i=k-d_2}^{k-d(k)-1}\delta^T(i)S_2\delta(i) - d\sum_{i=k-d(k)}^{k-d_1-1}\delta^T(i)S_2\delta(i)\right\} \\
&\leqslant \mathbb{E}\Big\{d^2\delta^T(k)S_2\delta(k)\Big\} \\
&-\mathbb{E}\Big\{\frac{d}{d(k)-d_1}\Big[\xi\big(k-d(k)\big)-\xi(k-d_1)\Big]^T S_2\Big[\xi\big(k-d(k)\big)-\xi(k-d_1)\Big]\Big\} \\
&-\mathbb{E}\Big\{\frac{d}{d_2-d(k)}\Big[\xi\big(k-d(k)\big)-\xi(k-d_2)\Big]^T S_2\Big[\xi\big(k-d(k)\big)-\xi(k-d_2)\Big]\Big\} \\
&\leqslant \mathbb{E}\Big\{d^2\delta^T(k)S_2\delta(k)\Big\} \\
&-\mathbb{E}\Big\{\begin{bmatrix}\xi(k-d_1)-\xi\big(k-d(k)\big)\\ \xi\big(k-d(k)\big)-\xi(k-d_2)\end{bmatrix}^T\begin{bmatrix}S_2 & Z_2\\ \star & S_2\end{bmatrix}\begin{bmatrix}\xi(k-d_1)-\xi\big(k-d(k)\big)\\ \xi\big(k-d(k)\big)-\xi(k-d_2)\end{bmatrix}\Big\}.
\end{aligned}
\tag{11.13}
$$

If $d(k)=d_1$ or $d(k)=d_2$, (11.13) still holds because $x\big(k-d(k)\big)-x(k-d_1)=0$ or $x\big(k-d(k)\big)-x(k-d_2)=0$. It follows from (11.11) and (11.13) that

$$
\mathbb{E}\Big\{\Delta V(k)\Big\} = \mathbb{E}\Big\{\varsigma^T(k)\bar{\Upsilon}(k)\varsigma(k)\Big\} < 0,
\tag{11.14}
$$

where

$$
\varsigma(k) \triangleq \Big[\,\xi^T(k)\ \xi^T(k-d_1)\ \xi^T(k-d_2)\ \xi^T\big(k-d(k)\big)\Big]^T,
$$

and

$$
\begin{aligned}
\bar{\Upsilon}(k) &\triangleq \begin{bmatrix} \bar{\Upsilon}_{11}(k) & \Upsilon_{12} & 0 & \bar{\Upsilon}_{14}(k) \\ \star & \Upsilon_{22}(k) & \Upsilon_{23} & \Upsilon_{24} \\ \star & \star & \Upsilon_{33}(k) & \Upsilon_{34} \\ \star & \star & \star & \bar{\Upsilon}_{44}(k) \end{bmatrix}, \\
\bar{\Upsilon}_{11}(k) &\triangleq \hat{A}^T(k)P(k+1)\hat{A}(k) - P(k) + Q_1(k) + Q_2(k) + (1+d)Q_3(k) - S_1 \\
&\quad + d_1^2\big(\hat{A}^T(k) - I\big)S_1\big(\hat{A}(k) - I\big) + d^2\big(\hat{A}^T(k) - I\big)S_2\big(\hat{A}(k) - I\big) \\
&\quad + \hat{B}_{\bar{w}}^T(k)P(k+1)\hat{B}_{\bar{w}}(k) + d_1^2\hat{B}_{\bar{w}}^T(k)S_1\hat{B}_{\bar{w}}(k) + d^2\hat{B}_{\bar{w}}^T(k)S_2\hat{B}_{\bar{w}}(k), \\
\bar{\Upsilon}_{14}(k) &\triangleq \hat{A}^T(k)P(k+1)\hat{A}_d(k) + d_1^2\big(\hat{A}^T(k) - I\big)S_1\hat{A}_d(k) \\
&\quad + d^2\big(\hat{A}^T(k) - I\big)S_2\hat{A}_d(k), \\
\bar{\Upsilon}_{44}(k) &\triangleq \hat{A}_d^T(k)P(k+1)\hat{A}_d(k) - Q_3\big(k - d(k)\big) + d_1^2\hat{A}_d^T(k)S_1\hat{A}_d(k) \\
&\quad + d^2\hat{A}_d^T(k)S_2\hat{A}_d(k) + Z_2^T + Z_2 - 2S_2.
\end{aligned} \tag{11.15}
$$

We can conclude $\bar{\Upsilon}(k) < 0$ from (11.6) and (11.9). Therefore, it can be seen that the dynamic error system (11.5) has mean-square asymptotical stability when $w(k) = 0$.

Next, we are going to verify the $\mathcal{H}_\infty$ performance for the concerned error system. Define the following index:

$$
J(k) \triangleq \mathbb{E}\left\{\sum_{i=0}^{k-1}\Big(e^T(k)e(k) - \gamma^2 w^T(k)w(k)\Big)\right\}.
$$

Based on the zero initial state $V(0) = 0$, we can get

$$
\begin{aligned}
J(k) &\leqslant \mathbb{E}\left\{\sum_{i=0}^{k-1}\Big(e^T(k)e(k) - \gamma^2 w^T(k)w(k)\Big) + V(k) - V(0)\right\} \\
&= \mathbb{E}\left\{\sum_{i=0}^{k-1}\Big(e^T(k)e(k) - \gamma^2 w^T(k)w(k) + \Delta V(k)\Big)\right\} \\
&= \mathbb{E}\left\{\sum_{i=0}^{k-1}\Big[\eta^T(i)\tilde{\Upsilon}(i)\eta(i)\Big]\right\}.
\end{aligned}
$$

For nonzero $w(k) \in \ell_2[0, \infty)$ and $k > 0$, $\eta(k) \triangleq \big[\,\xi^T(k)\ w^T(k)\,\big]^T$, it is easy to see $J(k) < 0$ on account of (11.9), which implies the resulting error system in (11.5) satisfies a particular $\mathcal{H}_\infty$ performance. Thus, the proof is accomplished.

Remark 11.5 It can be seen that the proposed conditions in Theorem 11.4 contain the noise input $w(k)$ and Gaussian white noise sequence $\bar{w}(k)$. Nevertheless, the existing research results in [211, 249] either not consider $w(k)$ or not $\bar{w}(k)$. In fact, among some practical engineering applications, there exist many unknown cases and external disturbances, introducing the noise input $w(k)$ and Gaussian white noise

$\bar{w}(k)$ can be more applicable in actual situations. Therefore, it is significative to give the solution of $\mathcal{H}_\infty$ reduced-order filter for fuzzy stochastic systems in this chapter.

Notice that when $w(k) = 0$ and $\bar{w}(k)$ is not considered, the fuzzy delayed system in (11.1) can be converted as

$$x(k+1) = \sum_{i=1}^{r} h_i\big(\theta(k)\big)\Big[A_i x(k) + A_{di} x\big(k - d(k)\big)\Big]. \tag{11.16}$$

Then, the corresponding stability condition can be obtained as follow.

Corollary 11.6 *For the given positive integers d_1 and d_2, the fuzzy system in (11.16) has the asymptotical stability if there are matrices $\mathcal{P}_i > 0$, $\mathcal{Q}_{1i} > 0$, $\mathcal{Q}_{2i} > 0$, $\mathcal{Q}_{3i} > 0$, $\mathcal{S}_1 > 0$, $\mathcal{S}_2 > 0$, $\mathcal{Z}_2$ and $\mathcal{W}$, which satisfy the following conditions for $i \in \mathbb{S}$:*

$$\hat{\Upsilon}_{ilstu} < 0, \tag{11.17}$$

$$\begin{bmatrix} \mathcal{S}_2 & \mathcal{Z}_2 \\ \star & \mathcal{S}_2 \end{bmatrix} \geqslant 0, \tag{11.18}$$

where

$$\hat{\Upsilon}_{ilstu} \triangleq \begin{bmatrix} \Upsilon_{11i} & \Upsilon_{12} & 0 & 0 & \hat{\Upsilon}_{15i} \\ \star & \Upsilon_{22t} & \Upsilon_{23} & \Upsilon_{24} & 0 \\ \star & \star & \Upsilon_{33s} & \Upsilon_{34} & 0 \\ \star & \star & \star & \Upsilon_{44u} & \hat{\Upsilon}_{45i} \\ \star & \star & \star & \star & \hat{\Upsilon}_{55l} \end{bmatrix},$$

$$\hat{\Upsilon}_{15i} \triangleq \Big[\hat{A}_i^T \mathcal{P}_i \ (\hat{A}_i^T - I)\mathcal{W} \ (\hat{A}_i^T - I)\mathcal{W}\Big],$$

$$\hat{\Upsilon}_{45i} \triangleq \Big[\hat{A}_{di}^T \mathcal{P}_i \ \hat{A}_{di}^T \mathcal{W} \ \hat{A}_{di}^T \mathcal{W}\Big],$$

$$\hat{\Upsilon}_{55l} \triangleq diag\left\{-\mathcal{P}_l \ d_1^{-2}(\mathcal{S}_1 - \mathcal{W} - \mathcal{W}^T) \ d^{-2}(\mathcal{S}_2 - \mathcal{W} - \mathcal{W}^T)\right\},$$

and the definitions are given in Theorem 11.4.

Proof Employing the similar ways as the proof process in Theorem 11.4, it is simple to prove that Corollary 11.6 can ensure the fuzzy delayed system to be asymptotically stable. And here, the detailed process of proof is ignored.

11.3.2 Reduced-Order Filter Design

In this part, $\mathcal{H}_\infty$ reduced-order filtering conditions for fuzzy time-varying delay systems are established based on the convex linearization method, and the results are shown as below.

Theorem 11.7 *Consider the system (11.2), there is an admissible reduced-order filter as (11.4), which ensures the dynamic filtering system (11.5) is mean-square asymptotically stable and possesses a given $\mathcal{H}_\infty$ performance index $\gamma > 0$, if there are positive scalars d_1, d_2 and proper matrices $P_1 > 0$, $\bar{Q}_{1i} > 0$, $\bar{Q}_{2i} > 0$, $\bar{Q}_{3i} > 0$, $\bar{S}_1 > 0$, $\bar{S}_2 > 0$, $\bar{Z}_2$, $\mathcal{Q}$, $\mathcal{A}$, $\mathcal{B}$, $\mathcal{L}$, $\mathcal{D}$, W_1, W_2, which satisfy the following requirements for $i, t, s, u, l \in \mathbb{S}$:*

$$\Gamma_{istu} < 0, \tag{11.19}$$

$$\begin{bmatrix} S_{21} & S_{22} & Z_{21} & Z_{22} \\ \star & S_{24} & Z_{23} & Z_{24} \\ \star & \star & S_{21} & S_{22} \\ \star & \star & \star & Z_{24} \end{bmatrix} \geqslant 0, \tag{11.20}$$

where

$$\Gamma_{istu} \triangleq \begin{bmatrix} \Gamma_{11i} & \Gamma_{12} & 0 & 0 & 0 & \Gamma_{16i} \\ \star & \Gamma_{22t} & \Gamma_{23} & \Gamma_{24} & 0 & 0 \\ \star & \star & \Gamma_{33s} & \Gamma_{34} & 0 & 0 \\ \star & \star & \star & \Gamma_{44u} & 0 & \Gamma_{46i} \\ \star & \star & \star & \star & \Gamma_{55} & \Gamma_{56i} \\ \star & \star & \star & \star & \star & \Gamma_{66} \end{bmatrix},$$

with

$$\bar{Q}_{1i} \triangleq \begin{bmatrix} Q_{1i1} & Q_{1i2} \\ \star & Q_{1i4} \end{bmatrix}, \ \bar{Q}_{2i} \triangleq \begin{bmatrix} Q_{2i1} & Q_{2i2} \\ \star & Q_{2i4} \end{bmatrix}, \ \bar{Q}_{3i} \triangleq \begin{bmatrix} Q_{3i1} & Q_{3i2} \\ \star & Q_{3i4} \end{bmatrix},$$

$$\bar{S}_1 \triangleq \begin{bmatrix} S_{11} & S_{12} \\ \star & S_{14} \end{bmatrix}, \quad \bar{S}_2 \triangleq \begin{bmatrix} S_{21} & S_{22} \\ \star & S_{24} \end{bmatrix}, \quad \bar{Z}_2 \triangleq \begin{bmatrix} Z_{21} & Z_{22} \\ Z_{23} & Z_{24} \end{bmatrix},$$

$$\Gamma_{11i} \triangleq \begin{bmatrix} \Gamma_{11i1} & \Gamma_{11i2} \\ \star & \Gamma_{11i4} \end{bmatrix}, \ \Gamma_{12} \triangleq \bar{S}_1, \ \Gamma_{23} \triangleq \bar{Z}_2, \ \Gamma_{24} \triangleq \begin{bmatrix} S_{21} - Z_{21} & S_{22} - Z_{22} \\ S_{22}^T - Z_{23} & S_{24} - Z_{24} \end{bmatrix},$$

$$\Gamma_{11i1} \triangleq -P_1 + Q_{1i1} + Q_{2i1} + (1+d)Q_{3i1} - S_{11}, \ \Gamma_{55} \triangleq -\gamma^2 I,$$

$$\Gamma_{11i2} \triangleq -\mathcal{H}\mathcal{Q} + Q_{1i2} + Q_{2i2} + (1+d)Q_{3i2} - S_{12},$$

$$\Gamma_{11i4} \triangleq -\mathcal{Q}^T + Q_{1i4} + Q_{2i4} + (1+d)Q_{3i4} - S_{14},$$

$$\Gamma_{22t} \triangleq \begin{bmatrix} -Q_{1t1} - S_{11} - S_{21} & -Q_{1t2} - S_{12} - S_{22} \\ \star & -Q_{1t4} - S_{14} - S_{24} \end{bmatrix},$$

$$\Gamma_{33s} \triangleq \begin{bmatrix} -Q_{2s1} - S_{21} & -Q_{2s2} - S_{22} \\ \star & -Q_{2s4} - S_{24} \end{bmatrix}, \ \Gamma_{34} \triangleq \begin{bmatrix} S_{21} - Z_{21}^T & S_{22} - Z_{23}^T \\ S_{22}^T - Z_{22}^T & S_{24} - Z_{24}^T \end{bmatrix},$$

$$\Gamma_{44u} \triangleq \begin{bmatrix} -Q_{3u1} - S_{21} - S_{21}^T + Z_{21} + Z_{21}^T & -Q_{3u2} - 2S_{22} + Z_{22} + Z_{23}^T \\ \star & -Q_{3u4} - S_{24} - S_{24}^T + Z_{24} + Z_{24}^T \end{bmatrix},$$

$$\Gamma_{16i} \triangleq \begin{bmatrix} \Gamma_{16i1} & \Gamma_{16i2} & \Gamma_{16i2} & \Gamma_{16i4} & \Gamma_{16i5} & \Gamma_{16i5} & \Gamma_{16i5} & \Gamma_{16i6} \end{bmatrix},$$

$$\Gamma_{46i} \triangleq \begin{bmatrix} \Gamma_{46i1} & \Gamma_{46i2} & \Gamma_{46i2} & \Gamma_{46i4} & 0 & 0 & 0 & 0 \end{bmatrix},$$

$$\Gamma_{56i} \triangleq \begin{bmatrix} \Gamma_{56i1} \ \Gamma_{56i2} \ \Gamma_{56i2} \ \Gamma_{56i4} \ 0\,0\,0\,0 \end{bmatrix},$$

$$\Gamma_{66} \triangleq diag\left\{ \Gamma_{661} \ \Gamma_{662} \ \Gamma_{663} \ -I \ \Gamma_{661} \ \Gamma_{662} \ \Gamma_{663} \ -I \right\},$$

$$\Gamma_{16i1} \triangleq \begin{bmatrix} A_i^T P_1 + C_i^T \mathcal{B}^T \mathcal{H}^T & A_i^T \mathcal{H}\mathcal{Q} + C_i^T \mathcal{B}^T \\ \mathcal{A}^T \mathcal{H}^T & \mathcal{A}^T \end{bmatrix},$$

$$\Gamma_{16i2} \triangleq \begin{bmatrix} A_i^T W_1 + C_i^T \mathcal{B}^T \mathcal{H}^T - W_1 & A_i^T W_2 + C_i^T \mathcal{B}^T - W_2 \\ \mathcal{A}^T \mathcal{H}^T - \mathcal{Q}^T \mathcal{H}^T & \mathcal{A}^T - \mathcal{Q}^T \end{bmatrix},$$

$$\Gamma_{16i5} \triangleq \begin{bmatrix} B_{\bar{w}i}^T W_1 + D_{\bar{w}i}^T \mathcal{B}^T \mathcal{H}^T & B_{\bar{w}i}^T W_2 + D_{\bar{w}i}^T \mathcal{B}^T \\ 0 & 0 \end{bmatrix}, \quad \Gamma_{16i4} \triangleq \begin{bmatrix} L_i^T - C_i^T \mathcal{D}^T \\ -\mathcal{L}^T \end{bmatrix},$$

$$\Gamma_{46i1} \triangleq \begin{bmatrix} A_{di}^T P_1 + C_{di}^T \mathcal{B}^T \mathcal{H}^T & A_{di}^T \mathcal{H}\mathcal{Q} + C_{di}^T \mathcal{B}^T \\ 0 & 0 \end{bmatrix}, \quad \Gamma_{16i6} \triangleq \begin{bmatrix} -D_{\bar{w}i}^T \mathcal{D}^T \\ 0 \end{bmatrix},$$

$$\Gamma_{46i2} \triangleq \begin{bmatrix} A_{di}^T W_1 + C_{di}^T \mathcal{B}^T \mathcal{H}^T & A_{di}^T W_2 + C_{di}^T \mathcal{B}^T \\ 0 & 0 \end{bmatrix}, \quad \Gamma_{46i4} \triangleq \begin{bmatrix} L_{di}^T - C_{di}^T \mathcal{D}^T \\ 0 \end{bmatrix},$$

$$\Gamma_{56i1} \triangleq \begin{bmatrix} B_i^T P_1 + D_i^T \mathcal{B}^T \mathcal{H}^T & B_i^T \mathcal{H}\mathcal{Q} + D_i^T \mathcal{B}^T \end{bmatrix}, \quad \Gamma_{56i4} \triangleq E_i^T - D_i^T \mathcal{D}^T,$$

$$\Gamma_{56i2} \triangleq \begin{bmatrix} B_i^T W_1 + D_i^T \mathcal{B}^T \mathcal{H}^T & B_i^T W_2 + D_i^T \mathcal{B}^T \end{bmatrix}, \quad \Gamma_{661} \triangleq \begin{bmatrix} -P_1 & -\mathcal{H}\mathcal{Q} \\ \star & -\mathcal{Q}^T \end{bmatrix},$$

$$\Gamma_{662} \triangleq d_1^{-2} \begin{bmatrix} S_{11} - W_1 - W_1^T & S_{12} - W_2 - \mathcal{H}\mathcal{Q} \\ \star & S_{14} - \mathcal{Q} - \mathcal{Q}^T \end{bmatrix},$$

$$\Gamma_{663} \triangleq d^{-2} \begin{bmatrix} S_{21} - W_1 - W_1^T & S_{22} - W_2 - \mathcal{H}\mathcal{Q} \\ \star & S_{24} - \mathcal{Q} - \mathcal{Q}^T \end{bmatrix}.$$

Furthermore, if the previously mentioned conditions have feasible solutions (P_1, $\mathcal{A}$, $\mathcal{B}$, $\mathcal{L}$, $\mathcal{D}$, $\mathcal{Q}$, W_1, W_2, Q_{1i1}, Q_{1i2}, Q_{1i4}, Q_{2i1}, Q_{2i2}, Q_{2i4}, Q_{3i1}, Q_{3i2}, Q_{3i4}, S_{11}, S_{12}, S_{14}, S_{21}, S_{22}, S_{24}, Z_{21}, Z_{22}, Z_{23}, Z_{24}), then the matrices of designed reduced-order filter can be formulated as

$$\begin{bmatrix} A_f & B_f \\ L_f & D_f \end{bmatrix} = \begin{bmatrix} \mathcal{Q}^{-1} & 0 \\ 0 & I \end{bmatrix} \begin{bmatrix} \mathcal{A} & \mathcal{B} \\ \mathcal{L} & \mathcal{D} \end{bmatrix}. \tag{11.21}$$

Proof In consideration of Corollary 11.6, it is not difficult to conclude the system (11.5) is mean-square asymptotically stable and satisfies the particular $\mathcal{H}_\infty$ performance γ if there exist some matrices $P_i > 0$, $Q_{1i} > 0$, $Q_{2i} > 0$, $Q_{3i} > 0$, $S_1 > 0$, $S_2 > 0$, Z_2 and W, which satisfy the condition below:

$$\hat{\Upsilon}_{ilstu} < 0, \tag{11.22}$$

where

$$\hat{\Upsilon}_{ilstu} \triangleq \begin{bmatrix} \Upsilon_{11i} & \Upsilon_{12} & 0 & 0 & 0 & \hat{\Upsilon}_{16i} \\ \star & \Upsilon_{22t} & \Upsilon_{23} & \Upsilon_{24} & 0 & 0 \\ \star & \star & \Upsilon_{33s} & \Upsilon_{34} & 0 & 0 \\ \star & \star & \star & \Upsilon_{44u} & 0 & \hat{\Upsilon}_{46i} \\ \star & \star & \star & \star & \Upsilon_{55} & \hat{\Upsilon}_{56i} \\ \star & \star & \star & \star & \star & \hat{\Upsilon}_{66l} \end{bmatrix},$$

$$\hat{\Upsilon}_{16i} \triangleq \left[\hat{A}_i^T P_i \; (\hat{A}_i^T - I)W \; (\hat{A}_i^T - I)W \; \hat{L}_i^T \; B_{\bar{w}i}^T W \; B_{\bar{w}i}^T W \; B_{\bar{w}i}^T W \; \hat{F}_i^T \right],$$

$$\hat{\Upsilon}_{46i} \triangleq \left[\hat{A}_{di}^T P_i \; \hat{A}_{di}^T W \; \hat{A}_{di}^T W \; \hat{L}_{di}^T \; 0\,0\,0\,0 \right],$$

$$\hat{\Upsilon}_{56i} \triangleq \left[\hat{B}_i^T P_i \; \hat{B}_i^T W \; \hat{B}_i^T W \; \hat{E}_i^T \; 0\,0\,0\,0 \right],$$

$$\hat{\Upsilon}_{66l} \triangleq diag\left\{ -P_l \; d_1^{-2}(S_1 - W - W^T) \; d^{-2}(S_2 - W - W^T) \; -I \; -P_l \; d_1^{-2}(S_1 - W - W^T) \; d^{-2}(S_2 - W - W^T) \; -I \right\}.$$

Let P_i in (11.22) be partitioned uniformly to

$$P_i \triangleq \begin{bmatrix} P_1 & \mathcal{H}P_2 \\ \star & P_3 \end{bmatrix}, \tag{11.23}$$

where $\mathcal{H} = \left[I_{r\times r} \; 0_{r\times(n-r)} \right]^T$, $P_1 \in \mathbb{R}^{n\times n}$, $P_3 \in \mathbb{R}^{r\times r}$ and $P_2 \in \mathbb{R}^{r\times r}$. We assume P_2 is nonsingular, in order to prove this, define the matrix $U_i \triangleq P_i + \beta V$ $(\beta > 0)$ and

$$V \triangleq \begin{bmatrix} 0_{n\times n} & \mathcal{H} \\ \star & 0_{r\times r} \end{bmatrix}, \quad U_i \triangleq \begin{bmatrix} U_1 & \mathcal{H}U_2 \\ \star & U_3 \end{bmatrix}. \tag{11.24}$$

Considering $P_i > 0$, we have $U_i > 0$ for $\beta > 0$. Hence, it is simple to prove U_2 is nonsingular with an arbitrarily small $\beta > 0$, and (11.24) is feasible with P_i rather than U_i. Without loss of generality, P_2 is assumed to be nonsingular subject to U_2.

Based on the points discussed above, introduce some relevant matrices as follows:

$$\begin{cases} J \triangleq \begin{bmatrix} I & 0 \\ 0 & P_3^{-1}P_2^T \end{bmatrix}, & W \triangleq \begin{bmatrix} W_1 & W_2 P_2^{-T} P_3 \\ (\mathcal{H}P_2)^T & P_3 \end{bmatrix}, \\ S_1 \triangleq J^{-T}\bar{S}_1 J^{-1}, & S_2 \triangleq J^{-T}\bar{S}_2 J^{-1}, \\ Q_{1i} \triangleq J^{-T}\bar{Q}_{1i} J^{-1}, & Q_{2i} \triangleq J^{-T}\bar{Q}_{2i} J^{-1}, \\ \mathcal{Q} \triangleq P_2 P_3^{-1} P_2^T, & Q_{3i} \triangleq J^{-T}\bar{Q}_{3i} J^{-1}, \\ Z_2 \triangleq J^{-T}\bar{Z}_2 J^{-1}, & \end{cases} \tag{11.25}$$

and

$$\begin{bmatrix} A_f & B_f \\ L_f & D_f \end{bmatrix} \triangleq \begin{bmatrix} P_2^{-1} & 0 \\ 0 & I \end{bmatrix} \begin{bmatrix} \mathcal{A} & \mathcal{B} \\ \mathcal{L} & \mathcal{D} \end{bmatrix} \begin{bmatrix} P_2^{-T}P_3 & 0 \\ 0 & I \end{bmatrix}. \tag{11.26}$$

Then it yields

$$
\begin{aligned}
J^T\hat{A}_i^T P_i J &\triangleq \begin{bmatrix} A_i^T P_1 + C_i^T \mathcal{B}^T \mathcal{H}^T & A_i^T \mathcal{H}\mathcal{Q} + C_i^T \mathcal{B}^T \\ \mathcal{A}^T \mathcal{H}^T & \mathcal{A}^T \end{bmatrix}, \\
J^T\hat{A}_{di}^T P_i J &\triangleq \begin{bmatrix} A_{di}^T P_1 + C_{di}^T \mathcal{B}^T \mathcal{H}^T & A_{di}^T \mathcal{H}\mathcal{Q} + C_{di}^T \mathcal{B}^T \\ 0 & 0 \end{bmatrix}, \\
\hat{B}_i^T P_i J &\triangleq \begin{bmatrix} B_i^T P_1 + D_i^T \mathcal{B}^T \mathcal{H}^T & B_i^T \mathcal{H}\mathcal{Q} + D_i^T \mathcal{B}^T \end{bmatrix}, \\
J^T\hat{A}_i^T W J &\triangleq \begin{bmatrix} A_i^T W_1 + C_i^T \mathcal{B}^T \mathcal{H}^T & A_i^T W_2 + C_i^T \mathcal{B}^T \\ \mathcal{A}^T \mathcal{H}^T & \mathcal{A}^T \end{bmatrix}, \\
J^T\hat{A}_{di}^T W J &\triangleq \begin{bmatrix} A_{di}^T W_1 + C_{di}^T \mathcal{B}^T \mathcal{H}^T & A_{di}^T W_2 + C_{di}^T \mathcal{B}^T \\ 0 & 0 \end{bmatrix}, \\
J^T\hat{B}_{\bar{w}i}^T W J &\triangleq \begin{bmatrix} \hat{B}_{\bar{w}i}^T W_1 + \hat{D}_{\bar{w}i}^T \mathcal{B}^T \mathcal{H}^T & \hat{B}_{\bar{w}i}^T W_2 + \hat{D}_{\bar{w}i}^T \mathcal{B}^T \\ 0 & 0 \end{bmatrix}, \\
\hat{B}_i^T W J &\triangleq \begin{bmatrix} B_i^T W_1 + D_i^T \mathcal{B}^T \mathcal{H}^T & B_i^T W_2 + D_i^T \mathcal{B}^T \end{bmatrix}, \\
J^T\hat{L}_i^T &\triangleq \begin{bmatrix} L_i^T - C_i^T \mathcal{D}^T \\ -\mathcal{L}^T \end{bmatrix}, \quad J^T P_i J \triangleq \begin{bmatrix} P_1 & \mathcal{H}\mathcal{Q} \\ \mathcal{Q}^T \mathcal{H}^T & \mathcal{Q}^T \end{bmatrix}, \\
J^T W J &\triangleq \begin{bmatrix} W_1 & W_2 \\ \mathcal{Q}^T \mathcal{H}^T & \mathcal{Q}^T \end{bmatrix}, \quad J^T\hat{L}_{di}^T \triangleq \begin{bmatrix} L_{di}^T - C_{di}^T \mathcal{D}^T \\ 0 \end{bmatrix}, \\
J^T\hat{F}^T &\triangleq \begin{bmatrix} -D_{\bar{w}i}^T \mathcal{D}^T \\ 0 \end{bmatrix}, \quad \hat{E}^T \triangleq E_i^T - D_i^T \mathcal{D}^T.
\end{aligned} \tag{11.27}
$$

Making the congruence transformations to (11.7) and (11.22) by diag (J, J) and diag $(J, J, J, J, I, J, J, J, I, J, J, J, I)$, separately, it can be readily seen that (11.19) and (11.20) hold from (11.25)–(11.27). Consequently, the dynamic filtering system in (11.5) is mean-square asymptotically stable and possesses a particular $\mathcal{H}_\infty$ performance level γ.

In addition, the reduced-order filter parameters (A_f, B_f, L_f, D_f) are rewritten as (11.26), which signifies that $P_2^{-T} P_3$ makes a similarity transformation to obtain the state-space form of the designed filter. Without loss of generality, set $P_2^{-T} P_3 = I$, then it follows (11.21), which is useful in constructing the filter in (11.4). Thus, it completes the proof.

Remark 11.8 The matrix $\mathcal{H}$ is denoted in Theorems 11.7, which plays a vital part in designing the reduced-order filter and it is employed as a namely order reduction element. Furthermore, if $\mathcal{H}$ becomes an unit matrix, the obtained results in Theorem 11.7 can be converted to a full-order case, which is easier compared with the reduced-order case. In the following, we will introduce the results in full-order case.

Corollary 11.9 *Consider the fuzzy system in (11.2) and the condition in (11.20). There exists a reduced-order filter as (11.4) so that the resulting system (11.5) can be guaranteed to be mean-square asymptotically stable subject to the specific $\mathcal{H}_\infty$ error performance sense γ, if there are positive scalars d_1, d_2 and suitable matrices $P_1 > 0$, $\bar{Q}_{1i} > 0$, $\bar{Q}_{2i} > 0$, $\bar{Q}_{3i} > 0$, $\bar{S}_1 > 0$, $\bar{S}_2 > 0$, $\bar{Z}_2$, $\mathcal{Q}$, $\mathcal{A}$, $\mathcal{B}$, $\mathcal{L}$, $\mathcal{D}$, W_1, W_2, which satisfy the following condition for $i, t, s, u, l \in \mathbb{S}$:*

$$\tilde{\Gamma}_{istu} < 0, \tag{11.28}$$

where

$$\tilde{\Gamma}_{istu} \triangleq \begin{bmatrix} \tilde{\Gamma}_{11i} & \Gamma_{12} & 0 & 0 & 0 & \tilde{\Gamma}_{16i} \\ \star & \Gamma_{22t} & \Gamma_{23} & \Gamma_{24} & 0 & 0 \\ \star & \star & \Gamma_{33s} & \Gamma_{34} & 0 & 0 \\ \star & \star & \star & \Gamma_{44u} & 0 & \tilde{\Gamma}_{46i} \\ \star & \star & \star & \star & \Gamma_{55} & \tilde{\Gamma}_{56i} \\ \star & \star & \star & \star & \star & \tilde{\Gamma}_{66} \end{bmatrix},$$

with

$$\begin{aligned}
\tilde{\Gamma}_{11i} &\triangleq \begin{bmatrix} \Gamma_{11i1} & \tilde{\Gamma}_{11i2} \\ \star & \Gamma_{11i4} \end{bmatrix}, \quad \tilde{\Gamma}_{661} \triangleq \begin{bmatrix} -P_1 & -\mathcal{Q} \\ \star & -\mathcal{Q}^T \end{bmatrix}, \\
\tilde{\Gamma}_{11i2} &\triangleq -\mathcal{Q} + Q_{1i2} + Q_{2i2} + (1+d)Q_{3i2} - S_{12}, \\
\tilde{\Gamma}_{16i} &\triangleq \begin{bmatrix} \tilde{\Gamma}_{16i1} & \tilde{\Gamma}_{16i2} & \tilde{\Gamma}_{16i2} & \Gamma_{16i4} & \tilde{\Gamma}_{16i5} & \tilde{\Gamma}_{16i5} & \tilde{\Gamma}_{16i5} & \Gamma_{16i6} \end{bmatrix}, \\
\tilde{\Gamma}_{46i} &\triangleq \begin{bmatrix} \tilde{\Gamma}_{46i1} & \tilde{\Gamma}_{46i2} & \tilde{\Gamma}_{46i2} & \Gamma_{46i4} & 0 & 0 & 0 & 0 \end{bmatrix}, \\
\tilde{\Gamma}_{56i} &\triangleq \begin{bmatrix} \tilde{\Gamma}_{56i1} & \tilde{\Gamma}_{56i2} & \tilde{\Gamma}_{56i2} & \Gamma_{56i4} & 0 & 0 & 0 & 0 \end{bmatrix}, \\
\tilde{\Gamma}_{66} &\triangleq diag\left\{ \tilde{\Gamma}_{661} \; \tilde{\Gamma}_{662} \; \tilde{\Gamma}_{663} \; -I \; \tilde{\Gamma}_{661} \; \tilde{\Gamma}_{662} \; \tilde{\Gamma}_{663} \; -I \right\}, \\
\tilde{\Gamma}_{16i1} &\triangleq \begin{bmatrix} A_i^T P_1 + C_i^T \mathcal{B}^T & A_i^T \mathcal{Q} + C_i^T \mathcal{B}^T \\ \mathcal{A}^T & \mathcal{A}^T \end{bmatrix}, \\
\tilde{\Gamma}_{16i2} &\triangleq \begin{bmatrix} A_i^T W_1 + C_i^T \mathcal{B}^T - W_1 & A_i^T W_2 + C_i^T \mathcal{B}^T - W_2 \\ \mathcal{A}^T - \mathcal{Q}^T & \mathcal{A}^T - \mathcal{Q}^T \end{bmatrix}, \\
\tilde{\Gamma}_{16i5} &\triangleq \begin{bmatrix} B_{\bar{w}i}^T W_1 + D_{\bar{w}i}^T \mathcal{B}^T & B_{\bar{w}i}^T W_2 + D_{\bar{w}i}^T \mathcal{B}^T \\ 0 & 0 \end{bmatrix}, \\
\tilde{\Gamma}_{46i1} &\triangleq \begin{bmatrix} A_{di}^T P_1 + C_{di}^T \mathcal{B}^T & A_{di}^T \mathcal{Q} + C_{di}^T \mathcal{B}^T \\ 0 & 0 \end{bmatrix}, \\
\tilde{\Gamma}_{46i2} &\triangleq \begin{bmatrix} A_{di}^T W_1 + C_{di}^T \mathcal{B}^T & A_{di}^T W_2 + C_{di}^T \mathcal{B}^T \\ 0 & 0 \end{bmatrix}, \\
\tilde{\Gamma}_{56i1} &\triangleq \begin{bmatrix} B_i^T P_1 + D_i^T \mathcal{B}^T & B_i^T \mathcal{Q} + D_i^T \mathcal{B}^T \end{bmatrix}, \\
\tilde{\Gamma}_{56i2} &\triangleq \begin{bmatrix} B_i^T W_1 + D_i^T \mathcal{B}^T & B_i^T W_2 + D_i^T \mathcal{B}^T \end{bmatrix}, \\
\tilde{\Gamma}_{662} &\triangleq d_1^{-2} \begin{bmatrix} S_{11} - W_1 - W_1^T & S_{12} - W_2 - \mathcal{Q} \\ \star & S_{14} - \mathcal{Q} - \mathcal{Q}^T \end{bmatrix}, \\
\tilde{\Gamma}_{663} &\triangleq d^{-2} \begin{bmatrix} S_{21} - W_1 - W_1^T & S_{22} - W_2 - \mathcal{Q} \\ \star & S_{24} - \mathcal{Q} - \mathcal{Q}^T \end{bmatrix},
\end{aligned}$$

where the definitions are given in Theorem 11.7.

Proof Applying the similar methods as Theorem 11.7, it is easy to conclude that the considered filtering system is mean-square asymptotically stable and satisfies the presupposed $\mathcal{H}_\infty$ performance sense γ in Corollary 11.9. Here, the detailed proof is omitted.

11.4 Illustrative Example

In this section, we give three examples to demonstrate the feasibility and superiority of our proposed design scheme.

Example 11.10 *(Conservative comparison analysis)* In the first example, we are going to compare the obtained results in Corollary 11.6 with some existing results. Consider the T-S fuzzy delayed system as (11.16), and relevant system matrices are chosen as follows:

$$x(k+1) = \sum_{i=1}^{2} h_i\big(\theta(k)\big)\Big[A_i x(k) + A_{di} x\big(k - d(k)\big)\Big],$$

and

$$A_1 = \begin{bmatrix} -0.291 & 1 \\ & 0\ 0.95 \end{bmatrix}, \quad A_{d1} = \begin{bmatrix} 0.012 & 0.014 \\ & 0\ 0.015 \end{bmatrix},$$
$$A_2 = \begin{bmatrix} -0.1 & 0 \\ 1 & -0.2 \end{bmatrix}, \quad A_{d2} = \begin{bmatrix} 0.01 & 0 \\ 0.01 & 0.015 \end{bmatrix}.$$

It is easily seen from Table 11.1 that the maximum allowable values of d_2 can be 62 when $d_1 = 12$. In addition, as for the identical d_1, the allowable upper bound of d_2 is larger than some existing results in [249, 298], that is to say, our presented scheme improves the stability standard and reduces the computation complexity.

Table 11.2 clearly shows the comparative results of the complexity analysis in the form as the quantity of used decision variables. It can be seen from Table 11.2 that the approach we propose possesses much fewer free variables demanded than the obtained results in [249, 298]. The quantity of variables in [298] requires $\frac{nr}{2}\big[n(6m^2 + 30m + 61) + 2m + 7\big]$ whereas it requires $\frac{nr}{2}\big[9n + nr + r + 5\big]$ in this chapter. Therefore, our presented method can be more favorable computationally, particularly when the time-delay is taken into account.

Example 11.11 *(Reduced-order filter analysis)* Consider the fuzzy delayed system under stochastic perturbation in (11.2) as follows:

Table 11.1 Maximum allowable values of d_2 for different values d_1

Different Methods	$d_1 = 3$	$d_1 = 5$	$d_1 = 10$	$d_1 = 12$
Theorem 3 of [298]	24	26	31	33
Corollary 1 of [249]	26	28	33	35
Corollary 11.6	53	55	60	62

Table 11.2 Allowable upper bound d_2 when $d_1 = 12$

Different Methods	Upper delay bound	Number of variables
Theorem 3 of [298]	33	804
Corollary 1 of [249]	35	56
Corollary 11.6	62	29

$$
\begin{aligned}
x(k+1) &= \sum_{i=1}^{2} h_i\big(\theta(k)\big)\Big[A_i x(k) + A_{di} x\big(k - d(k)\big) + B_i w(k) + B_{\bar{w}i} x(k)\bar{w}(k)\Big],\\
y(k) &= \sum_{i=1}^{2} h_i\big(\theta(k)\big)\Big[C_i x(k) + C_{di} x\big(k - d(k)\big) + D_i w(k) + D_{\bar{w}i} x(k)\bar{w}(k)\Big],\\
z(k) &= \sum_{i=1}^{2} h_i\big(\theta(k)\big)\Big[L_i x(k) + L_{di} x\big(k - d(k)\big) + E_i w(k)\Big],
\end{aligned}
$$

with

$$
A_1 = \begin{bmatrix} 0.3 & 0 & -0.1 \\ 0 & 0.3 & 0 \\ -0.1 & 0 & -0.3 \end{bmatrix}, \quad A_{d1} = \begin{bmatrix} 0.02 & 0 & 0 \\ 0 & -0.02 & 0 \\ 0 & 0 & 0.02 \end{bmatrix},
$$

$$
C_1 = \begin{bmatrix} 0.13 & 0 & -0.4 \\ 0.2 & 0 & 0 \\ 0 & 0 & 0.1 \end{bmatrix}, \quad C_{d1} = \begin{bmatrix} -0.01 & 0 & 0 \\ 0 & -0.02 & 0 \\ 0 & 0 & 0.02 \end{bmatrix},
$$

$$
B_{\bar{w}1} = \begin{bmatrix} 0.1 & 0 & 0 \\ 0 & -0.1 & 0 \\ 0 & 0 & 0.1 \end{bmatrix}, \quad D_{\bar{w}1} = \begin{bmatrix} 0.01 & 0 & 0 \\ 0 & -0.01 & 0 \\ 0 & 0 & 0.01 \end{bmatrix},
$$

$$
B_1 = \begin{bmatrix} 0.3 \\ 0.6 \\ 0.3 \end{bmatrix}, \quad D_1 = \begin{bmatrix} 0.5 \\ 0 \\ 0 \end{bmatrix}, \quad L_1 = \begin{bmatrix} 0.2 & 0.1 & 0 \end{bmatrix},
$$

$$
L_{d1} = \begin{bmatrix} -0.01 & 0 & 0 \end{bmatrix}, \quad E_1 = \begin{bmatrix} 0.1 \end{bmatrix},
$$

$$A_2 = \begin{bmatrix} 0.3 & 0 & 0.1 \\ 0 & -0.3 & 0 \\ 0.1 & 0 & -0.3 \end{bmatrix}, \; A_{d2} = \begin{bmatrix} 0.02 & 0 & 0 \\ 0 & 0.02 & 0 \\ 0 & 0 & 0.02 \end{bmatrix},$$

$$C_2 = \begin{bmatrix} 0.13 & 0 & 0.4 \\ 0.2 & 0 & 0 \\ 0 & 0 & -0.1 \end{bmatrix}, \; C_{d2} = \begin{bmatrix} 0.01 & 0 & 0 \\ 0 & -0.02 & 0 \\ 0 & 0 & 0.02 \end{bmatrix},$$

$$B_{\bar{w}2} = \begin{bmatrix} 0.1 & 0 & 0 \\ 0 & 0.1 & 0 \\ 0 & 0 & 0.1 \end{bmatrix}, \; D_{\bar{w}2} = \begin{bmatrix} 0.01 & 0 & 0 \\ 0 & 0.01 & 0 \\ 0 & 0 & 0.01 \end{bmatrix},$$

$$B_2 = \begin{bmatrix} 1 \\ 0.4 \\ 0.5 \end{bmatrix}, \; D_2 = \begin{bmatrix} 0.3 \\ 0 \\ 0 \end{bmatrix}, \; L_2 = \begin{bmatrix} 0.1 & 0.1 & 0 \end{bmatrix},$$

$$L_{d2} = \begin{bmatrix} 0.01 & 0 & 0 \end{bmatrix}, \; E_2 = \begin{bmatrix} 0.2 \end{bmatrix}.$$

Case 1. First, study the full-order filtering case, the minimised feasible γ is calculated as $\gamma^* = 0.1628$ via resolving the conditions (11.20) and (11.28) in Corollary 11.9. Moreover, the relevant parameters of the proposed full-order filter are obtained as

$$A_f = \begin{bmatrix} 0.7964 & -0.2809 & -0.0285 \\ -0.0041 & 0.7491 & 0.0158 \\ 0.0024 & -0.1071 & 0.6203 \end{bmatrix}, \; B_f = \begin{bmatrix} -1.6865 & 2.6378 & -6.2271 \\ -0.8167 & 2.2458 & -4.7485 \\ -0.3748 & 3.0907 & 0.0936 \end{bmatrix},$$

$$L_f = \begin{bmatrix} 0 & -0.0051 & -0.0010 \end{bmatrix}, \; D_f = \begin{bmatrix} 0.3689 & 0.7766 & 1.6557 \end{bmatrix}. \tag{11.29}$$

Case 2. Next, study the reduced-order filtering case, the minimised feasible γ is calculated as $\gamma^* = 0.1652$ via solving the conditions (11.19) and (11.20) in Theorem 11.7. Correspondingly, the parameters of the reduced-order filter are computed as

$$A_f = \begin{bmatrix} 0.8167 & 0.0109 \\ 0.0022 & 0.8192 \end{bmatrix}, \; B_f = \begin{bmatrix} 0.1144 & -0.1785 & 1.6873 \\ -0.1357 & 0.1536 & -1.1485 \end{bmatrix},$$

$$L_f = \begin{bmatrix} 0.0957 & -0.1929 \end{bmatrix} \times 10^{-3}, D_f = \begin{bmatrix} 0.3660 & 0.7886 & 1.6288 \end{bmatrix}. \tag{11.30}$$

Set the zero initial condition for the concerned system, which means $x(k) = 0$, $\hat{x}(k) = 0$. And $\bar{w}(k)$ denotes the typical Gaussian white noise with zero mean and unit variance. Besides, the membership functions are given by

$$h_1(\theta(k)) = \frac{1}{2}\Big[1 - 4x_1(k) - x_1(k - d(k))\Big],$$
$$h_2(\theta(k)) = \frac{1}{2}\Big[1 + 4x_1(k) + x_1(k - d(k))\Big].$$

The external disturbance $w(k)$ is set to

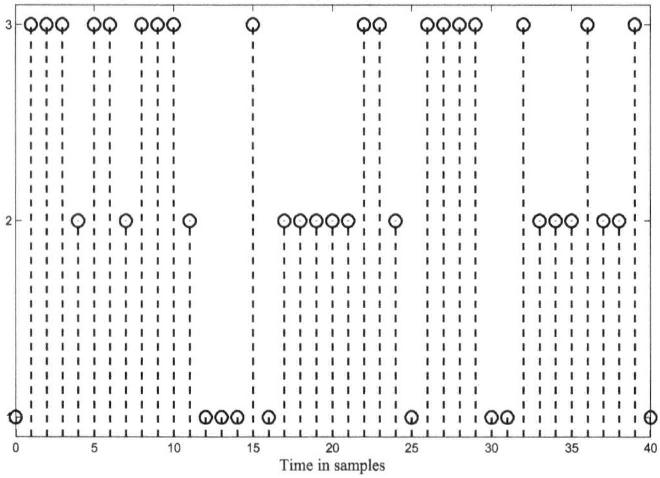

Fig. 11.2 Random time-varying delay $d(k)$

$$w(k) = 5\sin(-0.2k)\sin k.$$

Utilizing the MATLAB toolbox, simulation results of designed $\mathcal{H}_\infty$ filters are shown in Figs. 11.2, 11.3, 11.4, 11.5, 11.6, 11.7. Thereinto, the time-delay $d(k)$ changes stochastically from $d_1 = 1$ to $d_2 = 3$, which is plotted in Fig. 11.2, Fig. 11.3 depicts the signal $z(k)$ and its estimation $\hat{z}(k)$ of full-order filters (Case 1), and the estimation error $e(k)$ between $z(k)$ and $\hat{z}(k)$ in full-order case is displayed as Fig. 11.4. Next, on the basis of the same initial condition and disturbance signal, the signal $z(k)$ and its estimation $\hat{z}(k)$ of reduced-order filter (Case 2) are drawn in Fig. 11.5. And the estimation error $e(k)$ between them is plotted in Fig. 11.6. In addition, it can be observed that $e(k)$ increases when the dimension of reduced-order filter decreases. At last, Fig. 11.7 displays the estimation errors combined with the above-mentioned both cases.

Example 11.12 *(The inverted pendulum system)* In the third example, we consider the inverted pendulum system in [132, 329]. It is plotted schematically in Fig. 11.8 and we can see that there exists nonlinear characteristic in the upright position. Thus, we firstly construct the nonlinear dynamics of the inverted pendulum system via some physical rules, then we apply an efficient fuzzy approximation strategy to regulate the nonlinear dynamics. Finally, we can obtain the following results by the means of reduced-order filter design methods.

As for notational simplicity, the notation "(t)" in the system descriptions is omitted, in other words, y is equal to $y(t)$. And some relevant parameters of the inverted pendulum system are shown in Table 11.3. The dynamics of the considered system is described as

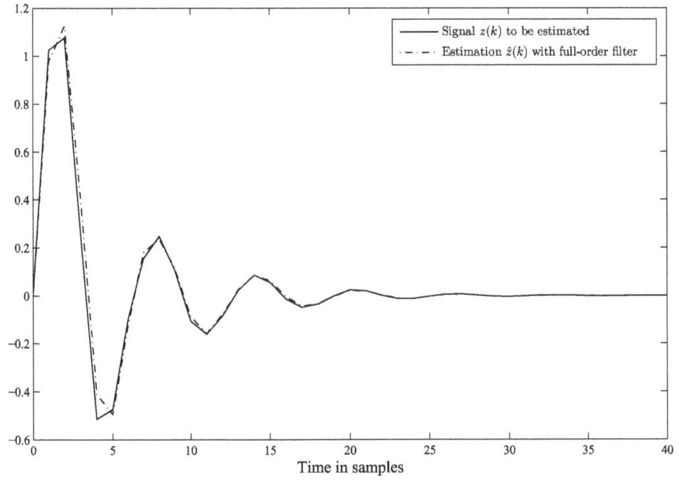

Fig. 11.3 Signal $z(k)$ and its estimations for the full-order case

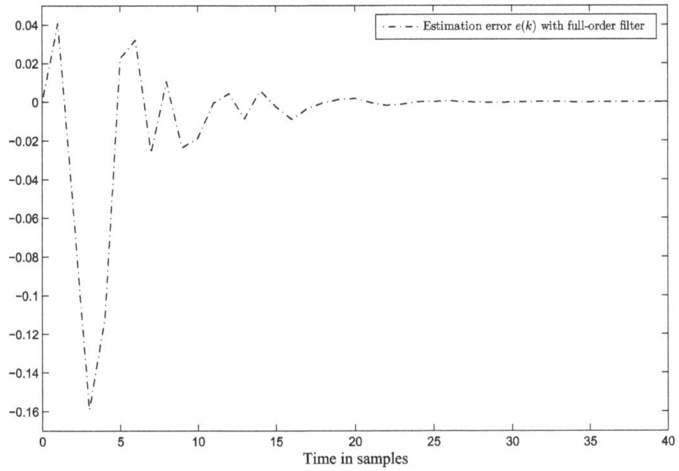

Fig. 11.4 Estimation error $e(k)$ between signal $z(k)$ and $\hat{z}(k)$ in Case 1

$$M\frac{d^2y}{dt^2} + m\frac{d^2}{dt^2}(y + l\sin\theta) = F_{\varpi} - F_r + u,$$
$$m\frac{d^2}{dt^2}(y + l\sin\theta)\cdot l\cos\theta = mgl\sin\theta,$$

where $F_r(t) = g_r\dot{y}(t-d) + c_r\dot{y}(t)$, with F_r denoting the resultant force produced by the damper and delayed resonator, $F_{\varpi}(t)$ denotes the position-dependent stochastic perturbation due to the rough road.

The relevant state variables are given by $x_1 = y$, $x_2 = \theta$, $x_3 = \dot{y}$, $x_4 = \dot{\theta}$. Then the system state-space equations can be obtained as follows:

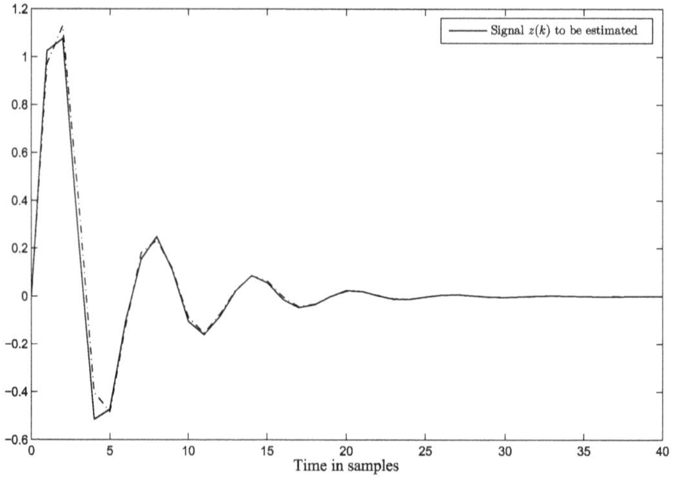

Fig. 11.5 Signal $z(k)$ and its estimations for the reduced-order case

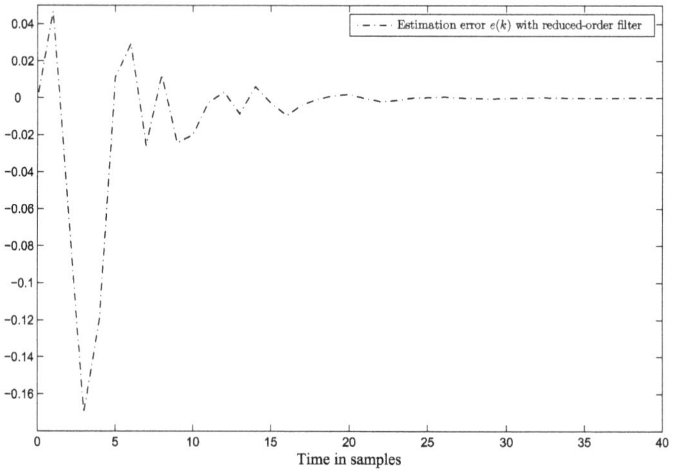

Fig. 11.6 Estimation error $e(k)$ between signal $z(k)$ and $\hat{z}(k)$ in Case 2

$$
\begin{aligned}
\dot{x}_1 &= x_3, \\
\dot{x}_2 &= x_4, \\
\dot{x}_3 &= \frac{-mg\sin x_2}{M\cos x_2} - \frac{c_r x_3 + g_r x_3(t-d) - F_{\varpi} - u}{M}, \\
\dot{x}_4 &= \frac{(M+m)g\sin x_2}{Ml\cos^2 x_2} + \frac{x_4^2 \sin x_2}{\cos x_2} + \frac{c_r x_3 + g_r x_3(t-d) - F_{\varpi} - u}{Ml\cos x_2}.
\end{aligned}
$$

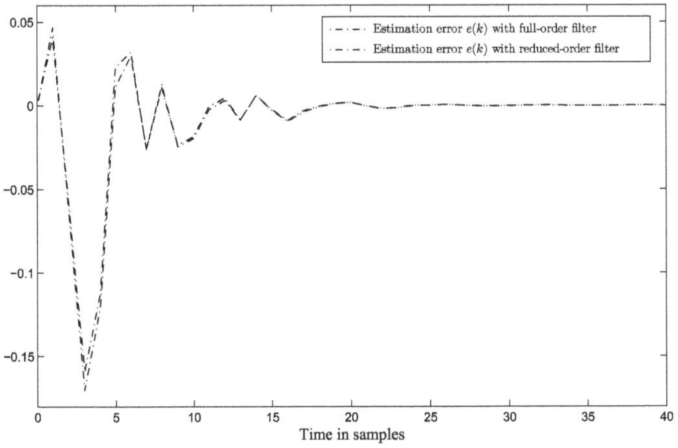

Fig. 11.7 Estimation error $e(k)$ between signal $z(k)$ and $\hat{z}(k)$ in Case 1–2

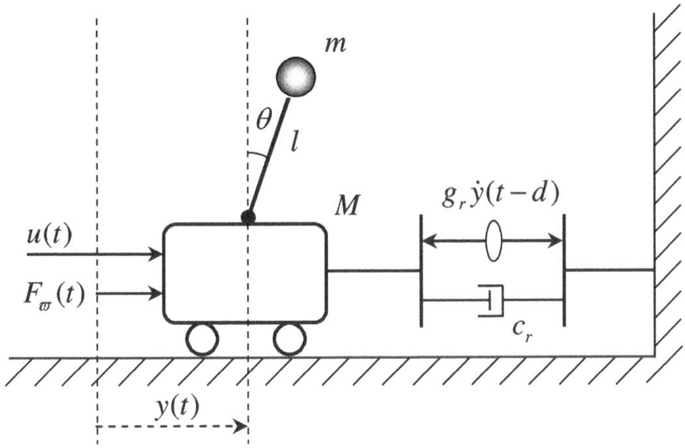

Fig. 11.8 Inverted pendulum on a cart with delayed resonator

To design the useful reduced-order filter, we will establish the T-S fuzzy model of inverted pendulum system via the approximation strategy.

(i) When x_2 is close to zero, the dynamics can be simplified to

$$\begin{aligned}
\dot{x}_1 &= x_3, \\
\dot{x}_2 &= x_4, \\
\dot{x}_3 &= \frac{-mgx_2}{M} - \frac{c_r x_3 + g_r x_3(t-d) - F_{\varpi} - u}{M}, \\
\dot{x}_4 &= \frac{(M+m)gx_2}{Ml} + \frac{c_r x_3 + g_r x_3(t-d) - F_{\varpi} - u}{Ml}.
\end{aligned}$$

Table 11.3 The parameter values of the inverted pendulum system

Parameter	Value	Description
M	1.378 kg	Mass of the cart
m	0.051 kg	Mass of the pendulum
l	0.325 m	Length of the pendulum
g	9.8 m/s^2	Acceleration of gravity
g_r	0.7 kg/s	Coefficient of the delayed resonator
g_ϖ	1.2 kg/s	Coefficient of the stochastic perturbation
c_r	5.98 kg/s	Coefficient of the damper
$\theta(t)$		Angle the pendulum from the upright position
$y(t)$		Displacement of the cart
$d(t)$		Time-varying delay
$u(t)$		Force acted on the cart

(ii) When x_2 is close to γ $(0 < |\gamma| < 1.57\ rad)$, the dynamics can be simplified to

$$
\begin{aligned}
\dot{x}_1 &= x_3, \\
\dot{x}_2 &= x_4, \\
\dot{x}_3 &= \frac{-mg\beta x_2}{M\alpha} - \frac{c_r x_3 + g_r x_3(t-d) - F_\varpi - u}{M}, \\
\dot{x}_4 &= \frac{(M+m)g\beta x_2}{Ml\alpha^2} + \frac{c_r x_3 + g_r x_3(t-d) - F_\varpi - u}{Ml\alpha},
\end{aligned}
$$

where $\gamma = 0.52rad$, $\alpha = \cos\gamma$ and $\beta = (\sin\gamma)/\gamma$. Employing the local approximated method in fuzzy partition spaces, the T-S fuzzy model with two rules can be formulated. And utilizing the Euler first-order approximation strategy, the discrete-time stochastic delayed fuzzy model can be obtained:

• **Plant Form**:

Rule 1: IF $x_2(k)$ is μ_1 and about 0, THEN

$$x(k+1) = A_1 x(k) + A_{d1} x(k - d(k)) + B_{11} u(k) + E_1 x(k)\bar{w}(k).$$

Rule 2: IF $x_2(k)$ is μ_2 and near γ, THEN

$$x(k+1) = A_2 x(k) + A_{d2} x(k - d(k)) + B_{12} u(k) + E_2 x(k)\bar{w}(k),$$

where $\bar{w}(k)$ stands for the stochastic perturbation generated by $F_\varpi(t)$.

Consider the above mentioned T-S fuzzy model, some involved system parameters are described by

$$A_1 = \begin{bmatrix} 0 & 0 & T & 0 \\ 0 & 0 & 0 & T \\ 0 & -\frac{Tmg}{M} & -\frac{Tc_r}{M} & 0 \\ 0 & \frac{T(M+m)g}{Ml} & \frac{Tc_r}{Ml} & 0 \end{bmatrix}, \quad A_{d1} = \begin{bmatrix} 0 & 0 & 0 & 0 \\ 0 & 0 & 0 & 0 \\ 0 & 0 & \frac{-Tg_r}{M} & 0 \\ 0 & 0 & \frac{Tg_r}{Ml} & 0 \end{bmatrix}, \quad B_{11} = \begin{bmatrix} 0 \\ 0 \\ \frac{T}{M} \\ \frac{-T}{Ml} \end{bmatrix},$$

$$A_2 = \begin{bmatrix} 0 & 0 & T & 0 \\ 0 & 0 & 0 & T \\ 0 & -\frac{Tmg\beta}{M\alpha} & -\frac{Tc_r}{M} & 0 \\ 0 & \frac{T(M+m)g\beta}{Ml\alpha^2} & \frac{Tc_r}{Ml\alpha} & 0 \end{bmatrix}, \quad A_{d2} = \begin{bmatrix} 0 & 0 & 0 & 0 \\ 0 & 0 & 0 & 0 \\ 0 & 0 & \frac{-Tg_r}{M} & 0 \\ 0 & 0 & \frac{Tg_r}{Ml\alpha} & 0 \end{bmatrix}, \quad B_{12} = \begin{bmatrix} 0 \\ 0 \\ \frac{T}{M} \\ \frac{-T}{Ml\alpha} \end{bmatrix},$$

$$E_1 = \begin{bmatrix} 0 & 0 & 0 & 0 \\ 0 & 0 & 0 & 0 \\ 0 & \frac{Tg_{\varpi}}{M} & 0 & 0 \\ 0 & \frac{-Tg_{\varpi}}{Ml} & 0 & 0 \end{bmatrix}, \quad E_2 = \begin{bmatrix} 0 & 0 & 0 & 0 \\ 0 & 0 & 0 & 0 \\ 0 & \frac{Tg_{\varpi}}{M} & 0 & 0 \\ 0 & \frac{-Tg_{\varpi}}{Ml\alpha} & 0 & 0 \end{bmatrix},$$

where T denotes the sampling time. Provided that $|x_2(k)|$ is smaller than $|\gamma|$, the fuzzy basis functions are chosen as

$$h_1\big(x_2(k)\big) = 1 - \frac{|x_2(k)|}{|\gamma|}, \quad h_2\big(x_2(k)\big) = \frac{|x_2(k)|}{|\gamma|}.$$

On the basis of the values shown in Table 11.3 and $T = 0.025s$, we have

$$A_1 = \begin{bmatrix} 0 & 0 & 0.0250 & 0 \\ 0 & 0 & 0 & 0.0250 \\ 0 & -0.0091 & -0.1085 & 0 \\ 0 & 0.7817 & 0.3338 & 0 \end{bmatrix}, \quad A_{d1} = \begin{bmatrix} 0 & 0 & 0 & 0 \\ 0 & 0 & 0 & 0 \\ 0 & 0 & -0.0127 & 0 \\ 0 & 0 & 0.0391 & 0 \end{bmatrix},$$

$$A_2 = \begin{bmatrix} 0 & 0 & 0.0250 & 0 \\ 0 & 0 & 0 & 0.0250 \\ 0 & -0.0100 & -0.1085 & 0 \\ 0 & 0.9954 & 0.3855 & 0 \end{bmatrix}, \quad A_{d2} = \begin{bmatrix} 0 & 0 & 0 & 0 \\ 0 & 0 & 0 & 0 \\ 0 & 0 & -0.0127 & 0 \\ 0 & 0 & 0.0451 & 0 \end{bmatrix},$$

$$E_1 = \begin{bmatrix} 0 & 0 & 0 & 0 \\ 0 & 0 & 0 & 0 \\ 0 & 0.0581 & 0 & 0 \\ 0 & -0.1786 & 0 & 0 \end{bmatrix}, \quad B_{11} = \begin{bmatrix} 0 \\ 0 \\ 0.0181 \\ -0.0558 \end{bmatrix},$$

$$E_2 = \begin{bmatrix} 0 & 0 & 0 & 0 \\ 0 & 0 & 0 & 0 \\ 0 & 0.0581 & 0 & 0 \\ 0 & -0.2063 & 0 & 0 \end{bmatrix}, \quad B_{12} = \begin{bmatrix} 0 \\ 0 \\ 0.0181 \\ -0.0645 \end{bmatrix},$$

and

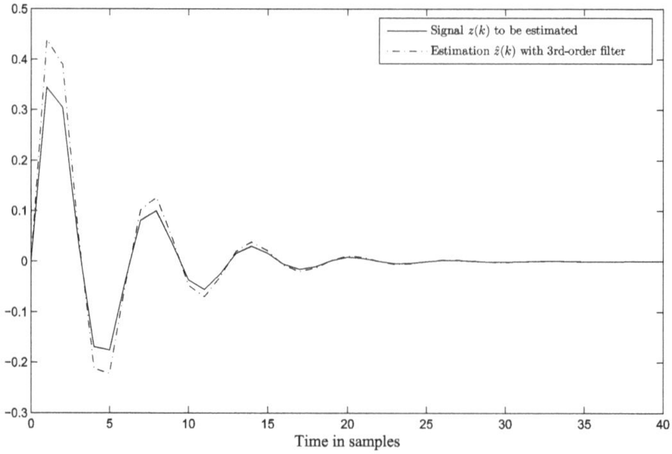

Fig. 11.9 Signal $z(k)$ and its estimations of the inverted pendulum system

$$h_1(x_2(k)) = 1 - 1.91\left|x_2(k)\right|, \quad h_2(x_2(k)) = 1.91\left|x_2(k)\right|.$$

Set the zero initial states, and $\bar{w}(k)$ denotes the Gaussian white noise. Besides, the external disturbance $u(k)$ is set to

$$u(k) = 5e^{(-0.2k)}sin(k).$$

Solving the conditions in Theorem 11.7, we can get the solutions as follows:

$$A_f = \begin{bmatrix} 0.8508 & 0.0098 & 0.0001 \\ -0.0396 & 0.8040 & -0.0001 \\ -0.5433 & -0.1951 & 0.8184 \end{bmatrix},$$

$$B_f = \begin{bmatrix} -0.0028 & -0.3520 & -0.0286 & 0.0029 \\ 0.0072 & 1.0635 & 0.0418 & -0.0215 \\ 0.0687 & 14.6230 & 0.7135 & -1.2580 \end{bmatrix},$$

$$L_f = \begin{bmatrix} 0.0062 & 0.0006 & 0.0001 \end{bmatrix},$$

$$D_f = \begin{bmatrix} 2.5436 & 2.9546 & 10.3485 & 2.5427 \end{bmatrix}.$$

The simulation results in this example are plotted in Figs. 11.9, 11.10. Figure 11.9 shows the state $z(k)$ and its estimation, and the estimation error $e(k)$ between $z(k)$ and $\hat{z}(k)$ is depicted in Fig. 11.10. It is easy to indicate that our proposed reduced-order filter let the inverted pendulum system obtain the desirable performance availably.

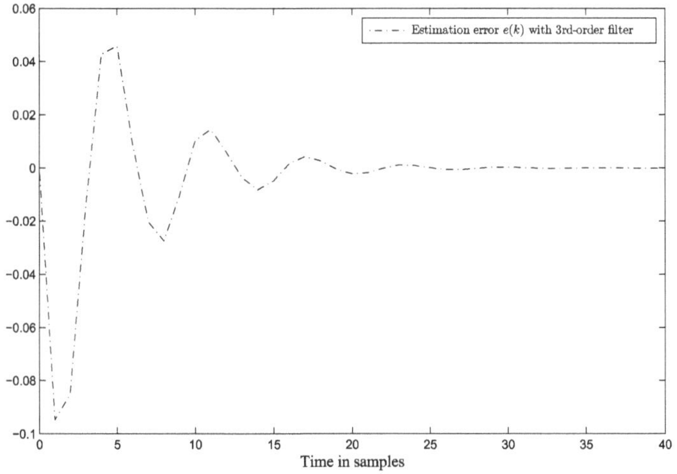

Fig. 11.10 Estimation error $e(k)$ between signal $z(k)$ and $\hat{z}(k)$ of the inverted pendulum system

11.5 Conclusion

This chapter investigated the reduced-order filtering issue for nonlinear systems with stochastic perturbations through T-S fuzzy modelling. Using a new fuzzy Lyapunov function and reciprocally convex method, sufficient conditions were established to ensure that the filtering system was mean-square asymptotically stable with $\mathcal{H}_\infty$ performance. Furthermore, the feasible conditions of reduced-order filtering were set using the convex linearization method. In this manner, the convex optimization issue could be readily resolved using standard numerical software. In the end, we presented illustrative examples to verify the feasibility and validity of the developed scheme.

Part IV
Event-Triggered Fuzzy Control Application

Chapter 12
Dissipative Event-Triggered Fuzzy Control of Truck-Trailer Systems

12.1 Introduction

This chapter aims to resolve the dissipative fuzzy control issue for fuzzy dynamic systems by the means of an event-triggered approach, which is adopted to reduce the transmissions and ensure the stability of the closed-loop system. Furthermore, the dissipative performance is considered in the design of the fuzzy controller to ensure that the resulting closed-loop system is asymptotically stable and strictly (X, Y, Z)-θ-*dissipative*. Based on the fuzzy model, the stability is analysed by using the Lyapunov stability theory. In addition, the designed controller is compactly described in the form of LMIs. An illustrative example is presented to validate the proposed design approach.

12.2 System Description and Preliminaries

12.2.1 Truck-Trailer Model

In this chapter, a truck-trailer model in [204] is considered and its schematic diagram is displayed in Fig. 12.1:

$$\begin{cases} \dot{x}_1(t) = -\dfrac{\bar{v}\bar{t}}{L\bar{t}_0}x_1(t) + \dfrac{\bar{v}\bar{t}}{l\bar{t}_0}u(t) + 0.1w_1(t) + 0.1w_2(t), \\ \dot{x}_2(t) = \dfrac{\bar{v}\bar{t}}{L\bar{t}_0}x_1(t) + 0.1w_1(t) + 0.1w_2(t), \\ \dot{x}_3(t) = -\dfrac{\bar{v}\bar{t}}{\bar{t}_0}\sin\left[x_2(t) + \dfrac{\bar{v}\bar{t}}{2L}x_1(t)\right], \end{cases} \tag{12.1}$$

X. Su et al., *Intelligent Control, Filtering and Model Reduction Analysis for Fuzzy-Model-Based Systems*, Studies in Systems, Decision and Control 385,
https://doi.org/10.1007/978-3-030-81214-0_12

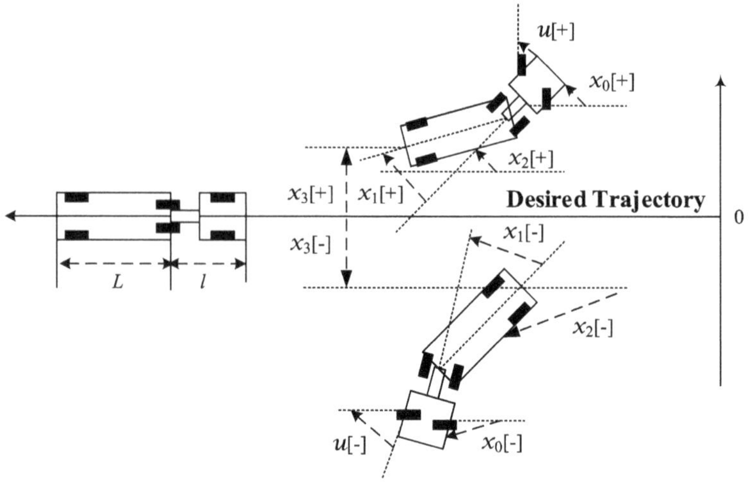

Fig. 12.1 Truck trailer model and its coordinate system

Table 12.1 The parameters of the truck-trailer model

Parameter	Value	Description
L	5.5 m	Length of the trailer
l	2.8 m	Length of the truck
$\bar{v}$	−1.0 m/s	Reversing constant speed
$\bar{t}$	2.0 s	Sampling time
$\bar{t}_0$	0.5 s	Initial time
$\bar{g}$	$10^{-2}/\pi$	Acceleration due to gravity

where $x_1(t)$ stands for the angle difference between the truck and trailer, $x_2(t)$ stands for the angle of the trailer, and $x_3(t)$ stands for the vertical position of rear end of the trailer. $\bar{v}$ and $w(t)$ represent the steering angel of the truck and the external disturbance, separately. Some relevant model matrices are given in Table 12.1. We concentrate on designing an efficient event-driven controller to ensure the resulting fuzzy system satisfies the asymptotic stability and strict (X, Y, Z)-θ-*dissipative* performance.

Denote $\zeta(t) = x_2(t) + \left(\frac{\bar{v}\bar{t}}{2L}\right) x_1(t)$ and choose the fuzzy basis functions as

$$h_1\Big(\zeta(t)\Big) = \begin{cases} \dfrac{\sin\Big(\zeta(t)\Big) - \bar{g}\zeta(t)}{\zeta(t)(1-\bar{g})}, & \text{if } \zeta(t) \neq 0, \\ 1, & \text{if } \zeta(t) = 0, \end{cases}$$

$$h_2\Big(\zeta(t)\Big) = 1 - \begin{cases} \dfrac{\sin\Big(\zeta(t)\Big) - \bar{g}\zeta(t)}{\zeta(t)(1-\bar{g})}, & \text{if } \zeta(t) \neq 0, \\ 1, & \text{if } \zeta(t) = 0. \end{cases}$$

In order to approximate the truck trailer system, we consider the T-S fuzzy model as follows:

Plant Rule 1: **IF** $\zeta(t)$ is close to 0, **THEN**

$$\dot{x}(t) = A_1x(t) + C_1w(t) + B_1u(t).$$

Plant Rule 2: **IF** $\zeta(t)$ is close to $\pm\pi$, **THEN**

$$\dot{x}(t) = A_2x(t) + C_2w(t) + B_2u(t),$$

where $x(t) = \begin{bmatrix} x_1(t) \\ x_2(t) \\ x_3(t) \end{bmatrix}$, $w(t) = \begin{bmatrix} w_1(t) \\ w_2(t) \end{bmatrix}$, and

$$A_1 = \begin{bmatrix} -\dfrac{\bar{v}\bar{t}}{L\bar{t}_0} & 0 & 0 \\ \dfrac{\bar{v}\bar{t}}{L\bar{t}_0} & 0 & 0 \\ \dfrac{\bar{v}^2\bar{t}^2}{2L\bar{t}_0} & \dfrac{\bar{v}\bar{t}}{\bar{t}_0} & 0 \end{bmatrix}, A_2 = \begin{bmatrix} -\dfrac{\bar{v}\bar{t}}{L\bar{t}_0} & 0 & 0 \\ \dfrac{\bar{v}\bar{t}}{L\bar{t}_0} & 0 & 0 \\ \dfrac{\bar{g}\bar{v}^2\bar{t}^2}{2L\bar{t}_0} & \dfrac{\bar{g}\bar{v}\bar{t}}{\bar{t}_0} & 0 \end{bmatrix}, B_1 = B_2 = \begin{bmatrix} \dfrac{\bar{v}\bar{t}}{l\bar{t}_0} \\ 0 \\ 0 \end{bmatrix},$$

$$C_1 = C_2 = \begin{bmatrix} 0.1 & 0.1 \\ 0.1 & 0.1 \\ 0 & 0 \end{bmatrix}.$$

12.2.2 T-S Fuzzy Systems

Introduce the nonlinear systems, which are approximated with the following T-S fuzzy models:

♦ **Plant Form**:

Rule i : IF $\zeta_1(t)$ is ϑ_{i1} and …and $\zeta_p(t)$ is ϑ_{ip}, THEN

$$\begin{cases} \dot{x}(t) = A_ix(t) + B_iu(t) + C_i\omega(t), \\ z(t) = D_ix(t) + E_iu(t) + F_i\omega(t), \end{cases} \tag{12.2}$$

where $x(t) \in \mathbb{R}^n$ denotes the system state vector; $u(t) \in \mathbb{R}^s$ denotes the control input; $z(t) \in \mathbb{R}^m$ represents the control output; $\omega(t) \in \mathbb{R}^q$ represents the exogenous disturbance belonging to $\mathcal{L}_2[0, \infty)$. $\vartheta_{ij}, i = 1, 2, \ldots, r, j = 1, 2, \ldots, p$, which stand for the fuzzy sets with r being the quantity of fuzzy-basis rules; A_i, B_i, C_i, D_i, E_i and F_i are some appropriately dimensioned matrices. And the fuzzy basis functions are expressed as $h_i\big(\zeta(t)\big) = \frac{v_i(\zeta(t))}{\sum_{i=1}^{r} v_i(\zeta(t))}$, $v_i\big(\zeta(t)\big) = \prod_{j=1}^{p} \vartheta_{ij}\big(\zeta_j(t)\big)$, where $\zeta(t) = \Big[\zeta_1(t), \zeta_2(t), \ldots, \zeta_p(t)\Big]$ and $\vartheta_{ij}\Big(\zeta_j(t)\Big)$ represent the premise variables and the grade of membership of $\zeta_j(t)$ in ϑ_{ij}. Provided that $v_i(\zeta(t)) \geqslant 0$, $i = 1, 2, \ldots, r$, $\sum_{i=1}^{r} v_i\Big(\zeta(t)\Big) > 0$ for all t, then $h_i\Big(\zeta(t)\Big) \geqslant$ for $i = 1, 2, \ldots, r$, and $\sum_{i=1}^{r} h_i\Big(\zeta(t)\Big) = 1$ for all t.

Thus, a complete expression of the dynamic T-S fuzzy model in (12.2) can be given by

$$\begin{cases} \dot{x}(t) = \sum_{i=1}^{r} h_i\Big(\zeta(t)\Big)\Big\{A_i x(t) + B_i u(t) + C_i \omega(t)\Big\}, \\ z(t) = \sum_{i=1}^{r} h_i\Big(\zeta(t)\Big)\Big\{D_i x(t) + E_i u(t) + F_i \omega(t)\Big\}. \end{cases} \tag{12.3}$$

For the promotion of further research, some assumptions are presented to analyze the event-based problem.

Assumption 12.1 The sensor is time-triggered and the sampling period is set as T, while the controller, zero-order holder (ZOH), and actuator are event-triggered. In this chapter, we do not consider the data-packet dropouts during the transmitting procedure.

Assumption 12.2 The holding interval time of ZOH is given by $t \in \big[t_k T + \tau_{t_k}, t_{k+1} T + \tau_{t_{k+1}}\big]$, with $t_k T + \tau_{t_k}$ being the instant when sampled data reach ZOH and τ_{t_k} being the network delay subject to an upper bound $\bar{\tau}$.

An event-driven condition is proposed to decide if the newly sampled signal can be transferred to the controller as follow:

$$\Big[x(kT) - x(t_k T)\Big]^T \Phi_1 \Big[x(kT) - x(t_k T)\Big] \leqslant \delta x^T(kT)\Phi_2 x(kT), \tag{12.4}$$

where $\Phi_1 > 0$, $\Phi_2 > 0$ denote the event-based weighting parameters, $x(kT)$ stands for the current sampled signal and $x(t_k T)$ stands for the latest transferred signal. The sampled signal can be stored and transmitted to the controller only when the sampled data exceeds the event-triggered condition in (12.4). For a finite positive integer q, which satisfies $t_{k+1} = t_k + q + 1$. Then the holding interval time of ZOH forms the set as $\Big[t_k T + \tau_{t_k}, t_{k+1} T + \tau_{t_{k+1}}\Big) = \bigcup_{n=0}^{q} \Lambda_{n,k}$, where $\Lambda_{n,k} = \Big[t_k T + nT + \tau_{i_k+n}, t_k T + (n+1)T + \tau_{i_k+n+1}\Big)$, $n = 0, 1, \ldots, q$.

The network time-delay is given by $h(t) = t - t_k T - nT$, $t \in \Lambda_{n,k}$, where $0 \leqslant h(t) \leqslant T + \bar{h} = h$. The transmission error between the present sampled signal and the latest transmission signal can be expressed as

$$e_k\left(s_kT\right)=x\left(s_kT\right)-x\left(t_kT\right), \tag{12.5}$$

where $s_kT=t_kT+nT$ denotes the sampling moment which starts with the present sampled time t_kT and finishes up with the next transmitted sampled time $t_{k+1}T$. Therefore, the event-driven condition is rewritten as

$$e_k^T\left(s_kT\right)\Lambda_1 e_k\left(s_kT\right)\leqslant \delta x^T\left(s_kT\right)\Lambda_2 x\left(s_kT\right). \tag{12.6}$$

Then we propose a state-feedback fuzzy controller as follow:

♦ **Fuzzy Controller Form**:

Rule j: IF $\zeta_1(t)$ is μ_{ij} and …and $\zeta_p(t)$ is μ_{ij}, THEN

$$u(t)=K_jx\left(t_kT\right), t\in\left[t_kT+h_{t_k}, t_{k+1}T+h_{t_{k+1}}\right), \tag{12.7}$$

where $K_j\in\mathbb{R}^{n\times n}$ represent the controller gains to be determined. It can be further described as

$$u(t)=\sum_{j=1}^{r}h_j\left(\zeta(t)\right)K_jx\left(t_kT\right), \tag{12.8}$$

where $t\in\left[t_kT+h_{t_k}, t_{k+1}T+h_{t_{k+1}}\right)$. On the basis of (12.3), (12.5), and (12.8), we can get the complete closed-loop fuzzy systems as below:

$$\begin{cases}\dot{x}(t)=\sum_{i=1}^{r}\sum_{j=1}^{r}h_i\left(\zeta(t)\right)h_j\left(\zeta(t)\right)\Big\{A_ix(t)+C_i\omega(t)\\ \qquad +B_iK_jx\left(t-h(t)\right)-B_iK_je_k\left(s_kT\right)\Big\},\\ z(t)=\sum_{i=1}^{r}\sum_{j=1}^{r}h_i\left(\zeta(t)\right)h_j\left(\zeta(t)\right)\Big\{D_ix(t)+F_i\omega(t)\\ \qquad +E_iK_jx\left(t-h(t)\right)-E_iK_je_k\left(s_kT\right)\Big\}.\end{cases} \tag{12.9}$$

Definition 12.1 For a scalar $\theta>0$, real matrices $X=X^T=-\hat{X}^T\hat{X}\leqslant 0$, Y and $Z=Z^T$, the concerned system (12.9) is asymptotically stable and strictly (X, Y, Z)–θ–*dissipative*, if

1. When $w(t) = 0$, (12.9) is asymptotically stable;
2. With the zero initial condition, the below condition is met:

$$\int_0^{\varphi} \left[z^T(t)Xz(t) + 2z^T(t)Yw(t) + w^T(t)Zw(t)\right]dt \geqslant \theta \int_0^{\varphi} \left[w^T(t)w(t)\right]dt, \tag{12.10}$$

for any $\varphi \geq 0$ and non-zero $w(t) \in \mathcal{L}_2[0, \infty)$.

Lemma 12.2 [207] *As for a positive definite matrix $\vartheta > 0$, and a differentiable function $\left\{x(u)|u \in [a, b]\right\}$, the following inequalities hold:*

$$\int_a^b \dot{x}^T(s)\vartheta\dot{x}(s)ds \geqslant \frac{1}{b-a}\Gamma_1^T\vartheta\Gamma_1 + \frac{3}{b-a}\Gamma_2^T\vartheta\Gamma_2 + \frac{5}{b-a}\Gamma_3^T\vartheta\Gamma_3,$$

$$\int_a^b\int_\theta^b \dot{x}^T(s)\vartheta\dot{x}(s)dsd\theta \geqslant 2\Gamma_4^T\vartheta\Gamma_4 + 4\Gamma_5^T\vartheta\Gamma_5,$$

$$\int_a^b\int_a^\theta \dot{x}^T(s)\vartheta\dot{x}(s)dsd\theta \geqslant 2\Gamma_6^T\vartheta\Gamma_6 + 4\Gamma_7^T\vartheta\Gamma_7,$$

$$\int_a^b \dot{x}^T(s)\vartheta\dot{x}(s)ds \geqslant \frac{1}{b-a}\left(\int_a^b x(s)ds\right)^T\vartheta\left(\int_a^b x(s)ds\right) + \frac{3}{b-a}\Gamma_8^T\vartheta\Gamma_8,$$

where

$$\Gamma_1 \triangleq x(b) - x(a),\ \Gamma_2 \triangleq x(b) + x(a) - \frac{2}{b-a}\int_a^b x(s)ds,$$

$$\Gamma_3 \triangleq \frac{6}{b-a}\int_a^b x(s)ds - \frac{12}{(b-a)^2}\int_a^b\int_\theta^b x(s)dsd\theta + x(b) - x(a),\ \Gamma_4 \triangleq x(b) - \frac{1}{b-a}\int_a^b x(s)ds,$$

$$\Gamma_5 \triangleq x(b) + \frac{2}{b-a}\int_a^b x(s)ds - \frac{6}{(b-a)^2}\int_a^b\int_\theta^b x(s)dsd\theta,$$

$$\Gamma_6 \triangleq x(a) - \frac{1}{b-a}\int_a^b x(s)ds,$$

$$\Gamma_7 \triangleq x(a) - \frac{4}{b-a}\int_a^b x(s)ds + \frac{6}{(b-a)^2}\int_a^b\int_\theta^b x(s)dsd\theta,$$

$$\Gamma_8 \triangleq \int_a^b x(s)ds - \frac{2}{b-a}\int_a^b\int_\theta^b x(s)dsd\theta.$$

12.3 Main Results

12.3.1 Dissipative Performance Analysis

This part will propose a design scheme to analyze the dissipative performance of T-S fuzzy systems in (12.9). To make the following description simpler, define some matrices as

$$\Psi(t) \triangleq col\left\{x(t),\ x(t-h),\ x(t-h(t)),\ \frac{1}{h(t)}\int_{t-h(t)}^{t} x(\alpha)d\alpha,\right.$$

$$\frac{1}{h-h(t)}\int_{t-h}^{t-h(t)} x(\alpha)d\alpha,\ \frac{2}{h^2(t)}\int_{-h(t)}^{0}\int_{t+\beta}^{t} x(\alpha)d\alpha d\beta,$$

$$\left.\frac{2}{(h-h(t))^2}\int_{-h}^{-h(t)}\int_{t+\beta}^{t-h(t)} x(\alpha)d\alpha d\beta,\ \dot{x}(t),\ e_k(s_kT),\ w(t)\right\},$$

$$\chi \triangleq \begin{bmatrix} \chi_{11} & \chi_{12} & \chi_{13} \\ \chi_{21} & \chi_{22} & \chi_{23} \\ \chi_{31} & \chi_{32} & \chi_{33} \end{bmatrix},\ O_{3\times 3} \triangleq \begin{bmatrix} 0 & 0 & 0 \\ 0 & 0 & 0 \\ 0 & 0 & 0 \end{bmatrix},\ \mathcal{I} \triangleq diag\{I, 3I, 5I\},$$

$$\Xi \triangleq \begin{bmatrix} \mathcal{I}\otimes R & \chi \\ \star & \mathcal{I}\otimes R \end{bmatrix},\ \Gamma \triangleq \Big[e_3 - e_2\ e_3 + e_2 - 2e_5\ e_3 - e_2 + 6e_5 - 6e_7$$

$$e_1 - e_3\ e_1 + e_3 - 2e_4\ e_1 - e_3 + 6e_4 - 6e_6\Big].$$

Theorem 12.3 *Given the scalars h, δ, θ, and real matrices $X = X^T = -\hat{X}^T\hat{X}$, Y, Z, if there are proper matrices $P > 0$, $Q > 0$, $R > 0$, $S > 0$, $T > 0$, $\Lambda_1 > 0$, $\Lambda_2 > 0$, Φ_1, Φ_2, χ, such that the following conditions are met:*

$$\frac{2}{r-1}\begin{bmatrix} \Omega_1^{ii} & \Omega_2^{ii} & \Omega_3^{ii} \\ \star & \Omega_4 & \Omega_5^{ii} \\ \star & \star & \Omega_6^{ii} \end{bmatrix} + \begin{bmatrix} \Omega_1^{ij} & \Omega_2^{ij} & \Omega_3^{ij} \\ \star & \Omega_4 & \Omega_5^{ij} \\ \star & \star & \Omega_6^{ij} \end{bmatrix} + \begin{bmatrix} \Omega_1^{ji} & \Omega_2^{ji} & \Omega_3^{ji} \\ \star & \Omega_4 & \Omega_5^{ji} \\ \star & \star & \Omega_6^{ji} \end{bmatrix} < 0, \quad (12.11)$$

$$\begin{bmatrix} \Omega_1^{ii} & \Omega_2^{ii} & \Omega_3^{ii} \\ \star & \Omega_4 & \Omega_5^{ii} \\ \star & \star & \Omega_6^{ii} \end{bmatrix} < 0, \quad (12.12)$$

$$F \triangleq \begin{bmatrix} \mathcal{I}\otimes R + \mathcal{I}\otimes S & \chi \\ \star & \mathcal{I}\otimes R + \mathcal{I}\otimes T \end{bmatrix} > 0, \quad (12.13)$$

where

$$\Omega_1^{ij} \triangleq \begin{bmatrix} \Omega_{11}^{ij} & \Omega_{12}^{ij} \\ \star & \Omega_{14}^{ij} \end{bmatrix}, \quad \Omega_2^{ij} \triangleq \begin{bmatrix} \Omega_{21}^{ij} & \Omega_{22}^{ij} \\ \Omega_{23}^{ij} & \Omega_{24}^{ij} \end{bmatrix},$$

$$\Omega_3^{ij} \triangleq \begin{bmatrix} -\Phi_1^T B_i K_j & \Phi_1^T C_j - D_i^T Y & D_i^T \hat{X}^T \\ 0 & 0 & 0 \\ 0 & -K_j^T E_i^T Y & K_j^T E_i^T \hat{X}^T \\ 0 & 0 & 0 \end{bmatrix},$$

$$\Omega_4 \triangleq \begin{bmatrix} \Omega_{41} & \Omega_{42} \\ \star & \Omega_{44} \end{bmatrix}, \quad \Omega_5^{ij} \triangleq \begin{bmatrix} O_{3\times 3} \\ \hline -\Phi_2^T B_i K_j \;\; \Phi_2^T C_j \;\; 0 \end{bmatrix},$$

$$\Omega_6^{ij} \triangleq \begin{bmatrix} -\Lambda_1 & K_j^T E_i^T Y & -K_j^T E_i^T \hat{X}^T \\ \star & -2F_i^T Y - Z + \theta & F_i^T \hat{X}^T \\ \star & \star & -I \end{bmatrix},$$

with

$$\Omega_{11}^{ij} \triangleq \begin{bmatrix} -9R - 6S + Q + 2\Phi_2^T A_i & M_1 \\ \star & -9R - 6T - Q \end{bmatrix},$$

$$\Omega_{12}^{ij} \triangleq \begin{bmatrix} M_2^{ij} & -24R - 6S \\ M_3 & M_4 \end{bmatrix}, \quad \Omega_{14}^{ij} \triangleq \begin{bmatrix} M_5 & M_6 \\ \star & -192R - 18S - 66T \end{bmatrix},$$

$$\Omega_{21}^{ij} \triangleq \begin{bmatrix} M_7 & 30R + 12S \\ 36R + 18T & (-1)^m \sum_{m=1}^{3} 6\chi_{m3} \end{bmatrix},$$

$$\Omega_{22}^{ij} \triangleq \begin{bmatrix} \sum\limits_{m=1}^{3} 6\chi_{3m}^T & P^T - \Phi_1^T + A_i^T \Phi_2 \\ -30R - 12T & 0 \end{bmatrix},$$

$$\Omega_{23}^{ij} \triangleq \begin{bmatrix} M_8 & M_9 \\ M_{11} & 180R + 24S + 48T \end{bmatrix}, \quad \Omega_{24}^{ij} \triangleq \begin{bmatrix} M_{10} & K_j^T B_i^T \Phi_2 \\ -12\chi_{32}^T + 36\chi_{33}^T & 0 \end{bmatrix},$$

$$\Omega_{41} \triangleq \begin{bmatrix} -192R - 18S - 66T & -12\chi_{23} + 36\chi_{33} \\ \star & -180R - 36T - 36S \end{bmatrix},$$

$$\Omega_{42} \triangleq \begin{bmatrix} 180R + 48T + 24S & 0 \\ -36\chi_{33} & 0 \end{bmatrix}, \quad \Omega_{44} \triangleq \begin{bmatrix} -180R - 36T - 36S & 0 \\ \star & M_{12} \end{bmatrix},$$

$$M_1 \triangleq \sum_{m=1}^{3} \left(\chi_{1m}^T - \chi_{2m}^T + \chi_{3m}^T\right), \quad M_7 \triangleq \sum_{m=1}^{3} \left(2\chi_{2m}^T - 6\chi_{3m}^T\right),$$

$$M_2^{ij} \triangleq \sum_{m=1}^{3} \left(-\chi_{1m}^T - \chi_{2m}^T - \chi_{3m}^T\right) + 3R + \Phi_1^T B_i K_j,$$

$$M_3 \triangleq -(1)^m \sum_{m=1}^{3} \left(\chi_{m1} - \chi_{m2} + \chi_{m3}\right) + 3R,$$

$$M_4 \triangleq -(1)^m \sum_{m=1}^{3} \left(2\chi_{m2} - 6\chi_{m3}\right),$$

$$M_5 \triangleq sym\left(\sum_{m=1}^{3} \left(\chi_{m1} - \chi_{m2} + \chi_{m3}\right)\right) - 18R - 6S - 6T + \delta\Lambda_2,$$

$$M_6 \triangleq \sum_{m=1}^{3} \left(2\chi_{m2} - 6\chi_{m3}\right) + 36R + 18T,$$

$$M_8 \triangleq (-1)^m \sum_{m=1}^{3} \left(2\chi_{2m}^T - 6\chi_{3m}^T\right) - 24R - 6S,$$

$$M_9 \triangleq \sum_{m=1}^{3} 6\chi_{m3} - 30R - 12T, \quad M_{10} \triangleq -(1)^m \sum_{m=1}^{3} 6\chi_{3m}^T + 30R + 12S,$$

$$M_{11} \triangleq -4\chi_{22}^T + 12\chi_{23}^T + 12\chi_{32}^T - 36\chi_{33}^T,$$

$$M_{12} \triangleq h^2 R + \frac{1}{2} h^2 (S + T) - 2\Phi_2^T,$$

then the system in (12.9) satisfies the asymptotic stability and strict (X, Y, Z)-θ-dissipative performance.

Proof The Lyapunov function is selected as

$$V(t) = \sum_{m=1}^{4} V_m(t), \tag{12.14}$$

where

$$V_1(t) \triangleq x^T(t)Px(t) + \int_{t-h}^{t} x^T(s)Qx(s)ds,$$

$$V_2(t) \triangleq h\int_{-h}^{0}\int_{t+\theta}^{t} \dot{x}^T(s)R\dot{x}(s)dsd\theta,$$

$$V_3(t) \triangleq \int_{-h}^{0}\int_{r}^{0}\int_{t+\theta}^{t} \dot{x}^T(s)S\dot{x}(s)dsd\theta dr,$$

$$V_4(t) \triangleq \int_{-h}^{0}\int_{-h}^{r}\int_{t+\theta}^{t} \dot{x}^T(s)T\dot{x}(s)dsd\theta dr.$$

It follows that

$$\dot{V}_1(t) = 2\dot{x}^T(t)Px(t) + x^T(t)Qx(t) - x^T(t-h)Qx(t-h), \quad (12.15)$$

$$\dot{V}_2(t) = h^2\dot{x}^T(t)R\dot{x}(t) - h\int_{t-h}^{t} \dot{x}^T(\alpha)R\dot{x}(\alpha)d\alpha, \quad (12.16)$$

$$\dot{V}_3(t) = \frac{1}{2}h^2\dot{x}^T(t)S\dot{x}(t) - \int_{-h}^{0}\int_{t+\beta}^{t} \dot{x}^T(\alpha)S\dot{x}(\alpha)d\alpha d\beta, \quad (12.17)$$

$$\dot{V}_4(t) = \frac{1}{2}h^2\dot{x}^T(t)T\dot{x}(t) - \int_{-h}^{0}\int_{t-h}^{t+\beta} \dot{x}^T(\alpha)T\dot{x}(\alpha)d\alpha d\beta. \quad (12.18)$$

Due to

$$-h\int_{t-h}^{t} \dot{x}^T(\alpha)R\dot{x}(\alpha)d\alpha = -h\int_{t-h}^{t-h(t)} \dot{x}^T(\alpha)R\dot{x}(\alpha)d\alpha - h\int_{t-h(t)}^{t} \dot{x}^T(\alpha)R\dot{x}(\alpha)d\alpha,$$

we have

$$
\begin{aligned}
&-h \int_{t-h}^{t-h(t)} \dot{x}^T(\alpha) R \dot{x}(\alpha) d\alpha \\
&\leqslant -\frac{h}{h-h(t)} \Psi^T(t) \Big\{ \Big(e_3 - e_2\Big) R \Big(e_3 - e_2\Big)^T \\
&+3\Big(e_3 + e_2 - 2e_5\Big) R \Big(e_3 + e_2 - 2e_5\Big)^T \\
&+5\Big(e_3 - e_2 + 6e_5 - 6e_7\Big) R \Big(e_3 - e_2 + 6e_5 - 6e_7\Big)^T \Big\} \Psi(t), \\
&-h \int_{t-h(t)}^{t} \dot{x}^T(\alpha) R \dot{x}(\alpha) d\alpha \\
&\leqslant -\frac{h}{h(t)} \Psi^T(t) \Big\{ \Big(e_1 - e_3\Big) R \Big(e_1 - e_3\Big)^T \\
&+3\Big(e_1 + e_3 - 2e_4\Big) R \Big(e_1 + e_3 - 2e_4\Big)^T \\
&+5\Big(e_1 - e_3 + 6e_4 - 6e_6\Big) R \Big(e_1 - e_3 + 6e_4 - 6e_6\Big)^T \Big\} \Psi(t).
\end{aligned}
$$

It is not difficult to see the following inequality is true:

$$
\begin{aligned}
&-\int_{-h}^{0} \int_{t+\beta}^{t} \dot{x}^T(\alpha) S \dot{x}(\alpha) d\alpha d\beta = -\int_{-h}^{-h(t)} \int_{t-h}^{t+\beta} \dot{x}^T(\alpha) T \dot{x}(\alpha) d\alpha d\beta \\
&-\int_{-h(t)}^{0} \int_{t-h(t)}^{t+\beta} \dot{x}^T(\alpha) T \dot{x}(\alpha) d\alpha d\beta - h(t) \int_{t-h}^{t-h(t)} \dot{x}^T(\alpha) T \dot{x}(\alpha) d\alpha.
\end{aligned}
$$

Thus,

$$
\begin{aligned}
&-\Big(h - h(t)\Big) \int_{t-h(t)}^{t} \dot{x}^T(\alpha) S \dot{x}(\alpha) d\alpha \\
&\leqslant -\Big(\frac{h}{h(t)} - 1\Big) \Psi^T(t) \Big\{ \Big(e_1 - e_3\Big) S \Big(e_1 - e_3\Big)^T \\
&+3\Big(e_1 + e_3 - 2e_4\Big) S \Big(e_1 + e_3 - 2e_4\Big)^T \\
&+5\Big(e_1 - e_3 + 6e_4 - 6e_6\Big) S \Big(e_1 - e_3 + 6e_4 - 6e_6\Big)^T \Big\} \Psi(t),
\end{aligned}
$$

$$
\begin{aligned}
&-h(t)\int_{t-h}^{t-h(t)}\dot{x}^T(\alpha)T\dot{x}(\alpha)d\alpha\\
&\leqslant -\frac{h(t)}{h-h(t)}\Psi^T(t)\Big\{\Big(e_3-e_2\Big)T\Big(e_3-e_2\Big)^T\\
&+3\Big(e_3+e_2-2e_5\Big)T\Big(e_3+e_2-2e_5\Big)^T\\
&+5\Big(e_3-e_2+6e_5-6e_7\Big)T\Big(e_3-e_2+6e_5-6e_7\Big)^T\Big\}\Psi(t).
\end{aligned}
$$

Denote $\alpha=\frac{h(t)}{h}$, $\beta=\frac{h-h(t)}{h}$. We can obtain

$$
\begin{aligned}
&-\frac{1}{\alpha}\Psi^T(t)\Big\{\Big(e_3-e_2\Big)\Big(T+R\Big)\Big(e_3-e_2\Big)^T\\
&+3\Big(e_3+e_2-2e_5\Big)\Big(T+R\Big)\Big(e_3+e_2-2e_5\Big)^T\\
&+5\Big(e_3-e_2+6e_5-6e_7\Big)\Big(T+R\Big)\\
&\Big(e_3-e_2+6e_5-6e_7\Big)^T\Big\}\Psi(t)\\
&-\frac{1}{\beta}\Psi^T(t)\Big\{\Big(e_1-e_3\Big)\Big(S+R\Big)\Big(e_1-e_3\Big)^T\\
&+3\Big(e_1+e_3-2e_4\Big)\Big(S+R\Big)\Big(e_1+e_3-2e_4\Big)^T\\
&+5\Big(e_1-e_3+6e_4-6e_6\Big)\Big(S+R\Big)\\
&\Big(e_1-e_3+6e_4-6e_6\Big)^T\Big\}\Psi(t)\\
&+\Psi^T(t)\Big\{\Big(e_3-e_2\Big)T\Big(e_3-e_2\Big)^T\\
&+3\Big(e_3+e_2-2e_5\Big)T\Big(e_3+e_2-2e_5\Big)^T\\
&+5\Big(e_3-e_2+6e_5-6e_7\Big)T\Big(e_3-e_2+6e_5-6e_7\Big)^T\Big\}\Psi(t)\\
&+\Psi^T(t)\Big\{\Big(e_1-e_3\Big)S\Big(e_1-e_3\Big)^T\\
&+3\Big(e_1+e_3-2e_4\Big)S\Big(e_1+e_3-2e_4\Big)^T\\
&+5\Big(e_1-e_3+6e_4-6e_6\Big)S\Big(e_1-e_3+6e_4-6e_6\Big)^T\Big\}\Psi(t)\\
&\leqslant -\Psi^T(t)\Gamma\Xi\Gamma^T\Psi(t).
\end{aligned}
\tag{12.19}
$$

On account of $S > 0$, $T > 0$, it has

$$
\begin{aligned}
&-\int_{-h}^{-h(t)}\int_{t+\beta}^{t-h(t)} \dot{x}^T(\alpha)S\dot{x}(\alpha)d\alpha d\beta \\
&\leqslant -\Psi^T(t)\Big\{2\Big(e_3-e_5\Big)S\Big(e_3-e_5\Big)^T \\
&+4\Big(e_3+2e_5-3e_7\Big)S\Big(e_3+2e_5-3e_7\Big)^T\Big\}\Psi(t), \qquad (12.20)
\end{aligned}
$$

$$
\begin{aligned}
&-\int_{-h(t)}^{0}\int_{t+\beta}^{t} \dot{x}^T(\alpha)S\dot{x}(\alpha)d\alpha d\beta \\
&\leqslant -\Psi^T(t)\Big\{2\Big(e_1-e_4\Big)S\Big(e_1-e_4\Big)^T \\
&+4\Big(e_1+2e_4-3e_6\Big)S\Big(e_1+2e_4-3e_6\Big)^T\Big\}\Psi(t), \qquad (12.21)
\end{aligned}
$$

$$
\begin{aligned}
&-\int_{-h}^{-h(t)}\int_{t-h}^{t+\beta} \dot{x}^T(\alpha)T\dot{x}(\alpha)d\alpha d\beta \\
&\leqslant -\Psi^T(t)\Big\{2\Big(e_2-e_5\Big)T\Big(e_2-e_5\Big)^T \\
&+4\Big(e_2-4e_5+3e_7\Big)T\Big(e_2-4e_5+3e_7\Big)^T\Big\}\Psi(t), \qquad (12.22)
\end{aligned}
$$

$$
\begin{aligned}
&-\int_{-h(t)}^{0}\int_{t-h(t)}^{t+\beta} \dot{x}^T(\alpha)T\dot{x}(\alpha)d\alpha d\beta \\
&\leqslant -\Psi^T(t)\Big\{2\Big(e_3-e_4\Big)T\Big(e_3-e_4\Big)^T \\
&+4\Big(e_3-4e_4+3e_6\Big)T\Big(e_3-4e_4+3e_6\Big)^T\Big\}\Psi(t). \qquad (12.23)
\end{aligned}
$$

Given suitably dimensioned matrices Φ_1^T and Φ_2^T, the condition below satisfies:

$$0 = \sum_{i=1}^{r}\sum_{j=1}^{r} h_i\Big(\zeta(t)\Big)h_j\Big(\zeta(t)\Big)\Big[x^T(t)\Phi_1^T + \dot{x}^T(t)\Phi_2^T\Big]$$
$$\Big[-\dot{x}(t) + A_i x(t) + B_i K_j x\Big(t - h(t)\Big) - B_i K_j e_k(s_k T) + C_i w(t)\Big]. \tag{12.24}$$

Introduce the index

$$\Theta(t) \triangleq -z^T(t)Xz(t) - 2z^T(t)Yw(t) - w^T(t)Zw(t) + \theta w^T(t)w(t).$$

On the basis of (12.6) and (12.15)–(12.24), we have

$$\dot{V}(t) + \Theta(t) \leqslant \sum_{i=1}^{r}\sum_{j=1}^{r} h_i\Big(\zeta(t)\Big)h_j\Big(\zeta(t)\Big)\left\{\Psi^T(t)\begin{bmatrix}\Omega_1^{ij} & \Omega_2^{ij} & \Omega_3^{ij}\\ \star & \Omega_4 & \Omega_5^{ij}\\ \star & \star & \Omega_6^{ij}\end{bmatrix}\Psi(t)\right\}.$$

Considering (12.11)–(12.12), it yields

$$\dot{V}(t) + \Theta(t) < 0. \tag{12.25}$$

Based on the zero initial condition, for any φ, it can be easily seen that

$$\int_0^{\varphi} \Theta(t)dt \leq \int_0^{\varphi} (\Theta(t) + \dot{V}(t))dt < 0,$$

which implies

$$\int_0^{\varphi}\Big[z^T(t)Xz(t) + 2z^T(t)Yw(t) + w^T(t)Zw(t)\Big]dt \geqslant \theta\int_0^{\varphi}\Big[w^T(t)w(t)\Big]dt.$$

At this point, the proof is completed.

12.3.2 Fuzzy Controller Design

Theorem 12.4 *Given the scalars h, δ, θ, ξ, β_1, β_2, and real matrices $X = X^T = -\hat{X}^T\hat{X}$, Y, Z, if there exist $\bar{P} > 0$, $\bar{Q} > 0$, $\bar{R} > 0$, $\bar{S} > 0$, $\bar{T} > 0$, $\bar{\Lambda}_1 > 0$, $\bar{\Lambda}_2 > 0$, $\bar{\chi} = (\bar{\chi}_{mn})_{3\times 3}$, such that the following conditions satisfy:*

$$\frac{2}{r-1}\begin{bmatrix} \Psi_1^{ii} & \Psi_2^{ii} & \Psi_3^{ii} \\ \star & \Psi_4 & \Psi_5^{ii} \\ \star & \star & \Psi_6^{ii} \end{bmatrix} + \begin{bmatrix} \Psi_1^{ij} & \Psi_2^{ij} & \Psi_3^{ij} \\ \star & \Psi_4 & \Psi_5^{ij} \\ \star & \star & \Psi_6^{ij} \end{bmatrix} + \begin{bmatrix} \Psi_1^{ji} & \Psi_2^{ji} & \Psi_3^{ji} \\ \star & \Psi_4 & \Psi_5^{ji} \\ \star & \star & \Psi_6^{ji} \end{bmatrix} < 0, \quad (12.26)$$

$$\begin{bmatrix} \Psi_1^{ii} & \Psi_2^{ii} & \Psi_3^{ii} \\ \star & \Psi_4 & \Psi_5^{ii} \\ \star & \star & \Psi_6^{ii} \end{bmatrix} < 0, \quad (12.27)$$

$$\bar{F} \triangleq \begin{bmatrix} \mathcal{I} \otimes \bar{R} + \mathcal{I} \otimes \bar{S} & \bar{\chi} \\ \star & \mathcal{I} \otimes \bar{R} + \mathcal{I} \otimes \bar{T} \end{bmatrix} > 0, \quad (12.28)$$

where

$$\Psi_1^{ij} \triangleq \begin{bmatrix} \Psi_{11}^{ij} & \Psi_{12}^{ij} \\ \star & \Psi_{14}^{ij} \end{bmatrix}, \quad \Psi_2^{ij} \triangleq \begin{bmatrix} \Psi_{21}^{ij} & \Psi_{22}^{ij} \\ \Psi_{23}^{ij} & \Psi_{24}^{ij} \end{bmatrix},$$

$$\Psi_3^{ij} \triangleq \begin{bmatrix} -\beta_1 B_i \Omega_j & \beta_1 C_j - \xi^T D_i^T Y & \xi^T \\ 0 & 0 & 0 \\ 0 & -\Omega_j^T E_i^T Y & \Omega_j^T E_i^T \hat{X}^T \\ 0 & 0 & 0 \end{bmatrix},$$

$$\Psi_4 \triangleq \begin{bmatrix} \Psi_{41} & \Psi_{42} \\ \star & \Psi_{44} \end{bmatrix}, \quad \Psi_5^{ij} \triangleq \left[\frac{O_{3\times 3}}{-\beta_2 B_i \Omega_j \;\; \beta_2 C_j \;\; 0} \right],$$

$$\Psi_6^{ij} \triangleq \begin{bmatrix} -\bar{\Lambda}_1 & \Omega_j^T E_i^T Y & -\Omega_j^T E_i^T \hat{X}^T \\ \star & -2F_i^T Y - Z + \theta & F_i^T \hat{X}^T \\ \star & \star & -I \end{bmatrix},$$

with

$$\Psi_{11}^{ij} \triangleq \begin{bmatrix} -9\bar{R} - 6\bar{S} + \bar{Q} + 2\beta_2 A_i \xi & N_1 \\ \star & -9\bar{R} - 6\bar{T} - \bar{Q} \end{bmatrix},$$

$$\Psi_{12}^{ij} \triangleq \begin{bmatrix} N_2^{ij} & -24\bar{R} - 6\bar{S} \\ N_3 & N_4 \end{bmatrix}, \quad \Psi_{14}^{ij} \triangleq \begin{bmatrix} N_5 & N_6 \\ \star & -192\bar{R} - 18\bar{S} - 66\bar{T} \end{bmatrix},$$

$$\Psi_{21}^{ij} \triangleq \begin{bmatrix} N_7 & 30\bar{R} + 12\bar{S} \\ 36\bar{R} + 18\bar{T} & (-1)\sum_{m=1}^{3} 6\bar{\chi}_{m3} \end{bmatrix},$$

$$\Psi_{22}^{ij} \triangleq \begin{bmatrix} \sum_{m=1}^{3} 6\bar{\chi}_{3m}^T & \bar{P}^T - \beta_1 \xi + \beta_2 \xi^T A_i^T \\ -30\bar{R} - 12\bar{T} & 0 \end{bmatrix},$$

$$\Psi_{23}^{ij} \triangleq \begin{bmatrix} N_8 & N_9 \\ N_{11} & 180\bar{R} + 24\bar{S} + 48\bar{T} \end{bmatrix}, \quad \Psi_{24}^{ij} \triangleq \begin{bmatrix} N_{10} & \beta_2 \Omega_j^T B_i^T \\ -12\bar{\chi}_{32}^T + 36\bar{\chi}_{33}^T & 0 \end{bmatrix},$$

$$\Psi_{41} \triangleq \begin{bmatrix} -192\bar{R} - 18\bar{S} - 66\bar{T} & -12\bar{\chi}_{23} + 36\bar{\chi}_{33} \\ \star & -180\bar{R} - 36\bar{T} - 36\bar{S} \end{bmatrix},$$

$$\Psi_{42} \triangleq \begin{bmatrix} 180\bar{R} + 48\bar{T} + 24\bar{S} & 0 \\ -36\bar{\chi}_{33} & 0 \end{bmatrix}, \quad \Psi_{44} \triangleq \begin{bmatrix} -180\bar{R} - 36\bar{T} - 36\bar{S} & 0 \\ \star & N_{12} \end{bmatrix},$$

$$N_1 \triangleq \sum_{m=1}^{3} \left(\bar{\chi}_{1m}^T - \bar{\chi}_{2m}^T + \bar{\chi}_{3m}^T\right), \quad N_4 \triangleq -(1)^m \sum_{m=1}^{3} \left(2\bar{\chi}_{m2} - 6\bar{\chi}_{m3}\right),$$

$$N_2^{ij} \triangleq \sum_{m=1}^{3} \left(-\bar{\chi}_{1m}^T - \bar{\chi}_{2m}^T - \bar{\chi}_{3m}^T\right) + 3\bar{R} + \beta_1 B_i \Omega_j,$$

$$N_3 \triangleq -(1)^m \sum_{m=1}^{3} \left(\bar{\chi}_{m1} - \bar{\chi}_{m2} + \bar{\chi}_{m3}\right) + 3\bar{R},$$

$$N_5 \triangleq sym\left(\sum_{m=1}^{3} \left(\bar{\chi}_{m1} - \bar{\chi}_{m2} + \bar{\chi}_{m3}\right)\right) - 18\bar{R} - 6\bar{S} - 6\bar{T} + \delta\bar{\Lambda}_2,$$

$$N_6 \triangleq \sum_{m=1}^{3} \left(2\bar{\chi}_{m2} - 6\bar{\chi}_{m3}\right) + 36\bar{R} + 18\bar{T}, \quad N_7 \triangleq \sum_{m=1}^{3} \left(2\bar{\chi}_{2m}^T - 6\bar{\chi}_{3m}^T\right),$$

$$N_8 \triangleq -(1)^m \sum_{m=1}^{3} \left(2\bar{\chi}_{2m}^T - 6\bar{\chi}_{3m}^T\right) - 24\bar{R} - 6\bar{S},$$

$$N_9 \triangleq \sum_{m=1}^{3} 6\bar{\chi}_{m3} - 30\bar{R} - 12\bar{T}, \quad N_{10} \triangleq -(1)^m \sum_{m=1}^{3} 6\bar{\chi}_{3m}^T + 30\bar{R} + 12\bar{S},$$

$$N_{11} \triangleq -4\bar{\chi}_{22}^T + 12\bar{\chi}_{23}^T + 12\bar{\chi}_{32}^T - 36\bar{\chi}_{33}^T, \quad N_{12} \triangleq h^2\bar{R} + \frac{1}{2}h^2\left(\bar{S} + \bar{T}\right) - 2\beta_2\xi,$$

the system in (12.9) is asymptotically stable in strict (X, Y, Z)-θ-dissipativity. In addition, the proposed fuzzy controller parameters can be further given by

$$K_j = \Omega_j \xi^{-1}. \tag{12.29}$$

Proof Denote

$$\begin{aligned} &\bar{R} \triangleq \xi^T R \xi, \quad \bar{S} \triangleq \xi^T S \xi, \quad \bar{T} \triangleq \xi^T T \xi, \quad \bar{P} \triangle \xi^T P \xi, \quad \bar{Q} \triangleq \xi^T Q \xi, \\ &\bar{\Lambda}_1 \triangleq \xi^T \Lambda_1 \xi, \bar{\Lambda}_2 \triangleq \xi^T \Lambda_2 \xi, \quad \Phi_1^T \triangleq \beta_1 \xi^{-T}, \quad \Phi_2^T \triangleq \beta_2 \xi^{-T}, \\ &\bar{\chi}_{mn} \triangleq \xi^T \chi_{mn} \xi, \quad m = 1, 2, 3, \quad n = 1, 2, 3, \end{aligned}$$

and $\Upsilon \triangleq diag\left\{\xi\ \xi\ \xi\ \xi\ \xi\ \xi\ \xi\ \xi\ \xi\ I\ I\right\}$, $\hat{\Upsilon} \triangleq diag\left\{\xi\ \xi\ \xi\ \xi\ \xi\ \xi\right\}$.

It is obvious to get $\bar{F} = \hat{\Upsilon}^T F \hat{\Upsilon}$, that is to say, (12.13) holds under the circumstance in (12.28). Then pre- and post-multiplying (12.11)–(12.12) with Υ^T and its transpose, separately, we have (12.26)–(12.27). Therefore, this completes the proof.

12.4 Simulation Results

Some simulation results are shown in this part to illustrate the feasibility of presented fuzzy control scheme, which has been mentioned in Sect. 12.2.1. Figures 12.2, 12.3, 12.4 plot the state trajectories of the open-loop system, we can see from them that the original system is unstable.

Therefore, we aim to present a valid event-based fuzzy controller to ensure the overall system is asymptotically stable in strict (X, Y, Z)-θ-$dissipative$ sense. By solving the conditions in Theorem 12.4, the fuzzy controller gains are computed as

$$K_1 = \left[\, 1.0519\ -0.3633\ 0.0364 \,\right], \quad K_2 = \left[\, 0.5766\ -0.1740\ 0.0321 \,\right].$$

Moreover, the event-driven matrices can be obtained as

$$W_1 = \begin{bmatrix} 622.6731 & -276.8543 & 36.0584 \\ -276.8543 & 134.5283 & -16.8499 \\ 36.0584 & -16.8499 & 2.1960 \end{bmatrix},$$
$$W_2 = \begin{bmatrix} 0.2090 & -0.0640 & 0.0078 \\ -0.0640 & 0.0481 & -0.0018 \\ 0.0078 & -0.0018 & 0.0011 \end{bmatrix}.$$

Set the initial condition of considered truck trailer as $x(0) = \begin{bmatrix} 0.16 \\ -0.1 \\ 0.16 \end{bmatrix}$ and the disturbance as $w(t) = \begin{bmatrix} \sin(0.1t)\exp(-0.1t) \\ \sin(0.1t)\exp(-0.1t) \end{bmatrix}$.

The event-driven released instants figure is depicted in Fig. 12.5. The state responses of the obtained closed-loop system are plotted in Figs. 12.6, 12.7, 12.8. Utilizing the designed fuzzy controller, the system states of resulting dynamic system can be converged to zero. Consequently, it can be readily to see our proposed fuzzy controller can stabilize the concerned truck trailer system.

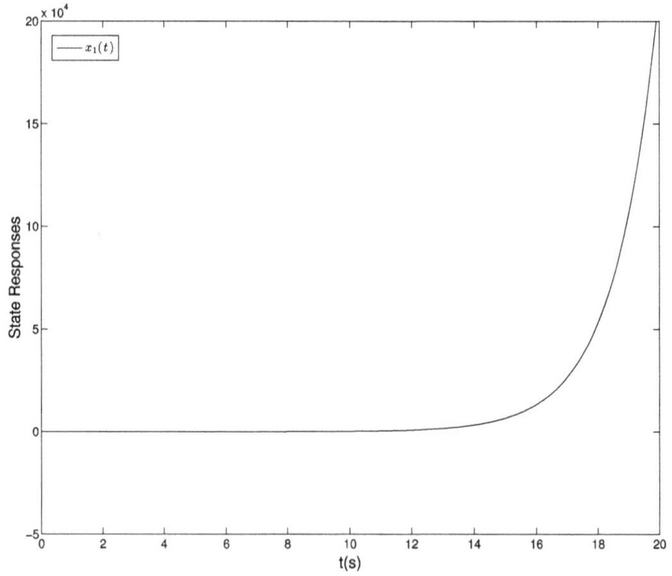

Fig. 12.2 State response $x_1(t)$ of open-loop system

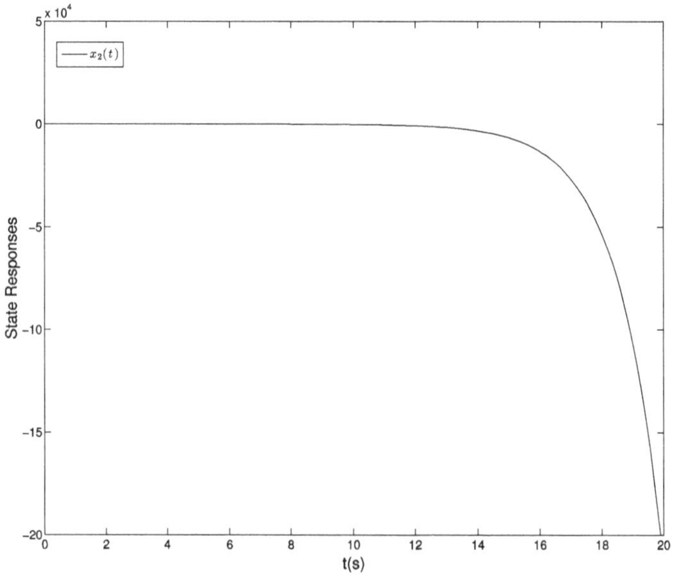

Fig. 12.3 State response $x_2(t)$ of open-loop system

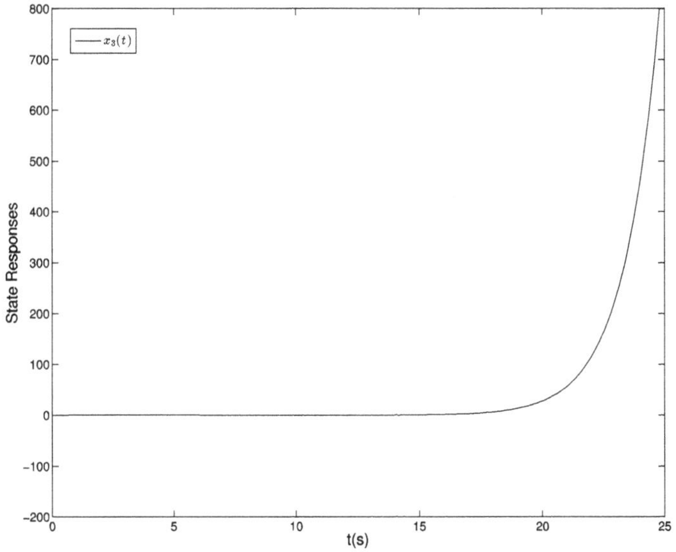

Fig. 12.4 State response $x_3(t)$ of open-loop system

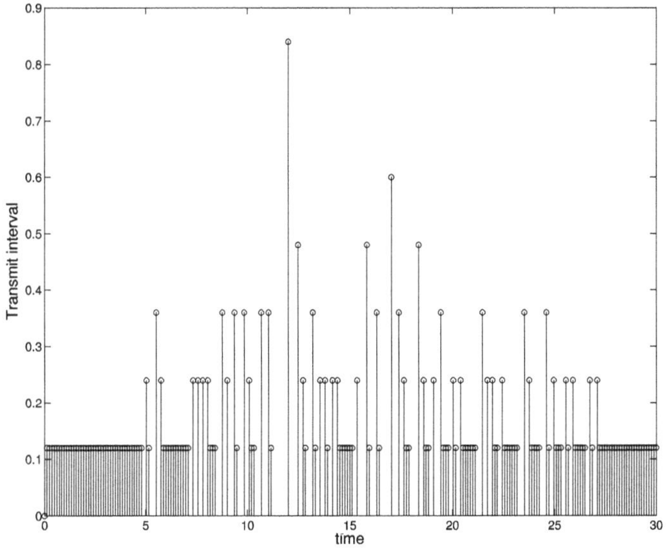

Fig. 12.5 The event-triggering release instants and intervals

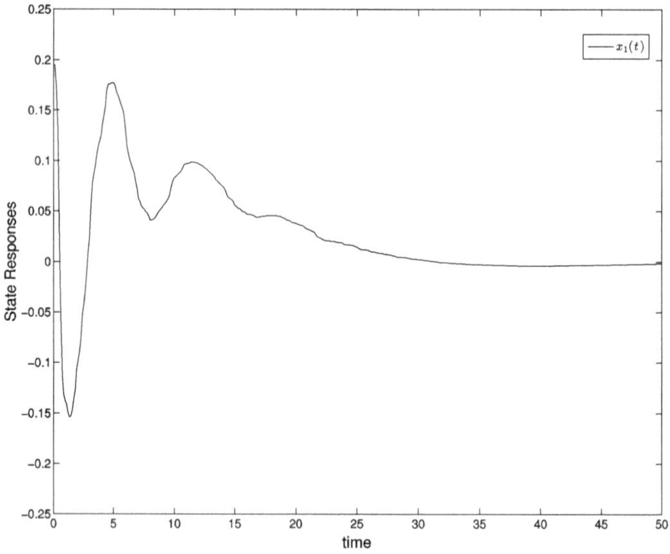

Fig. 12.6 State response $x_1(t)$ of fuzzy control system

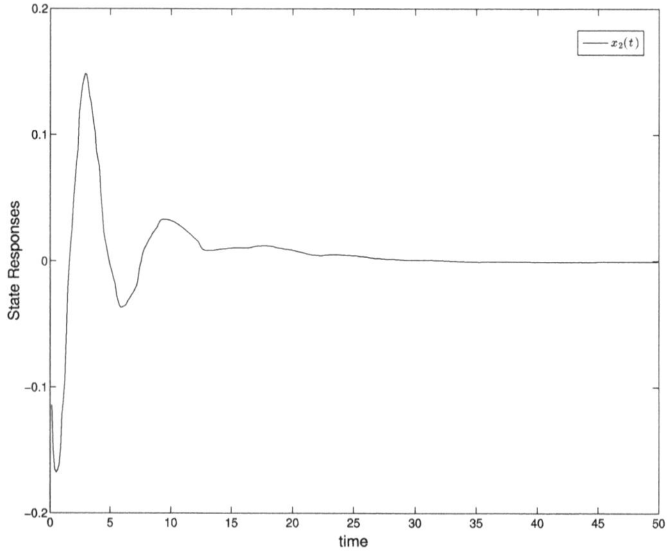

Fig. 12.7 State response $x_2(t)$ of fuzzy control system

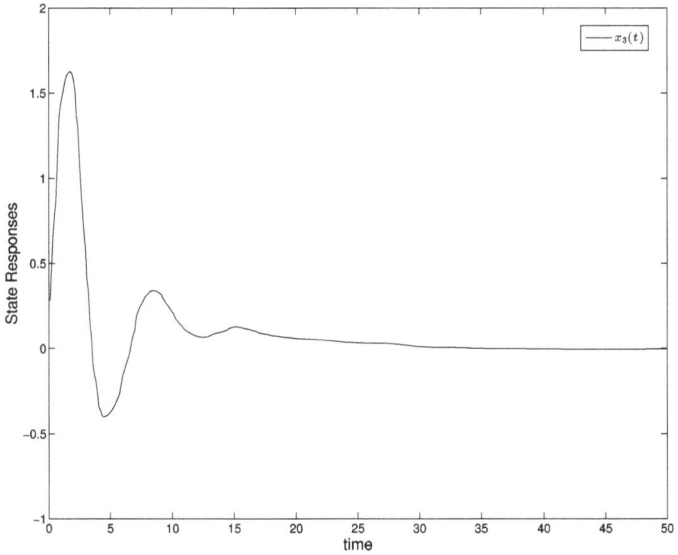

Fig. 12.8 State response $x_3(t)$ of fuzzy control system

12.5 Conclusion

In this chapter, the dissipative control problem of fuzzy logic systems based on the event-triggered strategy was examined. To reduce the unnecessary use of limited communication resources and network bandwidth, an efficient event-triggered scheme was introduced.And sufficient conditions of the fuzzy controller design, which could ensure the resulting closed-loop system was asymptotically stable subject to the specific dissipative property, were established. Finally, a truck-trailer model was presented as an illustrative example to verify the applicability of the developed design method.

Chapter 13
Event-Triggered Fuzzy Control of Inverted Pendulum Systems

13.1 Introduction

This chapter addresses the event-based fuzzy control design and its application to an inverted pendulum system. First, the time-varying delays in the inverted pendulum system and event-based method are considered in the system stability analysis. The interval of the time-delay is segmented to l non-uniform sub-intervals by applying an efficient delay-partition approach. The information in every subinterval is handled using the reciprocally convex technique. The stability conditions of the inverted pendulum model are established to be less conservative than existing research achievements. The reduction in the conservativeness is more notable when the number of delay partitions l is smaller. In addition, through the PDC rule, feasible conditions for the designed event-triggering fuzzy controller are derived for the considered inverted pendulum system.

13.2 System Description and Preliminaries

13.2.1 Inverted Pendulum System

Consider the nonlinear inverted pendulum model as studied in [132], which is plotted in Fig. 13.1, and its schematic diagram is shown in Fig. 13.2. It can be easily seen that the pendulum in the upright position is an unstable equilibrium with the nonlinearity.

In order to let the subsequent description be more convenient, some parameters of the inverted pendulum system are provided in Table 13.1.

The notation "(t)" in system variables is omitted, in other words, y means $y(t)$. Utilizing some physical analysis rules, the dynamics of inverted pendulum model is given by

X. Su et al., *Intelligent Control, Filtering and Model Reduction Analysis for Fuzzy-Model-Based Systems*, Studies in Systems, Decision and Control 385,
https://doi.org/10.1007/978-3-030-81214-0_13

Fig. 13.1 Inverted pendulum system with a delayed resonator

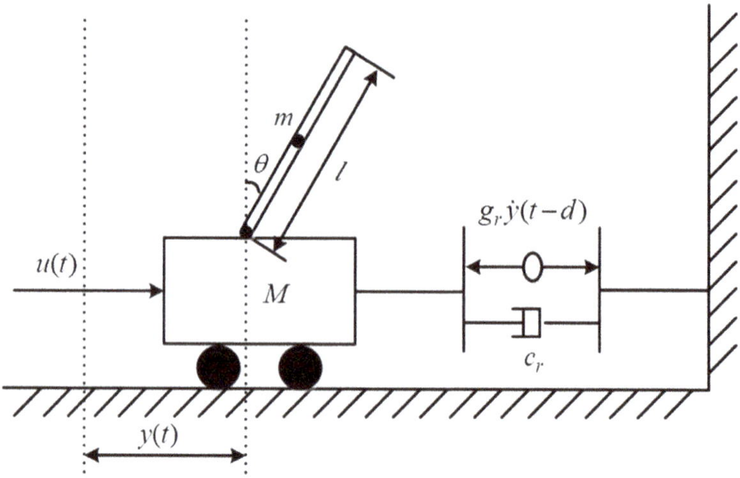

Fig. 13.2 Inverted pendulum system

$$M\frac{d^2y}{dt^2} + m\frac{d^2}{dt^2}(y + l\sin\theta) = u - F_r,$$
$$m\frac{d^2}{dt^2}(y + l\sin\theta)\cdot l\cos\theta = mgl\sin\theta,$$

where $F_r = g_r\dot{y}(t-d) + c_r\dot{y}$ with F_r denoting the resultant force produced by the damper and delayed resonator. And the state variables are set as

$$x_1 = y, \quad x_2 = \theta, \quad x_3 = \dot{y}, \quad x_4 = \dot{\theta}.$$

Table 13.1 The parameter values of the inverted pendulum system

Parameter	Value	Description
M	1.378 kg	Mass of the cart
m	0.051 kg	Mass of the pendulum
l	0.325 m	Length of the pendulum
g	9.8 m/s^2	Acceleration of gravity
g_r	0.7 kg/s	Coefficient of the delayed resonator
c_r	5.98 kg/s	Coefficient of the damper
$\theta(t)$		Angle from the upright position
$y(t)$		Displacement of the cart
$d(t)$		Time-varying delay
$u(t)$		Force acted on the cart

Therefore, we can obtain the state-space equations of resulting nonlinear system as follows:

$$\begin{aligned}\dot{x}_1 &= x_3,\\ \dot{x}_2 &= x_4,\\ \dot{x}_3 &= \frac{-mg\sin x_2}{M\cos x_2} - \frac{c_r x_3 + g_r x_3(t-d) - u}{M},\\ \dot{x}_4 &= \frac{(M+m)g\sin x_2}{Ml\cos^2 x_2} + \frac{x_4^2\sin x_2}{\cos x_2} + \frac{c_r x_3 + g_r x_3(t-d) - u}{Ml\cos x_2}.\end{aligned}$$

Then we establish the following T-S fuzzy model of considered inverted pendulum system with an approximation approach.

♦ **Plant Form:**

Rule 1: IF $x_2(k)$ is μ_1 and close to 0 rad, THEN

$$x(k+1) = A_1 x(k) + A_{d1} x\big(k - d(k)\big) + B_1 u(k).$$

Rule 2: IF $x_2(k)$ is μ_2 and close to γ rad or $-\gamma$ rad, THEN

$$x(k+1) = A_2 x(k) + A_{d2} x\big(k - d(k)\big) + B_2 u(k).$$

The corresponding parameters are formulated as

$$A_1 = \begin{bmatrix} 0 & 0 & T & 0 \\ 0 & 0 & 0 & T \\ 0 & -\frac{Tmg}{M} & -\frac{Tc_r}{M} & 0 \\ 0 & \frac{T(M+m)g}{Ml} & \frac{Tc_r}{Ml} & 0 \end{bmatrix}, \quad B_1 = \begin{bmatrix} 0 \\ 0 \\ \frac{T}{M} \\ -\frac{T}{Ml} \end{bmatrix},$$

$$A_2 = \begin{bmatrix} 0 & 0 & T & 0 \\ 0 & 0 & 0 & T \\ 0 & -\frac{Tmg\beta}{M\alpha} & -\frac{Tc_r}{M} & 0 \\ 0 & \frac{T(M+m)g\beta}{Ml\alpha^2} & \frac{Tc_r}{Ml\alpha} & 0 \end{bmatrix}, \quad B_2 = \begin{bmatrix} 0 \\ 0 \\ \frac{T}{M} \\ -\frac{T}{Ml\alpha} \end{bmatrix},$$

$$A_{d1} = \begin{bmatrix} 0 & 0 & 0 & 0 \\ 0 & 0 & 0 & 0 \\ 0 & 0 & -\frac{Tg_r}{M} & 0 \\ 0 & 0 & \frac{Tg_r}{Ml} & 0 \end{bmatrix}, \quad A_{d2} = \begin{bmatrix} 0 & 0 & 0 & 0 \\ 0 & 0 & 0 & 0 \\ 0 & 0 & -\frac{Tg_r}{M} & 0 \\ 0 & 0 & \frac{Tg_r}{Ml\alpha} & 0 \end{bmatrix}.$$

13.2.2 T-S Fuzzy System

In this part, we present the T-S fuzzy system and fuzzy controller to investigate the above-mentioned nonlinear control issue.

♦ **Plant Form:**
Rule i: IF $\theta_1(k)$ is μ_{i1} and $\theta_2(k)$ is μ_{i2} and $\cdots$ and $\theta_p(k)$ is μ_{ip}, THEN

$$\begin{aligned} x(k+1) &= A_i x(k) + A_{di} x\big(k - d(k)\big) + B_i u(k), \ i = 1, 2, \ldots, r, \\ x(k) &= \varphi(k), \quad k = -d_M, -d_M + 1, \ldots, 0, \end{aligned} \tag{13.1}$$

where $x(k) \in \mathbb{R}^n$ denotes the state vector; $u(k) \in \mathbb{R}^m$ denotes the control input; $\theta(k) = \big[\theta_1(k), \theta_2(k), \cdots, \theta_{p-1}(k), \theta_p(k)\big]$, sometimes represented by θ, stand for premise variables; $\mu_{i1}, \ldots, \mu_{ip}$ denote the fuzzy sets; $\varphi(k)$ implies the initial condition. A_i, A_{di} and B_i are known properly dimensioned matrices, where $i \in \mathbb{S} \triangleq \{1, 2, \ldots, r\}$, with r being the quantity of fuzzy rules. The time-delay $d(k)$ satisfies $0 < d_m \leqslant d(k) \leqslant d_M < \infty$, where d_m, d_M stand for the minimum and maximum bounds of time-delays, separately. Partitioning the time-delay interval non-uniformly to l fractions yields

$$\big[d_m, d_M\big] = \bigcup_{i=1}^{l} \big[d_{i-1}, d_i\big], \quad i = 1, 2, \ldots, l-1.$$

Denote $d_0 = d_m$, $d_l = d_M$, and η_i stands for the length of sub-interval, which means $\eta_i = d_i - d_{i-1}$ with $d_{-1} = 0$.

Provided that $u(k)$ is independent of $\theta(k)$, then the defuzzification of T-S fuzzy system is governed by

$$x(k+1) = \sum_{i=1}^{r} h_i(\theta)\left[A_i x(k) + A_{di} x\left(k - d(k)\right) + B_i u(k)\right],$$

where $h_i(\theta) = \nu_i(\theta) / \sum_{i=1}^{r} \nu_i(\theta)$ and $\nu_i(\theta) = \prod_{j=1}^{p} \mu_{ij}(\theta_j)$, with $\mu_{ij}(\theta_j)$ standing for the grade of membership of θ_j in μ_{ij}, it follows $\nu_i(\theta) \geqslant 0$, $h_i(\theta) \geqslant 0$, $i \in \mathbb{S}$ and $\sum_{i=1}^{r} h_i(\theta) = 1$. A complete presentation of the aforementioned system is inferred as

$$x(k+1) = A(k)x(k) + A_d(k)x\big(k - d(k)\big) + B(k)u(k), \tag{13.2}$$

where

$$A(k) \triangleq \sum_{i=1}^{r} h_i(\theta) A_i, \quad A_d(k) \triangleq \sum_{i=1}^{r} h_i(\theta) A_{di}, \quad B(k) \triangleq \sum_{i=1}^{r} h_i(\theta) B_i.$$

It is assumed that $\theta(k)$ of the fuzzy system can be available, then we design the fuzzy-basis-dependent controller as follow with a PDC method:

♦ **Fuzzy Controller:**
Rule i: IF $\theta_1(k)$ is μ_{i1} and $\theta_2(k)$ is μ_{i2} and $\cdots$ and $\theta_p(k)$ is μ_{ip}, THEN

$$u(k) = K_i x(k), \quad i \in \mathbb{S},$$

where K_i stands for the fuzzy controller gain to be designed later, and

$$u(k) = \sum_{i=1}^{r} h_i(\theta) K_i x(k), \tag{13.3}$$

that is $u(k) = K(k)x(k)$, where

$$K(k) = \sum_{i=1}^{r} h_i(\theta) K_i. \tag{13.4}$$

On account of the fuzzy controller (13.3), we can get the whole closed-loop system as

$$x(k+1) = \sum_{i=1}^{r}\sum_{j=1}^{r} h_i(\theta) h_j(\theta) \times \left[A_i x(k) + A_{di} x\left(k - d(k)\right) + B_i K_j x(k)\right], \tag{13.5}$$

which implies

$$x(k+1) = [A(k) + B(k)K(k)]\,x(k) + A_d(k)x\,(k - d(k))\,. \tag{13.6}$$

Remark 13.1 It is obvious to see the obtained conditions are comparatively conservative due to time-varying delay $d(k)$. Utilizing the delay partition method and reciprocally convex strategy, delay information of every subinterval are handled well to reduce the conservatism. In addition, the event-driven fuzzy controller design approach is also promoted. Actually, the reciprocally convex method plays an efficient part in dealing with the time-delay information of every delay subinterval for concerned fuzzy systems.

13.3 Fuzzy Controller Design

13.3.1 Stability of the Nonlinear Inverted Pendulum Systems

For the sake of notational simplification, partition the time-delay interval to two parts as $l = 2$. Define some relevant notations as follows:

$$\begin{cases} P(k) \triangleq \sum\limits_{i=1}^{r} h_i(\theta)P_i, & Q_m(k) \triangleq \sum\limits_{i=1}^{r} h_i(\theta)Q_{mi}, \\ R(k) \triangleq \sum\limits_{i=1}^{r} h_i(\theta)R_i, & Z_n(k) \triangleq \sum\limits_{i=1}^{r} h_i(\theta)Z_{ni}, \\ S_n(k) \triangleq \sum\limits_{i=1}^{r} h_i(\theta)S_{ni}, & \bar{S}_n(k) \triangleq \sum\limits_{i=1}^{r} h_i(\theta)\bar{S}_{ni}, \\ \bar{R}(k) \triangleq \sum\limits_{i=1}^{r} h_i(\theta)\bar{R}_i, & m = 0, 1, 2, 3,\ n = 1, 2. \end{cases}$$

Theorem 13.2 *Given some positive integers d_0, d_1, d_2, the system in (13.6) is asymptotically stable if there are matrices $P_i > 0$, $Q_{0i} > 0$, $Q_{1i} > 0$, $Q_{2i} > 0$, $Q_{3i} > 0$, $R_i > 0$, $\bar{R}_i > 0$, $S_{1i} > 0$, $S_{2i} > 0$, $\bar{S}_{1i} > 0$, $\bar{S}_{2i} > 0$, Z_{1i} and Z_{2i} such that for $i \in \mathbb{S}$,*

$$\Omega_1(k) < 0, \tag{13.7}$$

$$\Omega_2(k) < 0, \tag{13.8}$$

$$\begin{bmatrix} \bar{S}_1(k) & Z_1(k) \\ \star & \bar{S}_1(k) \end{bmatrix} \geqslant 0, \tag{13.9}$$

$$\begin{bmatrix} \bar{S}_2(k) & Z_2(k) \\ \star & \bar{S}_2(k) \end{bmatrix} \geqslant 0, \tag{13.10}$$

$$S_1(k) - \bar{S}_1(t) \geqslant 0, \tag{13.11}$$

$$S_2(k) - \bar{S}_2(t) \geqslant 0, \tag{13.12}$$

$$R(k) - \bar{R}(t) \geqslant 0, \tag{13.13}$$

where

$$\Omega_1(k) \triangleq \begin{bmatrix} \Gamma_{11}(k) & \Gamma_{12}(k) & 0 & 0 & \Gamma_{15}(k) \\ \star & \Gamma_{22}(k) & \Gamma_{23}(k) & 0 & \Gamma_{25}(k) \\ \star & \star & \Gamma_{33}(k) & 0 & \Gamma_{35}(k) \\ \star & \star & \star & \Gamma_{44}(k) & 0 \\ \star & \star & \star & \star & \Gamma_{55}(k) \end{bmatrix},$$

$$\Omega_2(k) \triangleq \begin{bmatrix} \Gamma_{11}(k) & \Gamma_{12}(k) & 0 & 0 & \Gamma_{15}(k) \\ \star & \Pi_{22}(k) & 0 & 0 & 0 \\ \star & \star & \Pi_{33}(k) & \Pi_{34}(k) & \Pi_{35}(k) \\ \star & \star & \star & \Pi_{44}(k) & \Pi_{45}(k) \\ \star & \star & \star & \star & \Pi_{55}(k) \end{bmatrix},$$

with

$$\begin{aligned}
\Gamma_{11}(k) &\triangleq A^T(k)P(k+1)A(k) - \bar{R}(k-d_0) + Q_0(k) + Q_1(k) \\
&\quad + Q_2(k) - P(k) + (1+\eta_1+\eta_2)Q_3(k) + \mathcal{A}(k)^T\mathcal{S}(k)\mathcal{A}(k), \\
\Gamma_{12}(k) &\triangleq \bar{R}(k-d_0), \; \Gamma_{25}(k) \triangleq \bar{S}_1(k) - Z_1(k), \\
\Gamma_{33}(k) &\triangleq -Q_1(k-d_1) - \bar{S}_1(k), \; \Gamma_{23}(k) \triangleq Z_1(k), \\
\Gamma_{22}(k) &\triangleq -Q_0(k-d_0) - \bar{R}(k-d_0) - \bar{S}_1(k), \\
\Gamma_{15}(k) &\triangleq A^T(k)P(k+1)A_d(k) + \mathcal{A}(k)^T\mathcal{S}(k)A_d(k), \\
\Gamma_{35}(k) &\triangleq \bar{S}_1(k) - Z_1^T(k), \; \Gamma_{44}(k) \triangleq -Q_2(k-d_2), \\
\Gamma_{55}(k) &\triangleq \mathcal{Q}(k) - 2\bar{S}_1(k) + Z_1(k) + Z_1^T(k), \\
\Pi_{22}(k) &\triangleq -Q_0(k-d_0) - \bar{R}(k-d_0), \\
\Pi_{33}(k) &\triangleq -Q_1(k-d_1) - \bar{S}_2(k), \; \mathcal{A}(k) \triangleq A(k) - I, \\
\Pi_{44}(k) &\triangleq -Q_2(k-d_2) - \bar{S}_2(k), \; \Pi_{34}(k) \triangleq Z_2(k), \\
\Pi_{35}(k) &\triangleq \bar{S}_2(k) - Z_2(k), \; \Pi_{45}(k) \triangleq \bar{S}_2(k) - Z_2^T(k), \\
\Pi_{55}(k) &\triangleq \mathcal{Q}(k) - 2\bar{S}_2(k) + Z_2(k) + Z_2^T(k), \\
\mathcal{S}(k) &\triangleq d_0^2R(k) + \eta_1^2S_1(k) + \eta_2^2S_2(k), \\
\mathcal{Q}(k) &\triangleq A_d^T(k)\left[P(k+1) + \mathcal{S}(k)\right]A_d(k) - Q_3(k-d(k)).
\end{aligned}$$

Proof The Lyapunov function is chosen as $V(k) \triangleq \sum_{i=1}^{5} V_i(k)$, where $V_i(k)$, $i = 1, 2, \ldots, 5$, with

$$
\begin{aligned}
V_1(k) &\triangleq x^T(k)P(k)x(k), \quad \delta(k) \triangleq x(k+1) - x(k), \\
V_2(k) &\triangleq \sum_{i=k-d_0}^{k-1} x^T(i)Q_0(i)x(i) + \sum_{i=k-d_1}^{k-1} x^T(i)Q_1(i)x(i) + \sum_{i=k-d_2}^{k-1} x^T(i)Q_2(i)x(i), \\
V_3(k) &\triangleq \sum_{i=k-d(k)}^{k-1} x^T(i)Q_3(i)x(i) + \sum_{j=-d_1+1}^{-d_0} \sum_{i=k+j}^{k-1} x^T(i)Q_3(i)x(i) \\
&\quad + \sum_{j=-d_2+1}^{-d_1} \sum_{i=k+j}^{k-1} x^T(i)Q_3(i)x(i), \\
V_4(k) &\triangleq d_0 \sum_{j=-d_0}^{-1} \sum_{i=k+j}^{k-1} \delta^T(i)R(i)\delta(i), \\
V_5(k) &\triangleq \sum_{l=1}^{2} \eta_l \sum_{j=-d_l}^{-d_{l-1}-1} \sum_{i=k+j}^{k-1} \delta^T(i)S_l(i)\delta(i), \\
\xi(k) &= \left[x^T(k) \; x^T(k-d_0) \; x^T(k-d_1) \; x^T(k-d_2) \; x^T(k-d(k)) \right]^T.
\end{aligned}
$$

Then we can obtain

$$
\begin{aligned}
\Delta V_1(k) &= x^T(k+1)P(k+1)x(k+1) - x^T(k)P(k)x(k), \\
\Delta V_2(k) &= x^T(k)\big[Q_0(k) + Q_1(k) + Q_2(k)\big]x(k) \\
&\quad - x^T(k-d_2)Q_2(k-d_2)x(k-d_2) \\
&\quad - x^T(k-d_0)Q_0(k-d_0)x(k-d_0) \\
&\quad - x^T(k-d_1)Q_1(k-d_1)x(k-d_1), \\
\Delta V_3(k) &\leqslant (1+\eta_1+\eta_2)x^T(k)Q_3(k)x(k) \\
&\quad - x^T\big(k-d(k)\big)Q_3\big(k-d(k)\big)x\big(k-d(k)\big), \\
\Delta V_4(k) &= d_0^2\delta^T(k)R(k)\delta(k) - d_0 \sum_{i=k-d_0}^{k-1} \delta^T(i)R(i)\delta(i), \\
\Delta V_5(k) &= \sum_{l=1}^{2} \left\{ \eta_l^2\delta^T(k)S_l(k)\delta(k) - \eta_l \sum_{i=k-d_l}^{k-d_{l-1}-1} \delta^T(i)S_l(i)\delta(i) \right\}.
\end{aligned}
\tag{13.14}
$$

Due to $R(k) - \bar{R}(t) \geqslant 0$, $\Delta V_4(k)$ can be further rewritten as

$$
\begin{aligned}
\Delta V_4(k) &\leqslant d_0^2\delta^T(k)R(k)\delta(k) - [x(k) - x(k-d_0)]^T \times \bar{R}(k-d_0) \\
&\quad \times [x(k) - x(k-d_0)].
\end{aligned}
\tag{13.15}
$$

Case 1. When $d_0 \leqslant d(k) < d_1$, considering (13.11), $\Delta V_5(k)$ can be given by

$$\Delta V_5(k) \leqslant \eta_1^2 \delta^T(k) S_1(k)\delta(k) + \eta_2^2 \delta^T(k) S_2(k)\delta(k) - \begin{bmatrix} x(k-d(k)) - x(k-d_0) \\ x(k-d_1) - x(k-d(k)) \end{bmatrix}^T \begin{bmatrix} \bar{S}_1(k) & Z_1(k) \\ \star & \bar{S}_1(k) \end{bmatrix} \times \begin{bmatrix} x(k-d(k)) - x(k-d_0) \\ x(k-d_1) - x(k-d(k)) \end{bmatrix}. \tag{13.16}$$

When $d(k) = d_0$, (13.16) still holds true because $x\big(k - d(k)\big) - x(k - d_0) = 0$. Combining (13.14), (13.15) and (13.16), for $d_0 \leqslant d(k) < d_1$, it follows that $\Delta V(k) = \xi^T(k)\Omega_1(k)\xi(k) < 0$.

Case 2. When $d_1 \leqslant d(k) \leqslant d_2$, due to $S_2(k) - \bar{S}_2(t) \geqslant 0$, $\Delta V_5(k)$ can be given by

$$\Delta V_5(k) \leqslant \eta_1^2 \delta^T(k) S_1(k)\delta(k) + \eta_2^2 \delta^T(k) S_2(k)\delta(k) - \begin{bmatrix} x(k-d_1) - x(k-d(k)) \\ x(k-d(k)) - x(k-d_2) \end{bmatrix}^T \begin{bmatrix} \bar{S}_2(k) & Z_2(k) \\ \star & \bar{S}_2(k) \end{bmatrix} \times \begin{bmatrix} x(k-d_1) - x(k-d(k)) \\ x(k-d(k)) - x(k-d_2) \end{bmatrix}. \tag{13.17}$$

When $d(k) = d_1$ or $d(k) = d_2$, the condition in (13.17) is still met because $x\big(k - d(k)\big) - x(k - d_1) = 0$ or $x\big(k - d(k)\big) - x(k - d_2) = 0$. On the basis of (13.14), (13.15) and (13.17), for $d_1 \leqslant d(k) \leqslant d_2$, we can get $\Delta V(k) < 0$. It can be easily concluded that the closed-loop system (13.6) satisfies the asymptotical stability. Then, the proof is accomplished.

Remark 13.3 The proposed stability conditions in Theorem 13.2 are in terms of parameter-dependent inequalities, which can be regarded as a non-convex feasibility issue. In order to resolve the problem, the mentioned inequalities in (13.7)–(13.13) are converted to finite constraints, which result in finding the feasible solution with a standard optimization toolbox. Then the stability conditions in the form of LMIs are presented as below.

Theorem 13.4 *Given some positive integers d_0, d_1, d_2, the system in (13.6) is asymptotically stable if there are matrices $P_i > 0$, $Q_{Ni} > 0$, $N = 0, 1, 2, 3$, $R_i > 0$, $\bar{R}_i > 0$, $S_{mi} > 0$, $\bar{S}_{mi} > 0$, $m = 1, 2$, Z_{1i} and Z_{2i}, such that for $i, j, t, s, u, o, l \in \mathbb{S}$,*

$$
\begin{aligned}
\Omega_{1iisltou} &< 0, \\
\Omega_{1ijsltou} + \Omega_{1jisltou} &< 0, \quad 1 \leqslant i < j \leqslant r, \\
\Omega_{2iisltou} &< 0, \\
\Omega_{2ijsltou} + \Omega_{2jisltou} &< 0, \quad 1 \leqslant i < j \leqslant r, \\
\begin{bmatrix} \bar{S}_{mi} & Z_{mi} \\ \star & \bar{S}_{mi} \end{bmatrix} &\geqslant 0, \\
S_{1i} - \bar{S}_{1j} &\geqslant 0, \\
S_{2i} - \bar{S}_{2j} &\geqslant 0, \\
R_i - \bar{R}_j &\geqslant 0,
\end{aligned}
$$

where

$$
\Omega_{1ijsltou} \triangleq \begin{bmatrix} \Gamma_{11ijlt} & \Gamma_{12t} & 0 & 0 & \Gamma_{15ijl} \\ \star & \Gamma_{22it} & \Gamma_{23i} & 0 & \Gamma_{25i} \\ \star & \star & \Gamma_{33is} & 0 & \Gamma_{35i} \\ \star & \star & \star & \Gamma_{44o} & 0 \\ \star & \star & \star & \star & \Gamma_{55ijul} \end{bmatrix},
$$

$$
\Omega_{2ijsltou} \triangleq \begin{bmatrix} \Gamma_{11ijlt} & \Gamma_{12t} & 0 & 0 & \Gamma_{15ijl} \\ \star & \Pi_{22t} & 0 & 0 & 0 \\ \star & \star & \Pi_{33is} & \Pi_{34i} & \Pi_{35i} \\ \star & \star & \star & \Pi_{44io} & \Pi_{45i} \\ \star & \star & \star & \star & \Pi_{55ijul} \end{bmatrix},
$$

with

$$
\begin{aligned}
\Gamma_{11ijlt} &\triangleq -P_i + A_j^T P_l A_j + \mathcal{A}_j^T \mathcal{S}_i \mathcal{A}_j + Q_{0i} + Q_{1i} + Q_{2i} \\
&\quad +(1 + \eta_1 + \eta_2) Q_{3i} - \bar{R}_t, \\
\mathcal{S}_i &\triangleq d_0^2 R_i + \eta_1^2 S_{1i} + \eta_2^2 S_{2i}, \ \mathcal{A}_j \triangleq A_j - I, \\
\Gamma_{25i} &\triangleq \bar{S}_{1i} - Z_{1i}, \ \Gamma_{15ijl} \triangleq \mathcal{A}_j^T \mathcal{S}_i A_{dj} + A_j^T P_l A_{dj}, \\
\Gamma_{22it} &\triangleq -Q_{0t} - \bar{R}_t - \bar{S}_{1i}, \ \Gamma_{35i} \triangleq S_{1i} - Z_{1i}^T, \\
\Gamma_{23i} &\triangleq Z_{1i}, \ \Gamma_{12t} \triangleq \bar{R}_t, \ \Gamma_{33is} \triangleq -Q_{1s} - \bar{S}_{1i}, \\
\Gamma_{44o} &\triangleq -Q_{2o}, \ \Gamma_{55ijul} \triangleq \mathcal{Q}_{ijul} + Z_{1i} + Z_{1i}^T - 2\bar{S}_{1i}, \\
\Pi_{22t} &\triangleq -Q_{0t} - \bar{R}_t, \ \Pi_{33is} \triangleq -Q_{1s} - \bar{S}_{2i}, \\
\Pi_{35i} &\triangleq \bar{S}_{2i} - Z_{2i}, \ \Pi_{44io} \triangleq -Q_{2o} - \bar{S}_{2i}, \\
\Pi_{34i} &\triangleq Z_{2i}, \ \Pi_{55ijul} \triangleq \mathcal{Q}_{ijul} - 2\bar{S}_{2i} + Z_{2i} + Z_{2i}^T, \\
\Pi_{45i} &\triangleq \bar{S}_{2i} - Z_{2i}^T, \ \mathcal{Q}_{ijul} \triangleq A_{dj}^T (P_l + \mathcal{S}_i) A_{dj} - Q_{3u}.
\end{aligned}
$$

These results can be obtained with the similar methods using in Theorem 13.2. For the sake of reducing the system conservatism, partitioning the time-delay to l

sub-intervals, then we construct the Lyapunov function as

$$
\begin{aligned}
V(k) &\triangleq x^T(k)P(k)x(k) + \sum_{j=0}^{l}\sum_{i=k-d_j}^{k-1} x^T(i)Q_j(i)x(i) \\
&+ \sum_{i=k-d(k)}^{k-1} x^T(i)Q_3(i)x(i) + \sum_{q=1}^{l}\sum_{j=-d_q+1}^{-d_{q-1}}\sum_{i=k+j}^{k-1} x^T(i)Q_3(i)x(i) \\
&+ d_0 \sum_{j=-d_0}^{-1}\sum_{i=k+j}^{k-1} \delta^T(i)R(i)\delta(i) + \sum_{q=1}^{l}\eta_q \sum_{j=-d_q}^{-d_{q-1}-1}\sum_{i=k+j}^{k-1} \delta^T(i)S_q(i)\delta(i).
\end{aligned}
$$

Using for reference the proof in Theorems 13.2 and 13.4, we can obtain the following delay-based stability results.

Theorem 13.5 *Given the positive integers $d_0, \ldots, d_{l-1}, d_l$ and $l \geqslant 2$, the system in (13.6) is asymptotically stable if there are matrices $P_i > 0$, $Q_{Ni} > 0$, $N = 0, 1, 2, \ldots, l+1$, $R_i > 0$, $\bar{R}_i > 0$, $S_{mi} > 0$, $\bar{S}_{mi} > 0$, and Z_{mi}, $m = 1, 2, \ldots, l$, such that for $i, j, t, s, u, o, f \in \mathbb{S}$,*

$$
\begin{aligned}
\Omega_{miisftou} &< 0, \\
\Omega_{mijsftou} + \Omega_{mjisftou} &< 0, \quad 1 \leqslant i < j \leqslant r, \\
\begin{bmatrix} \bar{S}_{mi} & Z_{mi} \\ \star & \bar{S}_{mi} \end{bmatrix} &\geqslant 0, \\
S_{mi} - \bar{S}_{mj} &\geqslant 0, \quad m = 1, 2, \ldots, l, \\
R_i - \bar{R}_j &\geqslant 0,
\end{aligned}
$$

where

$$
\Omega_{mijsftou} \triangleq \begin{bmatrix}
\Gamma_{11ijft} & \Gamma_{12t} & 0 & 0 & \ldots & \Gamma_{1(l+3)ijf} \\
\star & \Gamma_{22it} & \Gamma_{23i} & 0 & \ldots & \Gamma_{2(l+3)i} \\
\star & \star & \Gamma_{33io} & \Gamma_{34i} & \ldots & \Gamma_{3(l+3)i} \\
\star & \star & \star & \Gamma_{44io} & \ldots & \Gamma_{4(l+3)i} \\
\star & \star & \star & \star & \ddots & \vdots \\
\star & \star & \star & \star & \star & \Gamma_{(l+3)(l+3)ijuf}
\end{bmatrix},
$$

with

$$\Gamma_{(m+2)(l+3)i} \triangleq \bar{S}_{mi} - Z_{mi}^T, \quad \Gamma_{12t} \triangleq \bar{R}_t,$$

$$\Gamma_{(m+1)(m+2)i} \triangleq Z_{mi}, \quad \bar{S}_i \triangleq d_0^2 R_i + \sum_{N=1}^{l} \eta_N^2 S_{Ni},$$

$$\begin{aligned}
\Gamma_{(m+2)(m+2)io} &\triangleq -Q_{mo} - \bar{S}_{mi} \ (m \geqslant 2), \\
\Gamma_{(m+1)(m+1)io} &\triangleq -Q_{(m-1)o} - \bar{S}_{mi} \ (m \geqslant 2), \\
\Gamma_{1(l+3)ijf} &\triangleq A_j^T P_f A_{dj} + \mathcal{A}_j^T \bar{S}_i A_{dj}, \\
\Gamma_{(l+3)(l+3)ijuf} &\triangleq A_{dj}^T \left(P_f + \bar{S}_i\right) A_{dj} - 2\bar{S}_{mi} + Z_{mi} + Z_{mi}^T - Q_{(l+1)u}, \\
\Gamma_{(m+1)(l+3)i} &\triangleq \bar{S}_{mi} - Z_{mi}, \\
\Gamma_{11ijft} &\triangleq A_j^T P_f A_j - P_i - \bar{R}_t + Q_{(l+1)i} + \mathcal{A}_j^T \bar{S}_i \mathcal{A}_j \\
&\quad + \sum_{N=0}^{l} Q_{Ni} + \sum_{N=1}^{l} \eta_N Q_{(l+1)i}, \\
\Gamma_{22it} &\triangleq \begin{cases} -Q_t - \bar{R}_t - \bar{S}_{1i}, & \text{if } m = 1, \\ -Q_t - \bar{R}_t, & \text{if } m \geqslant 2, \end{cases} \\
\Gamma_{44io} &\triangleq \begin{cases} -Q_{2o}, & \text{if } m = 1, \\ \Gamma_{(m+2)(m+2)io}, & \text{if } m \geqslant 2. \end{cases}
\end{aligned}$$

Example 13.6 (Conservativeness analysis) The relevant system matrices are selected as follows:

$$A_1 = \begin{bmatrix} -0.291 & 1 \\ 0 & 0.95 \end{bmatrix}, \quad A_{d1} = \begin{bmatrix} 0.012 & 0.014 \\ 0 & 0.015 \end{bmatrix},$$

$$A_2 = \begin{bmatrix} -0.1 & 0 \\ 1 & -0.2 \end{bmatrix}, \quad A_{d2} = \begin{bmatrix} 0.01 & 0 \\ 0.01 & 0.015 \end{bmatrix}.$$

Here, we aim to obtain the maximum allowable values of delay bound based on Theorem 13.5, and the corresponding results are displayed in Table 13.2. It can be observed from Table 13.2 that the maximum allowable values d_M can be 34 and 40 when $l = 1$ and $l = 2$, separately. In addition, as for the same d_m, the maximum allowable values of d_M are much bigger than the results in [82, 249, 298]. When l increases, the allowable upper bound of d_M improves, that is to say, the conservativeness reduces when l increases. Table 13.3 presents the comparison with the number of decision variables, from which we can observe that the approach developed in this chapter has less free variables than these in [249, 298]. Such as the variable number in [298] is $\frac{nr}{2}\left[n(6m^2 + 30m + 61) + 2m + 7\right]$, while it is $nr(2nl + 2n + l + 2)$ in Theorem 13.5 of our obtained results.

Remark 13.7 From the obtained stability conditions, we can see the delay information can be properly handled with a reciprocally convex method. Therefore, the minimum value of time-delay is not demanded to be zero or one in our proposed approach, which is more applicable in many practical physical models.

Table 13.2 Maximum allowable values d_M for different values d_m

Different methods	$d_m = 3$	$d_m = 5$	$d_m = 10$	$d_m = 12$
Theorem 13.2 of [82]	14	16	20	21
Theorem 13.5 of [298]	24	26	31	33
Corollary 1 of [249]	26	28	33	35
Theorem 13.5, $l = 1$	25	27	32	34
Theorem 13.5, $l = 2$	31	33	38	40

Table 13.3 Allowable upper bound d_M when $d_m = 12$

Different methods	Upper bound	Number of variables
Theorem 13.2 of [82]	21	61
Theorem 13.5 of [298]	33	804
Corollary 1 of [249]	35	56
Theorem 13.5, $l = 1$	34	44
Theorem 13.5, $l = 2$	40	64

Remark 13.8 To verify the superiority of the developed scheme, the time-delay is divided into two parts, which means $l = 2$. Then this case is extended to arbitrary individual parts, which can illustrate clearly the reduction in conservatism. Example 13.12 indicates sufficient conditions in Theorem 13.5 are much less conservative than the results in [82, 249, 298]. Furthermore, it can be concluded that when the partition delay sub-intervals are gradually thinner, the reduced conservatism of proposed results is more obvious.

13.3.2 Fuzzy Control of Inverted Pendulum Systems

In this part, we are going to propose an efficient fuzzy controller as (13.3) such that the overall closed-loop system in (13.5) is stable.

Theorem 13.9 *For the given scalars $\lambda_1, \lambda_2, \ldots, \lambda_{l+4}, d_0, d_1, \ldots, d_l, l \geqslant 2$, the controller in (13.3) is designed to ensure the system in (13.5) is asymptotically stable if there are matrices $\bar{R} > 0$, $\bar{P}_i > 0$, $\bar{Q}_N > 0$, $N = 0, 1, \ldots, l$, $\bar{S}_m > 0$, $\bar{Z}_m$, $m = 1, 2, \ldots, l$, X and G_i such that for $i, j \in \mathbb{S}$,*

$$\Upsilon_{iis} + \Upsilon_{miis} < 0, \tag{13.18}$$

$$\Upsilon_{ijs} + \Upsilon_{mijs} + \Upsilon_{jis} + \Upsilon_{mjis} < 0, \ 1 \leqslant i < j \leqslant r, \tag{13.19}$$

$$\begin{bmatrix} \bar{S}_m & \bar{Z}_m \\ \star & \bar{S}_m \end{bmatrix} \geqslant 0, \ m = 1, 2, \ldots, l, \tag{13.20}$$

where

$$\begin{aligned}
\Upsilon_{ijs} \triangleq\ & \bar{M}\bar{W}_{Aij} + \bar{W}_{Aij}^T\bar{M}^T + W_{l+4}^T\bar{P}_s W_{l+4} - W_1^T\bar{P}_i W_1 \\
& + \sum_{N=0}^{l}\left(W_{N+1}^T\bar{Q}_N W_{N+1} - W_{N+2}^T\bar{Q}_N W_{N+2}\right) \\
& + \sum_{N=1}^{l}\eta_N^2\left(W_{l+4} - W_1\right)^T\bar{S}_N\left(W_{l+4} - W_1\right) \\
& + d_0^2(W_{l+4} - W_1)^T\bar{R}(W_{l+4} - W_1) \\
& - \left(W_1 - W_2\right)^T\bar{R}\left(W_1 - W_2\right), \\
\Upsilon_{mijs} \triangleq\ & \begin{bmatrix} W_{m+1} - W_{l+3} \\ W_{m+2} - W_{l+3} \end{bmatrix}^T \begin{bmatrix} \bar{S}_m & \bar{Z}_m \\ \star & \bar{S}_m \end{bmatrix} \begin{bmatrix} W_{l+3} - W_{m+1} \\ W_{l+3} - W_{m+2} \end{bmatrix} \\
& - \sum_{N=1, N\neq m}^{l}\left(W_{N+1} - W_{N+2}\right)^T\bar{S}_N\left(W_{N+1} - W_{N+2}\right), \\
\bar{W}_{Aij} \triangleq\ & \left[\, A_iX + B_iG_j \ 0_{n,(l+1)n} \ A_{di}X \ -X \,\right], \\
W_\omega \triangleq\ & \left[\, 0_{n,(\omega-1)n} \ I_n \ 0_{n,(l+4-\omega)n} \,\right], (\omega = 1, \ldots, l+4), \\
\bar{M} \triangleq\ & \left[\, \lambda_1 I_n \ \ldots \ \lambda_{l+4} I_n \,\right]^T.
\end{aligned}$$

Furthermore, the controller gains K_i in (13.5) are described as

$$K_i = G_i X^{-1}, \quad i \in \mathbb{S}. \tag{13.21}$$

Proof Denote $Z = X^{-T}$ and $E \triangleq \mathrm{diag}\{Z, Z, \ldots, Z\}$. The Lyapunov function is given by

$$\begin{aligned}
V(k) \triangleq\ & x^T(k)P(k)x(k) + \sum_{i=k-d_0}^{k-1} x^T(i)Q_0x(i) + \sum_{j=1}^{l}\sum_{i=k-d_j}^{k-d_{j-1}-1} x^T(i)Q_jx(i) \\
& + d_0\sum_{j=-d_0}^{-1}\sum_{i=k+j}^{k-1}\xi(i+1)^T R\xi(i+1) \\
& + \sum_{q=1}^{l}\eta_q\sum_{j=-d_q}^{-d_{q-1}-1}\sum_{i=k+j}^{k-1}\xi(i+1)^T S_q\xi(i+1),
\end{aligned}$$

where $\xi(i+1) \triangleq x(i+1) - x(i)$. It is not difficult to get the following results in similar ways as these in Theorem 13.5:

$$\Omega_{iits} + \Omega_{miits} < 0, \tag{13.22}$$

$$\Omega_{ijts} + \Omega_{mijts} + \Omega_{jits} + \Omega_{mjits} < 0, \ 1 \leqslant i < j \leqslant r, \tag{13.23}$$

$$\begin{bmatrix} S_m & Z_m \\ \star & S_m \end{bmatrix} \geqslant 0, \ m = 1, \ldots, l, \tag{13.24}$$

where

$$\begin{aligned}
W_{Ajt} &\triangleq \begin{bmatrix} A_j + B_j K_t & 0_{n,(l+1)n} & A_{dj} & -I_n \end{bmatrix}, \\
\Omega_{ijts} &\triangleq M W_{Ajt} + W_{Ajt}^T M^T - W_1^T P_i W_1 + W_{l+4}^T P_s W_{l+4} \\
&\quad + \sum_{N=0}^{l} \left(W_{N+1}^T Q_N W_{N+1} - W_{N+2}^T Q_N W_{N+2} \right) \\
&\quad + \sum_{N=1}^{l} \eta_N^2 \left(W_{l+4} - W_1 \right)^T S_N \left(W_{l+4} - W_1 \right) \\
&\quad + d_0^2 \left(W_{l+4} - W_1 \right)^T R \left(W_{l+4} - W_1 \right) \\
&\quad - \left(W_1 - W_2 \right)^T R \left(W_1 - W_2 \right), \\
\Omega_{mijts} &\triangleq \begin{bmatrix} W_{m+1} - W_{l+3} \\ W_{m+2} - W_{l+3} \end{bmatrix}^T \begin{bmatrix} S_m & Z_m \\ \star & S_m \end{bmatrix} \begin{bmatrix} W_{l+3} - W_{m+1} \\ W_{l+3} - W_{m+2} \end{bmatrix} \\
&\quad - \sum_{N=1, N \neq m}^{l} \left(W_{N+1} - W_{N+2} \right)^T S_N \left(W_{N+1} - W_{N+2} \right).
\end{aligned}$$

Pre- and post-multiplying (13.18), (13.19) with E, E^T separately, then we can obtain

$$\begin{aligned}
E \left(\Upsilon_{iis} + \Upsilon_{miis} \right) E^T &< 0, \\
E \left(\Upsilon_{ijs} + \Upsilon_{mijs} \right) E^T + E \left(\Upsilon_{jis} + \Upsilon_{mjis} \right) E^T &< 0, \\
1 \leqslant i < j &\leqslant r,
\end{aligned}$$

where

$$
\begin{aligned}
E\Upsilon_{mijs}E^T \triangleq & -\begin{bmatrix} W_{l+3}-W_{m+1} \\ W_{l+3}-W_{m+2} \end{bmatrix}^T \begin{bmatrix} Z\bar{S}_m Z^T & Z\bar{Z}_m Z^T \\ \star & Z\bar{S}_m Z^T \end{bmatrix} \begin{bmatrix} W_{l+3}-W_{m+1} \\ W_{l+3}-W_{m+2} \end{bmatrix} \\
& - \sum_{N=1, N\neq m}^{l} (W_{N+1} - W_{N+2})^T Z\bar{S}_N Z^T (W_{N+1} - W_{N+2}), \\
E\Upsilon_{ijs}E^T \triangleq & \, E\bar{M}\bar{W}_{Aij}E^T + \left(E\bar{M}\bar{W}_{Aij}E^T\right)^T + W_{l+4}^T Z\bar{P}_s Z^T W_{l+4} \\
& + \sum_{N=1}^{l} \eta_N^2 (W_{l+4} - W_1)^T Z\bar{S}_N Z^T (W_{l+4} - W_1) \\
& + \sum_{N=0}^{l} \left(W_{N+1}^T Z\bar{Q}_N Z^T W_{N+1} - W_{N+2}^T Z\bar{Q}_N Z^T W_{N+2}\right) \\
& - W_1^T Z\bar{P}_i Z^T W_1 - (W_1 - W_2)^T Z\bar{R}Z^T (W_1 - W_2) \\
& + d_0^2 (W_{l+4} - W_1)^T Z\bar{R}Z^T (W_{l+4} - W_1).
\end{aligned}
$$

Define some relevant matrices as below:

$$
\begin{aligned}
S_N &\triangleq Z\bar{S}_N Z^T, \quad R \triangleq Z\bar{R}Z^T, \quad Q_N \triangleq Z\bar{Q}_N Z^T, \\
P_i &\triangleq Z\bar{P}_i Z^T, \quad Z_m \triangleq Z\bar{Z}_m Z^T, \quad M \triangleq E\bar{M}.
\end{aligned}
$$

And it yields

$$
\begin{aligned}
E\bar{M}\bar{W}_{Aij}E^T &= \left[\lambda_1 Z^T \; \lambda_2 Z^T \; \ldots \; \lambda_{l+4} Z^T\right]^T \\
& \quad \times \left[A_i X + B_i G_j \; 0_{n,(l+1)n} \; A_{di} X \; -X\right] E^T \\
&= M \left[A_j + B_j K_t \; 0_{n,(l+1)n} \; A_{dj} \; -I_n\right].
\end{aligned}
$$

Then, by the means of pre- and post-multiplying with E and E^T, it can be seen that (13.18) and (13.19) hold if the conditions in (13.22) and (13.23) are satisfied. In a similar way, we can get (13.24) via pre- and post-multiplying (13.20) with E and E^T. Therefore, the proof is completed. ■

Remark 13.10 Employing a newly introduced matrix E, which implies our proposed sufficient conditions do not need involving the fuzzy-dependent Lyapunov matrix in system matrices. Actually, there is no requirement that the matrix used must be positive and symmetrical. Furthermore, it is easily observed that the designed fuzzy controller gains can be obtained smoothly by the standard numerical software.

13.3.3 Event-Triggered Fuzzy Control

In this part, a useful event-triggered control strategy is given and the inverted pendulum model is applied to demonstrate the validity of our developed method. First,

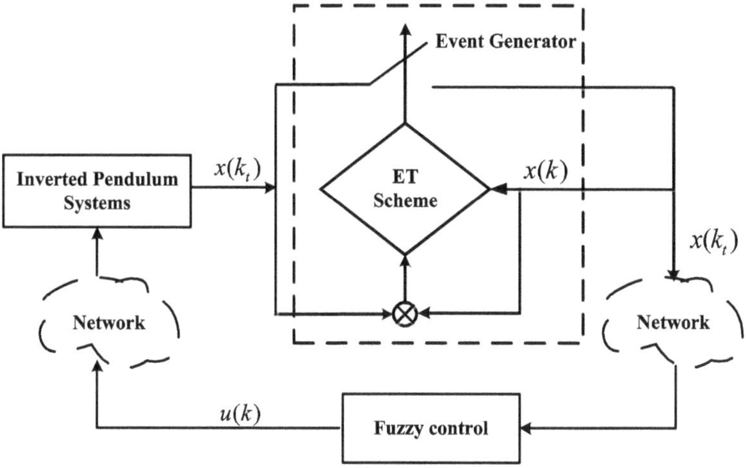

Fig. 13.3 Structure of the event-triggered control system

consider the following discrete-time fuzzy model:

$$x(k+1) = A(k)x(k) + B(k)u(k). \tag{13.25}$$

The relevant delay analysis approach in ([253]) is employed to design the event-based controller for considered inverted pendulum system. Therefore, as for $k \in [k_t, k_{t+1} - 1]$, whether the present sampling data $x(k_t)$ at the triggering moment k_t can be transferred to the event-based controller, which depends on the following condition:

$$[x(k) - x(k_t)]^T \, \Omega \, [x(k) - x(k_t)] \leqslant \varepsilon_i x^T(k_t)\Omega x(k_t), \tag{13.26}$$

where $\Omega > 0$ is to be decided subsequently, and ε_i is known parameter. The structure of introduced event-based control system is displayed as Fig. 13.3.

Provided that the network-induced delay τ_{k_t} satisfies $\tau_m \leqslant \tau_{k_t} \leqslant \tau_M$, where τ_m and τ_M denote the minimum and maximum bounds of time-varying delays, separately. Next, take the following two cases into account.

Case 1: When $k_t + 1 + \tau_M \geqslant k_{t+1} + \tau_{k_{t+1}} - 1$, introduce a function $h(k)$ as

$$h(k) = k - k_t, \; k \in \left[k_t + \tau_{k_t}, k_{t+1} + \tau_{k_{t+1}} - 1\right],$$

which yields

$$\tau_m \leq \tau_{k_t} \leqslant h(k) \leqslant k_{t+1} - k_t + \tau_{k_{t+1}} - 1 \leqslant 1 + \tau_M.$$

Case 2: When $k_t + 1 + \tau_M < k_{t+1} + \tau_{k_{t+1}} - 1$, two intervals are considered as follows:

$$\left[k_t + \tau_{k_t}, k_t + \tau_M\right], \left[k_t + \tau_M + n, k_t + \tau_M + n + 1\right],$$

where $n \in Z_+, n \geqslant 1$, and $\tau_{k_t} \leqslant \tau_M$, it is easy to see there exists d satisfying

$$k_t + d + \tau_M < k_{t+1} + \tau_{k_{t+1}} - 1 \leqslant k_t + d + \tau_M + 1.$$

In addition, $k_t + n = k_{t+1}$, $x(k_t)$ and $x(k_t + n)$ with $n = 1, 2, \ldots, d$, which satisfies the condition in (13.26). Denote

$$\begin{cases} I_0 = \left[k_t + \tau_{k_t}, k_t + \tau_M + 1\right), \\ I_n = \left[k_t + \tau_M + n, k_t + \tau_M + n + 1\right), \\ I_d = \left[k_t + \tau_M + d, k_{t+1} + \tau_{k_{t+1}} - 1\right], \end{cases}$$

where $n = 1, 2, \ldots, d - 1$, then it follows that

$$\left[k_t + \tau_{k_t}, k_{t+1} + \tau_{k_{t+1}} - 1\right] = \bigcup_{i=0}^{d} I_i.$$

$h(k)$ is given by

$$h(k) = \begin{cases} k - k_t, & k \in I_0, \\ k - k_t - n, & k \in I_n, \quad n = 1, 2, \ldots, d - 1, \\ k - k_t - d, & k \in I_d, \end{cases}$$

we can obtain

$$\begin{cases} \tau_{k_t} \leqslant h(k) \leq 1 + \tau_M = h_2, & k \in I_0, \\ \tau_{k_t} \leqslant h(k) \leq h_2, & k \in I_n, \\ \tau_{k_t} \leqslant h(k) \leq h_2, & k \in I_d. \end{cases} \tag{13.27}$$

Thus, it yields $0 \leqslant \tau_{k_t} \leqslant h(k) \leqslant h_2$.

In Case 1, for $k \in \left[k_t + \tau_{k_t}, k_{t+1} + \tau_{k_{t+1}} - 1\right]$, denote $e_i(k) = 0$.

In Case 2, denote

$$e_i(k) = \begin{cases} 0, & k \in I_0, \\ x(k_t) - x(k_t + n), & k \in I_n, \\ x(k_t) - x(k_t + d), & k \in I_d. \end{cases}$$

On the basis of event-triggering strategy in (13.26), for $k \in [k_t + \tau_{k_t}, k_{t+1} + \tau_{k_{t+1}} - 1]$, we have

$$e_i^T(k)\Omega e_i(k) \leqslant \varepsilon_i x^T\big(k - h(k)\big)\Omega x\big(k - h(k)\big). \tag{13.28}$$

Considering $h(k)$ and $e_i(k)$, a compact closed-loop system is described by

$$x(k+1) = A_i x(k) + B_i K_j x(k - h(k)) + B_i K_j e_i(k). \tag{13.29}$$

On account of Theorem 13.9, the following results are presented for designing the event-driven fuzzy controller.

Theorem 13.11 *Given scalars ε_i, $\lambda_1, \lambda_2, \ldots, \lambda_{l+5}$, $\tau_0, \tau_1, \ldots, \tau_l$, $l \geqslant 2$, the fuzzy controller as (13.3) is designed to ensure the system (13.29) is asymptotically stable if there are matrices $\bar{\Omega} > 0$, $\bar{R} > 0$, $\bar{P}_i > 0$, $\bar{Q}_N > 0$, $N = 0, 1, \ldots, l$, $\bar{S}_m > 0$, $\bar{Z}_m$, $m = 1, 2, \ldots, l$, X and G_i such that for $i, j \in \mathbb{S}$,*

$$\bar{\Upsilon}_{iis} + \bar{\Upsilon}_{miis} < 0, \tag{13.30}$$

$$\bar{\Upsilon}_{ijs} + \bar{\Upsilon}_{mijs} + \bar{\Upsilon}_{jis} + \bar{\Upsilon}_{mjis} < 0, \ 1 \leqslant i < j \leqslant r, \tag{13.31}$$

$$\begin{bmatrix} \bar{S}_m & \bar{Z}_m \\ \star & \bar{S}_m \end{bmatrix} \geqslant 0, \ m = 1, 2, \ldots, l, \tag{13.32}$$

where

$$\begin{aligned}
\bar{\Upsilon}_{ijs} \triangleq\ & \bar{M}\bar{W}_{Aij} + \bar{W}_{Aij}^T \bar{M}^T + W_{l+5}^T \bar{P}_s W_{l+5} - W_1^T \bar{P}_i W_1 \\
& + \sum_{N=0}^{l} \left(W_{N+1}^T \bar{Q}_N W_{N+1} - W_{N+2}^T \bar{Q}_N W_{N+2} \right) \\
& + \sum_{N=1}^{l} \eta_N^2 \left(W_{l+5} - W_1 \right)^T \bar{S}_N \left(W_{l+5} - W_1 \right) \\
& + \tau_0^2 (W_{l+5} - W_1)^T \bar{R} (W_{l+5} - W_1) \\
& - (W_1 - W_2)^T \bar{R} (W_1 - W_2) \\
& + \varepsilon_i W_{l+3}^T \bar{\Omega} W_{l+3} - W_{l+4}^T \bar{\Omega} W_{l+4}, \\
\bar{\Upsilon}_{mijs} \triangleq\ & \begin{bmatrix} W_{m+1} - W_{l+3} \\ W_{m+2} - W_{l+3} \end{bmatrix}^T \begin{bmatrix} \bar{S}_m & \bar{Z}_m \\ \star & \bar{S}_m \end{bmatrix} \begin{bmatrix} W_{l+3} - W_{m+1} \\ W_{l+3} - W_{m+2} \end{bmatrix} \\
& - \sum_{N=1, N \neq m}^{l} (W_{N+1} - W_{N+2})^T \bar{S}_N (W_{N+1} - W_{N+2}), \\
\bar{W}_{Aij} \triangleq\ & \begin{bmatrix} A_i X & 0_{n,(l+1)n} & B_i G_j & B_i G_j & -X \end{bmatrix}, \\
\bar{M} \triangleq\ & \begin{bmatrix} \lambda_1 I_n & \lambda_2 I_n & \ldots & \lambda_{l+5} I_n \end{bmatrix}^T, \\
W_\omega \triangleq\ & \begin{bmatrix} 0_{n,(\omega-1)n} & I_n & 0_{n,(l+5-\omega)n} \end{bmatrix}, \ (\omega = 1, \ldots, l+5).
\end{aligned}$$

Furthermore, the parameters of designed controller K_i can be formulated as

$$K_i = G_i X^{-1}, \quad i \in \mathbb{S}.$$

Proof The aforementioned results can be obtained referring to the similar methods as the proof in Theorem 13.9. ■

13.4 Simulation Results

To confirm our previous theoretical analysis, the fuzzy control and event-driven control methods are applied to stabilize the inverted pendulum model, separately. And the corresponding simulation results are shown as below.

Example 13.12 (Fuzzy control of the inverted pendulum model)
It follows from Sect. 13.2.1 that

$$\alpha = \cos\gamma, \quad \beta = (\sin\gamma)/\gamma,$$

where T represents the sampling time. With

$$\left|x_2(k)\right| < |\gamma|,$$

the fuzzy basis functions are given by

$$h_1\big(x_2(k)\big) = 1 - \frac{\left|x_2(k)\right|}{|\gamma|}, \quad h_2\big(x_2(k)\big) = \frac{\left|x_2(k)\right|}{|\gamma|}.$$

On account of the parameters given in Table 13.3, set

$$T = 0.3s, \quad \gamma = 0.52\text{rad (that is, } \gamma = 30^\circ),$$

then

$$A_1 = \begin{bmatrix} 0 & 0 & 0.3 & 0 \\ 0 & 0 & 0 & 0.3 \\ 0 & -0.1088 & -1.3019 & 0 \\ 0 & 9.3809 & 4.0058 & 0 \end{bmatrix}, \quad B_1 = \begin{bmatrix} 0 \\ 0 \\ 0.2177 \\ -0.6699 \end{bmatrix},$$

$$A_{d1} = \begin{bmatrix} 0 & 0 & 0 & 0 \\ 0 & 0 & 0 & 0 \\ 0 & 0 & -0.1524 & 0 \\ 0 & 0 & 0.4689 & 0 \end{bmatrix}, \quad A_{d2} = \begin{bmatrix} 0 & 0 & 0 & 0 \\ 0 & 0 & 0 & 0 \\ 0 & 0 & -0.1524 & 0 \\ 0 & 0 & 0.5414 & 0 \end{bmatrix},$$

$$A_2 = \begin{bmatrix} 0 & 0 & 0.3 & 0 \\ 0 & 0 & 0 & 0.3 \\ 0 & -0.12 & -1.3019 & 0 \\ 0 & 11.9442 & 4.6255 & 0 \end{bmatrix}, \quad B_2 = \begin{bmatrix} 0 \\ 0 \\ 0.2177 \\ -0.7735 \end{bmatrix},$$

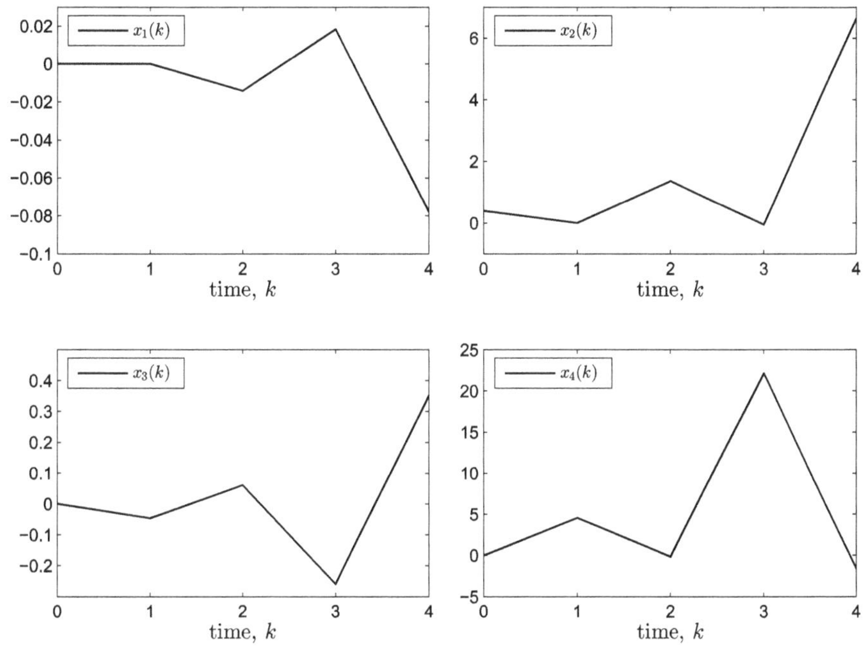

Fig. 13.4 States of the inverted pendulum system without the fuzzy controller

and $h_1(x_2(k)) = 1 - 1.9|x_2(k)|, \quad h_2(x_2(k)) = 1.9|x_2(k)|$.

The initial condition is set to $\varphi(k) = \begin{bmatrix} 0 & 0.4 & 0 & 0 \end{bmatrix}^T$, and the partition parts $l = 2$. The simulation results are displayed in Figs. 13.4, 13.5 and 13.6. From Fig. 13.4, we can see the original system is unstable based on the results in the Theorem 13.5. Solving the conditions in Theorem 13.9, we can get

$$\begin{aligned} G_1 &= \begin{bmatrix} 0.0671 \; -0.2023 \; -0.9314 \; 0.0893 \end{bmatrix}, \\ G_2 &= \begin{bmatrix} 0.0613 \; -0.2251 \; -0.9318 \; 0.0819 \end{bmatrix}, \\ K_1 &= \begin{bmatrix} 0.0320 \; 12.8026 \; 5.9830 \; -0.3147 \end{bmatrix}, \\ K_2 &= \begin{bmatrix} 0.0267 \; 14.3016 \; 5.9859 \; -0.2817 \end{bmatrix}. \end{aligned}$$

It can be seen from Fig. 13.5 that the states of closed-loop system are converged to zero via the developed state-feedback fuzzy controller. Figure 13.6 depicts the control results of original nonlinear system. It is easily indicated our designed fuzzy controller stabilizes the inverted pendulum model well.

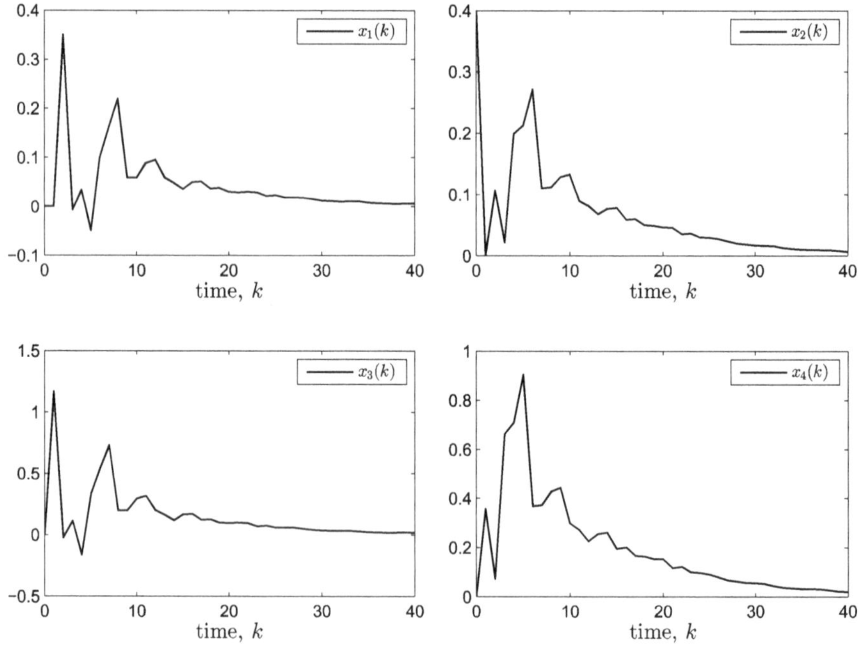

Fig. 13.5 Fuzzy control of the controlled model

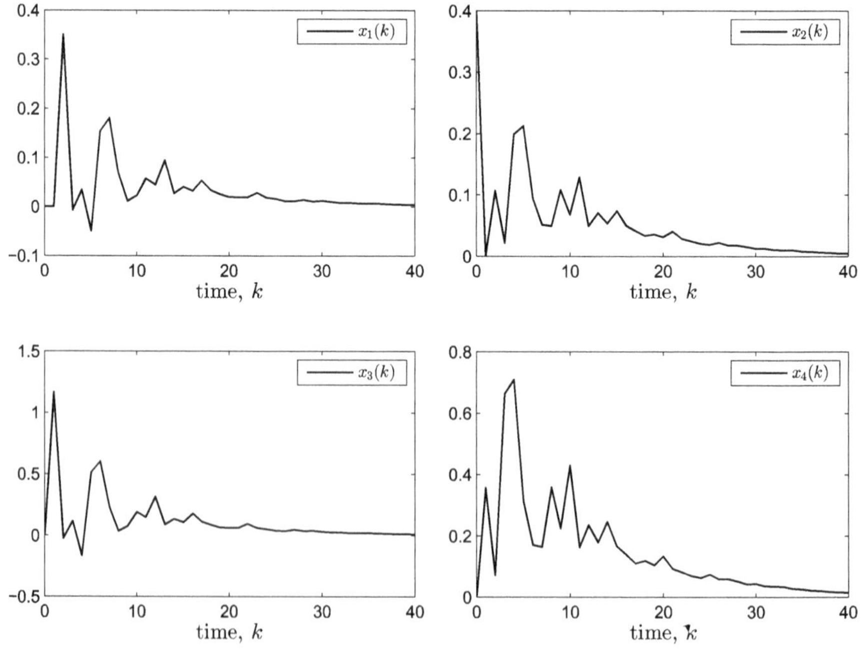

Fig. 13.6 States of the closed-loop inverted pendulum system with fuzzy control

Table 13.4 Relationship of the triggered parameter, trigger times and transmission rates

Triggered parameter	Trigger times	Transmission rates (%)
0	40	100
0.01	28	70
0.1	26	65
0.3	19	47.5

Example 13.13 (Event-triggered fuzzy control of the inverted pendulum model)
Utilizing the same parameters of the inverted pendulum model in Example 13.12, $\varphi(k) = [0 \quad 0.4 \quad 0 \quad 0]^T$, the time-varying delay $d(k)$ stochastically changes from $d_m = 1$ to $d_M = 3$, and $l = 2$. In consideration of the event-triggered scheme as (13.26), when ε_i takes different values, the relationship between the trigger parameter, triggered instants and transmission rates can be obtained, which is exhibited in Table 13.4. Apparently, it can be concluded from Table 13.4 that the bigger the trigger parameter is, the longer the sample interval holds.

Resolving the LMIs in Theorem 13.11, we can obtain the desired controller parameters as below:

$$\begin{aligned} G_1 &= \begin{bmatrix} 0.0130 \ -0.0780 \ 0.4312 \ -0.6452 \end{bmatrix}, \\ G_2 &= \begin{bmatrix} 0.0104 \ 0.0503 \ 0.2805 \ -0.3629 \end{bmatrix}, \\ K_1 &= \begin{bmatrix} -0.0210 \ 0.0793 \ -0.4739 \ 0.4709 \end{bmatrix}, \\ K_2 &= \begin{bmatrix} -0.0152 \ -0.1111 \ -0.3284 \ 0.2606 \end{bmatrix}. \end{aligned}$$

The corresponding simulation results are shown in Figs. 13.7, 13.8 and 13.9. Among them, Fig. 13.7 plots the triggering release instants and release intervals, which shows that only 28 times are triggered during the whole simulation time of 40 times. It is implied that the proposed event-driven strategy can effectively ease the communication burden. Figure 13.8 shows that our proposed design scheme let the states of resulting closed-loop system converge to zero. Figure 13.9 provides the event-based fuzzy control for the concerned inverted pendulum model.

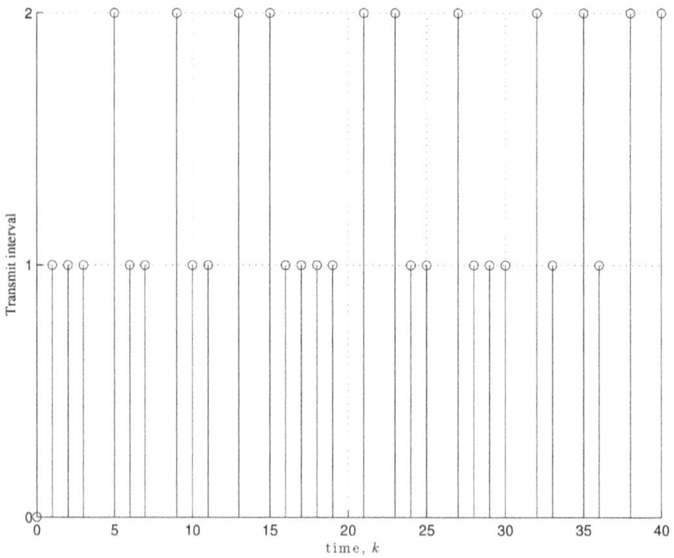

Fig. 13.7 The event-triggering release instants and release intervals

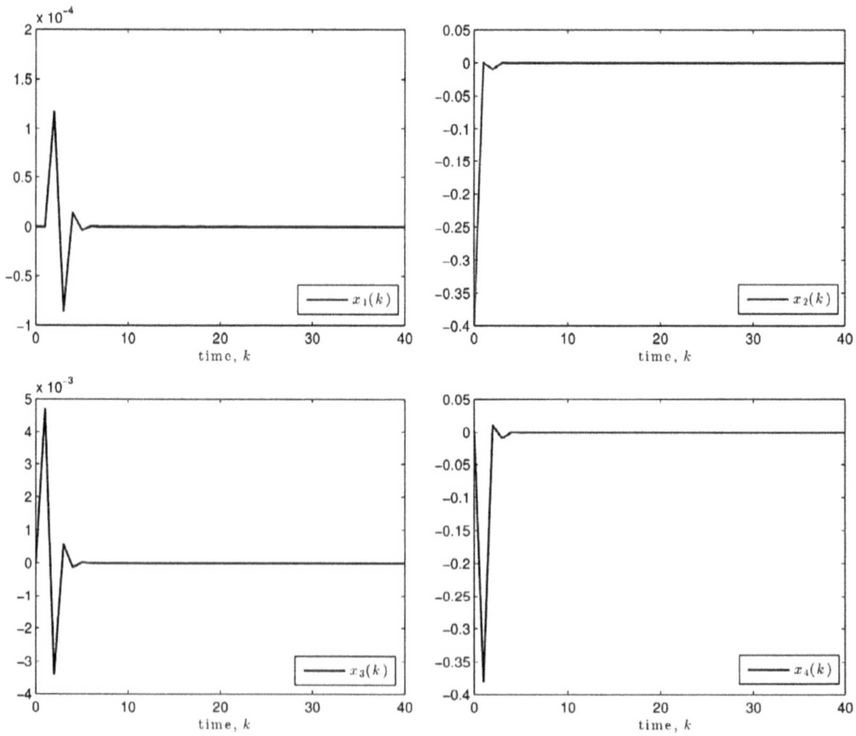

Fig. 13.8 Event-triggered states of the fuzzy control model

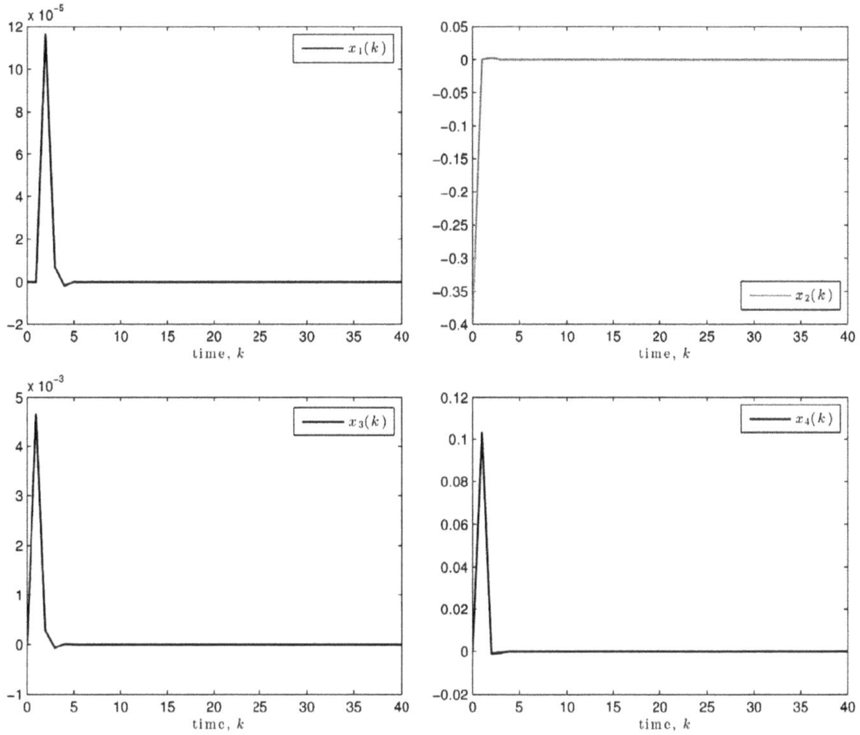

Fig. 13.9 Event-triggered fuzzy control of the inverted pendulum system

13.5 Conclusion

In this chapter, the event-based fuzzy control problem for nonlinear systems was addressed, and the inverted pendulum system was illustrated as a potential application. First, several delay-dependent stability conditions were established by combining the useful delay-partition method and reciprocally convex approach. Subsequently, utilizing the PDC strategy and event-based schemes, a fuzzy controller and an event-driven controller were constructed to stabilize the overall closed-loop system. At last, some simulation results have been presented to demonstrate the validity and applicability of the developed event-triggered fuzzy control scheme.

Chapter 14
Conclusion and Further Work

This chapter presents the concluding remarks and specifies the avenues for future research based on the present work.

14.1 Conclusion

This book focused on the stabilization analysis and synthesis, output-feedback controller design, fault detection, and reduced-order model approximation for dynamic fuzzy systems. Furthermore, several relevant research problems were discussed.

1. The stability and stabilization issues for continuous-time T-S fuzzy systems with time-varying delays were considered. Using the delay-partitioning approach and reciprocally convex method, new sufficient conditions, which could reduce the conservativeness against that of the existing results, were established. Next, an effective primary domain controller was developed to stabilize the fuzzy closed-loop systems.
2. The DOF Hankel-norm controller design issue was examined for T-S fuzzy stochastic systems. First, the Hankel-norm controller design scheme was established using certain additional matrix variables, which decoupled the Lyapunov functions and rendered the controller design feasible. Sufficient conditions with less conservativeness were obtained through the use of fuzzy-basis-dependent Lyapunov functions. Furthermore, a full-order OFC problem could be transformed to an optimization problem through the parameter transformation.
3. The DOFC issue for T-S fuzzy switched systems was investigated. The ADT technique was applied to stabilize the switched systems exponentially with an arbitrary switching rule. Next, we constructed a piecewise Lyapunov function and derived the sufficient conditions to ensure that the corresponding closed-loop system was exponentially stable with a specific $\mathcal{L}_2$–$\mathcal{L}_\infty$ performance level γ. Moreover, the feasible conditions of the fuzzy-rule-dependent DOFC were

X. Su et al., *Intelligent Control, Filtering and Model Reduction Analysis for Fuzzy-Model-Based Systems*, Studies in Systems, Decision and Control 385,
https://doi.org/10.1007/978-3-030-81214-0_14

derived through linearization, which could be promptly resolved using the standard toolbox.

4. The dissipativity-based filtering issue was resolved for T-S fuzzy switched systems with stochastic perturbations. Using the ADT approach and piecewise Lyapunov functions, sufficient conditions were established to ensure the mean-square exponential stability with dissipative performance for the corresponding filtering error dynamics. Furthermore, the solvable conditions for the dissipativity-based fuzzy filter were derived through linearization.
5. Weighted $\mathcal{H}_\infty$ fault detection filtering has been a key focus for nonlinear switched systems with stochastic disturbances, and the corresponding issue was addressed using the fuzzy-rule-dependent approach. Based on the piecewise Lyapunov functions and ADT method, sufficient conditions were established such that the mean-square exponential stability with weighted $\mathcal{H}_\infty$ performance for the dynamic error system could be ensured. Moreover, the solvable conditions of the proposed filter were derived based on the linearization procedure.
6. A reliable filtering technique with a dissipative performance for discrete-time T-S fuzzy delayed systems was designed. First, the sufficient conditions were formulated to ensure the asymptotical stability and strict dissipativity for the dynamic filtering system by using the reciprocally convex method. The reliable filter design issue could be transformed to a convex optimization issue.
7. The reduced-order model approximation issue subject to a prescribed system performance sense was considered for nonlinear hybrid stochastic switched systems through T-S fuzzy modelling by using the following steps: (1) Designed a reduced-order model to approximate the given high-order hybrid switched system; (2) Obtained the resulting dynamic error system by using the piecewise blending Lyapunov function and ADT method; the system was ensured to be mean-square exponentially stable with a particular performance index. In addition, the feasible conditions and corresponding parameters of designed reduced-order models were set based on the projection approach and CCL algorithm by solving the sequential minimization problem.
8. The model approximation issue for T-S fuzzy systems with time-varying delay was examined. Asymptotic stability conditions with less conservativeness were established for the resulting error system by using the reciprocally convex method. Moreover, the reduced-order model was developed using the projection strategy, which could not only approximate the original system with a specific $\mathcal{H}_\infty$ performance level, but also facilitate the analysis and synthesis of complex high-order systems.
9. The model approximation method for fuzzy switched systems with stochastic disturbances was established. By employing the piecewise Lyapunov function and ADT technique, sufficient conditions were derived to ensure the mean-square exponential stability and Hankel-norm performance for the resulting dynamic system. Furthermore, the relevant solvable conditions of the designed reduced-order models were set using an efficient linearization procedure method.
10. The reduced-order filtering issue for nonlinear systems with stochastic perturbations was examined through T-S fuzzy modelling. Using a new fuzzy Lyapunov

function and reciprocally convex method, sufficient conditions were established to ensure that the filtering system was mean-square asymptotically stable with $\mathcal{H}_\infty$ performance. Furthermore, the feasible conditions of reduced-order filtering were set using the convex linearization method. In this manner, the convex optimization issue could be readily resolved using standard numerical software.

11. The problem of the dissipative control of fuzzy logic systems based on the event-triggered strategy was examined. To reduce the unnecessary use of limited communication resources and network bandwidth, an efficient event-triggered scheme was introduced. Based on this analysis, sufficient conditions of the fuzzy controller design, which could ensure that the resulting closed-loop system was asymptotically stable subject to the specific dissipative property, were established. Finally, the truck-trailer model was presented as an illustrative example to verify the applicability of the developed design method.
12. The event-based fuzzy control problem for nonlinear systems was addressed, and the inverted pendulum system was illustrated as a potential application. First, several delay-dependent stability conditions were established by combining the useful delay-partition method and reciprocally convex approach. Subsequently, utilizing the PDC strategy and event-based schemes, a fuzzy controller and an event-driven controller were constructed to stabilize the overall closed-loop system.

14.2 Further Work

The presented research can be expanded in several directions:

1. The stability analysis and controller design issues for different fuzzy systems, such as stochastic fuzzy systems, Markov jumping fuzzy systems, switched fuzzy systems, and two-dimensional fuzzy systems can be further examined. In the case of switched fuzzy systems subject to multiple delays and stochastic perturbation, several valid schemes (such as event-based fault tolerance control approaches, switched quadratic Lyapunov functions, piecewise Lyapunov functions, and ADT method) presented in this book can be used to resolve the analysis and design issues for switched fuzzy systems. The research results of such systems are challenging and meaningful from both theoretical and practical perspectives.
2. NCSs have been broadly utilized in many industrial systems, such as for factory automation, grid-connected photovoltaic generation, autonomic mobile robots, and cascaded H-bridge converters. Nevertheless, communication networks involve several challenges, for instance, in the modelling, analysis and synthesis of NCSs, including network-induced delays, data-packets dropouts, finite network bandwidth aspects, and quantization. Future work may consider the application of fuzzy systems to approximate nonlinear NCSs in the uniform frame and focus on the stability analysis, controller design, and fault detection filtering problems.

3. The results pertaining to the stability analysis for fuzzy systems with time-delay considered in this book are conservative to a certain extent. Many advanced approaches, such as the free-weight matrix approach, delay-partition approach, small-gain-based input-output approach, and reciprocally convex approach, can be applied to reduce the conservativeness pertaining to the time-delay. Moreover, these approaches can be used to reduce the conservativeness in sampling controller design. Thus, the above-mentioned approaches can help design the controller and filter for dynamic fuzzy systems with time delays.

References

1. Ahn, C.K.: Input-to-state stable nonlinear filtering for a class of continuous-time delayed nonlinear systems. Int. J. Control **86**(6), 1179–1185 (2013)
2. Ahammed, A.K.I., Azeem, M.F.: Robust stabilization and control of Takagi-Sugeno fuzzy systems with parameter uncertainties and disturbances via state feedback and output feedback. Int. J. Fuzzy Syst. **21**(8), 2556–2574 (2019)
3. Alif, A., Darouach, M., Boutayeb, M.: Design of robust $\mathcal{H}_\infty$ reduced-order unknown-input filter for a class of uncertain linear neutral systems. IEEE Trans. Autom. Control **55**(1), 6–19 (2010)
4. Ali, C., Mohammed, C., Peng, S., Naceur, B.B.: Fuzzy fault detection filter design for T-S fuzzy systems in the finite-frequency domain. IEEE Trans. Fuzzy Syst. **25**(5), 1051–1061 (2017)
5. Assawinchaichote, W., Nguang, S.K.: $\mathcal{H}_\infty$ filtering for fuzzy dynamic systems with D stability constraints. IEEE Trans. Circuits Syst. I Fundam. Theory Appl. **50**(11), 1503–1508 (2003)
6. Assawinchaichote, W., Nguang, S.K.: Fuzzy $\mathcal{H}_\infty$ output feedback control design for singularly perturbed systems with pole placement constraints: an LMI approach. IEEE Trans. Fuzzy Syst. **14**(3), 361–371 (2006)
7. Assawinchaichote, W., Nguang, S.K., Shi, P.: Output feedback control design for uncertain fuzzy singularly perturbed systems: an LMI approaches. Automatica **40**(12), 2147–2152 (2004)
8. Ali, M.S., Narayanan, G., Shekher, V., Alsulami, H., Saeed, T.: Dynamic stability analysis of stochastic fractional-order memristor fuzzy BAM neural networks with delay and leakage terms. Appl. Math. Comput. (2020). https://doi.org/10.1016/j.amc.2019.124896
9. Aberkane, S., Ponsart, J.C., Rodrigues, M., Sauter, D.: Output feedback control of a class of stochastic hybrid systems. Automatica **44**(5), 1325–1332 (2008)
10. Ahn, C.K., Shi, P., Basin, M.V.: Two-dimensional dissipative control and filtering for Roesser model. IEEE Trans. Autom. Control **60**(7), 1745–1759 (2015)
11. Ahn, C.K., Shi, P., Li, H.: $\mathcal{H}_2$ output-feedback control with finite multiple measurement information. IEEE Trans. Autom. Control **63**(8), 2588–2595 (2018)
12. Ahn, C.K., Shi, P., Wu, L.: Receding horizon stabilization and disturbance attenuation for neural networks with time-varying delay. IEEE Trans. Cybern. **45**(12), 2680–2692 (2015)
13. Amini, F., Vahdani, R.: Fuzzy optimal control of uncertain dynamic characteristics in tall buildings subjected to seismic excitation. J. Vib. Control **14**(12), 1843–1867 (2008)
14. Ali, M.S., Vadivel, R., Saravanakumar, R.: Event-triggered state estimation for Markovian jumping impulsive neural networks with interval time-varying delays. Int. J. Control **92**(2), 270–290 (2019)

X. Su et al., *Intelligent Control, Filtering and Model Reduction Analysis for Fuzzy-Model-Based Systems*, Studies in Systems, Decision and Control 385,
https://doi.org/10.1007/978-3-030-81214-0

15. Arslan, E., Vadivel, R., Ali, M.S., Arik, S.: Event-triggered $\mathcal{H}_\infty$ filtering for delayed neural networks via sampled-data. Neural Netw. **91**, 11–21 (2017)
16. An, J., Wen, G., Lin, C.: New results on a delay-derivative-dependent fuzzy $\mathcal{H}_\infty$ filter design for T-S fuzzy systems. IEEE Trans. Fuzzy Syst. **19**(4), 770–779 (2011)
17. Aslam, M.S., Zhang, B.Y., Zhang, Y.J., Zhang, Z.Q.: Extended dissipative filter design for T-S fuzzy systems with delays multiple time delays. ISA Trans. **80**, 22–34 (2018)
18. Briat, C.: Convergence and equivalence results for the Jensen's inequality: application to time-delay and sampled-data systems. IEEE Trans. Autom. Control **56**(7), 1660–1665 (2011)
19. Beyhan, S.: Affine T-S fuzzy model-based estimation and control of Hindmarsh-Rose neuronal model. IEEE Trans. Syst. Man Cybern. Syst. **47**(8), 2342–2350 (2017)
20. Baillieul, J., Antsaklis, P.J.: Control and communication challenges in networked real-time systems. Proc. IEEE **95**(1), 9–28 (2007)
21. Bingul, Z., Cook, G.E., Strauss, A.M.: Application of fuzzy logic to spatial thermal control in fusion welding. IEEE Trans. Ind. Appl. **36**(6), 1523–1530 (2000)
22. Battistelli, G., Chisci, L., Selvi, D.: A distributed Kalman filter with event-triggered communication and guaranteed stability. Automatica **93**, 75–82 (2018)
23. Bolender, M., Doman, D.: A nonlinear longitudinal dynamical model of an air-breathing hypersonic vehicle. J. Spacecr. Rockets **44**(2), 374–387 (2007)
24. Boukezzoula, R., Galichet, S., Foulloy, L.: Observer-based fuzzy adaptive control for a class of nonlinear systems: real-time implementation for a robot wrist. IEEE Trans. Control Syst. Technol. **12**(3), 340–351 (2004)
25. Boyd, S., Ghaoui, L.E., Feron, E., Balakrishnan, V.: Linear Matrix Inequalities in Systems and Control Theory. SIAM, Philadelphia, PA (1994)
26. Baigzadehnoe, B., Rezaie, B., Rahmani, Z.: Fuzzy-model-based fault detection for nonlinear networked control systems with periodic access constraints and Bernoulli packet dropouts. Appl. Soft Comput. **80**, 465–474 (2019)
27. Bai, Y., Zhuang, H.Q., Roth, Z.S.: Fuzzy logic control to suppress noises and coupling effects in a laser tracking system. IEEE Trans. Control Syst. Technol. **13**(1), 113–121 (2005)
28. Choi, H.H.: Robust stabilization of uncertain fuzzy systems using variable structure system approach. IEEE Trans. Fuzzy Syst. **16**(3), 715–724 (2008)
29. Chang, X.-H.: Robust nonfragile $\mathcal{H}_\infty$ filtering of fuzzy systems with linear fractional parametric uncertainties. IEEE Trans. Fuzzy Syst. **20**(6), 1001–1011 (2012)
30. Chadli, M., Abdo, A., Ding, S.X.: Robust nonfragile $\mathcal{H}_\infty$ filtering of fuzzy systems with linear fractional parametric uncertainties. IEEE Trans. Fuzzy Syst. **20**(6), 1001–1011 (2013)
31. Cheng, J., Ahn, C.K., Karimi, H.R., Cao, J., Qi, W.: An event-based asynchronous approach to Markov jump systems with hidden mode detections and missing measurements. IEEE Trans. Syst. Man Cybern. Syst. **49**(9), 1749–1758 (2019)
32. Choi, H.D., Ahn, C.K., Karimi, H.R., Lim, M.T.: Filtering of discrete-time switched neural networks ensuring exponential dissipative and $\updownarrow_2 - \updownarrow_\infty$ Performances. IEEE Trans. Cybern. **47**(10), 3195–3207 (2017)
33. Choi, H.D., Ahn, C.K., Shi, P., Lim, M.T., Song, M.K.: $\mathcal{L}_2 - \mathcal{L}_\infty$ filtering for Takagi-Sugeno fuzzy neural networks based on Wirtinger-type inequalities. Neurocomputing **153**, 117–125 (2015)
34. Choi, H.D., Ahn, C.K., Shi, P., Wu, L., Lim, M.T.: Dynamic output-feedback dissipative control for T-S fuzzy systems with time-varying input delay and output constraints. IEEE Trans. Fuzzy Syst. **25**(3), 511–526 (2017)
35. Chen, Z., Ding, S.X., Luo, H., Zhang, K.: An alternative data-driven fault detection scheme for dynamic processes with deterministic disturbances. J. Frankl. Inst. **354**(1), 556–570 (2017)
36. Cai, J., Ferdowsi, H., Arangapani, J.: A flexible terminal approach to sampled-data exponentially synchronization of Markovian neural networks with time-varying delayed signals. Automatica **66**, 122–131 (2016)
37. Chadli, M., Guerra, T.M.: LMI solution for robust static output feedback control of discrete Takagi-Sugeno fuzzy models. IEEE Trans. Fuzzy Syst. **20**(6), 1160–1165 (2012)

38. Cho, S., Gao, Z., Moan, T.: Model-based fault detection, fault isolation and fault-tolerant control of a blade pitch system in floating wind turbines. Renew. Energy **120**, 306–321 (2018)
39. Chen, G., Gao, Y., Zhu, S.: Finite-time dissipative control for stochastic interval systems with time-delay and Markovian switching. Appl. Math. Comput. **310**, 169–181 (2017)
40. Chadli, M., Karimi, H.R.: Robust observer design for unknown inputs Takagi-Sugeno models. IEEE Trans. Fuzzy Syst. **21**(1), 158–164 (2013)
41. Chen, B., Liu, X.P.: Delay-dependent robust $\mathcal{H}_\infty$ control for T-S fuzzy systems with time delay. IEEE Trans. Fuzzy Syst. **13**(4), 544–556 (2005)
42. Chen, B., Lin, C., Liu, X.P., Liu, K.F.: Observer-based adaptive fuzzy control for a class of nonlinear delayed systems. IEEE Trans. Syst. Man Cybern. Syst. **46**(1), 27–36 (2016)
43. Chung, L.Y., Lien, C.H., Yu, K.W., Chen, J.D.: Robust $\mathcal{H}_\infty$ filtering for discrete switched systems with interval time-varying delay. Signal Process. **94**, 661–669 (2014)
44. Chen, Y.-J., Ohtake, H., Tanaka, K., Wang, W.-J.: Relaxed stabilization criterion for T-S fuzzy systems by minimum-type piecewise-Lyapunov-function-based switching fuzzy controller. IEEE Trans. Fuzzy Syst. **20**(6), 1166–1173 (2012)
45. Cheng, J., Park, J.H., Karimi, H.R., Shen, H.: Model-based fault detection, estimation, and prediction for a class of linear distributed parameter systems. IEEE Trans. Fuzzy Syst. **48**(4), 2232–2244 (2018)
46. Chen, G.R., Pham, T.T., Weiss, J.J.: Fuzzy modeling of control-systems. IEEE Trans. Aerosp. Electron. Syst. **31**(1), 414–429 (1995)
47. Cattivelli, F.S., Sayed, A.H.: Diffusion strategies for distributed Kalman filtering and smoothing. IEEE Trans. Autom. Control **55**(9), 2069–2084 (2010)
48. Campos, V.C.S., Souza, F.O., Torres, L.A.B., Palhares, R.M.: New stability conditions based on piecewise fuzzy Lyapunov functions and tensor product transformations. IEEE Trans. Fuzzy Syst. **21**(4), 748–760 (2013)
49. Chen, B.S., Tseng, C.S., Uang, H.J.: Mixed $\mathcal{H}_2/\mathcal{H}_\infty$ fuzzy output feedback control design for nonlinear dynamic systems: an LMI approach. IEEE Trans. Fuzzy Syst. **8**(3), 249–265 (2000)
50. Chen, T., Wang, C., Hill, D.J.: Fault detection for a class of uncertain sampled-data systems using deterministic learning. IEEE Trans. Cybern. (2020). https://doi.org/10.1109/TCYB.2019.2963259
51. Chang, X.H., Yang, G.H.: Nonfragile $\mathcal{H}_\infty$ filter design for T-S fuzzy systems in standard form. IEEE Trans. Ind. Electron. **61**(7), 3448–3458 (2014)
52. Chen, X., Yuan, P.: Event-triggered generalized dissipative filtering for delayed neural networks under aperiodic DoS jamming attacks. IEEE Trans. Ind. Electron. **400**, 467–479 (2020)
53. Dolk, V., Borgers, D., Heemels, W.: Output-based and decentralized dynamic event-triggered control with guaranteed $\mathcal{L}_p$-gain performance and zeno-freeness. IEEE Trans. Autom. Control **62**(1), 33–49 (2017)
54. Dong, S., Fang, M., Shi, P., Wu, Z., Zhang, D.: Dissipativity-based control for fuzzy systems with asynchronous modes and intermittent measurements. IEEE Trans. Cybern. **50**(6), 2389–2399 (2020)
55. Ding, L., He, Y., Wu, M., Zhang, Z.: A novel delay partitioning method for stability analysis of interval time-varying delay systems. J. Frankl. Inst. Eng. Appl. Math. **354**(2), 1209–1219 (2017)
56. Du, B., Lam, J., Shu, Z., Wang, Z.: A delay-partitioning projection approach to stability analysis of continuous systems with multiple delay components. IET Control Theory Appl. **3**(4), 383–390 (2009)
57. Du, D., Jiang, B., Shi, P., Zhou, S.: $\mathcal{H}_\infty$ filtering of discrete-time switched systems with state delays via switched lyapunov function approach. IEEE Trans. Autom. Control **52**(8), 1520–1525 (2007)
58. Daafouz, J., Riedinger, P., Iung, C.: Stability analysis and control synthesis for switched systems: a switched Lyapunov function approach. IEEE Trans. Autom. Control **47**(11), 1883–1887 (2002)

59. Dovzan, D., Skrjanc, I.: Fuzzy space partitioning based on hyperplanes defined by eigenvectors for Takagi-Sugeno fuzzy model identification. IEEE Trans. Ind. Electron. **67**(6), 5144–5153 (2020)
60. Du, S., Wu, M., Chen, X., Cao, W.: An intelligent control strategy for iron ore sintering ignition process based on the prediction of ignition temperature. IEEE Trans. Ind. Electron. **67**(2), 1233–1241 (2020)
61. Dong, H., Wang, Z., Ho, D.W.C., Gao, H.: Robust $\mathcal{H}_\infty$ fuzzy output-feedback control with multiple probabilistic delays and multiple missing measurements. IEEE Trans. Fuzzy Syst. **18**(4), 712–725 (2010)
62. Deng, X.H., Xu, T., Wang, R.: Risk evaluation model of highway tunnel portal construction based on BP fuzzy neural network. Comput. Intell. Neurosci. (2018). https://doi.org/10.1155/2018/8547313
63. Dong, J., Yang, G., Zhang, H.: Stability analysis of T-S fuzzy control systems by using set theory. IEEE Trans. Fuzzy Syst. **23**(4), 827–841 (2015)
64. Ding, S.X., Zhang, P., Yin, S., Ding, E.L.: An integrated design framework of fault-tolerant wireless networked control systems for industrial automatic control applications. IEEE Trans. Ind. Inform. **9**(1), 462–471 (2013)
65. Elahi, A., Alfi, A.: Finite-time stability analysis of uncertain network-based control systems under random packet dropout and varying network delay. Nonlinear Dyn. **91**(1), 713–731 (2018)
66. Eltag, K., Aslam, M.S., Chen, Z.R.: Functional observer-based T-S fuzzy systems for quadratic stability of power system synchronous generator. Int. J. Fuzzy Syst. **22**(1), 172–180 (2020)
67. El-Kasri, C., Hmamed, A., Tadeo, F.: Reduced-order $\mathcal{H}_\infty$ filters for uncertain 2-D continuous systems, via LMIs and polynomial matrices. Circuits Syst. Signal Process. **33**(4), 1189–1214 (2014)
68. Feng, C.: A design scheme of variable structure adaptive control for uncertain dynamic systems. Automatica **32**(4), 561–567 (1996)
69. Feng, G.: Robust $\mathcal{H}_\infty$ filtering of fuzzy dynamic systems. IEEE Trans. Aerosp. Electron. Syst. **41**(2), 658–670 (2005)
70. Feng, G.: A survey on analysis and design of model-based fuzzy control systems. IEEE Trans. Fuzzy Syst. **14**(5), 676–697 (2006)
71. Fayek, A.R.: Fuzzy logic and fuzzy hybrid techniques for construction engineering and management.J. Constr. Eng. Manag. **146**(7) (2020)
72. Fei, Z., Gao, H., Shi, P.: New results on stabilization of Markovian jump systems with time delay. Automatica **45**(10), 2300–2306 (2009)
73. Feng, Z., Lam, J.: Robust reliable dissipative filtering for discrete delay singular systems. Signal Process. **92**(12), 3010–3025 (2012)
74. Feng, Z., Lam, J., Yang, G.-H.: Optimal partitioning method for stability analysis of continuous/discrete delay systems. Int. J. Robust Nonlinear Control **25**(4), 559–574 (2015)
75. Filev, D., Syed, F.U.: Applied intelligent systems: blending fuzzy logic with conventional control. Int. J. Gen. Syst. **39**(4), 395–414 (2010)
76. Gahinet, P., Apkarian, P.: A linear matrix inequality approach to H_∞ control. Int. J. Robust Nonlinear Control **4**(4), 421–448 (1994)
77. Garcis, E., Antsaklis, P.J.: Model-based event-triggered control for systems with quantization and time-varying network delays. IEEE Trans. Autom. Control **58**(2), 422–434 (2013)
78. Gupta, R.A., Chow, M.-Y.: Networked control system: overview and research trends. IEEE Trans. Ind. Electron. **57**(7), 2527–2535 (2010)
79. Gao, H., Chen, T., Wang, L.: Robust fault detection with missing measurements. Int. J. Control **81**(5), 804–819 (2008)
80. Ghazali, M.R., Ibrahim, Z., Suid, M., Saealal, M., Tumari, M.: Single input fuzzy logic controller for flexible joint manipulator. Int. J. Innov. Comput. Inf. Control **12**(1), 181–191 (2016)
81. Gao, Y., Li, H., Chadli, M., Lam, H.K.: Static output-feedback control for interval type-2 discrete-time fuzzy systems. Complexity **21**(3), 74–88 (2016)

82. Gao, H., Liu, X., Lam, J.: Stability analysis and stabilization for discrete-time fuzzy systems with time-varying delay. IEEE Trans. Syst. Man Cybern. Part B Cybern. **39**(2), 306–317 (2009)
83. Guo, H., Qiu, J., Tian, H., Gao, H.: Fault detection of discrete-time T-S fuzzy affine systems based on piecewise Lyapunov functions. J. Frankl. Inst. **351**(7), 3633–3650 (2014)
84. Goyal, P., Redmann, M.: Time-limited $\mathcal{H}_2 - optimal$ model order reduction. Appl. Math. Comput. **355**, 184–197 (2019)
85. Gilles, T., Reine, T., Ali, C.: Design and comparison of robust nonlinear controllers for the lateral dynamics of intelligent vehicles. Trans. Intell. Transp. Syst. **17**(3), 796–809 (2016)
86. Gilles, T., Reine, T., Ali, C.: Event-based fault detection for T-S fuzzy systems with packet dropouts and (x, v)-dependent noises. Signal Process. **138**, 211–219 (2017)
87. Gagliardi, G., Tedesco, F., Casavola, A.: A LPV modeling of turbocharged spark-ignition automotive engine oriented to fault detection and isolation purposes. J. Frankl. Inst. **355**(14), 6710–6745 (2018)
88. Gao, H., Zhao, Y., Chen, T.: $\mathcal{H}_\infty$ fuzzy control of nonlinear systems under unreliable communication links. IEEE Trans. Fuzzy Syst. **17**(2), 265–278 (2009)
89. Higham, D.: An algorithmic introduction to numerical simulation of stochastic differential equations. SIAM Rev. **43**, 525–546 (2001)
90. Karimi, H.R.: Robust delay-dependent $\mathcal{H}_\infty$ control of uncertain time-delay systems with mixed neutral, discrete, and distributed time-delays and Markovian switching parameters. IEEE Trans. Circuits Syst. I Regul. Pap. **58**(8), 1910–1923 (2011)
91. Karimi, H.R.: A sliding mode approach to $\mathcal{H}_\infty$ synchronization of master-slave time-delay systems with Markovian jumping parameters and nonlinear uncertainties. J. Frankl. Inst. **349**(4), 1480–1496 (2012)
92. Heydari, A.: Optimal triggering of networked control systems. IEEE Trans. Neural Netw. Learn. Syst. **29**(7), 3011–3021 (2018)
93. Husek, P.: Monotonic smooth Takagi-Sugeno fuzzy systems with fuzzy sets with compact support. IEEE Trans. Fuzzy Syst. **27**(3), 605–611 (2019)
94. Ho, W.-H., Chen, S.-H., Chou, J.-H.: Observability robustness of uncertain fuzzy-model-based control systems. Int. J. Innov. Comput. Inf. Control **9**(2), 805–819 (2013)
95. Hameed, A., Elmadbouly, I., Abdo, I.: Sensor and actuator fault-hiding reconfigurable control design for a four-tank system benchmark. Int. J. Innov. Comput. Inf. Control **11**(2), 679–690 (2015)
96. Huang, H., Huang, T., Cao, Y.: Reduced-order filtering of delayed static neural networks with Markovian jumping parameters. IEEE Trans. Neural Netw. Learn. Syst. **29**(11), 5606–5618 (2018)
97. Hmamed, A., El Kasri, C., Tissir, E.H., Alvarez, T., Tadeo, F.: Robust $\mathcal{H}_\infty$ filtering for uncertain 2-D continuous systems with delays. Int. J. Innov. Comput. Inf. Control **9**(5), 2167–2183 (2013)
98. Hua, C., Liu, G., Guan, X.: Switching regulation based stabilisation of discrete-time 2D switched systems with stable and unstable modes. IET Control Theory Appl. **12**(7), 953–960 (2018)
99. Hill, D.J., Moylan, P.J.: Dissipative dynamical systems: basic input-output and state properties. J. Frankl. Inst. **309**(5), 327–357 (1980)
100. Huang, D., Nguang, S.K.: State feedback control of uncertain networked control systems with random time delays. IEEE Trans. Autom. Control **53**(3), 829–834 (2008)
101. Hespanha, J.P., Naghshtabrizi, P., Xu, Y.: A survey of recent results in networked control systems. Proc. IEEE **95**(1), 138–162 (2007)
102. Harirchi, F., Ozay, N.: Guaranteed model-based fault detection in cyber-physical systems: a model invalidation approach. Automatica **93**, 476–488 (2018)
103. He, X., Wang, Z., Zhou, D.: Robust fault detection for networked systems with communication delay and data missing. Automatica **45**(11), 2634–2639 (2009)
104. Hu, S., Yue, D., Xie, X., Du, Z.: Event-triggered $\mathcal{H}_\infty$ stabilization for networked stochastic systems with multiplicative noise and network-induced delays. Inf. Sci. **299**(9), 178–197 (2015)

105. Hua, M., Zheng, D., Deng, F.: Partially mode-dependent $\mathcal{L}_2 - \mathcal{L}_\infty$ filtering for discrete-time nonhomogeneous Markov jump systems with repeated scalar nonlinearities. Inf. Sci. **451**, 223–239 (2018)
106. Hong, Y., Zhang, H., Zheng, Q.: Asynchronous $\mathcal{H}_\infty$ filtering for switched T-S fuzzy systems and its application to the continuous stirred tank reactor. Int. J. Fuzzy Syst. **20**(5), 1470–1482 (2018)
107. Iftar, A.: Robust optimal control of power systems. Int. J. Power Energy Syst. **36**(2), 71–75 (2016)
108. Ishii, H., Francis, B.A.: Limited Data Rate in Control Systems with Networks, p. 275. Springer, Berlin, Germany (2002)
109. Il Lee, W., Park, P.: Second-order reciprocally convex approach to stability of systems with interval time-varying delays. Appl. Math. Comput. **229**, 245–253 (2014)
110. Ibrahim, O., Yahaya, N., Saad, N., Ahmed, K.Y.: Development of observer state output feedback for phase-shifted full bridge DC-DC converter control. IEEE Access **5**, 18143–18154 (2017)
111. Ishihara, T., Zheng, L.A.: LQG/LTR procedure using reduced-order Kalman filters. Int. J. Control **92**(3), 461–475 (2019)
112. Jayawardhana, B.: Towards a generic constructive nonlinear control design tool using relaxed control. IEEE Trans. Autom. Control **60**(12), 3293–3298 (2015)
113. Jiang, L., Qi, R.: Adaptive actuator fault compensation for discrete-time T-S fuzzy systems with multiple input-output delays. Int. J. Innov. Comput. Inf. Control **12**(4), 1043–1058 (2016)
114. Kishida, M.: Encrypted control system with quantiser. IET Control Theory Appl. **13**(1), 146–151 (2019)
115. Ksantini, M., Hammami, M.A., Delmotte, F.: On the global exponential stabilization of Takagi-Sugeno fuzzy uncertain systems. Int. J. Innov. Comput. Inf. Control **11**(1), 281–294 (2015)
116. Kim, H.J., Koo, G.B., Park, J.B., Joo, Y.H.: Decentralized sampled-data $\mathcal{H}_\infty$ fuzzy filter for nonlinear large-scale systems. Fuzzy Sets Syst. **273**, 68–86 (2015)
117. Kumar, J., Kumar, V., Rana, K.P.S.: Design of robust fractional order fuzzy sliding mode PID controller for two link robotic manipulator system. J. Intell. Fuzzy Syst. **35**(5), 5301–5315 (2018)
118. Kazachek, N., Lokhin, V., Manko, S., Romanov, M.: Research of periodic oscillations in control systems with fuzzy controllers. Int. J. Innov. Comput. Inf. Control **11**(3), 985–997 (2015)
119. Kazachek, N., Lokhin, V., Manko, S., Romanov, M.: Application of fuzzy control-system to industrial processes. Automatica **13**(3), 235–242 (1977)
120. Karer, G., Music, G., Skrjanc, I., Zupancic, B.: Model predictive control of nonlinear hybrid systems with discrete inputs employing a hybrid fuzzy model. J. Intell. Robot. Syst. **50**(3), 297–319 (2007)
121. Kim, H.J., Park, J.B., Joo, Y.H.: An improved $\mathcal{H}_\infty$ fuzzy filter for nonlinear sampled-data systems. Int. J. Control Autom. Syst. **15**(3), 1394–1404 (2017)
122. Kavikumar, R., Sakthivel, R., Kwon, O.M., Kaviarasan, B.: Reliable non-fragile memory state feedback controller design for fuzzy Markov jump systems. Nonlinear Anal. Hybrid Syst. (2020). https://doi.org/10.1016/j.nahs.2019.100828
123. Kumar, S.V., Sakthivel, R., Sathishkumar, M., Anthoni, S.M.: Finite time passive reliable filtering for fuzzy systems with missing measurements. J. Dyn. Syst. Meas. Control Trans. ASME **140**(8) (2018)
124. Karami, A., Yazdanpanah, M.J., Moshiri, B.: Robust switching signal estimation for a class of uncertain nonlinear switched systems. Int. J. Control **92**(5), 1094–1102 (2019)
125. Liberzon, D.: Switching in Systems and Control. Birkhauser, Boston (2003)
126. Lam, H.-K.: LMI-based stability analysis for fuzzy-model-based control systems using artificial T-S fuzzy model. IEEE Trans. Syst. Man Cybern. Part B Cybern. **19**(3), 505–513 (2011)
127. Lam, H.-K.: Polynomial fuzzy-model-based control systems: stability analysis via piecewise-linear membership functions. IEEE Trans. Fuzzy Syst. **19**(3), 588–593 (2011)

128. Lam, H.-K.: Stabilization of nonlinear systems using sampled-data output-feedback fuzzy controller based on polynomial fuzzy-model-based control approach. IEEE Trans. Syst. Man Cybern. Part B Cybern. **42**(1), 258–267 (2012)
129. Li, S., Ahn, C.K., Xiang, Z.: Sampled-data adaptive output feedback fuzzy stabilization for switched nonlinear systems with asynchronous switching. IEEE Trans. Fuzzy Syst. **27**(1), 200–205 (2019)
130. Lin, T.-C., Chen, C.-H.: Robust fault-tolerant control for a biped robot using a recurrent cerebellar model articulation controller. IEEE Trans. Syst. Man Cybern. Part B Cybern. **37**(1), 110–123 (2007)
131. Lam, H.-K., Chan, E.W.S.: Stability analysis of sampled-data fuzzy-model-based control systems. Int. J. Fuzzy Syst. **10**(2), 129–135 (2008)
132. Landry, M., Campbell, S.A., Morris, K., Aguilar, C.O.: Dynamics of an inverted pendulum with delayed feedback control. SIAM J. Appl. Dyn. Syst. **4**(2), 333–351 (2005)
133. Li, H., Chen, Z., Wu, L., Lam, H.K., Du, H.: Event-triggered fault detection of nonlinear networked systems. IEEE Trans. Cybern. **47**(4), 1041–1052 (2017)
134. Linsenmayer, S., Dimarogonas, D.V., Allgoewer, F.: Periodic event-triggered control for networked control systems based on non-monotonic Lyapunov functions. Automatica **106**, 35–46 (2019)
135. Luo, S., Deng, F., Chen, W.: Dynamic event-triggered control for linear stochastic systems with sporadic measurements and communication delays. Automatica **107**, 86–94 (2019)
136. Li, L.L., Ding, S.X., Qiu, J.B., Yang, Y.: Real-time fault detection approach for nonlinear systems and its asynchronous T-S fuzzy observer-based implementation. IEEE Trans. Cybern. **47**(2), 283–294 (2017)
137. Liu, Y., Fang, F., Park, J.H.: Decentralized dissipative filtering for delayed nonlinear interconnected systems based on T-S fuzzy model. IEEE Trans. Fuzzy Syst. **27**(4), 790–801 (2019)
138. Li, X., Gao, H.: Reduced-order generalized $\mathcal{H}_\infty$ filtering for linear discrete-time systems with application to channel equalization. IEEE Trans. Signal Process. **62**(13), 3393–3402 (2014)
139. Li, X., Gao, H., Gu, K.: Delay-independent stability analysis of linear time-delay systems based on frequency discretization. Automatica **70**, 288–294 (2016)
140. Liu, Y., Guo, B., Park, J.H., Lee, S.: Event-based reliable dissipative filtering for T-S fuzzy systems with asynchronous constraints. IEEE Trans. Fuzzy Syst. **26**(4), 2089–2098 (2018)
141. Lee, D., Hu, J.: Periodic stabilization of discrete-time switched linear systems. IEEE Trans. Autom. Control **62**(7), 3382–3394 (2017)
142. Lin, T., Huang, F., Du, Z., Lin, Y.: Synchronization of fuzzy modeling chaotic time delay memristor-based Chua's circuits with application to secure communication. Int. J. Fuzzy Syst. **17**(2), 206–214 (2015)
143. Lam, H.-K., Lo, J.-C.: Output regulation of polynomial-fuzzy-model-based control systems. IEEE Trans. Fuzzy Syst. **21**(2), 262–274 (2013)
144. Lin, C.M., Li, H.Y.: Adaptive dynamic sliding-mode fuzzy CMAC for voice coil motor using asymmetric gaussian membership function. IEEE Trans. Ind. Electron. **61**(10), 5662–5671 (2014)
145. Li, M., Lin, H.-J.: Design and implementation of smart home control systems based on wireless sensor networks and power line communications. IEEE Trans. Ind. Electron. **62**(7), 4430–4442 (2015)
146. Li, L., Liu, X., Chai, T.: New approaches on $\mathcal{H}_\infty$ control of T-S fuzzy systems with interval time-varying delay. Fuzzy Sets Syst. **160**(12), 1669–1688 (2009)
147. Li, L., Lin, Z., Chai, Y., Cai, J.: Output feedback stabilization of two-dimensional fuzzy systems. Multidimens. Syst. Signal Process. **30**(4), 1731–1748 (2018). https://doi.org/10.1007/s11045-018-0625-x
148. Li, X., Mehran, K., Lam, H.-K., Xiao, B., Bao, Z.: Stability analysis of discrete-time positive polynomial-fuzzy-model-based control systems through fuzzy co-positive Lyapunov function with bounded control. IET Control Theory Appl. **14**(2), 233–243 (2020)
149. Lam, H.-K., Narimani, M.: Quadratic stability analysis of fuzzy-model-based control systems using staircase membership functions. IEEE Trans. Fuzzy Syst. **18**(1), 125–137 (2010)

150. Lugli, A.B., Neto, E.R., Henriques, J.P.C., Hervas, M.D.A., Santos, M.M.D., Justo, J.F.: Industrial application control with fuzzy systems. Int. J. Innov. Comput. Inf. Control **12**(2), 665–676 (2016)
151. Lam, H.-K., Narimani, M., Li, H., Liu, H.: Stability analysis of polynomial-fuzzy-model-based control systems using switching polynomial Lyapunov function. IEEE Trans. Fuzzy Syst. **21**(5), 800–813 (2013)
152. Lu, R., Peng, H., Liu, S., Xu, Y., Li, X.: Reliable $\ell_2 - \ell_\infty$ filtering for fuzzy Markov stochastic systems with sensor failures and packet dropouts. IET Control Theory Appl. **11**(14), 2195–2203 (2017)
153. Lee, D.H., Park, J.B., Joo, Y.H.: Approaches to extended non-quadratic stability and stabilisation conditions for discrete-time Takagi-Sugeno fuzzy systems. Automatica **47**(3), 534–538 (2011)
154. Li, S., Su, H., Ding, X.: Synchronized stationary distribution of hybrid stochastic coupled systems with applications to coupled oscillators and a Chua's circuits network. J. Frankl. Inst. Eng. Appl. Math. **355**(17), 8743–8765 (2018)
155. Liu, X., Su, X., Shi, P., Shen, C., Peng, Y.: Event-triggered sliding mode control of nonlinear dynamic systems. Automatica (2020). https://doi.org/10.1016/j.automatica.2019.108738
156. Li, F., Shi, P., Wu, L., Basin, M., Lim, C.: Quantized control design for cognitive radio networks modeled as nonlinear semi-Markovian jump systems. IEEE Trans. Ind. Electron. **62**(4), 2330–2340 (2015)
157. Lam, H.K., Tsai, S.H.: Stability analysis of polynomial-fuzzy-model-based control systems with mismatched premise membership functions. IEEE Trans. Fuzzy Syst. **22**(1), 223–229 (2014)
158. Li, Y., Tong, S., Liu, L., Feng, G.: Adaptive output-feedback control design with prescribed performance for switched nonlinear systems. Automatica **80**, 225–231 (2017)
159. LLamas, P.M.V., Vega, P.: Analytical fuzzy predictive control applied to wastewater treatment biological processes. Complexity (2019) https://doi.org/10.1155/2019/5720185
160. Li, Y., Voos, H., Darouach, M., Hua, C.: An application of linear algebra theory in networked control systems: stochastic cyber-attacks detection approach. IMA J. Math. Control Inf. **33**(4), 1081–1102 (2016)
161. Li, L., Wang, X., Lemmon, M.D.: Efficiently attentive event-triggered systems with limited bandwidth. IEEE Trans. Autom. Control **62**(3), 1491–1497 (2017)
162. Lin, C., Wang, Q.G., Lee, T.H., He, Y., Chen, B.: Observer-based $\mathcal{H}_\infty$ fuzzy control design for T-S fuzzy systems with state delays. Automatica **44**(3), 868–874 (2008)
163. Liu, J., Wu, C., Wang, Z., Wu, L.: Reliable filter design for sensor networks using type-2 fuzzy framework. IEEE Trans. Ind. Inform. **13**(4), 1742–1752 (2017)
164. Liu, J., Wang, J., Yang, G.: Reliable guaranteed variance filtering against sensor failures. IEEE Trans. Signal Process. **51**(5), 1403–1411 (2003)
165. Li, H., Wu, C., Feng, Z.: Fuzzy dynamic output-feedback control of non-linear networked discrete-time system with missing measurements. IET Control Theory Appl. **9**(3), 327–335 (2015)
166. Li, X., Xiang, Z., Karimi, H.R.: Asynchronously switched control of discrete impulsive switched systems with time delays. Inf. Sci. **249**, 132–142 (2013)
167. Liu, G., Xu, S., Park, J.H., Zhuang, G.: Reliable exponential $\mathcal{H}_\infty$ filtering for singular Markovian jump systems with time-varying delays and sensor failures. Int. J. Robust Nonlinear Control **28**(14), 4230–4245 (2018)
168. Liu, J., Xia, J., Tian, E., Fei, S.: Hybrid-driven-based $\mathcal{H}_\infty$ filter design for neural networks subject to deception attacks. Appl. Math. Comput. **320**, 158–174 (2018)
169. Lien, C.H., Yu, K.W., Chen, W.D., Wan, Z.L., Chung, Y.J.: Stability criteria for uncertain Takagi-Sugeno fuzzy systems with interval time-varying delay. IET Control Theory Appl. **1**(3), 764–769 (2007)
170. Liang, M., Yeap, T., Hermansyah, A., Rahmati, S.: Fuzzy control of spindle torque for industrial CNC machining. Int. J. Mach. Tools Manuf. **43**(14), 1497–1508 (2003)

171. Li, A., Yi, S., Wang, X.: New reliable $\mathcal{H}_\infty$ filter design for networked control systems with external disturbances and randomly occurring sensor faults. Neurocomputing **185**, 21–27 (2016)
172. Li, F., Zhang, X.: Delay-range-dependent robust $\mathcal{H}_\infty$ filtering for singular LPV systems with time variant delay. Int. J. Innov. Comput. Inf. Control **9**(1), 339–353 (2013)
173. Li, L., Zhong, L.: Generalised nonlinear $l_2 - l_\infty$ filtering of discrete-time Markov jump descriptor systems. Int. J. Control **87**(3), 653–664 (2014)
174. Lian, B., Zhang, Q., Li, J.: Integrated sliding mode control and neural networks based packet disordering prediction for nonlinear networked control systems. IEEE Trans. Neural Netw. Learn. Syst. **30**(8), 2324–2335 (2019)
175. Liu, L., Zhao, X., Niu, B., Wang, H., Xie, X.: Global output-feedback stabilisation of switched stochastic non-linear time-delay systems under arbitrary switchings. IET Control Theory Appl. **9**(2), 283–292 (2015)
176. Mahmoud, M.S.: Reliable decentralized control of interconnected discrete delay systems. Automatica **48**(5), 986–990 (2012)
177. Mao, X.R.: Stabilization of continuous-time hybrid stochastic differential equations by discrete-time feedback control. Automatica **49**(12), 3677–3681 (2013)
178. Mahmoud, M.S.: Adaptive control of nonlinear MIMO system with orthogonal endocrine intelligent controller. IEEE Trans. Cybern. (2020). https://doi.org/10.1109/TCYB.2020.2998505
179. Mir, U., Bhatti, Z.A.: Time triggered handoff schemes in cognitive radio networks: A survey. J. Netw. Comput. Appl. **102**, 71–85 (2018)
180. Mao, J.F., Cassandras, C.G.: Optimal control of multilayer discrete event systems with real-time constraint guarantees. IEEE Trans. Syst. Man, Cybern. Syst. **44**(10), 1425–1434 (2014)
181. Mary, P.M., Marimuthu, N.S.: Design of self-tuning fuzzy logic controller for the control of an unknown industrial process. IET Control Theory Appl. **3**(4), 428–436 (2009)
182. Mitchell, R., Kim, Y., El-Korchi, T., Cha, Y.J.: Wavelet-neuro-fuzzy control of hybrid building-active tuned mass damper system under seismic excitations. J. Vib. Control **19**(12), 1881–1894 (2013)
183. Ma, X., Djouadi, S.M., Li, H.: State estimation over a semi-Markov model based cognitive radio system. IEEE Trans. Wirel. Commun. **11**(7), 2391–2401 (2012)
184. Meng, X., Lam, J., Du, B., Gao, H.: A partial delay-partitioning approach to stability analysis of discrete-time systems. Automatica **46**(3), 610–614 (2009)
185. Morais, C.F., Palma, J.M., Peres, P.L.D., Oliveira, R.C.L.F.: An LMI approach for $\mathcal{H}_2$ and $\mathcal{H}_\infty$ reduced-order filtering of uncertain discrete-time Markov and Bernoulli jump linear systems. Automatica **95**, 463–471 (2018)
186. Mahmoud, M.S., Xia, Y.: Analysis and synthesis of fault-tolerant control systems. Wiley (2014)
187. Ma, Y., Yang, P., Zhang, Q.: Robust $\mathcal{H}_\infty$ control for uncertain singular discrete T-S fuzzy time-delay systems with actuator saturation. J. Frankl. Inst. Eng. Appl. Math. **353**(13), 3290–3311 (2016)
188. Nguang, S.K.: Robust nonlinear $\mathcal{H}_\infty$-output feedback control. IEEE Trans. Autom. Control **41**(7), 1003–1007 (1996)
189. Nasiri, A., Nguang, S.K., Swain, A., Almakhles, D.J.: Reducing conservatism in an $\mathcal{H}_\infty$ robust state-feedback control design of T-S fuzzy systems: a nonmonotonic approach. IEEE Trans. Fuzzy Syst. **26**(1), 386–390 (2018)
190. Nguang, S.K., Shi, P.: Nonlinear $\mathcal{H}_\infty$ filtering of sampled-data systems. Automatica **36**(2), 303–310 (2000)
191. Nguang, S.K., Shi, P.: Output feedback control of nonlinear systems under sampled measurements. Automatica **39**(12), 2169–2174 (2003)
192. Ning, Z., Wang, T., Song, X., Yu, J.: Fault detection of nonlinear stochastic systems via a dynamic event-triggered strategy. Signal Process. (2020). https://doi.org/10.1016/j.sigpro.2019.107283

193. Nguang, S.K., Shi, P., Ding, S.: Fault detection for uncertain fuzzy systems: an LMI approach. IEEE Trans. Fuzzy Syst. **15**(6), 1251–1262 (2007)
194. Ougli, A.E., Tidhaf, B.: Optimal type-2 fuzzy adaptive control for a class of uncertain nonlinear systems using an LMI approach. Int. J. Innov. Comput. Inf. Control **11**(3), 851–863 (2015)
195. Obuz, S., Klotz, J., Kamalapurkar, R., Dixon, W.: Unknown time-varying input delay compensation for uncertain nonlinear systems. Automatica **76**, 222–229 (2017)
196. Ogura, M., Preciado, V.M.: Stability of Markov regenerative switched linear systems. Automatica **69**, 169–175 (2016)
197. Pawlak, Z.: Al and intelligent industrial applications: the rough set perspective. Cybern. Syst. **31**(3), 227–252 (2000)
198. Park, P.: Markov chain model of fault-tolerant wireless networked control systems. Wirel. Netw. **25**(5), 2291–2303 (2018). https://doi.org/10.1007/s11276-017-1657-0
199. Pak, J.M., Ahn, C.K., Shi, P., Shmaliy, Y.S., Lim, M.T.: Distributed hybrid particle/FIR filtering for mitigating NLOS effects in TOA-based localization using wireless sensor networks. IEEE Trans. Ind. Electron. **64**(6), 5182–5191 (2017)
200. Petreczky, M., Bako, L., van Schuppen, J.H.: Realization theory of discrete-time linear switched systems. Automatica **49**, 3337–3344 (2013)
201. Proskurnikov, A.V., Fradkov, A.L.: Problems and methods of network control. Autom. Remote Control **77**(10), 1711–1740 (2016). https://doi.org/10.1134/S0005117916100015
202. Pan, J., Fei, S., Jaadri, A., Guerra, T.M.: Nonquadratic stabilisation of continuous T-S models: LMI solution for local approach. IEEE Trans. Fuzzy Syst. **20**(3), 594–602 (2012)
203. Precup, R.-E., Hellendoorn, H.: A survey on industrial applications of fuzzy control. Comput. Ind. **62**, 213–226 (2011)
204. Park, C., Kim, B., Lee, J.: Digital stabilization of fuzzy systems with time-delay and its application to backing up control of a truck-trailer. Int. J. Fuzzy Syst. **9**(1), 14–21 (2007)
205. Park, P.G., Ko, J.W., Jeong, C.: Reciprocally convex approach to stability of systems with time-varying delays. Automatica **47**(1), 235–238 (2011)
206. Shi, P., Li, F.: A survey on Markovian jump systems: modeling and design. Int. J. Control Autom. Syst. **13**(1), 1–16 (2015)
207. Park, P., Lee, W., Lee, S.: Auxiliary function-based integral inequalities for quadratic functions and their applications to time-delay systems. J. Frankl. Inst. **352**(4), 1378–1396 (2015)
208. Parriaux, J., Millérioux, G.: Nilpotent semigroups for the characterization of flat outputs of switched linear and LPV discrete-time systems. Syst. Control Lett. **62**, 679–685 (2013)
209. Park, J., Park, P.: Sampled-data control for continuous-time Markovian jump linear systems via a fragmented-delay state and its state-space model. J. Frankl. Inst. Eng. Appl. Math. **356**(10), 5073–5086 (2019)
210. Pu, X., Tan, X.: Stability of hybrid stochastic differential systems with switching and time delay. Stoch. Anal. Appl. **35**(3), 569–585 (2017)
211. Peng, T., Yang, X., Wu, L., Pang, B.: Reduced-order $\mathcal{L}_2$–$\mathcal{L}_\infty$ filtering for discrete-time T-S fuzzy systems with stochastic perturbation. Int. J. Syst. Sci. **46**(1), 179–192 (2015)
212. Shi, P., Yin, Y., Liu, F., Zhang, J.: Robust control on saturated Markov jump systems with missing information. Inf. Sci. **265**, 123–138 (2014)
213. Pang, B., Zhang, Q.L.: Stability analysis and observer-based controllers design for T-S fuzzy positive systems. Neurocomputing **275**, 1468–1477 (2018)
214. Qiu, J., Ding, S.X., Gao, H., Yin, S.: Fuzzy-model-based reliable static output feedback $\mathcal{H}_\infty$ control of nonlinear hyperbolic PDE systems. IEEE Trans. Fuzzy Syst. **24**(2), 388–400 (2016)
215. Qiu, J., Feng, G., Gao, H.: Fuzzy-model-based piecewise $\mathcal{H}_\infty$ static output-feedback controller design for networked nonlinear systems. IEEE Trans. Fuzzy Syst. **18**(5), 919–934 (2010)
216. Qiu, J., Gao, H., Ding, S.X.: Recent advances on fuzzy-model-based nonlinear networked control systems: a survey. IEEE Trans. Ind. Electron. **63**(2), 1207–1217 (2016)
217. Qiu, S.-B., Liu, X.-G., Wang, F.-X., Shu, Y.-J.: Robust stability analysis for uncertain recurrent neural networks with leakage delay based on delay-partitioning approach. Neural Comput. Appl. (10), 1–12 (2016). https://doi.org/10.1007/s00521-016-2670-4

218. Qiu, L., Shi, Y., Yao, F., Xu, G., Xue, B.: Network-based robust $\mathcal{H}_2/\mathcal{H}_\infty$ control for linear systems with two-channel random packet dropouts and time delays. IEEE Trans. Cybern. **45**(8), 1450–1462 (2015)
219. Rojas, O.J., Bao, J., Lee, P.L.: On dissipativity, passivity and dynamic operability of nonlinear processes. J. Process Control **18**(5), 515–526 (2008)
220. Ren, Y., Li, Q., Ding, D., Xie, X.: Dissipativity-preserving model reduction for Takagi-Sugeno fuzzy systems. IEEE Trans. Fuzzy Syst. **27**(4), 659–670 (2019)
221. Rosa, T.E., Morais, C.F., Oliveira, R.C.L.F.: New robust LMI synthesis conditions for mixed $\mathcal{H}_2/\mathcal{H}_\infty$ gain-scheduled reduced-order DOF control of discrete-time LPV systems. Int. J. Robust Nonlinear Control **28**(18), 6122–6145 (2018)
222. Rocha, R., Ruthiramoorthy, J., Kathamuthu, T.: Memristive oscillator based on Chua's circuit: stability analysis and hidden dynamics. Nonlinear Dyn. **88**(4), 2577–2587 (2017)
223. Ren, W., Xiong, J.: Vector-lyapunov-function-based input-to-state stability of stochastic impulsive switched time-delay systems. IEEE Trans. Autom. Control **64**(2), 654–669 (2019)
224. Sugeno, M.: Industrial Applications of Fuzzy Control. Elsevier, Amsterdam, The Netherlands (1985)
225. Selami, B.: Affine T-S fuzzy model-based estimation and control of Hindmarsh-Rose neuronal model. IEEE Trans. Syst. Man Cybern. Syst. **47**(8), 2342–2350 (2017)
226. Saravanakumar, R., Ali, M.S., Huang, H., Cao, J., Joo, Y.H.: Robust $\mathcal{H}_\infty$ state-feedback control for nonlinear uncertain systems with mixed time-varying delays. Int. J. Control Autom. Syst. **16**(1), 225–233 (2018)
227. Sharma, R., Bhasin, S., Gaur, P., Joshi, D.: A switching-based collaborative fractional order fuzzy logic controllers for robotic manipulators. Appl. Math. Model. **73**, 228–246 (2019)
228. Sim, K.-B., Byun, K.-S.: Harashima, F: Internet-based teleoperation of an intelligent robot with optimal two-layer fuzzy controller. IEEE Trans. Ind. Electron. **53**(4), 1362–1372 (2006)
229. Santin, I., Barbu, M., Pedret, C., Vilanova, R.: Fuzzy logic for plant-wide control of biological wastewater treatment process including greenhouse gas emissions. ISA Trans. **77**, 146–166 (2018)
230. Shi, D., Chen, T., Shi, L.: Event-triggered maximum likelihood state estimation. Automatica **50**(1), 247–254 (2014)
231. St-Onge, X.F., Cameron, J., Saleh, S., Scheme, E.J.: A symmetrical component feature extraction method for fault detection in induction machines. IEEE Trans. Ind. Electron. **66**(9), 7281–7289 (2019)
232. Slyn'ko, V.I., Denysenko, V.S.: The stability analysis of abstract Takagi-Sugeno fuzzy impulsive system. Fuzzy Sets Syst. **254**, 67–82 (2019)
233. Soliman, M.A., Hasanien, H.M., Azazi, H.Z., El-Kholy, E.E., Mahmoud, S.A.: An adaptive fuzzy logic control strategy for performance enhancement of a grid-connected PMSG-based wind turbine. IEEE Trans. Ind. Inform. **15**(6), 3163–3173 (2019)
234. Shi, Y., Huang, J., Yu, B.: Robust tracking control of networked control systems: application to a networked DC motor. IEEE Trans. Ind. Electron. **60**(12), 5864–5874 (2013)
235. Sugeno, M., Kang, G.T.: Structure identification of fuzzy model. Fuzzy Sets Syst. **28**(1), 15–33 (1988)
236. Spacapan, I., Kocijan, J., Bajd, T.: Simulation of fuzzy-logic-based intelligent wheelchair control system. J. Intell. Robot. Syst. **39**(2), 227–241 (2004)
237. Sun, J., Liu, G., Chen, J., Rees, D.: Improved delay-range-dependent stability criteria for linear systems with time-varying delays. Automatica **46**(2), 466–470 (2010)
238. Su, X., Liu, X., Song, Y.-D., Lam, H.K., Wang, L.: Reduced-order model approximation of fuzzy switched systems with pre-specified performance. Inf. Sci. **370**, 538–550 (2016)
239. Sun, K., Li, Y., Tong, S.: Fuzzy adaptive output feedback optimal control design for strict-feedback nonlinear systems. IEEE Trans. Syst. Man Cybern. Syst. **47**(1), 33–44 (2016)
240. Shi, P., Liu, M., Zhang, L.: Fault-tolerant sliding mode observer synthesis of Markovian jump systems using quantized measurements. IEEE Trans. Ind. Electron. **62**(9), 5910–5918 (2015)
241. Song, J., Niu, Y., Lam, J., Shu, Z.: A hybrid design approach for output feedback exponential stabilization of Markovian jump systems. IEEE Trans. Autom. Control **63**(5), 1404–14174 (2018)

242. Souza, F.D., de Oliveira, M.C., Palhares, R.M.: A simple necessary and sufficient LMI condition for the strong delay-independent stability of LTI systems with single delay. Automatica **89**, 407–410 (2018)
243. Shen, M., Park, J.H., Fei, S.: Event-triggered nonfragile $\mathcal{H}_\infty$ filtering of Markov jump systems with imperfect transmissions. Signal Process. **149**, 204–213 (2018)
244. Sowmiya, C., Raja, R., Zhu, Q., Rajchakit, G.: Further mean-square asymptotic stability of impulsive discrete-time stochastic BAM neural networks with Markovian jumping and multiple time-varying delays. J. Frankl. Inst. **356**(1), 561–591 (2019)
245. Senthilraj, S., Raja, R., Zhu, Q., Samidurai, R., Yao, Z.: New delay-interval-dependent stability criteria for static neural networks with time-varying delays. Neurocomputing **186**, 1–7 (2016)
246. Sato, K., Sato, H.: Structure preserving $\mathcal{H}_2$ optimal model reduction based on Riemannian trust-region method. IEEE Trans. Autom. Control **63**(2), 505–512 (2018)
247. Shi, P., Su, X., Li, F.: Dissipativity-based filtering for fuzzy switched systems with stochastic perturbation. IEEE Trans. Autom. Control **61**(6), 1694–1699 (2016)
248. Sakthivel, R., Sakthivel, R., Nithya, V., Selvaraj, P., Kwon, O.M.: Fuzzy sliding mode control design of Markovian jump systems with time-varying delay. Automatica **355**(14), 6353–6370 (2018)
249. Su, X., Shi, P., Wu, L., Basin, M.: Reliable filtering with strict dissipativity for T-S fuzzy time-delay systems. IEEE Trans. Cybern. **44**(12), 2470–2483 (2014)
250. Su, X., Shi, P., Wu, L., Song, Y.-D.: A novel control design on discrete-time Takagi-Sugeno fuzzy systems with time-varying delays. IEEE Trans. Fuzzy Syst. **21**(4), 655–671 (2013)
251. Samidurai, R., Sriraman, R., Zhu, S.: Leakage delay-dependent stability analysis for complex-valued neural networks with discrete and distributed time-varying delays. Neurocomputing **338**, 262–273 (2019)
252. Syed Ali, M., Vadivel, R., Saravanakumar, R.: Design of robust reliable control for T-S fuzzy Markovian jumping delayed neutral type neural networks with probabilistic actuator faults and leakage delays: an event-triggered communication scheme. ISA Trans. **77**, 30–48 (2018)
253. Shi, P., Wang, H., Lim, C.-C.: Network-based event-triggered control for singular systems with quantizations. IEEE Trans. Ind. Electron. **63**(2), 1230–1238 (2016)
254. Su, X., Wu, L., Shi, P.: Sensor networks with random link failures: distributed filtering for T-S fuzzy systems. IEEE Trans. Ind. Inform. **9**(3), 1739–1750 (2013)
255. Su, X., Wu, L., Shi, P., Chen, C.L.P.: Model approximation for fuzzy switched systems with stochastic perturbation. IEEE Trans. Fuzzy Syst. **23**(5), 1458–1473 (2015)
256. Su, X., Wu, L., Shi, P., Song, Y.-D.: H_∞ model reduction of Takagi-Sugeno fuzzy stochastic systems. IEEE Trans. Syst. Man Cybern. Part B Cybern. **42**(6), 1574–1585 (2012)
257. Sahoo, A., Xu, H., Jagannathan, S.: Adaptive neural network-based event-triggered control of single-input single-output nonlinear discrete-time systems. IEEE Trans. Neural Netw. Learn. Syst. **27**(1), 151–164 (2016)
258. Sahoo, A., Xu, H., Jagannathan, S.: Near optimal event-triggered control of nonlinear discrete-time systems using neurodynamic programming. IEEE Trans. Neural Netw. Learn. Syst. **27**(9), 1801–1815 (2016)
259. Sun, Y., Yu, J., Chen, Z., Xing, X.: Event-triggered filtering for nonlinear networked discrete-time systems. IEEE Trans. Ind. Electron. **62**(11), 7163–7170 (2015)
260. Sun, X., Zhang, Q.: Admissibility analysis for interval type-2 fuzzy descriptor systems based on sliding mode control. IEEE Trans. Cybern. **49**(8), 3032–3040 (2019)
261. Socha, L., Zhu, Q.: Exponential stability with respect to part of the variables for a class of nonlinear stochastic systems with Markovian switchings. Math. Comput. Simul. **155**, 2–14 (2019)
262. Shi, P., Zhang, Y., Chadli, M., Agarwal, R.: Mixed $\mathcal{H}_\infty$ and passive filtering for discrete fuzzy neural networks with stochastic jumps and time delays. IEEE Trans. Neural Netw. Learn. Syst. **27**(4), 903–909 (2016)
263. Truong, D.Q., Ahn, K.K.: Robust variable sampling period control for networked control systems. IEEE Trans. Ind. Electron. **62**(9), 5630–5643 (2015)

264. Tong, R.M., Beck, M.B., Latten, A.: Fuzzy control of the activated sludge wastewater treatment process. Automatica **16**(6), 695–701 (1980)
265. Tang, X., Deng, L., Qu, H.: Predictive control for networked interval type-2 T-S fuzzy system via an event-triggered dynamic output feedback scheme. IEEE Trans. Fuzzy Syst. **27**(8), 1573–1586 (2019)
266. Tanelli, M., Ferrara, A.: Enhancing robustness and performance via switched second order sliding mode control. IEEE Trans. Autom. Control **58**(4), 962–974 (2013)
267. Thumati, B.T., Feinstein, M.A., Jagannathan, S.: A model-based fault detection and prognostics scheme for Takagi-Sugeno fuzzy systems. IEEE Trans. Fuzzy Syst. **22**(4), 736–748 (2014)
268. Tsai, S., Jen, C.: $\mathcal{H}_\infty$ stabilization for polynomial fuzzy time delay system: a sum-of-squares approach. IEEE Trans. Fuzzy Syst. **26**(6), 3630–3644 (2018)
269. Tariq, M.F., Khan, A.Q., Abid, M., Mustafa, G.: Data-driven robust fault detection and isolation of three-phase induction motor. IEEE Trans. Ind. Electron. **66**(6), 4707–4715 (2019)
270. Tobenkin, M.M., Manchester, I.R., Megretski, A.: Convex parameterizations and fidelity bounds for nonlinear identification and reduced-order modelling. IEEE Trans. Autom. Control **62**(7), 3679–3686 (2017)
271. Tabuada, P., Pappas, G.J.: From nonlinear to Hamiltonian via feedback. IEEE Trans. Autom. Control **48**(8), 1439–1442 (2003)
272. Takagi, T., Sugeno, M.: Fuzzy identification of systems and its applications to modeling and control. IEEE Trans. Syst. Man Cybern. Part B Cybern. **15**(1), 116–132 (1985)
273. Tong, S.C., Sui, S., Li, Y.M.: Fuzzy adaptive output feedback control of MIMO nonlinear systems with partial tracking errors constrained. IEEE Trans. Fuzzy Syst. **23**(4), 729–742 (2015)
274. Talla, J., Streit, L., Peroutka, Z., Drabek, P.: Position-based T-S fuzzy power management for tram with energy storage system. IEEE Trans. Ind. Electron. **62**(5), 3061–3071 (2015)
275. Tanaka, K., Wang, H.O.: Fuzzy Control Systems Design and Analysis: A Linear Matrix Inequality Approach. Wiley, New York (2001)
276. Tian, E., Yue, D.: Reliable $\mathcal{H}_\infty$ filter design for T-S fuzzy model-based networked control systems with random sensor failure. Int. J. Robust Nonlinear Control **23**(1), 15–32 (2013)
277. Tang, L., Zhao, J.: Switched threshold-based fault detection for switched nonlinear systems with its application to Chua's circuit system. IEEE Trans. Circuits Syst. I Regul. Pap. **66**(2), 733–741 (2019)
278. Veillette, R., Medanic, J.B., Perkins, W.: Design of reliable control systems. IEEE Trans. Autom. Control **37**(3), 290–304 (1992)
279. Wang, H.: Application of intelligent materials in the control system. J. Comput. Theor. Nanosci. **12**(9), 2830–2836 (2015)
280. Wu, Y., Dong, J.: Fault detection for T-S fuzzy systems with partly unmeasurable premise variables. Fuzzy Sets Syst. **338**, 136–156 (2018)
281. Wu, Z., Dong, S., Shi, P., Su, H., Huang, T.: Reliable filtering of nonlinear Markovian jump systems: the continuous-time case. IEEE Trans. Syst. Man Cybern. Syst. **49**(2), 386–394 (2019)
282. Wang, Y., Ding, S.X., Xu, D., Shen, B.: An $\mathcal{H}_\infty$ fault estimation scheme of wireless networked control systems for industrial real-time applications. IEEE Trans. Control Syst. Technol. **22**(6), 2073–2086 (2014)
283. Wang, T., Gao, H., Qiu, J.: A combined adaptive neural network and nonlinear model predictive control for multirate networked industrial process control. IEEE Trans. Neural Netw. Learn. Syst. **27**(2), 416–425 (2016)
284. Wu, L., Ho, D.W.C.: Fuzzy filter design for $It\hat{o}$ stochastic systems with application to sensor fault detection. IEEE Trans. Fuzzy Syst. **17**(1), 233–242 (2009)
285. Wu, Y., Lu, R.: Event-based control for network systems via integral quadratic constraints. IEEE Trans. Circuits Syst. I Regul. Pap. **65**(4), 1386–1394 (2018)
286. Wang, L., Lam, H.K.: New stability criterion for continuous-time Takagi-Sugeno fuzzy systems with time-varying delay. IEEE Trans. Cybern. **49**(4), 1551–1556 (2019)

287. Wu, C., Liu, J., Jing, X., Li, H., Wu, L.: Adaptive fuzzy control for nonlinear networked control systems. IEEE Trans. Syst. Man Cybern. Syst. **47**(8), 2420–2430 (2017)
288. Wang, Y., Lim, C.C., Shi, P.: Adaptively adjusted event-triggering mechanism on fault detection for networked control systems. IEEE Trans. Cybern. **64**(6), 5203–5211 (2017)
289. Wang, J., Ma, S., Zhang, C., Fu, M.: Finite-time $\mathcal{H}_\infty$ filtering for nonlinear singular systems with nonhomogeneous Markov jumps. IEEE Trans. Cybern. **49**(6), 2133–2143 (2019)
290. Wu, Z., Park, J.H., Su, H., Song, B., Chu, J.: Reliable $\mathcal{H}_\infty$ filtering for discrete-time singular systems with randomly occurring delays and sensor failures. IET Control Theory Appl. **6**(14), 2308–2317 (2012)
291. Wang, M., Qiu, J., Chadli, M., Wang, M.: A switched system approach to exponential stabilization of sampled-data T-S fuzzy systems with packet dropouts. IEEE Trans. Cybern. **46**(12), 3145–3156 (2016)
292. Wang, T., Qiu, J., Fu, S., Ji, W.: Distributed fuzzy $\mathcal{H}_\infty$ filtering for nonlinear multirate networked double-layer industrial processes. IEEE Trans. Ind. Electron. **64**(6), 5203–5211 (2017)
293. Wang, C., Shen, Y.: Improved delay-dependent robust stability criteria for uncertain time delay systems. Appl. Math. Comput. **218**(6), 2880–2888 (2011)
294. Wu, L., Shi, P., Gao, H., Wang, J.: $\mathcal{H}_\infty$ model reduction for linear parameter-varying systems with distributed delay. Int. J. Control **82**(3), 408–422 (2009)
295. Wang, Y., Shi, P., Lim, C.-C.: Liu, Y: Event-triggered fault detection filter design for a continuous-time networked control system. IEEE Trans. Cybern. **46**(12), 3414–3426 (2016)
296. Wu, Z., Shi, P., Su, H., Chu, J.: Network-based robust passive control for fuzzy systems with randomly occurring uncertainties. IEEE Trans. Fuzzy Syst. **21**(5), 966–971 (2013)
297. Wu, Z., Shi, P., Su, H., Chu, J.: Sampled-data fuzzy control of chaotic systems based on T-S fuzzy model. IEEE Trans. Fuzzy Syst. **22**(1), 153–163 (2014)
298. Wu, L., Su, X., Shi, P., Qiu, P.: A new approach to stability analysis and stabilization of discrete-time T-S fuzzy time-varying delay systems. IEEE Trans. Syst. Man Cybern. Part B Cybern. **41**(1), 273–286 (2001)
299. Wang, H.O., Tanaka, K., Griffin, M.F.: An approach to fuzzy control of nonlinear systems: stability and design issues. IEEE Trans. Fuzzy Syst. **4**(1), 14–23 (1996)
300. Wu, X., Tang, Y., Cao, J., Mao, X.: Stability analysis for continuous-time switched systems with stochastic switching signals. IEEE Trans. Autom. Control **63**(9), 3083–3090 (2018)
301. Wang, C., Wu, M.: Hierarchical intelligent control system and its application to the sintering process. IEEE Trans. Ind. Inform. **9**(1), 190–197 (2013)
302. Wakaiki, M., Yamamoto, Y.: Stability analysis of sampled-data switched systems with quantization. Automatica **69**, 157–168 (2016)
303. Wu, L., Yang, X., Lam, H.K.: Dissipativity analysis and synthesis for discrete-time T-S fuzzy stochastic systems with time-varying delay. IEEE Trans. Fuzzy Syst. **22**(2), 380–394 (2014)
304. Wu, L., Yao, X., Zheng, W.-X.: Generalized $\mathcal{H}_2$ fault detection for two-dimensional Markovian jump systems. Automatica **48**(8), 1741–1750 (2012)
305. Wu, L., Zheng, W.: $\mathcal{L}_2$-$\mathcal{L}_\infty$ control of nonlinear fuzzy Ito stochastic delay systems via dynamic output feedback. IEEE Trans. Syst. Man Cybern. Part B Cybern. **39**(5), 1308–1315 (2009)
306. Wen, S., Zeng, Z., Huang, T.: Reliable $\mathcal{H}_\infty$ filter design for a class of mixed-delay Markovian jump systems with stochastic nonlinearities and multiplicative noises via delay-partitioning method. Int. J. Control Autom. Syst. **10**(4), 711–720 (2012)
307. Wu, L., Zheng, W.X., Gao, H.: Dissipativity-based sliding mode control of switched stochastic systems. IEEE Trans. Autom. Control **58**(3), 785–793 (2013)
308. Xu, S., Chen, T., Lam, J.: Robust $\mathcal{H}_\infty$ filtering for uncertain Markovian jump systems with mode-dependent time-delays. IEEE Trans. Autom. Control **48**(5), 900–907 (2003)
309. Xi, Z., Feng, G., Cheng, D., Lu, Q.: Nonlinear decentralized saturated controller design for power systems. IEEE Trans. Control Syst. Technol. **11**(4), 539–547 (2003)
310. Xia, M., Gupta, V., Antsaklis, P.J.: Networked state estimation over a shared communication medium. IEEE Trans. Control Syst. Technol. **62**(4), 1729–1741 (2017)
311. Xu, H., Jagannathan, S.: Neural network-based finite horizon stochastic optimal control design for nonlinear networked control systems. IEEE Trans. Neural Netw. Learn. Syst. **26**(3), 472–485 (2015)

312. Xiao, B., Lam, H.K., Li, H.: Stabilization of interval type-2 polynomial-fuzzy-model-based control systems. IEEE Trans. Fuzzy Syst. **25**(1), 205–217 (2017)
313. Xie, K., Lyu, Z., Liu, Z., Zhang, Y., Chen, C.L.P.: Adaptive neural quantized control for a class of MIMO switched nonlinear systems with asymmetric actuator dead-zone. IEEE Trans. Neural Netw. Learn. Syst. **31**(6), 1927–1941 (2020)
314. Xiao, Q., Lewis, F.L., Zeng, Z.: Event-based time-interval pinning control for complex networks on time scales and applications. IEEE Trans. Ind. Electron. **65**(11), 8797–8808 (2018)
315. Xia, W., Ma, Q., Lu, J., Zhuang, G.: Reliable filtering with extended dissipativity for uncertain systems with discrete and distributed delays. Int. J. Syst. Sci. **48**(12), 2644–2657 (2017)
316. Xue, M., Tang, Y., Wu, L., Qian, F.: Model approximation for switched genetic regulatory networks. IEEE Trans. Neural Netw. Learn. Syst. **29**(8), 3404–3417 (2018)
317. Xu, Q., Wong, P., Zhang, C., Xie, S., Yen, P.: Engineering applications of intelligent monitoring and control. Math. Probl. Eng. (2013). https://doi.org/10.1155/2013/564021
318. Xing, M., Xia, J., Huang, X., Shen, H.: On dissipativity-based filtering for discrete-time switched singular systems with sensor failures: a persistent dwell-time scheme. IET Control Theory Appl. **13**(12), 1814–1822 (2019)
319. Xiao, F., Xie, X., Jiang, Z., Sun, L., Wang, R.: Utility-aware data transmission scheme for delay tolerant networks. Peer-To-Peer Netw. Appl. **9**(5), 936–944 (2015). https://doi.org/10.1007/s12083-015-0354-y
320. Xiong, W., Yu, W., Lu, J., Yu, X.: Fuzzy modelling and consensus of nonlinear multiagent systems with variable structure. IEEE Trans. Circuits Syst. I Regul. Pap. **61**(4), 1183–1191 (2014)
321. Xie, X., Yue, D., Ma, T., Zhu, X.: Further studies on control synthesis of discrete-time T-S fuzzy systems via augmented multi-indexed matrix approach. IEEE Trans. Cybern. **44**(12), 2784–2791 (2014)
322. Xiao, F., Yang, X., Yang, M., Sun, L., Wang, R., Yang, P.: Surface coverage algorithm in directional sensor networks for 3D complex terrains. Tsinghua Sci. Technol. **21**(4), 397–406 (2016)
323. Yin, C., Cheng, Y., Huang, X., Zhong, S., Li, Y., Shi, K.: Delay-partitioning approach design for stochastic stability analysis of uncertain neutral-type neural networks with Markovian jumping parameters. Neurocomputing **207**, 437–449 (2016)
324. Yu, Y., Lam, H.K., Chan, K.Y.: T-S fuzzy-model-based output feedback tracking control with control input saturation. IEEE Trans. Fuzzy Syst. **26**(6), 3514–3523 (2018)
325. Yang, H., Li, H., Xia, Y., Li, L.: Nonuniform sampling Kalman filter for networked systems with Markovian packets dropout. J. Frankl. Inst. Eng. Appl. Math. **355**(10), 4218–4240 (2018)
326. Yan, S., Nguang, S.K., Shen, M., Zhang, G.: Event-triggered $\mathcal{H}_\infty$ control of networked control systems with distributed transmission delay. IEEE Trans. Autom. Control (2019). https://doi.org/10.1109/TAC.2019.2953460
327. Yin, Y., Shi, P., Liu, F., Teo, K.L., Lim, C.-C.: Robust filtering for nonlinear nonhomogeneous Markov jump systems by fuzzy approximation approach. IEEE Trans. Cybern. **45**(9), 1706–1716 (2015)
328. Yang, W., Tong, S.: Adaptive output feedback fault-tolerant control of switched fuzzy systems. Inf. Sci. **329**, 478–490 (2016)
329. Yang, X., Wu, L., H. K., Su, X.: Stability and stabilization of discrete-time T-S fuzzy systems with stochastic perturbation and time-varying delay. IEEE Trans. Fuzzy Syst. **22**(1), 124–138 (2014)
330. Yan, H., Xu, X., Zhang, H., Yang, F.: Distributed event-triggered $\mathcal{H}_\infty$ state estimation for T-S fuzzy systems over filtering networks. J. Frankl. Inst. **354**(9), 3760–3779 (2017)
331. Yang, G., Ye, D.: Adaptive reliable $\mathcal{H}_\infty$ filtering against sensor failures. IEEE Trans. Signal Process. **55**(7), 3161–3171 (2007)
332. Yang, H., Zhang, J., Jia, X., Li, S.: Non-fragile control of positive Markovian jump systems. J. Frankl. Inst. **356**(5), 2742–2758 (2019)
333. Yang, R., Zheng, W.: $\mathcal{H}_\infty$ filtering for discrete-time 2-D switched systems: an extended average dwell time approach. Automatica **98**, 302–313 (2018)

334. Yang, R., Zheng, W., Yu, Y.: Event-triggered sliding mode control of discrete-time two-dimensional systems in Roesser model. Automatica (2020). https://doi.org/10.1016/j.automatica.2020.108813
335. Yan, H., Zhang, H., Yang, F., Huang, C., Chen, S.: Distributed $\mathcal{H}_\infty$ filtering for switched repeated scalar nonlinear systems with randomly occurred sensor nonlinearities and asynchronous switching. IEEE Trans. Syst. Man Cybern. Syst. **48**(12), 2263–2270 (2018)
336. Zadeh, L.A.: Outline of a new approach to analysis of complex systems and decision processes. IEEE Trans. Syst. Man Cybern. SMC **3**(1), 28–44 (1973)
337. Zhang, D., Cai, W.J., Xie, L.H., Wang, Q.G.: Non-fragile distributed filtering for T-S fuzzy systems in sensor networks. IEEE Trans. Fuzzy Syst. **23**(5), 1883–1890 (2015)
338. Zhao, T., Dian, S.: State feedback control for interval type-2 fuzzy systems with time-varying delay and unreliable communication links. IEEE Trans. Fuzzy Syst. **26**(2), 951–966 (2018)
339. Zhao, L., Gao, H., Karimi, H.R.: Robust stability and stabilization of uncertain T-S fuzzy systems with time-varying delay: an input-output approach. IEEE Trans. Fuzzy Syst. **21**(5), 883–897 (2013)
340. Zhao, J., Hill, D.J.: Dissipativity theory for switched systems. IEEE Trans. Autom. Control **53**(4), 941–953 (2008)
341. Zhang, L., Hua, C., Cheng, G., Li, K., Guan, X.: Decentralized adaptive output feedback fault detection and control for uncertain nonlinear interconnected systems. IEEE Trans. Cybern. **50**(3), 935–945 (2020)
342. Zhang, C., Hu, J., Qiu, J., Chen, Q.: Event-triggered nonsynchronized $\mathcal{H}_\infty$ filtering for discrete-time T-S fuzzy systems based on piecewise lyapunov functions. IEEE Trans. Syst. Man Cybern. Syst. **47**(8), 2330–2341 (2016)
343. Zhang, X., Han, Q., Seuret, A., Gouaisbaut, F.: An improved reciprocally convex inequality and an augmented Lyapunov-Krasovskii functional for stability of linear systems with time-varying delay. Automatica **84**, 221–226 (2017)
344. Zhang, X., Han, Q., Yu, X.: Survey on recent advances in networked control systems. IEEE Trans. Ind. Inform. **12**(5), 1740–1752 (2016)
345. Zhang, Z., Liang, H., Wu, C., Ahn, C.K.: Adaptive Event-triggered output feedback fuzzy control for nonlinear networked systems with packet dropouts and actuator failure. IEEE Trans. Fuzzy Syst. **27**(9), 1793–1806 (2019)
346. Zhang, X., Lu, G., Zheng, Y.: Stabilization of networked stochastic time-delay fuzzy systems with data dropout. IEEE Trans. Fuzzy Syst. **16**(3), 798–807 (2008)
347. Zhao, J., Mili, L.: Sparse state recovery versus generalized maximum-likelihood estimator of a power system. IEEE Trans. Power Syst. **33**(1), 1104–1106 (2018)
348. Zamani, M., Mazo, M., Khaled, M., Abate, A.: Symbolic abstractions of networked control systems. IEEE Trans. Control Netw. Syst. **5**(4), 1622–1634 (2018)
349. Zhao, M., Peng, C., He, W., Song, Y.: Event-triggered communication for leader-following consensus of second-order multiagent systems. IEEE Trans. Cybern. **48**(6), 1888–1897 (2018)
350. Zhang, H., Qiu, Z., Xiong, L., Jiang, G.: Stochastic stability analysis for neutral-type Markov jump neural networks with additive time-varying delays via a new reciprocally convex combination inequality. Int. J. Syst. Sci. **50**(5), 970–988 (2019)
351. Zhang, Q., Qiao, L., Zhu, B., Zhang, H.: Dissipativity analysis and synthesis for a class of T-S fuzzy descriptor systems. IEEE Trans. Syst. Man Cybern. Syst. **47**(8), 1774–1784 (2017)
352. Zhang, J.H., Shi, P., Qiu, J.Q., Nguang, S.K.: A novel observer-based output feedback controller design for discrete-time fuzzy systems. IEEE Trans. Fuzzy Syst. **23**(1), 223–229 (2015)
353. Zhang, H.G., Shan, Q., Wang, Z.: Stability analysis of neural networks with two delay components based on dynamic delay interval method. IEEE Trans. Neural Netw. Learn. Syst. **28**(2), 259–267 (2017)
354. Zhang, H., Shi, Y., Wang, J., Chen, H.: A new delay-compensation scheme for networked control systems in controller area networks. IEEE Trans. Ind. Electron. **65**(9), 7239–7247 (2018)
355. Zhang, Z., Shi, Y., Zhang, Z., Yan, W.: New results on sliding-mode control for Takagi-Sugeno fuzzy multiagent systems. IEEE Trans. Cybern. **49**(5), 1592–1604 (2019)

356. Zhang, J., Xia, Y.: Design of $\mathcal{H}_\infty$ fuzzy controllers for nonlinear systems with random data dropouts. Optim. Control Appl. Methods **32**, 328–349 (2011)
357. Zhang, H., Wang, J.: State estimation of discrete-time Takagi-Sugeno fuzzy systems in a network environment. IEEE Trans. Cybern. **45**(8), 1525–1536 (2015)
358. Zhao, G., Wang, J.: Reset control systems with time-varying delay: delay-dependent stability and $\mathcal{L}_2$ gain performance improvement. IEEE Trans. Cybern. **17**(6), 2460–2468 (2015)
359. Zou, L., Wang, Z., Hu, J., Gao, H.: On $\mathcal{H}_\infty$ finite-horizon filtering under stochastic protocol: dealing with high-rate communication networks. IEEE Trans. Autom. Control **62**(9), 4884–4890 (2017)
360. Zhou, Q., Wu, C., Shi, P.: Observer-based adaptive fuzzy tracking control of nonlinear systems with time delay and input saturation. Fuzzy Sets Syst. **316**, 49–68 (2017)
361. Zhou, H., Zhang, Y., Li, W.: Synchronization of stochastic levy noise systems on a multi-weights network and its applications of Chua's circuits. IEEE Trans. Circuits Syst. I Regul. Pap. **66**(7), 2709–2722 (2019)
362. Zhang, Z., Zhang, Z., Zhang, H., Shi, P., Karimi, H.R.: Finite-time $\mathcal{H}_\infty$ filtering for T-S fuzzy discrete-time systems with time-varying delay and norm-bounded uncertainties. IEEE Trans. Fuzzy Syst. **23**(6), 2427–2434 (2015)
363. Zheng, Q., Zhang, H.: Mixed $\mathcal{H}_\infty$ and passive filtering for switched Takagi-Sugeno fuzzy systems with average dwell time. ISA Trans. **75**, 52–63 (2018)
364. Zhang, Z., Zhou, Q., Wu, C., Li, H.: Mixed $\mathcal{H}_\infty$ and passive filtering for switched Takagi-Sugeno fuzzy systems with average dwell time. Int. J. Fuzzy Syst. **20**(20), 390–402 (2018)

Lightning Source UK Ltd.
Milton Keynes UK
UKHW021309220822
407638UK00002B/10

9 783030 812164